The Spectroscopy of Semiconductors

SEMICONDUCTORS
AND SEMIMETALS
Volume 36

Semiconductors and Semimetals

A Treatise

Edited by

R. K. Willardson
CONSULTING PHYSICIST
SPOKANE, WASHINGTON

Albert C. Beer
CONSULTING PHYSICIST
COLUMBUS, OHIO

Eicke R. Weber
DEPARTMENT OF MATERIALS SCIENCE AND MINERAL ENGINEERING
UNIVERSITY OF CALIFORNIA AT BERKELEY

The Spectroscopy of Semiconductors

SEMICONDUCTORS
AND SEMIMETALS

Volume 36

Volume Editors

DAVID G. SEILER
SEMICONDUCTOR ELECTRONICS DIVISION
NATIONAL INSTITUTE OF STANDARDS AND TECHNOLOGY
GAITHERSBURG, MARYLAND

CHRISTOPHER L. LITTLER
DEPARTMENT OF PHYSICS
UNIVERSITY OF NORTH TEXAS
DENTON, TEXAS

ACADEMIC PRESS, INC.
Harcourt Brace Jovanovich, Publishers

Boston   San Diego   New York
London   Sydney   Tokyo   Toronto

This book is printed on acid-free paper. ⊛

COPYRIGHT © 1992 BY ACADEMIC PRESS, INC.
ALL RIGHTS RESERVED.
NO PART OF THIS PUBLICATION MAY BE REPRODUCED OR
TRANSMITTED IN ANY FORM OR BY ANY MEANS, ELECTRONIC
OR MECHANICAL, INCLUDING PHOTOCOPY, RECORDING, OR
ANY INFORMATION STORAGE AND RETRIEVAL SYSTEM, WITHOUT
PERMISSION IN WRITING FROM THE PUBLISHER.

ACADEMIC PRESS, INC.
1250 Sixth Avenue, San Diego, CA 92101-4311

United Kingdom Edition published by
ACADEMIC PRESS LIMITED
24-28 Oval Road, London NW1 7DX

LIBRARY OF CONGRESS CATALOGING-IN-PUBLICATION DATA

Semiconductors and semimetals.—Vol. 1—New York: Academic Press, 1966–

v.: ill.; 24 cm.

Irregular.
Each vol. has also a distinctive title.
Edited by R. K. Willardson, Albert C. Beer, and Eicke R. Weber
ISSN 0080-8784 = Semiconductors and semimetals

1. Semiconductors—Collected works. 2. Semimetals—Collected works.
I. Willardson, Robert K. II. Beer, Albert C. III. Weber, Eicke R.
QC610.9.S48   621.3815'2—dc19   85-642319
                                           AACR 2   MARC-S

Library of Congress   [8709]
ISBN 0-12-752136-4 (v. 36)

PRINTED IN THE UNITED STATES OF AMERICA
92 93 94 95 96 97   BC   9 8 7 6 5 4 3 2 1

Contents

LIST OF CONTRIBUTORS vii
PREFACE ix

Chapter 1 **Laser Spectroscopy of Semiconductors at Low Temperatures and High Magnetic Fields**

    D. Heiman

I.	Introduction	2
II.	Experimental Techniques	7
III.	GaAs Quantum Well Structures	14
IV.	Magnetic Semiconductors	52
	Acknowledgments	78
	References	79

Chapter 2 **Transient Spectroscopy by Ultrashort Laser Pulse Techniques**

    Arto V. Nurmikko

I.	Introduction	85
II.	Instrumentation for Time-Resolved Spectroscopy	88
III.	Time-Resolved Spectroscopy in Semiconductors: Approaches	96
IV.	Portfolio of Experimental Examples in Time-Resolved Spectroscopy	105
V.	Concluding Remarks	131
	References	133

Chapter 3 **Piezospectroscopy of Semiconductors**

    A. K. Ramdas and S. Rodriguez

I.	Introduction	137
II.	Elasticity, Symmetry, Latent Anisotropy, and Deformation Potentials	139

III.	Experimental Techniques	145
IV.	Valence and Conduction Bands of Elemental and Compound Semiconductors under External Stress	150
V.	Interband Transitions and Associated Excitons	166
VI.	Lyman Spectra of Shallow Donors and Acceptors	179
VII.	Zone-Center Optical Phonons	197
VIII.	Optical Phonons and Electronic States	206
IX.	Concluding Remarks	213
	Acknowledgments	215
	References	215

Chapter 4 Photoreflectance Spectroscopy of Microstructures

Orest J. Glembocki and Benjamin V. Shanabrook

I.	Introduction	222
II.	Microstructures	223
III.	Modulation Spectroscopy	228
IV.	Experimental Details	238
V.	Experimental Spectra	240
VI.	Conclusions	288
	References	289

Chapter 5 One- and Two-Photon Magneto-Optical Spectroscopy of InSb and $Hg_{1-x}Cd_xTe$

David G. Seiler, Christopher L. Littler, and Margaret H. Weiler

I.	Introduction	294
II.	Experimental Methods	329
III.	One-Photon Magnetospectroscopy	339
IV.	Two-Photon Absorption (TPA) and Two-Photon Magnetoabsorption (TPMA) Spectroscopy	389
	References	418

INDEX	429
CONTENTS OF PREVIOUS VOLUMES	437

List of Contributors

Numbers in parentheses indicate pages on which the authors' contributions begin.

OREST J. GLEMBOCKI (221), *Naval Research Laboratory, Code 6833, Washington, DC 20375*

D. HEIMAN (2), *Massachusetts Institute of Technology, Francis Bitter National Magnet Laboratory, Cambridge, Massachusetts 02139*

CHRISTOPHER L. LITTLER (293), *Department of Physics, University of North Texas, Denton, Texas 76203*

ARTO V. NURMIKKO (85), *Division of Engineering and Department of Physics, Brown University, Providence, Rhode Island 02912*

A. K. RAMDAS (137), *Department of Physics, Purdue University, West Lafayette, Indiana 47907*

S. RODRIQUEZ (137), *Department of Physics, Purdue University, West Lafayette, Indiana 47907*

DAVID G. SEILER (293), *Materials Technology Group, Semiconductor Electronic Division, National Institute of Standards and Technology, Building 225, Room A305, Gaithersburg, Maryland 20899*

BENJAMIN V. SHANABROOK (221), *Naval Research Laboratory, Code 6833, Washington, DC 20375*

MARGARET H. WEILER (293), *Loral Infrared & Imaging Systems, Mail Stop 146, Lexington, Massachusetts 02173-7393*

Preface

Spectroscopy has played a fundamental role in understanding and establishing the nature of atoms, molecules, and solids. Spectroscopy can be defined as the study of the emission, absorption, dispersion, and scattering of electromagnetic radiation by matter. Optical spectroscopy is a simple yet powerful technique for studying semiconductors. Various spectroscopic techniques have been, and continue to be, extremely fruitful in discovering and delineating a wide variety of phenomena exhibited by semiconductors and in characterizing their properties. Group theory and quantum mechanics are then needed to interpret properly these spectroscopic investigations. In addition, spectroscopy carried out under external perturbations (uniaxial stress, hydrostatic pressure, magnetic field, electric field, etc.) further advances our understanding of many areas of semiconductor physics: degeneracies, symmetries, impurities, defects, phonons, energy band structure features, time-resolved behavior, novel effects, etc. The strength of all of these spectroscopic methods in scientific and technological research is perhaps best emphasized by the diversity of their applications. This book covers work at the frontiers of magneto-optics, Raman scattering, photoluminescence, photoreflectance, piezospectroscopy, ultra-fast spectroscopy, and spectroscopy at extremely low temperatures and high magnetic fields. The tutorial approach in the chapters of this volume facilitates learning by newcomers to the field of semiconductor spectroscopy, while experts will profit by the state-of-the-art aspects. The extensive bibliographies in each chapter will give readers direct access to the wide variety of experimental and theoretical investigations that illustrate the power and sensitivity of spectroscopy as applied to semiconductors.

Since the discovery of the transistor and the refinement of crystal growth, spectroscopic techniques have been demonstrated to be among the most powerful and sensitive characterization methods for investigating semiconductors. This volume presents reviews of a number of major spectroscopic techniques extensively used to investigate bulk and artificially structured semiconductors. A strong emphasis is placed on work done during the last ten years and also upon major semiconductor systems and artificially structured materials such as GaAs, InSb, $Hg(1 - x)Cd(x)Te$ and MBE grown structures based on GaAs/AlGaAs materials. Both the novice and the expert in spectroscopy will profit from the descriptions and discussions of these methods, principles, and applications relevant for understanding and characterizing today's semiconductor structures. The present book consists of a number of chapters written by experts; each chapter describes a different facet of the spectroscopy of semiconductors.

The successful growth of new semiconductor materials and artificially structured materials such as superlattices, quantum wells, and modulation-doped heterojunction systems by molecular beam epitaxy has opened entirely new opportunities for using spectroscopic techniques to characterize their physical and chemical properties. The artificially structured materials are of scientific and technological interest because they exhibit properties that are significantly different from bulk. Reduced dimensionality and various layers (consisting of different semiconductors) produce novel properties such as band offsets, strain mismatch, quantum confined electronic states, and many-body effects. It is safe to say that future progress in materials growth and engineering of these structures will only come by making proper use of spectroscopic methods such as those presented in this volume.

The first chapter, written by D. Heiman (MIT Francis Bitter National Magnet Laboratory), reviews and discusses recent spectroscopic studies on GaAs-based quantum layered systems and magnetic semiconductors such as $Cd(1-x)Mn(x)Te$ using temperatures as low as 40 mK and magnetic fields up to 150 Tesla. High magnetic fields and low temperatures have been crucial to the discovery of new phenomena in semiconductors—notable examples are the integer and fractional quantum Hall effects (QHE). These effects were first observed in resistance measurements, rather than optically, mainly because of the difficulty in accessing these extreme environments with optics. This situation has changed in the past several years, due primarily to the use of fiber optics in spectroscopic measurements. As a result, Raman scattering and photoluminescence measurements can now be routinely performed down to $T = 0.04$ K in a dilution refrigerator and up to $B = 35$ T in dc magnets. Magneto-optics of magnetic semiconductors are important for studying a whole host of new phenomena arising from exchange interactions. An example of novel low-temperature spectroscopy is the observation of cooperative magnetic polarons in spin-flip Raman scattering experiments down to $T = 0.1$ K. Recent spectroscopy of 2D electron systems in high fields has concentrated on the effects of many-body electron–electron interactions. This has led to the identification of the unique roton-like many-body dispersion and metal-insulating transition in the integer-QHE, in addition to measurement of collective energies in the incompressible quantum fluid of the fractional-QHE. Currently, optical investigations are aimed toward understanding the predicted Wigner crystallization of electrons in high magnetic fields, as well as new phenomena in zero- and one-dimensional systems.

Spectroscopic studies of semiconductors by time-resolved methods in the picosecond and femtosecond time scales are presented in Chapter 2 by A. V. Nurmikko (Brown University). The major emphasis of this chapter is on summarizing the main technical features and virtues of ultrafast techniques with illustrations given in the context of basic contemporary phenomena in crystalline materials and heterostructures such as quantum wells and superlattices. Examples covered include dephasing processes of electrons and excitons, electron–phonon inelastic scattering, exciton–exciton and exciton-free carrier interactions, determination of exciton–polariton group velocity, optical Stark effect, electron transport in layered structures such as superlattices, and spin-related effects in magnetic semiconductors.

In solid-state spectroscopy, as in atomic and molecular spectroscopy, spectra recorded under an external perturbation provide deep insights into the degeneracy and symmetry of the initial and final states responsible for an optical transition. The focus of the third chapter of A. K. Ramdas and S. Rodriquez (Purdue University) is on uniaxial stress as an external perturbation. This type of perturbation is unique to solids and spectroscopy of this type is labeled piezospectroscopy. The theme of this chapter is piezospectroscopy as it has been performed in the field of semiconductor physics. Theoretical issues—theory of elasticity, symmetry arguments including group theory, and the quantitative theory in terms of a strain Hamiltonian (the so-called deformation potential theory)—are developed with sufficient detail and generality. The unified theoretical background thus is readily accessible for interpreting experimental results obtained in semiconductor physics with piezospectroscopy. A section on experimental techniques describes a typical apparatus used in this field.

In particular, results on the piezospectroscopy of interband and excitonic transitions are given; excitation spectra of donors and acceptors and zone center optical phonons observed in Raman spectroscopy are discussed extensively in Chapter 3. There is increasing appreciation that built-in strains in semiconductors play a significant role in the behavior of heterostructures; one section illustrates how special insights have been obtained with piezospectroscopy in this field. Coupled exitons of plasmons and phonons in polar semiconductors, and of interband transitions and zone center optical phonons in Si, are discussed as phenomena where uniaxial stress as a perturbation has provided a microscopic characterization.

O. J. Glembocki and B. V. Shanabrook (Naval Research Laboratory) discuss in Chapter 4 the application and flexibility of modulation spectroscopy, particularly electroreflectance and photoreflectance, to the study of artificially structured materials. They show how these modulation measurements have enhanced our understanding of the properties of artificially structured materials and of the modulation mechanisms involved in electromodulation spectroscopy. Using modulation spectroscopy, the derivative of the absorptivity or reflectivity with respect to some parameter is measured. The resulting spectra are sensitive to critical point transitions and have sharp derivative-like features which are characteristic of the material's structure and properties. Weak features that may be difficult to observe in the absolute absorption or reflection spectrum can be enhanced.

The overview of photoreflectance spectroscopy as applied to artificially structured materials will serve as a useful guide to the interested reader. This technique is one of the simplest room temperature optical probes of semiconductors. Although the technique is simple to implement, the measured line shapes are often complex and require careful analysis. The theoretical treatment provided by Glembocki and Shanabrook should serve as a guide in interpreting a wide variety of photoreflectance data. The chapter focuses on AlGaAs/GaAs materials and quantum wells, coupled wells, superlattices, and modulation doped heterojunctions. Applications to bulk materials such as GaAs and InP, as well as strained SiGe and InGaAs/GaAs structures, are also given.

The last chapter, written by D. G. Seiler, C. L. Littler, and M. H. Weiler (National Institute of Standards and Technology, University of North Texas, and Loral Infrared & Imaging Systems, respectively), focuses on a wide variety of magneto-optical phenomena that have been studied in InSb and $Hg(1-x)Cd(x)Te$ during the last decade. These two narrow-gap materials exhibit many interesting scientific features related to their energy band structures that can be studied using magneto-optical techniques. Technologically, they are two of the most important materials used in the construction of infrared detectors because of their sensitivity in the regions of the electromagnetic spectrum corresponding to the atmospheric windows at $3-5$ μm and $8-14$ μm. Spectroscopic techniques have been used extensively to study and characterize InSb and HgCdTe in the past. This chapter concentrates on those magneto-optical studies done that involve the use of lasers as light sources. Reviews of a wide variety of intraband and interband one-photon absorption phenomena and comprehensive reviews of all two-photon absorption work are presented. The emphasis is on phenomena observed during the last decade. Unique aspects presented here include the review of two-photon magneto-optical studies and of impurity- and defect-related magneto-optical spectroscopy. Numerous examples are given of the wide variety of laser-induced magneto-optical phenomena in InSb and HgCdTe—interband one- and two-photon magnetoabsorption, free-carrier magnetoabsorption (conduction-band and phonon-assisted cyclotron resonance, free-hole and combined resonance), and impurity- and defect-related transitions from shallow and deep levels. Experimental methods used to study magneto-optical effects are presented with novel features such as ac magnetic-field modulation techniques for derivative spectroscopy stressed.

The editors are indebted to their spouses for their encouragement and patience during the time this volume was being written. This volume was initiated while DGS was at the University of North Texas and as such it represents the contributions and views of the editors and is not to be construed as official NIST work.

David G. Seiler
Christopher L. Littler

CHAPTER 1

Laser Spectroscopy of Semiconductors at Low Temperatures and High Magnetic Fields

D. Heiman

MASSACHUSETTS INSTITUTE OF TECHNOLOGY
FRANCIS BITTER NATIONAL MAGNET LABORATORY
CAMBRIDGE, MASSACHUSETTS

I. INTRODUCTION 2
   1. *Two-Dimensional Electron Systems*. 3
   2. *Magnetic Semiconductors* 5
II. EXPERIMENTAL TECHNIQUES. 7
   3. *Apparatus for Generating Extreme Conditions* 7
      a. *Low-Temperature Apparatus*. 7
      b. *High-Field Magnets*. 9
      c. *High-Pressure Devices* 11
   4. *Spectroscopic Techniques* 11
   5. *Fiber-Optic Apparatus* 11
III. GaAs QUANTUM WELL STRUCTURES 14
   6. *Background on Two-Dimensional Systems* 14
      a. *Overview* 14
      b. *Basic Physics of the 2D Electron System*. 16
   7. *Band-gap Optical Transitions* 21
      a. *Basic Processes* 21
      b. *Optical Quantum Oscillations and Selection Rules* 21
      c. *Nonlinear Landau Level Tuning in Emission* 24
      d. *Spectral Blue-Shift Anomaly at $v = 1$* 26
      e. *Optical Analog of the Quantum Hall Effect* 28
      f. *Blue-Shifting Spectral Doublet Anomaly near $v = 2/3$* 30
      g. *Time-Resolved Photoluminescence at $v = 1$* 32
      h. *Summary of Optical Emission in the Quantum Hall Regime*. . . 34
   8. *Inelastic Light Scattering of 2D Electron and Hole Systems* 35
      a. *Spectroscopy of 2D Valence Subband States* 35
      b. *Excitations of the 2D Electron Gas in High Magnetic Fields* . . . 39
      c. *Magnetoroton Density of States*. 41
      d. *Regime of the Electron Solid* 43
   9. *Excitons in Quantum Wells and Strain-Layer Superlattices*. 44
      a. *Comparison of Magnetoexciton Theory and Experiment* 45
      b. *Exciton States in High Magnetic Fields* 46
      c. *Exciton Localization in a Magnetic Field* 48
      d. *Competition between Strain and Magnetic Field* 50

IV. MAGNETIC SEMICONDUCTORS 52
   10. *Background on Exchange Interactions.* 53
      a. *Magnetic-Ion Subsystem* 53
      b. *sp-d Exchange Interaction* 55
      c. *Bound Magnetic Polarons*. 56
   11. *High-Field Spin Alignment in* (Cd,Mn)Te. 58
      a. *Results to* 45 T 58
      b. *Results to* 150 T 60
   12. *Field-Induced Transitions within Magnetic-Ion Clusters.* 61
      a. *Magnetic-Ion Pairs* 62
      b. *Magnetic-Ion Triplets* 64
   13. *Bound Magnetic Polarons below* $T = 1$ K 66
   14. *Metal–Insulator Transition below* $T = 1$ K 69
   15. *Spectroscopy of* Fe *and Other Ions* 73
   ACKNOWLEDGMENTS 78
   REFERENCES 79

I. Introduction

The search for new semiconductor physics often requires extreme conditions of low temperatures or high magnetic fields. This chapter reviews recent studies of semiconductor systems that employ optical spectroscopy at temperatures as low as 40 mK and magnetic fields up to 1.5 megagauss (150 T). The main focus is on two categories of materials that show a variety of interesting effects:

(1) GaAs quantum-layered systems, including many-electron interactions;
(2) magnetic semiconductors, such as the wide-band-gap prototype $Cd_{1-x}Mn_xTe$.

One of the main obstacles for optical studies under these extreme conditions is the difficulty in making measurements in complex Dewars with optical windows. Now however, new techniques using fiber-optic light pipes allow access to these environments without the use of an optical Dewar. Photons can be routed through low-loss glass fibers as readily as electrons are pushed through wires. In the simplest case, the design of a fiber-optic probe is not much different from a cryogenic insert for electrical measurements—small, flexible, optical fibers are strung in the same way as conducting wires. Collimated light from tunable lasers is easily coupled into these fibers, even into small-diameter (5-μm) single-mode fibers. As an end result, these techniques have provided a breakthrough for laser spectroscopy, especially for temperatures below $T = 1$ K and magnetic fields above $B = 15$ T (Heiman *et al.*, 1989c).

1. Two-Dimensional Electron Systems

The integer and fractional quantum Hall effects (QHE) and the electron solid are examples of new physics that rely on both extremes of low temperatures and high magnetic fields (see Prange and Girvin, 1987). High magnetic fields are extremely useful because they effectively switch the two-dimensional (2D) electron system between a metal and an insulator in the region of the QHE. Although these novel effects were first observed in electrical transport experiments, in principle, optical measurements can be important and have unique advantages. For example, an inelastic optical process, such as simple absorption, by virtue of its single-quantum interaction is capable of probing *local* electron states. With optics it is also possible to examine the entire electron density of states, not just those states at the Fermi surface that are accessible in electrical transport. Optical studies are just beginning to be applied to 2D electron systems at low temperatures and high magnetic fields, and they have already revealed interesting single-particle effects and many-body electron–electron interactions. Many of these are discussed at length in Section III. For a complementary review of magneto-optical studies of confined electron systems see (Petrou and McCombe 1991), which contains sections on infrared studies, confined impurities, and electron–phonon interactions.

Our current knowledge of electronic states in two dimensions owes a great deal to laser spectroscopy. Optical studies often rely on high magnetic fields to measure properties like subband effective masses and excitonic interactions. For example, Landau-level energies are governed by the carrier effective mass. Most of the recent attention has been focused on GaAs–(Al, Ga)As quantum well structures because of their high-purity and lattice-matched heterostructures of GaAs and AlAs. Such materials are well suited for optical studies because the band-gap region (1.5 eV) is matched by tunable laser sources and high-efficiency detectors. Besides this, the quality of GaAs-based heterostructures continues to improve; electron mobilities in modulation-doped heterostructures now exceed 10^7 cm^2/V-sec, layers can be made with near monolayer smoothness, and layer thicknesses approach a single monolayer.

In addition to the 2D carrier confinement induced by the quantum well barriers, a large magnetic field applied perpendicular to the layers produces in-plane carrier localization as well. The field constrains the mobile electrons to move in cyclotron orbits having quantized angular momenta. This series of Landau levels leads to quantum oscillations in the 2D electron gas as a function magnetic field, which are readily observed in the electrical conductance. These quantum oscillations, such as those present in the integer QHE, reflect field-dependent changes in the degeneracy of the electron states

(Klitzing et al., 1980). To illustrate the effects of degeneracy, consider a cold 2D electron gas in zero magnetic field and neglect the spin degeneracy. Because of the Pauli exclusion principle, each electron must have a different energy, and consequently there is no degeneracy of energy levels along each coordinate direction. In the single-particle picture, a magnetic field localizes each electron to a dimension the size of the magnetic length (cyclotron radius). Now since this length varies as $1/\sqrt{B}$, a higher field allows a higher density of cyclotron orbits per unit area, thus increasing the Landau-level degeneracy. The magnetic quantum limit is defined when all electrons are degenerate at the same energy (when the level broadening is neglected).

The field-dependent degeneracy that shows up in the electrical conductivity also shows up in the optical spectra. A good example is found in the conduction-to-valence band luminescence spectrum associated with the free electrons in a 2D layer, which is described in detail in Section III 7a. At $B = 0$ there are transitions from electrons in the entire Fermi sea, giving rise to a broad spectral continuum with a width equal to the Fermi energy. In a magnetic field this continuum breaks up into a set of discrete spectral peaks, one for each Landau level that is populated with electrons.

One might ask why so much effort has been applied to the study of lower-dimensional electron systems. The most important feature, aside from the effects of "particle-in-a-box" quantum confinement, is the enhancement of many-body electron–electron interactions. (Ando, et al., 1982). To see this, consider screening of the Coulomb interaction between two charged particles caused by a third particle. The Coulomb interaction between the two particles acts through the electric field lines in 3D space and screening occurs when the third particle distorts these field lines. For a system of carriers confined to a 2D plane, the screening is greatly reduced because the electric-field lines lying out of the plane are much less disturbed. This leads to larger energies for many-body interactions in a 2D system compared with a 3D system. For example, a 2D electron system with carrier density of 10^{11} cm^{-2} has a typical Hartree energy of a few milli electron volts, while a 3D system with the same interparticle spacing has an energy an order of magnitude smaller. Experimental results are described in Sections III 7 and III 8.

At much higher magnetic fields and lower temperatures than those used to study the integer QHE, a high-mobility 2D electron system shows some remarkable many-body electron–electron interactions. Correlated motions of the electrons in a 2D layer give rise to a unique incompressible quantum fluid (Tsui et al., 1982; Laughlin, 1983; Prange and Girvin, 1987). The 2D electron system is incompressible in the sense that, in order to decrease the system density a "bubble" must be formed, where the bubble is equivalent to a quasiparticle excitation. At extremely high fields, when the kinetic energy of the electrons is further quenched, the 2D electron system should condense into a Wigner crystal or glass (Lam and Girvin, 1984). Because of its simplicity, magnetotransport has been the primary probe of these many-electron ground states. Now, a number of research groups are beginning to use optical probes such as photoluminescence and inelastic light scattering. These optical studies have uncovered new manifestations of the electron–electron interactions and are discussed in the following section.

2. MAGNETIC SEMICONDUCTORS

Another important system for high-magnetic-field and low-temperature investigations is that of magnetic semiconductors (see Nagaev, 1983). Materials like $Cd_{1-x}Mn_xTe$, also referred to as dilute magnetic or semimagnetic semiconductors, are generally II-VI semiconductors with a transition-metal ion substituting for a fraction of the column II cations (see Furdyna and Kossut, 1988; Aggarwal et al, 1987). Perhaps the most interesting property of this class of materials is the sp-d exchange interaction between magnetic ion spins and the spins of semiconductor carriers. As a result of this interaction, these materials exhibit many novel effects like giant Faraday rotation, magnetically reduced band gap, and magnetic polarons. In the elementary picture of the sp-d exchange, first a moderate applied magnetic field at low temperatures aligns the magnetic ion spins, then the exchange interaction is responsible for generating a much larger *exchange field* felt by the carrier spins. Thus, the field amplification gives rise to large Zeeman splittings of the semiconductor bands, which can be as large as 10^2 meV. This effect is observed in optical experiments either directly as conduction- or valence-intraband splittings or as changes in the conduction-to-valence interband energy. Note that the field amplification acts only on spin moments and not on the translational motion of carriers. The application of a magnetic field at liquid helium temperatures can reduce a 2-eV band gap by as much as 50 to 100 meV. Without this interaction, a typical wide-band-gap (1.5–3 eV) semiconductor would have a Zeeman splitting of order 1 meV.

Magnetic polarons are another fascinating manifestation of the sp-d exchange (see Nagaev, 1983; Wolff, 1988). Here, the spin of a carrier produces an exchange field that aligns the magnetic-ion spins, and this alignment gives an exchange field acting back to the carrier. The self-consistent exchange interaction acting in both directions, between carrier spin and ion spins, creates a ferromagnetic alignment of ion spins as the vicinity of the carrier. Bound magnetic polarons consist of carriers localized in donor-bound electrons and acceptor-bound holes, which are embedded in a dense background of magnetic ions. Typical polaron magnetic moments are in the range of 10 to 10^2 Bohr magnetons.

The most important probe of the wide range of phenomena arising from exchange interactions is, perhaps, magneto-optics. Although magnetic measurements like magnetization and susceptibility determine the average spin alignment of the magnetic ions, they are less useful for measuring either the enhanced Zeeman splitting or local effects due to bound magnetic polarons. In magneto-optics, one measures both the bulk and local responses of the magnetic-ion subsystem, thus probing a whole host of interesting physics resulting from the exchange interaction. Thus, optical measurements on wide-band-gap semiconductors have been responsible for the discovery of the giant Faraday rotation ≤ 10 deg/cm-G (Komarov et al., 1977), large Zeeman splitting ~ 10–10^2 meV (Gaj et al., 1978), enhanced effective Landé g-factors $\sim 10^2$ (Heiman et al., 1983), magnetically tunable band gaps and lasers $\Delta\lambda > 10$ nm (Heiman, 1983), and large bound magnetic polaron binding energies ~ 1–10^2 meV (Golnik et al., 1980).

In connection with such effects, magneto-optical experiments are necessary for the measurement of sp-d exchange energies. Here, one compares the optically determined Zeeman splitting with field-induced spin alignment, measured via magnetization or susceptibility. Typically, sp-d exchange energies are $1/4$ eV for the conduction band and -1 eV for the valence band. The exchange fields and splittings vary with temperature approximately as $1/T$. The field-induced band-gap reduction at $B = 10$ T might be 50 meV at liquid helium temperature and only 1 meV at room temperature. This fact dictates that the optical experiments be done at low temperatures and high magnetic fields.

The d-d exchange interaction acting between magnetic ions can also be probed with optical measurements. Faraday rotation has been used to measure the average spin alignment in applied magnetic fields to 150 T (Isaacs et al., 1990a). Although ion spins that are relatively isolated (having no nearest-neighbor spins) reach saturation by 10 T, coupled ions need much higher fields to reach saturation, $\sim 10^2$ T. Since the d-d exchange is antiferromagnetic, a given ion spin feels a large exchange field that is in opposition to the applied field. Measurements of the high-field spin alignment are important for modeling acceptor-bound magnetic polarons where internal fields reach 10^2 T (Ram-Mohan and Wolff, 1988).

High-field optical experiments are important for direct measurements of the d-d exchange energy. In samples with dilute concentrations of Mn^{2+} ions, 1–10%, a significant fraction of the ions are in isolated, nearest-neighbor pair clusters. Without an external field, the spins are antiferromagnetically coupled in opposing alignment at low temperature. For increasing applied field, the spin alignment along the field direction increases from 0 to 5 in integer steps. The fields at which these steps occur give an accurate value for the exchange constant J/k_B (see Shapira, 1990). For example, spin-flip Raman scattering measurements to 30 T have revealed the first two steps in (Cd,Mn)Se and the first three steps in (Cd,Mn)Te, giving $J/k_B = -7.5$ K and -6.1, respectively (Isaacs et al., 1988b; Heiman, 1990). Details can be found in Section IV 12.

II. Experimental Techniques

Techniques for laser spectroscopy in extreme environments are described in this section. The major topics covered are

(1) the generation of low temperatures and high magnetic fields;
(2) laser spectroscopies in these environments;
(3) novel approaches to spectroscopy that use optical fibers as light pipes.

3. APPARATUS FOR GENERATING EXTREME CONDITIONS

a. Low-Temperature Apparatus

Current methods for achieving low temperatures have been reviewed in a number of excellent books (Rose-Innes, 1973; Lounasmaa, 1974; White, 1979; Richardson and Smith, 1988). The Richardson and Smith book contains an up-to-date compilation of recent low-temperature techniques presented in a delightful "cookbook" approach. Here, I will only give an introductory discussion for simple cryogenic techniques that use pure liquid ^{4}He or pure ^{3}He, as well as ^{3}He–^{4}He dilution refrigeration. In all these techniques the cooling principle involves lowering the entropy of the liquid, and hence temperature, as helium atoms cross a phase boundary. At this boundary the vapor pressure varies exponentially with temperature, so as the pressure above the liquid is reduced by pumping, the temperature decreases. Since ^{3}He atoms are lighter, they have larger zero-point fluctuations and hence give a higher pressure at a given temperature. This translates into lower operating temperatures for ^{3}He than for ^{4}He, $T \sim 0.4$ K versus 1.3 K. In the dilution process, ^{3}He atoms are pulled from a ^{3}He-rich ^{3}He–^{4}He mixture into a dilute phase. Cooling to $T < 0.05$ K is thus achieved at the phase boundary between the two liquids.

The simplest technique for reducing sample temperatures below $T = 4.2$ K is by immersion in liquid ^{4}He. Direct contact between the sample and liquid usually holds the sample temperature near that of the bath for moderate incident laser powers (<0.1 W). This is sufficient for optical experiments where the light does not have to pass through the turbulent boiling liquid, as when the sample is placed within a millimeter of an optical fiber or window.

Otherwise, the boiling may be quenched in the superfluid phase by cooling below the lambda point at $T = 2.2$ K, by pumping down to a pressure below 38 mm Hg. The superfluid also provides an excellent thermal contact with the sample and keeps the sample temperature within a tenth of a degree for modest optical powers of 10^{-2} W. When the laser intensity on an absorbing surface is above 10^{-1} W/cm^2, local boiling occurs. The resulting turbid optical path can be overcome by lowering the liquid level to just below the laser beam, but with reduced cooling power. In this case, the sample temperature may be as much as a few kelvins above that of the bath at high optical powers. Temperatures as low as $T \sim 1.5$ K can easily be reached in a ^{4}He Dewar with optical windows by this simple pumping technique. Reaching these temperatures requires a large mechanical vacuum pump having a pumping speed of at least 20 L/sec. Of course, the base temperature depends on the heat load input into the liquid. One watt input corresponds to pumping away 1.3 L of liquid ^{4}He per hour, due to the latent heat of vaporization. When a coldfinger arrangement is used, in which the sample is surrounded by a vacuum, the only heat transfer is through contact to the cooled stage. Here, the sample can be considerably warmer ($\Delta T \sim 2$–10 K) than the copper coldfinger, depending on the amount of heating from the incident light.

Pumped liquid ^{3}He can generate temperatures between $T = 0.3$ K and $T = 1.8$ K. These lower operating temperatures come at the expense of lower cooling power at these lower temperatures. A simple ^{3}He cryostat consists of a ^{3}He chamber inside of a ^{4}He chamber, where a lower portion of the cryostat is thermally isolated by a vacuum jacket between the two chambers. When the ^{4}He surrounding the unisolated upper section is pumped below $T = 1.8$ K, ^{3}He gas condenses on this wall and flows down the wall to the isolated bottom section. A small mechanical vacuum pump on the ^{3}He allows the bath to be reduced to $T = 0.5$ K. Increased pumping, either by a higher-speed pump and a larger-diameter pumping tube or a cryopump located within the ^{4}He Dewar, can provide working temperatures as low as $T \sim 0.3$ K. Optical windows on this type of Dewar have two disadvantages. The additional ^{3}He chamber and vacuum jacket require the extra complexity of added windows, and the windows are a source of heat input. There is a particularly annoying problem with the use of ^{3}He in high magnetic fields. Above about 18 T there are large diamagnetic forces on the liquid that lead to large thermal gradients along the field direction. In the extreme case of large input heat, a bubble can develop in the liquid column and will not rise to the top in a restricted geometry due to the magnetic force. Thus care must be taken to allow for sufficient unrestricted fluid flow and reduced heat input.

The lowest temperature achieved for laser spectroscopy in high magnetic fields is below $T = 0.1$ K. These temperatures are generated by a dilution refrigerator containing a mixture of ^{3}He and ^{4}He. Heat is removed from the sample surface by collisions with helium atoms and is then extracted from the mixture surface where the ^{3}He is preferentially pumped out. The top-loading dilution refrigerator facility at the MIT National Magnet Laboratory has been outfitted with silica fibers for optical spectroscopy. Results of spectroscopic experiments indicate a lower cooling power of the sample for the dilution mixture compared with pure ^{3}He. Although the refrigerator cooling power was 0.4 W at $T = 0.1$ K, the sample cooling power was found to be one to two orders of magnitude lower, due to Kapitza boundary resistance. This is discussed further in Sections IV 13 and IV 14, which deal with spin-flip Raman scattering below $T = 1$ K. Although dilution refrigerators are usually expensive, it is possible to build a simple "weekend" cryostat insert (Anderson, 1970).

b. High-Field Magnets

High magnetic fields are produced in steady-state dc and in pulsed modes by a number of techniques (see Herlach and Miura, 1987; Herlach, 1985). In order of increasing field, these techniques include 18–20 T superconductor solenoid; 24-T Bitter-type copper resistive solenoid; 25-T polyhelix copper solenoid; 30–35 T hybrid (superconductor–resistive) solenoid; 50-T pulsed (10–100 msec) copper solenoid; 70-T pulsed copper–niobium matrix solenoid; 80-T pulsed (10–100 μsec) steel polyhelix solenoid; 200-T pulsed (1–10 μsec) exploding single-turn coil; and 350–1000-T pulsed single-turn imploding coil.

Commercially available superconducting magnets produce dc fields up to $B = 12$ T for Nb–Ti windings or $B = 18$ T for Nb$_3$Sn. Because of the large inductance of superconducting solenoids, the sweep time for ramping the field is usually long, ~ 10–20 min for full field. Bore diameters are from 3 to 5 cm and normally have access along the solenoid axis. When radial access to the magnet bore is desired, the solenoid is separated in the middle by a few centimeters, which reduces the maximum field by 2 to 4 T.

For higher dc fields, one must use dissipative copper conductors. So-called Bitter solenoids routinely provide fields to $B \approx 24$ T in a 3-cm bore. These magnets have conductors in the form of thin copper disks 30–40 cm in diameter that are able to withstand the large $\mathbf{J} \times \mathbf{B}$ Lorentz radial force. Typically, 10 MW (250 V at 40 KA) is needed for $B \sim 24$ T, while 5 MW gives $B \sim 19$ T. For a larger bore diameter of 5 cm, peak fields reach $B \sim 20$ T at 10 MW or $B \sim 15$ T at 5 MW. The Joule heating is removed by flowing water through the stack of disks at about 100 L/sec.

The largest dc fields, typically $B \approx 35$ T, are obtained with a hybrid configuration of a Bitter solenoid surrounded by a 1-m diameter superconducting solenoid. The current world record for dc fields is $B = 35.3$ T, made at the MIT National Magnet Laboratory (Guinness Book, 1991). This total was produced via a 24.8-T Bitter solenoid, 7.0-T superconducting solenoid, and 3.5-T holmium flux concentrators. A new hybrid magnet having a 13-T superconducting solenoid has just been built at MIT. Future hybrid magnets are expected to reach $B \sim 45$ T by using a 15-T superconducting solenoid and a 30-T multiple-helix resistive solenoid operating at 20 MW. Sections III and IV describe Raman scattering and photoluminescence measurements made in fields up to $B = 30$ T.

Pulsed magnets are currently required for fields above $B \sim 35$ T. The obvious disadvantages of pulsed magnets are (a) less integration time for data collection ($\leq 10^{-3}$ sec per constant field interval); (b) small bore sizes, 1 cm or less; (c) and eddy current heating induced by the large time-dependent magnetic flux. Optical measurements can have advantages over experiments requiring conducting wires in the bore, such as conductivity and magnetization measurements which are plagued by induced emf's in the wires leading out of the magnet.

A simple reinforced copper wire solenoid is capable of producing 10–100 msec pulses with peak fields of $B = 40$–50 T. In the late 1920s, Kapitza first used such a magnet, and, later, Foner developed the currently used design (Foner and Kolm, 1956). These solenoids have a few hundred turns of 3×3 mm^2 copper wire with a 1–2 cm bore. Typical energies in the 100–200 kJ range are stored in a capacitor bank, then discharged into the liquid-nitrogen-cooled solenoid. This type of magnet facility is now found in many countries (see Herlach and Miura, 1987). The field maximum in this type of magnet is not limited by Joule heating as in the Bitter solenoids, but is limited by the strength of the wire and its reinforcement. When the $J \times B$ force on the inner winding becomes excessive, the diameter stretches enough to damage the insulation and arcing results. Realizing this, Foner developed coils made with a remarkably strong Cu–Nb composite (Foner, 1986). This material has a tensile strength a few times larger than pure copper, yet the conductivity is similar to that of copper. A nominal 6:1 Cu–Nb composite billet is extruded into wire having a copper matrix with long ~ 10-nm-diameter niobium filaments. These magnets now reach fields as high as $B = 70$ T.

Destructive magnets produce fields larger than 100 T (Herlach, 1985). A single turn of a copper plate, 3 mm thick, 1 cm in diameter, and 1 cm wide, yields $B = 150$ T in a 6-μsec-long pulse. Normally, the violent explosion is radially outward, allowing the cryostat and sample to survive as many as 10 shots. Results of Faraday rotation measurements on (Cd,Mn)Te to $B = 150$ T, presented in Section IV 11b. show spin saturation typical of random dilute antiferromagnets (Isaacs et al., 1990a).

Optical measurements have been made in flux-compression devices to fields of $B = 350$ T, produced by electromagnetic compression (Cnare effect), and to $B = 1000$ T, using shaped explosive charges (see Herlach, 1985).

c. High-Pressure Devices

Uniaxial pressures of a few kilobars can be obtained in a simple mechanical clamp, while hyrostatic pressure bombs are useful up to about 10–20 kbar. The pressure bombs can be either prestressed outside the spectroscopic apparatus with a hydraulic press or stressed with isobarically frozen helium (Schirber, 1970). Spectroscopy at high pressures usually requires optical fibers because they easily tolerate the high pressures and can be inserted through small holes in the bombs (Ackermann *et al.*, 1985). Much higher pressures, in the 100-kbar range, are produced in diamond anvil presses.

4. Spectroscopic Techniques

There are a number of types of optical experiments having large light signals and thus require only simple light collecting systems. Among these are photoluminescence, photoluminescence excitation, reflectivity, transmission (absorption), and Faraday rotation. These techniques are easily compatible with optical fibers and are discussed further in the next section.

More complex experiments include Raman scattering, Brillouin scattering, photoreflectivity, and optically detected magnetic resonance. The first two techniques, involving inelastic light scattering, usually have weak optical signals and are more difficult to detect. Because of the additional sensitivity required, care must be taken to collect a large solid angle of the inelastically scattered light. Also, the small ratio of elastic to inelastic light ($\sim 10^{-5} - 10^{-10}$) and finite spectral rejection requires one to avoid collecting the elastically scattered light coming from the sample surfaces and optics. In extreme cases where samples have very small scattering signals, due to a small light penetration depth (~ 10 nm) found in metals and superconductors, it is useful to have Dewar windows with antireflection coatings.

The techniques of photoreflectivity will not be covered here, since they are reviewed in another chapter. However, a photoreflectivity apparatus employing fiber-optic light pipes has been used in high-field experiments on GaAs quantum well systems (Zheng *et al.*, 1988a).

5. Fiber-Optic Apparatus

Simple single-fiber apparatuses have been used for photoluminescence measurements at low temperatures, where they eliminate the need for a

Dewar with optical windows (Ackermann et al., 1985; Heiman, 1985; Sakaki et al., 1985). The main advantage of fiber-optic (FO) light pipes is their utility in reaching environments having restricted optical access. In fact, FO systems are often the only practical method available for optical experiments in dilution refrigerators, small-bore pulsed magnets, and high-pressure cells in magnets. Several FO apparatuses have been developed for optical experiments in high magnetic fields at the MIT National Magnet Laboratory (Heiman, 1985; Isaacs and Heiman, 1987). In addition, silica FO cables transmit over a wide wavelength range, 0.3 to 3 μm, have negligible loss over laboratory distances and connect with uncomplicated optical couplers. A general review of fiber optics for spectroscopy can be found in Heiman et al. (1989c).

Single-, dual-, and multiple-fiber systems are illustrated in this section. In cases where the input light is easily discriminated from the output light, by a spectrometer or polarizer, a single fiber can be used for both input and output. In other cases, the collected light must be discriminated from the laser light by separate fibers for the input and output light.

A single-light-pipe apparatus was made using a 600-μm silicone-clad silica FO (Heiman, 1985). For Photoluminescence (PL) experiments, the output light from the FO cable is first imaged onto the spectrometer slit, while the laser excitation light is focused onto this output end at a moderate angle (10°) to the optical centerline. In this way, laser light travels down the fiber to the sample, then both the PL signal and the reflected laser light return to the spectrometer entrance slit. For materials having a quantum efficiency greater than 10^{-2}, the PL signal is typically smaller than the residual excitation light by a factor of 10^{-4} to 10^{-5}. A single-grating spectrometer is usually adequate to reject the stray laser light.

Large Faraday rotations are conveniently monitored by placing at the end of a FO a "sandwich" containing a sample between a linear polarizer and mirror (Heiman et al., 1987a). Laser light is first polarized by the polarizer, then passes through the sample twice before being analyzed by the same polarizer. The output signal monitored by a photodiode has a sinusoidal variation with changing magnetic field. (There is also a small constant background arising from surface reflections.) Each peak represents 180° rotation. When less than a few rotations are observed, it may be necessary to fit the signal to a cosine curve.

Figure 1 shows a two-fiber apparatus for Raman scattering (Isaacs and Heiman, 1987). A similar apparatus was used for spin-flip Raman scattering down to $T = 0.1$ K and up to $B = 30$ T, as described in Section IV. Light from the input fiber is collimated by a graded refractive index (GRIN) lens, after which a mirror then reflects the light onto the sample. A small lens collects the inelastically scattered light and focuses it into the output fiber. Angle-dependent scattering measurements were also made by placing an ad-

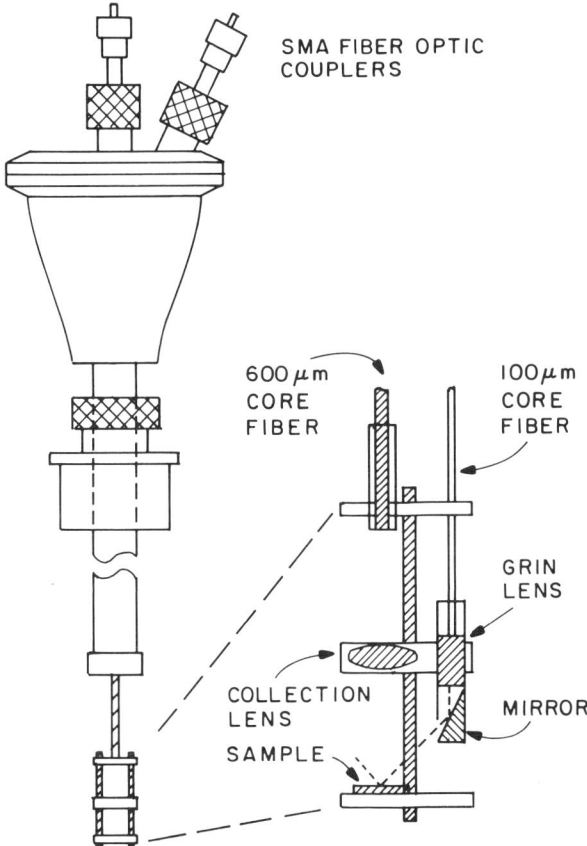

FIG. 1. Diagram of fiber-optic apparatus for Raman scattering and photoluminescence. Input fiber (100-μm core) on right has graded refractive index (GRIN) lens at the bottom to collimate the excitation light, in addition to a mirror to deflect the excitation light onto sample. Scattered light from the sample is focused onto the output fiber (600-μm core) with a 5 mm focal length lens.

ditional input assembly in a forward scattering geometry, aimed 20° away from the output direction (Issacs *et al.*, 1990b).

Polarization selection is easily produced by placing plastic Polaroid polarizers near the sample. However, in the near-infrared region at 0.8 μm, available polarizers lose their polarization at low temperatures. The only polarizer found to work in this application consists of a glass matrix with microscopic silver threads aligned along one direction, available from Corning Glass (Heiman *et al.*, 1989c).

III. GaAs Quantum Well Structures

6. BACKGROUND ON TWO-DIMENSIONAL SYSTEMS

a. Overview

Laser spectroscopy in high magnetic fields is a powerful technique for studying many properties unique to 2D systems. For example, the present understanding of many-body interactions among electrons (or holes) confined to 2D layers in modulation-doped heterostructures owes a great deal to optical measurements. One of the most exciting aspects of 2D electron systems is the enhanced many-body interactions compared with equivalent 3D systems (see Schmitt-Rink *et al.*, 1989). These interactions lead to variations in the screening of the 2D electron gas in the quantum Hall effect (QHE) regime, and as will be shown here, can be directly probed with optical measurements. In addition, optics is probably the most useful technique for studying the effect of quantum confinement on energy levels in undoped materials.

Among the variety of effects observed using optics on GaAs heterostructures, the following are covered in this section.

(1) Integer QHE: Band-gap emission spectra show optical analogs of quantum oscillations, and time-resolved measurements reveal increased recombination at $v = 1$ due to localization.

(2) Fractional QHE: Optical anomalies associated with the fractional quantum Hall effect (FQHE) are manifested as spectral doublets or discontinuous spectral shifts in emission spectra.

(3) Extreme quantization: Inelastic light scattering reveals a dramatic collapse of the intersubband energy.

(4) Collective-mode dispersion: The magnetoroton dispersion, associated with inter–Landau level transitions, was first observed through inelastic light scattering.

(5) Valence subbands: Measurement of intravalence band transitions has mapped out the complex mixture of light- and heavy-hole valence subbands.

(6) Exciton localization: Tunneling and localization of excitons in weak potential fluctuations are observed in high magnetic fields.

Perhaps the simplest and most widely used optical probe is the technique of photoluminescence spectroscopy. In undoped materials the photoexcited electrons and valence band holes bind to form simple 2D excitons. When the 2D quantum well is doped with electrons, the emission originates from the recombination of photoexcited holes with electrons in the 2D electron layer. In this way the emission spectra can directly probe not only single-particle effects but also many-body electron–electron interactions. When a magnetic field is applied perpendicular to a 2D quantum well containing mobile carriers, their motion becomes quantized in Landau orbits. As a consequence, the emission spectrum develops into discrete peaks corresponding to each occupied Landau level. In the regime of the QHE, the emission peak arising from the lowest Landau level does not show a simple linear increase in the transition energy with increasing magnetic field. Instead, at integer and fractional Landau filling factors there appear striking shifts and intensity variations in the spectral peak. This will be described in detail in the following section.

In addition to high magnetic fields, low temperatures are essential for investigating weaker many-electron processes in 2D electron systems. Figure 2 illustrates step-back or modulation doping in single and multiple quantum

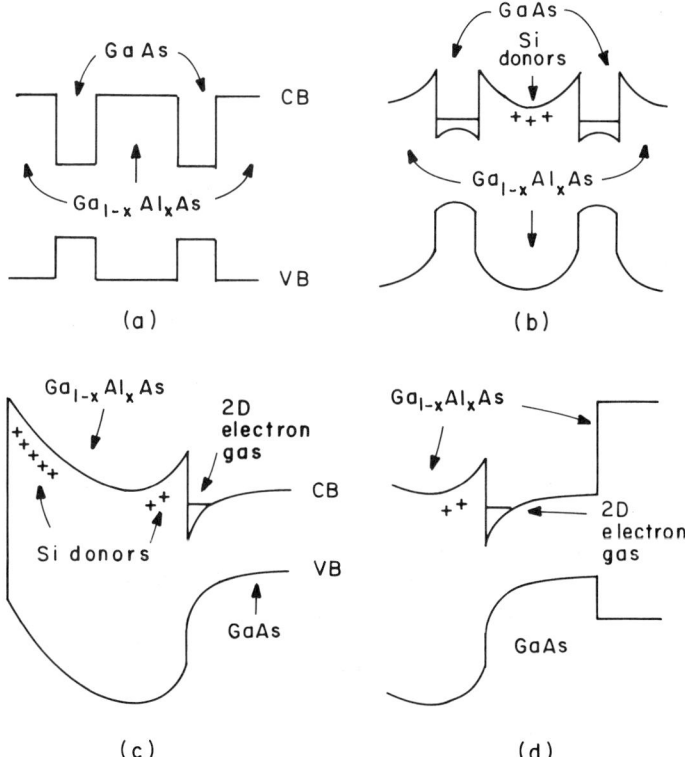

FIG. 2. Diagram of two-dimensional semiconductor heterostructures, where upper and lower lines show conduction and valence bands. Sequential layers of GaAs, $Ga_{1-x}Al_xAs$, and $Ga_{1-x}As{:}Si$ are deposited on a GaAs:Cr substrate. An undoped superlattice is shown in (a). Modulation-doped structures shown are (b) multiple quantum well; (c) single heterostructure; and (d) one-side-doped single quantum well. The curvatures are produced by the electric fields of the separated electrons and ionized donors.

well structures and a single interface. The Si donor ions are spatially removed from the electron well, allowing electron–electron interactions to dominate over electron-impurity scattering. Under these conditions, correlated motions of the electrons in the 2D layer can give rise to a unique incompressible fluid (Laughlin, 1983). The FQHE is found at high fields and low temperatures as a manifestation of the condensation of the 2D electron gas into a quantum fluid. (Tsui et al., 1982; reviews in Prang and Girvin, 1987) Excitations of the ground state of this incompressible fluid appear as quasiparticles having a fractional charge (Laughlin, 1983). These quasiparticles can be considered as defects in the perfect quantum fluid, induced through changes in the filling factor, and they also appear at local structural defects and impurities. The FQHE is observed in magnetoresistance experiments as plateaus in the Hall resistance and minima in the diagonal resistivity at fractional values of the Laudau-level filling factors having an odd denominator (Stormer et al., 1984). Mobilities high enough to reach the FQHE regime, greater than 100,000 cm^2/V-sec GaAs, are only possible with modulation doping.

Optical detection of the FQHE was first reported for Si MIS (metal–insulator–semiconductor) structures. The recombination emission spectra showed a small temperature-dependent shift at $v = 7/3$ for $T = 1.5$ K (Kukushkin and Timofeev, 1986). A temperature-dependent *spectral doublet* in the emission spectra was first observed in GaAs at $v = 2/3$ (Heiman, 1987c; Goldberg et al., 1988a; Heiman et al., 1988b). Following this, spectral doublets were reported for Si at $v = 7/3$ and $8/3$ (Kukushkin and Timofeev, 1988). Spectral shifts and intensity minima at integer and fractional Landau filling have been seen in very high mobility ($>2 \times 10^6$ cm^2/V-sec) GaAs quantum wells (Goldberg et al., 1990; Turberfield et al., 1990; Buhmann et al., 1990).

Perhaps the most novel use of optical studies of many-electron interactions is the observation of the peculiar collective-mode dispersion of inter-Landau transitions, referred to as the magnetoroton (Pinczuk et al., 1988). The electron-electron interactions in a 2D electron gas are responsible for a small shift ($\sim$meV) in the energy between Landau levels. Charge density excitations give rise to a shift that first increases from zero for increasing wave vector, then decreases to a local minimum before increasing again (Kallin and Halperin, 1984, 1985; MacDonald, 1985). This local minimum in the dispersion curve is similar to the roton minimum found in liquid helium. See Section III 3 for more details.

b. Basic Physics of the 2D Electron System

First it will be useful to characterize some basic properties of a 2D electron system, including Landau quantization of the 2D electron gas in high magnetic fields which is a single-particle effect.

A 2D quantum well usually consists of heterostructures of two materials that have an offset in the conduction or valence bands (see Fig. 2), such as two thick slabs of $Ga_{0.7}Al_{0.3}As$ with a thin layer of GaAs sandwiched between them. The lower conduction band in the GaAs creates a well between two barriers having a height of about 1/3 eV. Since the electron motion is severely limited in the direction perpendicular to the layers, the confinement creates discrete kinetic energy levels. The lowest confinement subband, in the limit where the confinement energy is small compared with the barrier height (infinite well model), is pushed up by

$$E_{co} = \frac{\hbar^2 k_z^2}{2m^*}, \qquad (1)$$

where m^* is the effective mass, the confinement wave vector $k_z = \pi/l_z$, and l_z is the well thickness. Thus,

$$E_{co} = \frac{\hbar^2 \pi^2}{2m^* l_z^2} \qquad (2a)$$

$$= \frac{3.8}{(m^*/m_0)(l_z/100 \text{ Å})^2} \text{ meV}. \qquad (2b)$$

For the conduction band in GaAs the effective-mass ratio is $m^*/m_0 = 0.068$ (m_0 is the free-electron mass), and

$$E_{co} = \frac{55}{(l_z/10 \text{ nm})^2} \text{ meV}. \qquad (3)$$

This formula works well only for $l_z > 10$ nm. The excited-state subbands, $M = 0, 1...$, are at $(M + 1)^2 E_{co}$; however, these levels are usually not small compared with the barrier height and one must use a finite barrier model.

In the absence of an applied magnetic field, Fermi statistics limits the number of electrons per kinetic energy state to the spin degeneracy of two. At $B = 0$, electrons in a 2D layer have a constant density of states and fill the lowest conduction subband up to the Fermi energy, given by

$$E_F = \frac{\pi \hbar^2 n}{m^*}, \qquad (4)$$

where n is the 2D electron density. For GaAs,

$$E_F = \frac{n}{10^{11} \text{ cm}^{-2}} \times 3.5 \text{ meV}. \qquad (5)$$

In cases where the Fermi level is above excited subbands, these populated subbands must be taken into account.

An applied magnetic field breaks down the translational invariance that gave rise to the subband, localizing the electron motion to cyclotron orbits. Now since there are many equivalent cyclotron orbits, a large number of electrons have equivalent energies. This produces the large degeneracy of electron states in a magnetic field. In a 2D electron gas it is convenient to characterize the state of the system by the number of electrons that can occupy a given Landau level (cyclotron orbit). The Landau filling factor v is defined as the total number of levels occupied, including two spin levels per Landau level,

$$v \equiv \frac{n}{n_B}, \qquad (6)$$

where n is the 2D carrier density of the populated subband,

$$n_B = \frac{1}{2\pi l_C^2} = \frac{eB}{hc} \qquad (7)$$

is the number of states per unit area of a full Landau level, B is the magnetic field applied perpendicular to the plane of the 2D electron layer, and

$$l_C \equiv \left(\frac{\hbar c}{eB}\right)^{1/2} \qquad (8)$$

is the magnetic length (cyclotron radius) with radial electron density e^{-r^2/l_C^2}. The field required for $v = 1$ is simply

$$B_{v=1} = \frac{n}{10^{11} \text{ cm}^{-2}} \times 4.14 \text{ T}, \qquad (9)$$

and the magnetic length at a given field is

$$l_C = \left(\frac{B}{10 \text{ T}}\right)^{-1/2} \times 8.1 \text{ nm}, \qquad (10)$$

independent of mass. The quantum limit denotes the condition where all the electrons are in the lowest Landau level, and $v \leq 2$ since there are a pair of electrons with opposite spins. The extreme quantum limit occurs when $v \leq 1$, and all the electrons are in the lowest spin state of the lowest Landau level. In practice, however, the levels are often broader than the spin splitting. Then

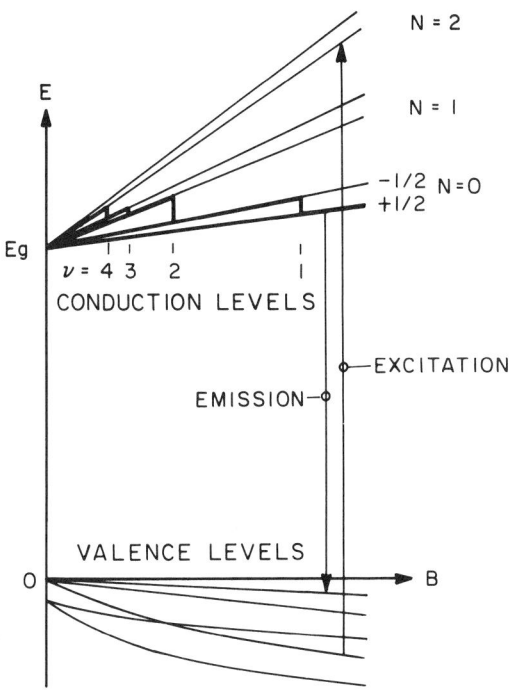

FIG. 3. Schematic diagram of conduction and valence subband energy levels E in an applied magnetic field B. The conduction levels have a small spin splitting of $g\mu_B B$ for the two z-components of electron spin ($\pm 1/2$), and the energies of the electron Landau levels are $(N + 1/2)\hbar\omega_C$. This simple picture comes from a noninteracting system and does not include many-body electron–electron interactions that can be a few milli electron volts. In a photoluminescence experiment an absorbed excitation photon lifts an electron from a valence subband to a high conduction subband, followed by relaxation of the electron (hole) to a lower (higher) level before recombination emission takes place.

$v \leq 1$ only implies that the number of spin-up electrons in the lowest Landau level is larger than the number of spin-down electrons.

A schematic diagram (fan diagram) of the 2D conduction-band Landau levels in GaAs is displayed in Fig. 3. These energies are given by

$$E = E_g + E_{CM} + \left(N + \frac{1}{2}\right)\hbar\omega_C \pm \frac{g\mu_B B}{2}, \qquad (11)$$

where E_g is the band gap, including many-body renormalization, E_{CM} is the energy of the Mth subband in the absence of a field, $N = 0, 1, 2\ldots$ is the Landau level number,

$$\hbar\omega_C = \frac{\hbar eB}{m^*c} \tag{12a}$$

$$= \frac{2\mu_B B}{m^*/m_0}, \tag{12b}$$

is the cyclotron energy, g is the bare (unenhanced) Landé g-factor for electrons in the absence of electron–electron interactions, $m^*/m_0 = 0.069$ at $B = 10$ T, and

$$\mu_B = 0.0579 \text{ meV/T} \tag{13}$$

is the Bohr magneton. For conduction–valence interband transitions, shown in Fig. 3, the valence subband mass is sometimes included by using the reduced mass of the electron and hole, replacing m^* by $(1/m_e^* + 1/m_h^*)^{-1}$. In practice, this is inaccurate for all valence levels except the lowest energy valence subband at high fields, because the combined effects of the magnetic field and the confinement mix the valence subbands in a complicated way, and the resulting highly nonparabolic bands leads to magnetic-field-dependent hole masses (Broido and Sham, 1985; Altarelli, 1985). However, in the range $B = 10–30$ T for a well with $l_z = 20$ nm, the upper hole level has a slope appropriate to a heavy hole $m_{hh}^*/m_0 = 0.45$ (compared with 0.19 for the light hole) (Goldberg et al., 1988c). Electron–hole Coulomb effects (excitons with binding energy E_X) have been neglected here. These are small for $n > 1 \times 10^{11}$ cm^{-2} and negligible for $n > 6 \times 10^{11}$ cm^{-2}. In undoped materials the low-field dependence $[(1/2)\hbar\omega_C \ll E_X]$ is quadratic in B and approaches the Landau level at high fields $[(1/2)\hbar\omega_C \gg E_X]$. Characteristic values for GaAs at $B = 10$ T, where $g = -0.44$, are

$$g\mu_B B = 0.3 \text{ meV} \tag{14}$$

and

$$\hbar\omega_C = 16.8 \text{ meV}, \tag{15}$$

It is clear that the spin splitting is almost negligible compared with the cyclotron splitting. Note that the spin is exaggerated in Fig. 3. Electron–electron interactions can alter these values by as much as a few meV, and are discussed in Section III 8c (Kallin and Halperin, 1984; 1985). With the single-particle energies varying linearly with B, shown in Fig. 3, the Fermi energy follows the heavy line, dropping down to a lower Landau level at even filling factors and to a lower spin level at odd filling factors.

7. BAND-GAP OPTICAL TRANSITIONS

a. Basic Processes

A photoluminescence emission process is shown in Fig. 3, where the absorption of a photon from the incident laser lifts an electron from a valence-subband level to a conduction-subband level. In most cases there is rapid energy relaxation of the electron and hole before recombination. The photoexcited hole in the valence subband quickly rises to the highest valence-subband level, and likewise the electron drops into the Fermi sea. Emission occurs when the photoexcited hole recombines with an electron in the Fermi sea. There are also processes where carriers recombine at impurity and defect sites, but these will not considered here (Kukushkin *et al.*, 1988).

The emission spectrum at $B = 0$ for a sample with electron density $n = 4 \times 10^{11}$ cm^{-2} is shown at the bottom of Fig. 4. Note that the spectral intensity does not go to zero abruptly below $E_0 = E_g + E_{C0}$, but decreases at a moderate rate due to the combined broadening of both conduction and valence subbands. For this material $E_0 = 1515$ meV, $E_F = 14$ meV, $l_z = 200$ Å, $E_{C0} = 16$ meV (this also includes the hole effective mass), making $E_g = 1515 - 16 = 1499$ meV, which is considerably smaller than the bulk band gap in GaAs of 1519.4 meV. This difference is due to the renormalization of the band gap and exciton corrections (Pinczuk *et al.*, 1984; Schmitt-Rink *et al.*, 1989). The intensity decrease for E increasing toward $E_0 + E_F$ is due to lack of large hole wave vectors to match the large electron wave vectors at the Fermi edge. Of particular importance is the cutoff above $E_0 + E_F$ where there are no available electrons.

b. Optical Quantum Oscillations and Selection Rules

When a magnetic field is applied perpendicular to the 2D electron layer, the broad Fermi sea shown in Fig. 4 develops into discrete bands separated by $\hbar\omega_C$. As B increases, $\hbar\omega_C$ increases and the bands move to higher energies. The highest occupied Landau level depopulates completely as it crosses the Fermi level. Actually, the highest Landau level ($N = 2$), seen in the upper spectrum of Fig. 4 ($B = 3$ T), should be empty since $v = 6$ occurs at $B = 2.7$ T, but level broadening is responsible for the partial occupation. The general behavior of the disappearance of the Landau peak at even filling factors confirms the picture of the Fermi level jumps represented in Fig. 3. Similar effects have been discussed by Perry *et al.* (1987).

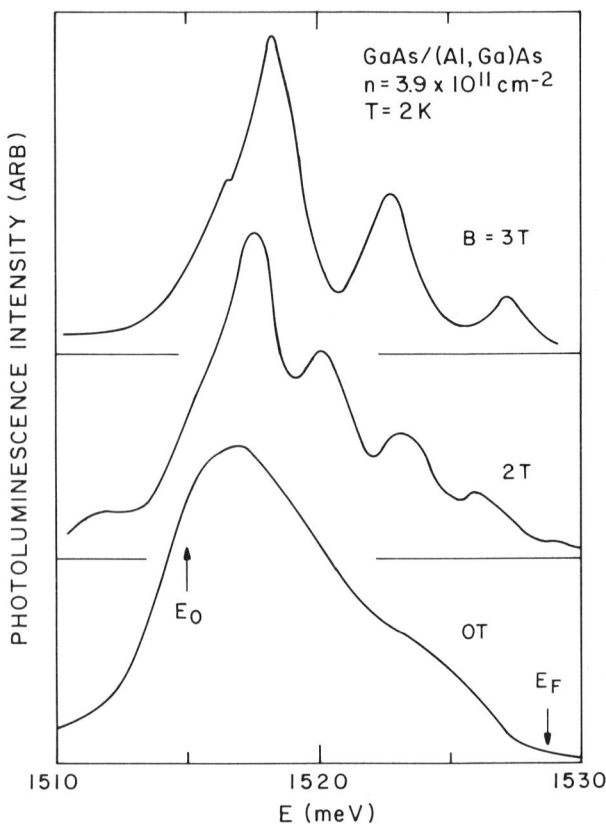

FIG. 4. Photoluminescence emission spectra of a 2D electron gas at small magnetic fields B. The modulation-doped multiple quantum well sample of GaAs–(Al,Ga)As had a carrier density $n = 3.9 \times 10^{11}$ cm^{-2} and mobility $\mu = 2 \times 10^5$ cm^2/V-sec. The bottom of the conduction subband is at E_0, and electrons are filled up to the Fermi energy E_F. The magnetic field breaks up the photoluminescence arising from the Fermi gas at $B = 0$ into spectral peaks associated with discrete Landau levels.

An inverse effect appears in the optical transmission spectrum—new absorption peaks suddenly *appear* at even values of v for increasing B (Goldberg et al., 1988b). As the Fermi energy drops down to a lower Landau level that level begins to empty, allowing a new absorption channel. The inverse nature arises because emission requires *occupied* states, while absorption requires *unoccupied* states. Figure 5 shows the onset of the $N = 1$ Landau level at $B = 4.5$ T and the $N = 0$ level at $B = 8$ T. The first-excited subband transition is seen at $E = 1543$ meV. This makes the energy difference between the ground and first-excited subbands 28 meV, considerably smaller than the infinite-barrier value of 50 meV.

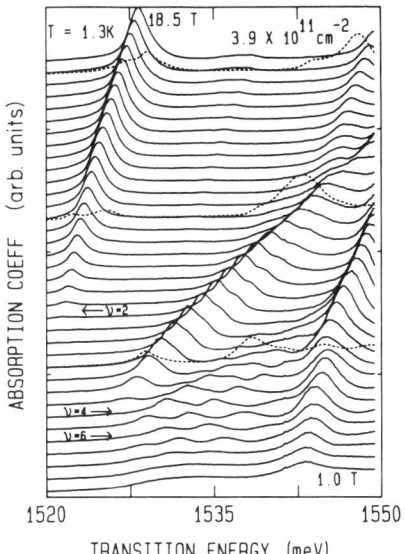

FIG. 5. Absorption spectra of a 2D gas at high magnetic fields, from 1 to 18.5 T in steps of 0.5 T. Note the emergence of new transitions at $v = 6$, 4, and 2, indicating that these Landau levels are no longer full at higher fields. The modulation-doped multiple quantum well GaAs–(Al,Ga)As sample had a carrier density of $n = 3.9 \times 10^{11}$ cm^{-2} and mobility $\mu = 2 \times 10^5$ cm^2/V-sec. (From Goldberg et al., 1988c.)

Identification of the conduction- and valence-subband states has been determined from polarization measurements of the optical transitions, as a function of applied magnetic field (Goldberg et al., 1988c, 1989; Heiman et al., 1989a). Figure 6 shows a set of three closely spaced absorption transitions associated with the two spin states of the lowest conduction Landau level. Calculations of the valence subbands were matched to the measured energies and optical polarizations. At high magnetic fields, the lowest energy transition in absorption, transition c, corresponds to that observed in optical emission, since the photoexcited hole rises up to the highest valence subband 0^+ ($m_j = +3/2$). The selection rule for emission only permits the hole to recombine with a spin-up 0^+ ($m_j = +1/2$) electron. (Spin-up electrons are the lowest energy state because of the negative g-factor.) Thus, recombination emission, denoted as $0^+ \to 0^+$, is polarized primarily σ_- at high fields.

It is important to note that it is not necessary to have a spin-polarized electron gas ($v \leq 1$) to obtain entirely circularly polarized emission, since the polarization may be determined by the spin-polarized hole. Furthermore, it is not possible to measure the conduction-band spin splitting directly, since selection rules prohibit transitions from both spin states to a single valence level.

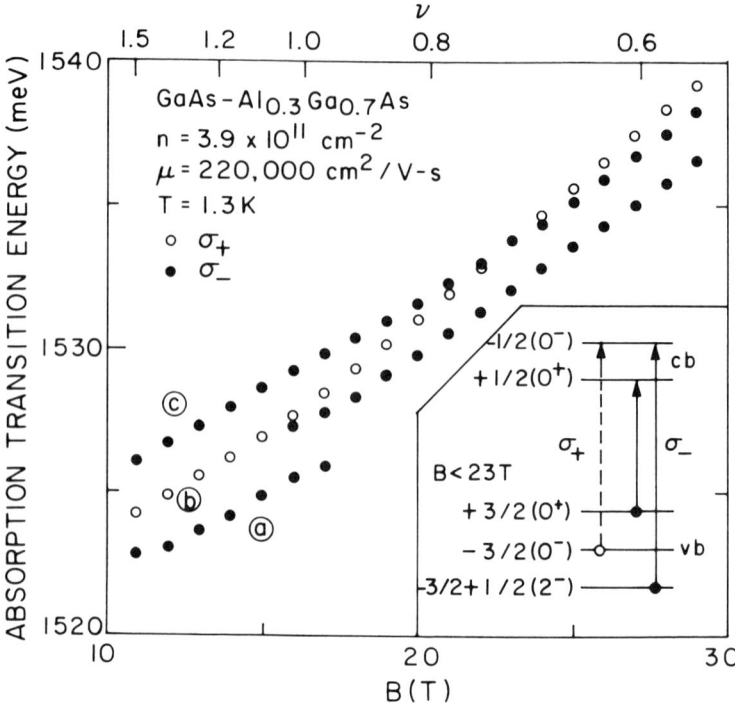

FIG. 6. Conduction–valence band transition energies of a 2D electron gas at high magnetic fields. Circles represent the spectral peaks from transmission measurements on a modulation-doped multiple quantum well sample. Closed (open) circles are for light circularly polarized σ_- (σ_+). The inset shows the participating levels for the three transitions (a, b, and c). (From Heiman et al., 1989a.)

c. Nonlinear Landau Level Tuning in Emission

In multiple quantum well (MQW) structures of GaAs–(Al,Ga)As that are modulation-doped with electrons, the interband emission spectra show striking *nonlinear* energy shifts with increasing B (see the review Perry et al., 1987). At low temperatures ($T < 10$ K) the energy of the valence- to conduction-subband transition is an oscillatory function of filling factor v. These features are not observed in the spectra of undoped quantum well systems and require a theory that contains electron–electron interactions. The electron self-energy may account for this behavior (Uenoyama and Sham, 1989).

Figure 7 shows the measured interband transition energies involving the $N = 0$ and $N = 1$ Landau level levels. The upper transition, $N = 1$, shows a nearly linear increase in energy for increasing B (The small change in slope near $B = 3$ T is attributed to a change from one valence subband to another

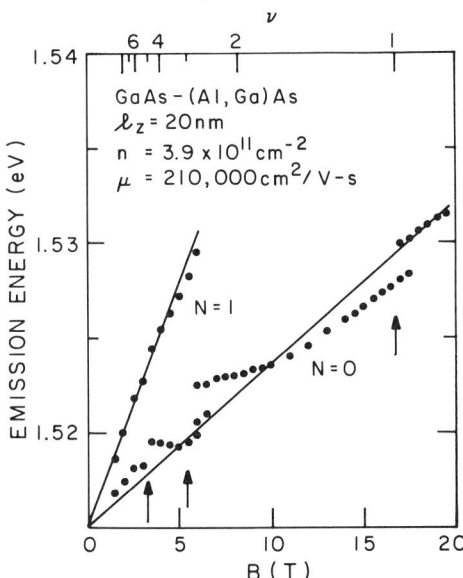

FIG. 7. Landau-level energies of a 2D electron gas at high magnetic fields, derived from spectral emission peaks in photoluminescence from a multiple quantum well sample. The arrows denote discontinuous spectral blue shifts at filling factors $v = 5, 3,$ and 1. Solid lines represent the expected energy of the two lowest Landau levels in a noninteracting electron gas.

where a crossing occurs.) For comparison, solid lines are drawn having slopes of $1/2$ and $(3/2)\hbar\omega_C$. In marked contrast, the lowest transition, $N = 0$, shows regions where the transition energy is nearly field-independent and even decreases slightly with increasing field near $v = 4$. In addition, there are a number of *discontinuities* as a function of filling factor. These deviations from linear-in-B energy shifts appear to be less apparent for lower-mobility samples, but have been observed in samples with mobilities as low as $\mu = 10,000 \text{ cm}^2/\text{V-sec}$.

These nonlinearities are related to changes in the screening of the 2D electron gas with changing filling factor. At even filling factors the cyclotron levels are fully occupied and screening is smallest. Here, electrons do not effectively screen out a Coulomb field because they cannot move without jumping the gap $\hbar\omega_C$ into an unccoupied Landau level at higher energy. However, between these even filling factors the screening is larger because the Landau level is only partially full. Those electrons provide screening because they are free to move into unoccupied states.

A simple picture that uses excitonic-like screening of the valence-subband hole cannot account for the nonlinearity. In this case, increased screening would lead to a lowering of the transition energy caused by buildup of charge around the hole. This would amount to oscillations in the transition energy between the $\hbar\omega_C/2$ line and below. The minimum energy lowering should be centered around even filling factors, while the maximum lowering would be between these values if the spin gap is neglected. This is not observed in the data. In fact, the data go substantially above the $\hbar\omega_C/2$ line. Next, the behavior at $\nu = 1$ is examined in more detail.

d. Spectral Blue-Shift Anomaly at $\nu = 1$

An optical blue-shift anomaly has been found in the interband optical spectrum near $\nu = 1$ in GaAs MQW structures (Heiman et al., 1988b). Closer examination of the data in Fig. 7 reveals a similar discontinuity near $\nu = 3$. There is even a substantial blue shift at $\nu = 5$; however, the linewidth is too broad to establish whether the shift is discontinuous. The appearance of a new spectral line at a higher transition energy is evident in many samples having mobilities greater than 10^5 cm^2/V-sec. The anomaly at $\nu = 1$ appears in both emission (Heiman et al., 1988b, 1989a) and absorption (Goldberg et al., 1989; Heiman et al., 1989a).

Figure 8 shows a set of emission spectra for a sample with $n = 3.9 \times 10^{11}$ cm^{-2} and $\mu = 2 \times 10^5$ cm^2/V-sec. The feature at low energy is identified as

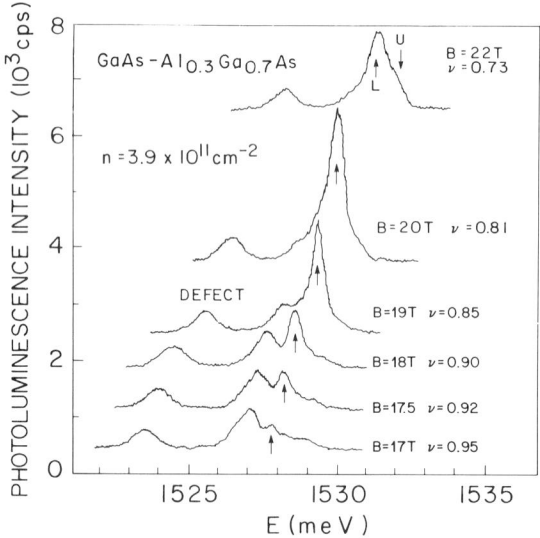

FIG. 8. Photoluminescence emission spectra of a 2D electron gas at high magnetic fields B. The arrows indicate a new peak that appears in the extreme quantum limit $\nu < 1$. The modulation-doped multiple quantum well sample had a carrier density $n = 3.9 \times 10^{11}$ cm^{-2}, mobility $\mu = 2 \times 10^5$ cm^2/V-sec, and temperature $T = 0.5$ K. (From Heiman et al., 1988b.)

recombination at a defect and will not be discussed, since it does not show anomalous behavior. The structure at higher energy is identified as recombination of an electron in the 2D gas with a photoexcited hole in the valence band, since this energy coincides with the absorption onset. Notice a new spectral peak, indicated by the arrow, that increases in intensity with increasing field. It first appears as a small shoulder on the high-energy side of the main intrinsic peak; then at high fields ($B > 18$ T) it becomes the dominant feature. Figure 9 displays the energies of the spectral peaks versus B. For increasing B, the intrinsic peak (solid circles) deviates to lower energy at $\nu = 1$ and becomes replaced by the upper peak. The magnitude of the spectral splitting is about 1 meV at this temperature ($T < 1$ K) and decreases for

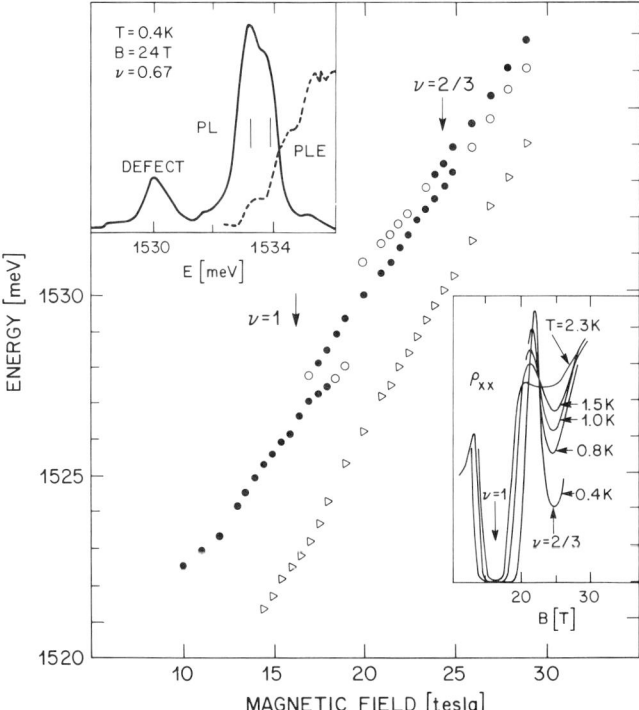

FIG. 9. Transition energies from photoluminescence spectra of a 2D electron system at high magnetic fields. The modulation-doped multiple quantum well sample of GaAs–(Al,Ga)As had a carrier density $n = 3.9 \times 10^{11}$ cm^{-2}, mobility $\mu = 2 \times 10^5$ cm^2/V-sec, and temperature $T = 0.5$ K. Open circles represent weaker intensities compared with the solid circles, and the open triangles show the position of the defect-related peak. Upper left inset shows the photoluminescence emission spectrum, and the PL excitation (PLE) which represents the absorption process. The lower right inset shows the ρ_{xx} magnetotransport under illumination at various temperatures. (From Heiman et al., 1988b.)

increasing T. Above $T \sim 10$ K the $v = 1$ anomaly disappears and the peak shifts continuously with increasing field. A more extensive study of the temperature dependence in absorption measurements has been reported (Goldberg et al., 1989). Similar blue-shifting doublets have also been observed in single quantum well structures that are modulation-doped on one side and have mobilities up to 3×10^6 cm^2/V-sec.

Recombination transitions within an electron gas depend significantly on the local environment of the minority carrier, the valence-subband hole. Screening of the positively charged hole redistributes the local change density of the 2D electron gas. When v moves away from an integer, the screening is enhanced because the electrons can change their state within the lowest Landau level. (Das Sarma, 1981; Ando and Murayama, 1985) The resulting modification of the local electron distribution effects the exchange and Coulomb corrections and the excitonic (vertex) corrections to the transition energy (Kallin and Halperin, 1984, 1985; Yang and Sham, 1987; Uenoyama and Sham, 1989; Katayama and Sham, 1989) as well as the oscillator strength (matrix element) of the transition. The observed spectral doublet anomaly at $v = 1$ appears to be a result of increased screening for $v < 1$. This is discussed further in the next section.

e. Optical Analog of the Quantum Hall Effect

Next, very high mobility, $\mu > 2 \times 10^6$ cm^2/V-sec, single-side-doped single quantum well structures are examined. As displayed in Fig. 2, the hole wave function is somewhat spatially separated from the electron wave function, in contrast to the multiple quantum well systems where the wave functions have a stronger overlap. The blue-shift anomaly in the emission spectra at $v = 1$ observed in the previous section develops into a plateau region where the emission peak energy varies slowly with changing B field in these single-side-doped structures (Goldberg et al., 1990). The range of field over which the plateau exists coincides almost exactly with the ρ_{XY} quantum Hall plateau observed in magnetotransport. The solid curves in the upper part of Fig. 10 display ρ_{XY}, and the open circles show the transition energy of the spectral PL peak. Note that the lower curves in each set, taken at $T = 0.4$ K, both show a plateau width of about 2 T. This correlation in plateau widths for the optics and transport even extends to the temperature dependence: as the temperature is raised to $T = 2.5$ K, the width of the plateau decreases to zero. It is also clear that over the plateau region the intensity of the emission is reduced, similar to the ρ_{XX} minimum in the magnetotransport. The intensity minimum also exists only over a region of temperature and magnetic field where the quantum Hall plateau in transport exists. Finally, note that the intensity minimum is much more robust than the spectral plateau; while the

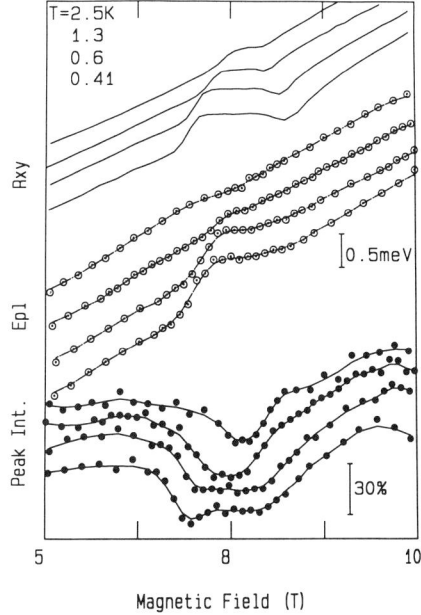

FIG. 10. Optical analogue of the integer quantum Hall effect at $v = 1$. Upper solid curves show ρ_{xy} quantum Hall plateaus in magnetotransport, open circles show pinning of the photoluminescence emission energy of the lowest Landau level, and the solid circles show minima in the emission peak intensity. The modulation-doped single quantum well sample of GaAs–(Al,Ga)As had a carrier density $n = 2.0 \times 10^{11}$ cm^{-2}, mobility $\mu = 3 \times 10^6$ cm^2/V-sec, and various temperatures. (From Goldberg et al., 1990.)

intensity minimum is present in all samples examined, the spectral plateau is well developed only in selected samples, and often it is not a flat plateau but has a finite $E(B)$ slope.

The observed optical analog to the Hall quantization has a similar origin as the quantum Hall effect itself; both appear when the Fermi level lies in the region of localized transport states. With changing magnetic field, as the Fermi level moves from a region of extended states to localized states the screening behavior of the 2D electron gas changes from metallic-like to insulating-like. At this transition the screening of the valence band hole is reduced because the electron gas can no longer react to the presence of the positively charged hole. This has the effect of altering the many-body energy corrections of the electron and hole. For electrons, the contributions from the screened-exchange and the Coulomb-hole interactions nearly cancel (Uenoyama and Sham, 1989; Katayama and Ando, 1989). The only remaining contribution comes from the Coulomb-hole interaction of the valence-band hole. Thus, as the screening changes with changing magnetic field, the spectral peak should also shift. Unfortunately, calculations that include the effects of localization are difficult and have not been completed at this time. Qualitatively, when there is a gap in the allowed energy states the spectral emission near integer filling factors is expected to blue-shift when the coulomb-hole interaction dominates, and red-shift when the vertex correction dominates (Goldberg, et al., 1991).

f. Blue-Shifting Spectral Doublet Anomaly near $v = 2/3$

Spectral doublets with blue-shift anomalies in the regime of the FQHE have been observed in GaAs–AlGaAs single and multiple quantum well structures. First, n-type modulation-doped multiple quantum well structures will be discussed. Simultaneous magnetotransport and recombination emission measurements were made in order to correlate the optical anomalies with the fractional effects.

The samples had carrier densities between $n = 1.4$ and 6×10^{11} cm^{-2} and mobilities of $\mu = 1.3–5 \times 10^5$ cm^2/V-sec, and all showed well-developed FQHE ρ_{XX} minima at either $v = 1/3$ or $2/3$. Well widths were approximately 20 nm, with 25 to 85 periods. A fiber-optic apparatus (Heiman et al., 1989c) was used to access the He^3 cryostat and high-field magnet, and when polarization information was needed a circular polarizer was placed next to the sample. The optical power at 1.58 eV was limited to 10^{-5} W over an area 10^{-2} cm^2 to avoid heating, which was monitored by the temperature-dependent minima in ρ_{XX} at $v = 2/3$, displayed in the inset of Fig. 9. Roughly, the electron temperature increased by 0.1 K for every 0.1 W of input power, between $T = 0.5$ and 1 K.

The anomaly at $v = 2/3$ is similar to the one at $v = 1$, where their common characteristic is the development of a new emission peak at *higher* energy for increasing field. The energies of the emission doublet components are plotted as a function of B in Fig. 9. The stronger component is represented by the solid circles. In the vicinity of $v = 2/3$ the splitting is ~ 0.6 meV and independent of field, and the two components persist over a wide field range, ~ 10 T, in contrast to the $v = 1$ anomaly. These two traits are not consistent with level crossing, such as resonant polaron coupling [cyclotron energy equal to longitudinal-optic (LO) phonon energy], which occurs at $B \approx 22$ T, or to crossing of the cyclotron level with an intersubband level. For example, polaron coupling should result in a sharp magnetic-field-dependent level splitting only over a narrow range of B, which is not observed.

The B- and T-dependencies of the spectral doublet at $v = 2/3$ have some peculiar characteristics. Although the doublet separation is nearly constant, the spectral weights of the doublet components vary rapidly. Figure 11 shows the emission spectra of the doublet, the larger structure at high energies, in addition to the smaller defect-related peak at lower energy. The vertical bars indicate the intensities and energies of the two components determined by curve-fitting the spectra to the sum of two Gaussian line shapes with equal linewidths. Notice that the higher-energy component increases relative to the lower-energy one for increasing B. This is evident at both temperatures, $T = 0.4$ and 0.8 K. However, there is a remarkable difference for the two temperatures — the magnetic fields corresponding to the equal-intensity crossovers are different. For $T = 0.4$ K this occurs at $B = 24.5$ K, while for $T = 0.8$ K the crossover occurs 2 T higher, at $B = 26.6$ T. (Another way to look at the temperature dependence is to plot the intensity ratio, I_L/I_U (lower divided by upper) at constant B for various temperatures.)

In the temperature dependence the key feature is the energy scale of the $v = 2/3$ optical anomaly, which is similar to that of the transport FQHE. The intensity ratio I_L/I_U has been studied for a range of temperatures $T = 0.4$–1.5 K, magnetic fields $B = 10$–30 T, as well as for densities $n = 1.8$–3.9×10^{11} cm^{-2}.

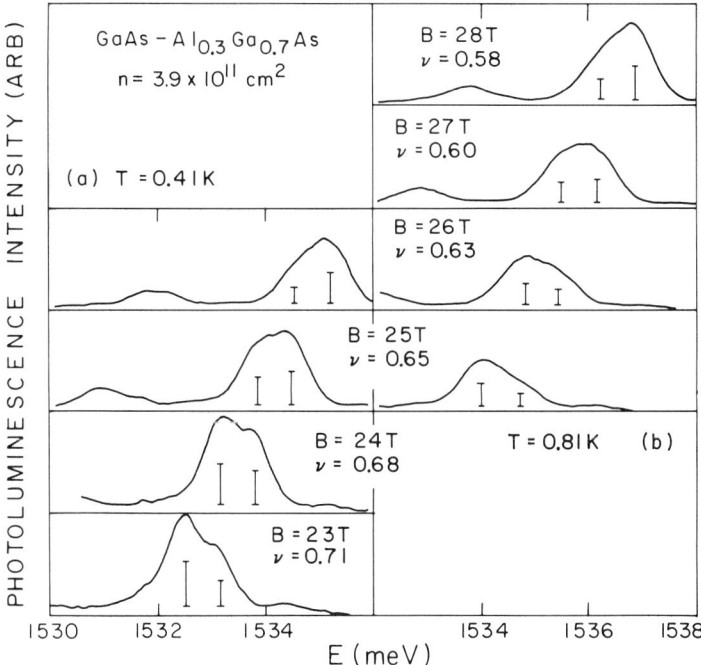

FIG. 11. Temperature dependence of the $v = 2/3$ spectral blue-shift anomaly in the emission spectra. At fixed B the higher energy dominates at low temperature. The vertical bars represent the intensity of each spectral component of the doublet, obtained by curve fitting. The modulation-doped multiple quantum well sample had a carrier density $n = 3.9 \times 10^{11}$ cm^{-2} and mobility $\mu = 2 \times 10^5$ cm^2/V-sec. (From Heiman et al., 1988b.)

(Heiman et al., 1988b). The characteristic optical energy has been determined from the intensity ratio I_L/I_U versus, $1/T$. Although the dependence is not strictly a straight line, the slope is similar to that of the magnetoresistance ρ_{XX} minima at $v = 2/3$. Magnetotransport has an activation energy of $\Delta E/2 = 0.44$ K, which is about a factor of 4 lower than in comparable mobility single hetrostructures (Boebinger et al., 1987). Although the optical results have a slope similar to the activation energy of the FQHE, it does not demonstrate that the optical intensity ratio is activated. It merely implies that they have the same energy scale. In fact, the slope of I_L/I_U is only weakly dependent on field between $B = 21$ and 29 T, whereas the activation energy of ρ_{XX} is maximum at $v = 2/3$ and decrease to zero for a change of field of ± 3 T.

Optical transmission measurements were made in the region of the $v = 2/3$ anomaly (Heiman et al., 1989a). In contrast to the emission, the absorption showed no measureable temperature-dependent energy shifts. Thus, the blue shift observed near $v = 2/3$ only appears in the final state interaction, unlike the $v = 1$ anomaly that appears in both the initial absorption and the final emission.

Higher-mobility single-side-doped samples, $\mu \sim 2 \times 10^6$ cm^2/V-sec, display an intensity minima and spectral blue shift at $v = 1/3$ and $2/3$ (Goldberg et al., 1990). The qualitative features of the intensity minima, including its temperature and field dependence, are similar to the minima at $v = 1$ described earlier, except with different temperature scales. The spectral blue shift differs from that found in the lower-mobility MQW sample—the blue shift persists only over a narrow field range, simultaneously with the ρ_{XX} magnetoresistance minima; and the spectral peak red-shifts to the original unshifted position for fields higher than the fractional filling factor. This may be a general behavior occuring at all fractional states; however, it has only been established at these two fractions.

The screening changes that led to the $v = 1$ anomaly are different for the case at fractional filling. Here, quasiparticles redistribute the charge in the vicinity of the valence-subband hole. Thus, the hole might be expected to be "dressed" with a quasiparticle defect. The response of the 2D electron charge distribution to the photoexcited hole might be treated like the cases of charged impurities and disorder (Zhang et al., 1985; Rezayi and Haldane, 1985; MacDonald et al., 1986). A quantitative model should take these into account.

g. Time-Resolved Photoluminescence at $v = 1$

Time evolution of the PL emission has been studied near the $v = 1$ optical anomaly (Dahl et al., 1992). In the experiment, a short pulse of light creates valence-band holes in excited subbands. These holes relax in energy during the PL rise time and are removed as they recombine with electrons in a time characteristic of PL decay time. Figure 12 shows an increase in both the rise and decay times for the emission peaks at $v = 1$. The rise time increases by 30–50% from the background decay time of 200 psec. This increase goes away at higher temperatures, $T > 3$ K, in correspondence with the disappearance of the quantum Hall effect itself.

The slowing down appears to be a direct result of the localization giving rise to the insulating state of the QHE. At $v = 1$, the Fermi level is in the middle of the spin gap, and the electrons cannot respond to either the valence band hole or the potential fluctuations of the remote ionized donors. This results in a small electron–hole overlap, in perhaps both the z-direction and xy-plane, and the reduced optical matrix element gives a lower probability for recombining and, hence, a longer recombination time. Such slowing down also explains the intensity minimum at $v = 1$. Since the hole stays around longer, it has a higher probability of recombining with electrons in excited conduction subbands. In fact, the intensity that is lost in the ground state is precisely made up for in the excited-state emission (Turberfield, et al., 1990;

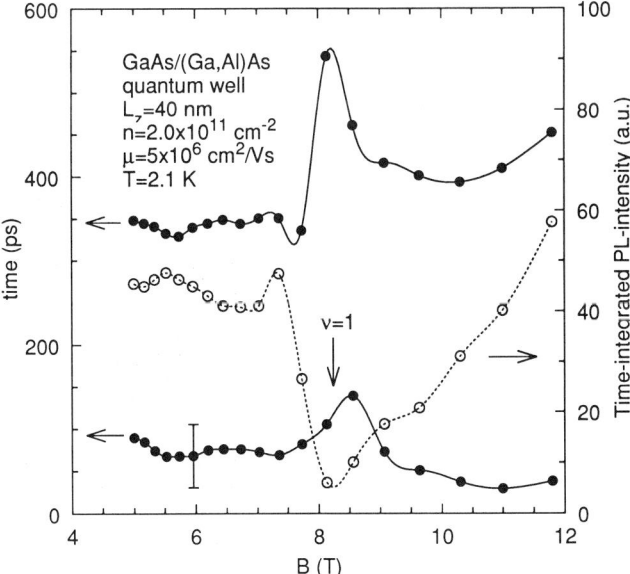

FIG. 12. Rise time and decay time from time-resolved photoluminescence measurements of the ground-state emission near $v = 1$ as a function of applied magnetic field B. The GaAs–AlGaAs single interface had a carrier density of $n = 2.0 \times 10^{11}$ cm^{-2} and an electron mobility of $\mu = 5 \times 10^6$ cm^2/V-sec.

Dahl, et al., 1992). This also points to the fact that the holes do not decay nonradiatively, on the nanosecond time scale, and there is merely an exchange of spectral intensity as the different electrons compete for the hole.

h. Summary of Optical Emission in the Quantum Hall Regime

Recombination photoluminescence has been shown to be a valuable probe of electron–electron interactions in 2D systems. This is especially true in GaAs heterostructures for a number of reasons: (1) the materials grown with state-of-the-art molecular beam epitaxy have exceedingly high purity, with electron mobilities above 10×10^6 cm^2/V-sec; (2) the quantum efficiency for optical recombination is near unity; (3) tunable laser sources and high-efficiency optical detectors are readily available; and (4) low-loss fiber-optic light pipes now allow efficient light transmission to small-bore high-field magnets and low-temperature cryostats and refrigerators.

It is by no means presumed that present experiments have established universal behavior. For example, some samples might show clear spectral doublets while others only show a spectral shift, even with equivalent carrier density and mobility. Materials can even become worse after being thermally cycled. A second problem is that of density changes with light illumination (see Section 8d). This is especially difficult when the changes occur on the time scale of many minutes to hours. For this reason, optical experiments should be carried out with simultaneous magnetotransport measured over the same region of the sample.

Recent experiments have already been applied to samples with mobilities near 8×10^6 cm^2/V-sec, in magnetic fields up to $B = 45$ T, and temperatures down to $T = 0.1$ K. From these measurements, novel features have been found, including the optical analog of the integer QHE, spectral doublets associated with the FQHE, and blue shifts at very high magnetic fields. At the present time, these effects can be attributed qualitatively to localization and screening by electrons and quasiparticles. In emission experiments the presence of the valence-band hole affects the dense electron system a great deal and must be included. Many of the theoretical ingredients already exist, but have not yet been combined and applied to the specific processes. In the regime of the integer QHE the interactions include screened-exchange and Coulomb-hole self-energies, vertex corrections (exciton effect), and self-consistent screening (Uenoyama and Sham, 1989; Katayama and Ando, 1989; Schmitt-Rink et al., 1989). At fractional filling factors the problem also includes the additional short-range interactions giving rise to the incompressible Laughlin fluid.

8. Inelastic Light Scattering of 2D Electron and Hole Systems

This section describes the utility of Raman inelastic light scattering at high magnetic fields for probing

(1) single-particle intersubband excitations to study the confinement subband states,
(2) collective modes of carriers confined in two dimensions, such as magnetoplasmons,
(3) many-body effects of high-mobility 2D electron systems; and
(4) 2D electron solid formation in high magnetic fields.

Inelastic light scattering on 2D electron systems has been reviewed in a number of publications (Pinczuk and Worlock, 1982; Worlock et al., 1983; Absteiter et al., 1986; Pinczuk and Heiman, 1987).

a. Spectroscopy of 2D Valence-Subband States

Inelastic light scattering on p-doped GaAs quantum wells has been useful for studying the complex character of the valence-subband states (Pinczuk and Heiman, 1987). These complexities result from quantum confinement effects on the degenerate valence-band states of the zincblende structure, which splits and mixes light- and heavy-hole valence-band states of the bulk. Calculations have shown substantial nonparabolicity and even electron-like, negative-mass hole states. A large magnetic field provides additional confinement in the 2D plane. This gives rise to *field-dependent* mixing of the valence-subband states. Effective-mass theories have been useful in predicting these properties and providing quantitative results (Broido and Sham, 1985; Altarelli, 1985; Bastard, 1985; Bangert and Landwehr, 1985; Sham, 1987). Comparisons between theory and experiments will be reviewed in this section.

To understand the results of laser spectroscopy of hole states at high fields, $B = 0$ spectra will be described first (see Pinczuk and Heiman, 1987; Pinczuk et al., 1986, 1987). In a manner similar to n-type modulation-doping, the (Al,Ga)As barriers have acceptor atoms only in the central region of the barrier. Thus, the free holes in the quantum well are spatially separated from the A^- ionized acceptors in the barrier layer. At moderate doping levels $p \sim 10^{11}-10^{12}$ cm^{-2}, the holes reside in the highest valence subband, the heavy-hole h_0 state. Figure 13 represents the resonant scattering process in p-type, modulation-doped quantum wells. Transitions can be made between

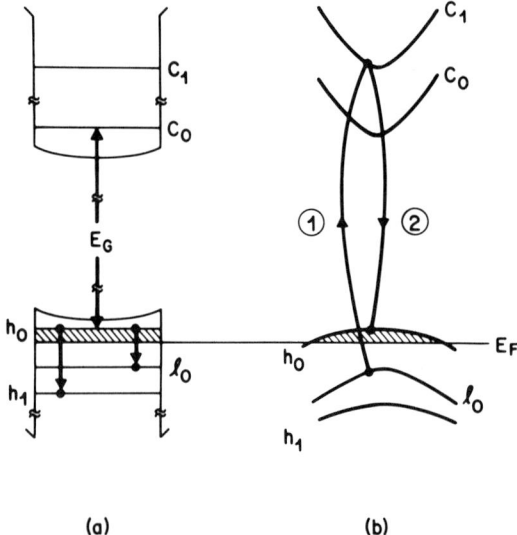

FIG. 13. Schematic diagram of the process of inelastic Raman light scattering of holes in a 2D system. Transitions are shown between conduction–valence subbands in a *p*-doped quantum well. Free holes occupy the shaded region between the uppermost heavy-hole valence subband and the Fermi level. Raman transitions are shown in (a) by the two arrows connecting occupied hole states and empty excited hole states. In (b) the resonance Raman process is shown where (1) the incident photon lifts an electron from valence level to the excited conduction subband, followed by (2) emission when the electron recombines with a hole in the Fermi sea. (From Pinczuk and Heiman, 1987.)

h_0 and both the lowest light-hole l_0 and first-excited heavy-hole h_1, denoted by downward-pointing arrows in Fig. 13a. The input laser energy is made resonant with the first-excited conduction-band state c_1 (Fig. 13b) to avoid the background from recombination emission associated with the lowest conduction subband c_0. Figure 14 shows the light scattering spectra for a multiple quantum well sample with a doping level of $p = 1.5 \times 10^{11}$ cm^{-2}. Two intersubband transitions can be identified, $h_0 \to l_0$ and $h_0 \to h_1$. The lower-energy transition $h_0 \to l_0$ appears as an onset of intensity followed by a gradual decline for higher-energy shifts. This arises because the two bands are not parallel and their separation increases for increasing *k*-vector. The higher energy peak denoted as $h_0 \to h_1$ is sharp because the initial and final states are parallel in *k*-space.

The picture of valence-subband transitions is quite a bit more complicated with a magnetic field applied perpendicular to the 2D layers (Heiman et al., 1987c). Each of the subband levels breaks up into a set of Landau levels.

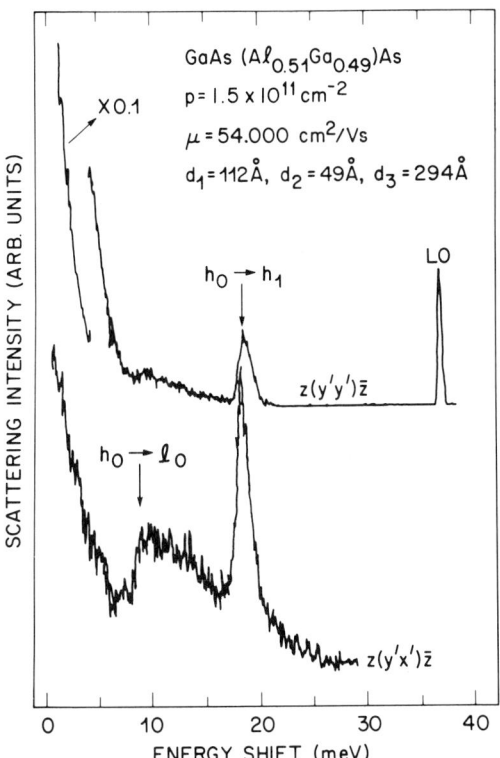

FIG. 14. Inelastic light scattering from free holes in a GaAs 2D system. Upper spectrum in a depolarized configuration shows the heavy to heavy-hole transition, while the lower polarized spectrum shows in addition the onset of the heavy to light-hole transition. (From Pinczuk and Heiman, 1987.).

Figure 15 shows the six lowest Landau levels associated with each h_0, l_0, and h_1 subband levels. These were obtained from envelope-function calculations that include mixing between the first 10 valence-subband levels (Altarelli, 1985). The Fermi level is shown by the heavy line on the uppermost Landau levels. Raman-allowed transitions originate from one of the filled levels and terminate on an empty level, with a change of Landau quantum number $\Delta L = 0, \pm 2$. Some of these transitions are indicated by vertical lines. Figure 16 displays the energies of the strongest peaks observed in the spectra. The two data points at $B = 0$ are the $h_0 \to l_0$ and $h_0 \to h_1$ transitions mentioned previously. The solid lines show the calculated transitions with *no adjustable parameters*; the only inputs were the sample parameters. Notice the good agreement between theory and experiment for the $1 \to 1$ ($h_0 \to l_0$) above $B = 5$ T. Relatively good agreement is also seen for the two higher-energy $h_0 \to h_1$

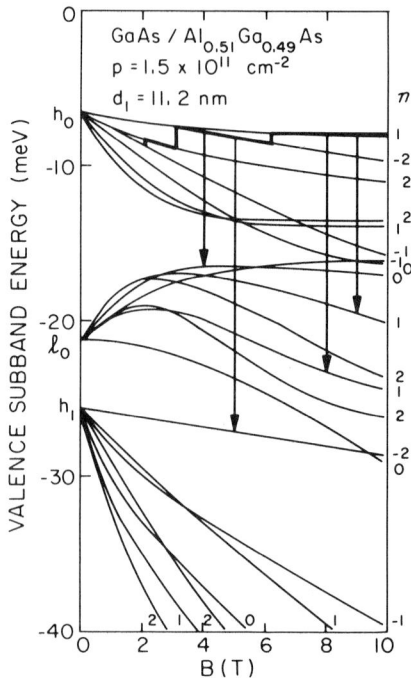

FIG. 15. Calculated valence-subband energies versus magnetic field for a GaAs quantum well. The bold curves at the top show the Fermi level for the hole gas, while the arrows indicate Raman-allowed transitions from occupied hole states to unfilled states. (From Heiman et al., 1987c.)

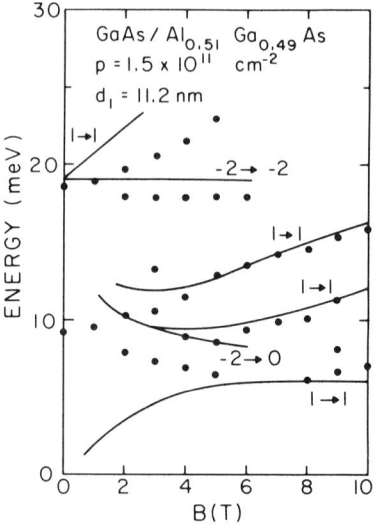

FIG. 16. Measured transition energies for a 2D system of holes in GaAs versus magnetic field. The solid curves were derived from calculations. (From Heiman et al., 1987c.)

transitions, except for a minor downward shift of the experimental data. This small discrepancy has been accounted for under more careful examination. It was observed that the energy of the $h_0 \to h_1$ transition varies slightly with laser excitation energy; monolayer well-width fluctuations change both the intersubband transition energy and the resonance condition (Goldberg et al., 1988b). Another striking observation is the disappearance of two strong lines, $-2 \to 0$ and $-2 \to -2$, above $B = 6$ T. These two transitions are expected to disappear abruptly at $B = 6.2$ T when the $L = -2$ Landau level becomes empty at $v = 1$. The upper peak at 18 meV is field-independent, characteristic of the calculated $-2 \to -2$ transition. The $L = -2$ lowest Landau levels of the h_0 and h_1 subbands do not mix with any other states and thus have linear tuning with magnetic field. Although the lowest energy peak only agrees qualitatively with the calculated $-2 \to 0$ transition, it is important to note that it has a strong shift to lower energy with increasing field. Such a decrease in transition energy is a verification of the *negative-mass curvature* introduced by subband mixing. These l_0 levels are seen to increase toward the conduction band for increasing field.

b. Excitations of the 2D Electron Gas in High Magnetic Fields

This section covers three types of excitations of the 2D electron gas in GaAs. Quasi-3D magnetoplasmons are collective excitations that couple separate 2D electron layers in *n*-type modulation-doped MQW systems. Single-particle excitations include both inter–Landau level transitions (cyclotron transitions) and combined Landau level–intersubband excitations. The first observations of resonant inelastic light scattering by 2D electrons in high magnetic fields were reported in 1981 (Worlock et al., 1981; Tien et al., 1982).

Inelastic light scattering obtained from a modulation-doped MQW structure containing 15 wells of 27 nm thickness, 98 nm superlattice period, carrier density $n = 8.2 \times 10^{11}$ cm^{-2}, and mobility $\mu = 1.3 \times 10^5$ cm^2/V-sec reveal the three afore-mentioned excitations. The spectral peaks become tractable when plotted as a function of perpendicular magnetic field, shown in Fig. 17. The straight line through the origin, denoted $\hbar\omega_C$, is due to a transition from an occupied Landau level just below the Fermi energy to the next higher Landau level just above the Fermi energy. This ensures that an electron makes a transition to an empty or partially empty state. From the slope of the line, the electron effective mass is determined for that field range. Here, $m^*/m_0 = 0.069$, which is a few percent larger than the $B = 0$ value due to conduction-band nonparabolicity.

The lower two lines in Fig. 17 labeled $E_{01} - \hbar\omega_C$ and $\hbar\omega_C - E_{01}$ are combined Landau level–intersubband excitations (Pinczuk et al., 1987; and Heiman, 1987; Abstreiter et al., 1986). The electrons undergo a combined

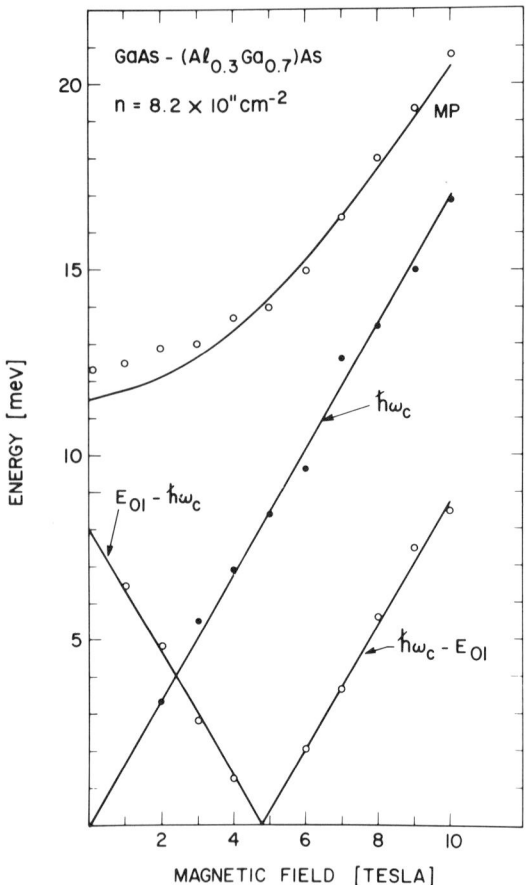

FIG. 17. Transition energies of a 2D electron gas at high magnetic fields, obtained from resonant inelastic light scattering. Note the inter–Landau level transition ($\hbar\omega_C$), the combined intersubband–inter–Landau level transitions ($E_{01}, \hbar\omega_C$), and quasi-3D magnetoplasmon. The sample was an n-type modulation-doped multiple quantum well structure with 15 layers of 27-nm-thick wells and 98-nm period, mobility $\mu = 1.3 \times 10^5$ cm^2/V-sec, and $T = 2$ K. (From Pinczuk and Heiman, 1987.)

change of Landau level and subband state. Such combined transitions were only measured previously in far-infrared optical absorption experiments in which *tilted* magnetic fields cause coupling between in-plane and normal to the plane motions (Schlesinger et al., 1985). In the light scattering experiment the combined transitions are observed with magnetic fields *normal* to the plane. Here, the new selection rules are attributed to mixing between subband and in-plane motions in the valence-band states that participate in the two virtual transitions of the resonant light scattering process (Pinczuk and Heiman, 1987). The intersubband energy for this sample is $E_{01} = 8$ meV.

In Fig. 17, MP labels the magnetoplasma mode of the multilayer structure. For a near-backscattering geometry, the normal component of scattering wave vector is $k_Z \approx 6 \times 10^5$ cm^{-1}, the distance between layers is $d \approx 1.0 \times 10^{-5}$ cm, making $k_Z d \approx 6 \sim 2\pi$. Thus, all of the 2D electron layers should oscillate in phase. The solid line is the magnetoplasma energy calculated from

$$\omega_P(B) = [\omega_P^2(0) + \omega_C^2]^{1/2}, \tag{16}$$

with

$$\omega_P^2(0) = \frac{4\pi n e^2}{\varepsilon m^* d}, \tag{17}$$

and $\varepsilon = 13$ is the background dielectric constant of GaAs. For $B > 4$ T, there is good agreement between measured and calculated energies. The discrepancies at lower field may be due to the interaction between the magnetoplasmon and the collective intersubband.

c. *Magnetoroton Density of States*

Electron-electron interactions in a 2D electron gas give rise to roton minima in the inter–Landau level mode dispersion curve. This is shown in the lower part of Fig. 18. The energies of these modes, magnetoplasmons and spin density excitations, have dispersion

$$\omega(q) = \omega_C + \Delta(q, B). \tag{18}$$

Hartree–Fock calculations of $\Delta(q)$ reveal roton-like minima. Instead of a monotonically increasing energy for increasing wave-vector, the energy dips down to a local minima near $2q_0 = 1/l_C$. The roton structure arises from the reduction of excitonic binding energy at large wave vectors between the electron in the excited Landau level and the hole left in the lower Landau level (Kallin and Halperin, 1984; MacDonald, 1985). Unfortunately, the relevant wave vectors, $q \geq 10^6$ cm^{-1}, are usually not accessible in infrared absorption or inelastic light scattering experiments, where in-plane wave vectors are $k < 10^4$ cm^{-1}. However, one can argue that the localization giving rise to the QHE and FQHE might break the wave vector conservation on the scale of the interparticle separation.

Recently, inelastic light scattering revealed the 2D roton density of states in GaAs (Pinczuk *et al.*, 1988; 1990). The fortunate breakdown of the $q \approx 0$ selection rule is evidently caused by impurities and disorder, allowing the $q \gg 0$ Landau-level transitions to couple to large wave vector excitations. Figure 18a displays results at $v = 2$ ($B = 7.4$ T) in the vicinity of $\hbar\omega_C$. The

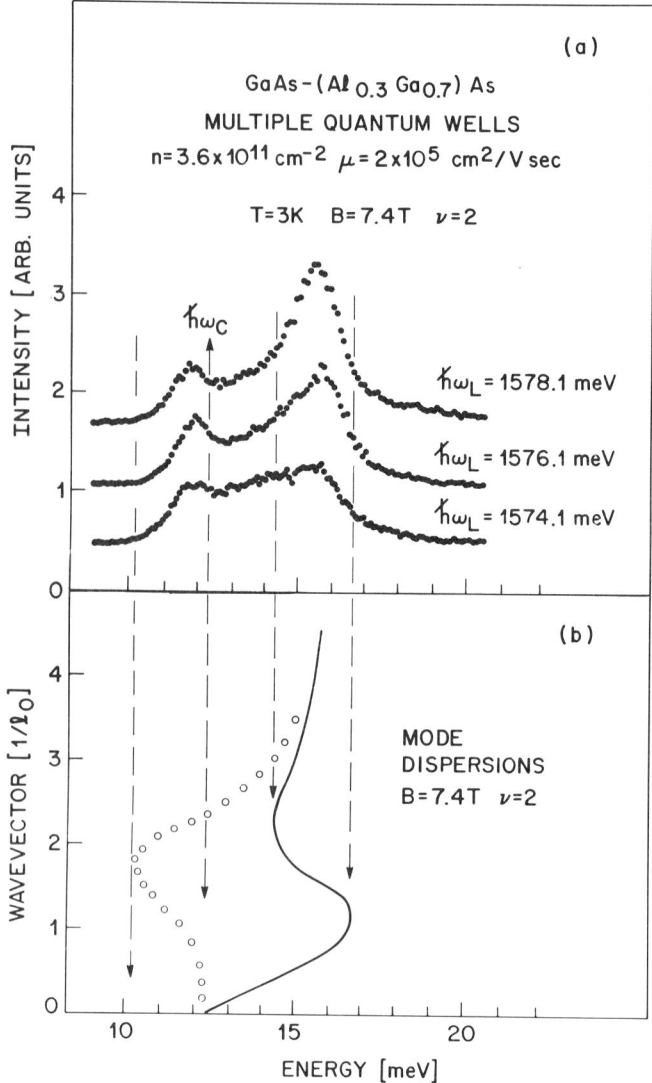

FIG. 18. Magnetoroton dispersion of inter–Landau level transitions in a 2D electron system. Inelastic light scattering in (a) shows spectral peaks corresponding to critical points in the magnetoroton dispersion curves in (b). Solid curve is from charge density excitations, and the open circles correspond to spin density excitations. (From Pinczuk et al., 1988.)

spectra have an onset a few milli electron volts below $\hbar\omega_C$ and a cutoff at about 5 meV above $\hbar\omega_C$. The spectra overlap with the calculated energies of the dispersion curves of the inter–Landau level excitations shown at the bottom. This identifies the lower-energy peak as due to spin density excitations and the higher-energy peak as magnetoplasmons, all at large wave vectors $q \geq 1/l_c$. Similar spectra were measured in multiple quantum well systems, as well as in single heterostructures (Pinczuk et al., 1988, 1990). The only requirement appears to be a high-mobility 2D electron gas; similar features were observed for mobilities ranging between 1.5×10^5 and 8×10^6 cm^2/V-sec.

Note that the spin-density wave lying below $\hbar\omega_c$ is direct evidence of *excitonic* binding, the Hartree–Fock interaction between the electron in excited Landau level and the "hole" in the lower Landau level. In addition, we note that the activation gap measured in magnetotransport only measures the infinite wave vector energy of the magnetoroton dispersion, thus, inelastic light scattering can give additional information.

Tilting the magnetic field relative to the 2D plane causes the Landau excitations to mix with the intersubband excitations. Thus, one no longer has subband levels that are independent of magnetic field and Landau levels that are classically dependent on magnetic field. Instead, there are two sets of magnetic-field-dependent hybrid modes. Now, due to the hybrid nature of the modes, one can observe the $q = 0$ excitation of the Landau-like mode in inelastic light scattering that was parity-forbidden for pure Landau-level transitions (Liao et al., 1992). For analyzing these results, a time-dependent Hartree–Fock model with a parabolic potential has been developed. An interesting result of this model is the prediction that the Landau-like spin density and charge density excitations are no longer degenerate as $q \to 0$. This generates an energy gap that is proportional to the density of states and is an increasing function of the tilt angle. Inelastic light scattering might provide a test of this predicted gap.

d. *Regime of the Electron Solid*

An exciting prospect is the possibility of using inelastic light scattering in the regimes of the FQHE (MacDonald, et al., 1985) and the electron solid, at higher fields and lower temperatures. For example, the shear mode of an electron solid might be observed directly in the Raman spectrum. The energy shift has been predicted to be about one meV (Côté and MacDonald, 1990).

In the first such measurements, the intersubband energy E_{10} in single interfaces has been found to collapse for filling factors $\nu < 1/2$ (Heiman et al., 1992). This is shown in Fig. 19 for electron density $n = 1.2 \times 10^{11}$ cm^{-2}. For increasing fields beyond $\nu = 1/2$ and $T = 0.5$ K, $E_{10}(B)$ decreases rapidly until saturation is reached at high fields. Another striking feature of the collapse is its strong temperature dependence, the onset of collapse moves to higher fields for higher temperatures and disappears above $T \sim 2$–3 K. This appears to be a universal effect—it is observed in both single interface and single quantum well structures over a wide range of electron density. The

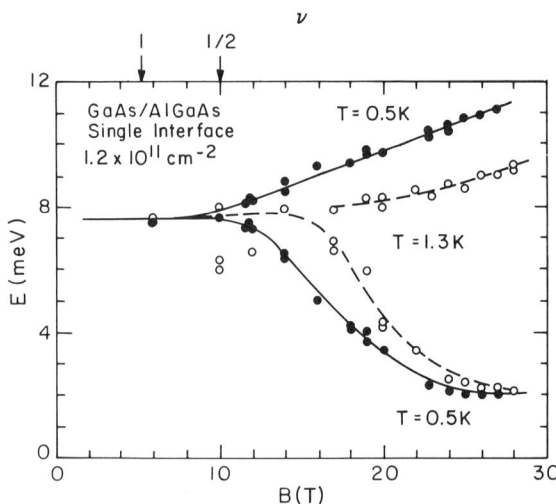

FIG. 19. Intersubband excitation energy E_{10} from inelastic light scattering versus applied magnetic field B at $T = 0.5$ K. The GaAs–AlGaAs single interface had a carrier density of $n = 1.2 \times 10^{11}$ cm^{-2} and electron mobility of $\mu \sim 5 \times 10^6$ cm^2/V-sec.

mechanism of the collapse results from optical pumping of the electrons. Magnetoresistance measurements under illumination show a simultaneous decrease in electron density. This correlation between E_{01} and n exists over a wide range of B-field, temperature, and optical power. There is good agreement between the measured E_{01} and the self-consistent calculation of E_{01} determined from the measured n. From this we conclude that electrons are pumped out of the 2D channel and returned to the (Al,Ga)As barrier region, perhaps back on to the ionized donors in the doping layer, (Heiman et al., 1992).

9. EXCITONS IN QUANTUM WELLS AND STRAIN-LAYER SUPERLATTICES

High magnetic fields are shown here to be a useful perturbation on excitons in undoped quantum wells and in overcoming strain offsets in strain-layer superlattices.

Interband magneto-optic experiments are a valuable probe of exciton states in undoped quantum wells, provided one can subtract the conduction-to valence-band transition energies properly. Normally, the conduction band is taken to be nearly parabolic, while the large nonparabolicity of the mixed valence subbands is considerably more difficult to evaluate. In spite of these difficulties, a number of useful studies of magnetoexcitons have been made and will be reviewed here.

a. Comparison of Magnetoexciton Theory and Experiment

Magnetoexcitons in undoped GaAs quantum wells have been studied by using excitation (Maan *et al.*, 1984), absorption (Tarucha *et al.*, 1984), emission (Ossau *et al.*, 1987), photoconductivity (Rogers *et al.*, 1987), reflectivity (Duggan, 1988), and photoreflectivity (Zheng *et al.*, 1988a) spectroscopy. Theoretical $k \cdot p$ models have been developed that use various basis states, subband Landau-level states (Yang and Sham, 1987), subband 2D hydrogenic states (Bauer and Ando, 1988), and a combination of both (Zheng *et al.*, 1989).

A comparison of results from excitation spectroscopy with theory (Yang and Sham, 1987) brings out a number of interesting points.

(1) The observable intraband optical transitions in a finite magnetic field are all *excitons*; unbound electron–hole pairs are not optically allowed in a finite field.
(2) The nonparabolic valence subbands lead to nonlinear field dependence for the Landau levels. This is in addition to the diamagnetic B^2 dependence associated with a hydrogenic-like exciton, discussed in the next section.
(3) The valence-subband coupling leads to additional lines in the magneto-optic spectra. In the presence of band coupling, the z-component of angular momentum m_J is not a good quantum number, and mixing occurs between states of different m_J. This leads to unusual ordering of the exciton states (1s, 2s, 2p, etc.) for the heavy- and light-hole (hh, lh) series.

The first two items imply that the exciton binding cannot be obtained by extrapolating the excited states to $B = 0$. This can be seen in Fig. 20, which compares the measured and calculated exciton spectral lines in a magnetic field. The dashed straight lines were drawn through the experimental points, while the solid curves were drawn through the calculated points. Notice that the $B = 0$ extrapolations are not very good, yet the overall correspondence at higher fields is much better. Similar effects occur in the other polarization (Yang and Sham, 1987). If $B = 0$ extrapolations are made, an erroneously large light-hole and small heavy-hole exciton binding energy is obtained.

The calculations also point out the unusual ordering of the exciton states as mentioned in (3). The magnetoexciton levels in Fig. 20 are ordered 1s-hh,

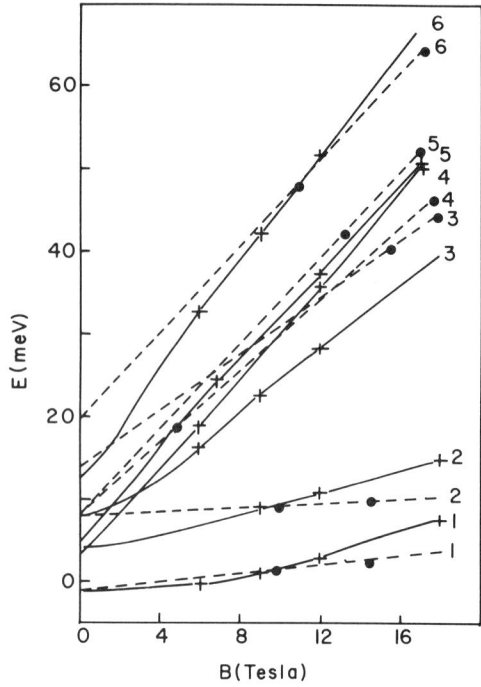

FIG. 20. Exciton energies (conduction–valence band transition energy minus bulk band gap) in a 2D GaAs quantum well at high magnetic fields. Crosses are the calculated points with smooth curves, and the solid points are from absorption measurements with σ_- circularly polarized light. (From Yang and Sham, 1987.)

3d-hh, 1s-lh, 2s-hh, 4d-hh, and 2s-lh. Instead of the second level being 1s-lh, the magnetic field mixes in a significant amount of an excited exciton state of the heavy hole. This is most easily understood by considering the E versus B diagram of the exciton states. An excited-hh state that is relatively flat can cross a lower-level lh state that has a much larger $E(B)$ dependence.

b. Exciton States in High Magnetic Fields

Photoreflection spectra of undoped GaAs reveal excitonic transitions in both bulk and quantum wells at room temperature, even at $B = 0$. Here, the thermal energy is larger than the exciton ionization energy, 4.2 meV in bulk and less than 10 meV in quantum wells wider than 10 nm. Figure 21 shows a plot of the prominent peaks in the room temperature photoreflectivity spectra as a function of magnetic field (Zheng et al., 1988b). The higher-energy transition is linear with a slope of approximately $(1/2)\hbar\omega_c$. This is associated

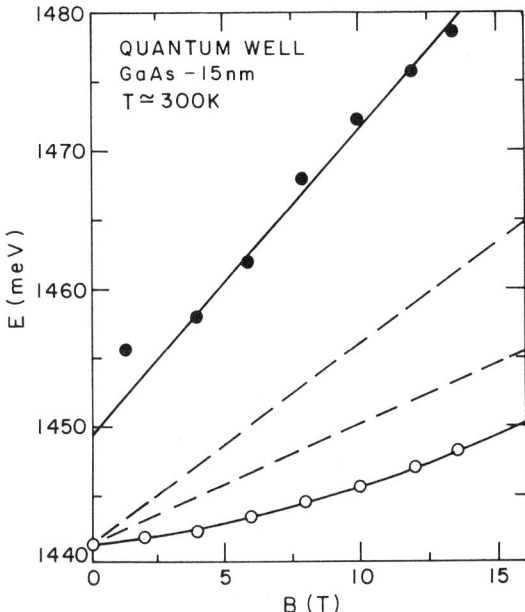

FIG. 21. Conduction–valence band transition energies in GaAs quantum well at high magnetic fields, from room temperature photoreflectivity measurements. Open circles and curve represent measured excitonic transitions, closed circles and line represent measured free-particle-like Landau transitions, and the dashed straight lines represent calculated Landau-level transitions. (From Zheng et al., 1988b.)

with a highly excited exciton state, a nearly free electron–hole pair. The lower-energy line has characteristic exciton features: a B^2 quadratic dependence at low field, giving way to a linear Landau-like behavior at a higher magnetic field.

To understand the quadratic dependence of the 1s ground state of the exciton, we look at the behavior of a hydrogenic system. A relevant parameter that measures the relative magnetic field strength is the ratio $\gamma = \hbar\omega_C/2E_B$, magnetic energy divided by exciton Rydberg energy $E_B = \hbar^2/2\mu^* a_B^2$, where μ^* is the reduced mass of the exciton and a_B is the effective Bohr radius. At low fields, $\gamma \ll 1$, the magnetic field is a perturbation on the exciton and the energy dependence is quadratic in B,

$$E \approx E_0 + \Delta E_X, \qquad (19)$$

where

$$\Delta E_X = \left(\frac{e^2 B^2}{8\mu^* c^2}\right)\langle(x^2 + y^2)\rangle \qquad (20)$$

The last factor is averaged over the wave function for the system. In three dimensions it is $2a_B$, and for 2D it is $(3/8)a_B$, thus

$$\Delta E_X = \frac{1}{2} E_B \gamma^2 (1.01 - 0.34\gamma) \quad \text{for 3D and } \gamma \leq 1 \tag{21}$$

and

$$\Delta E_X = \frac{3}{32} E_B \gamma^2 \quad \text{for 2D.} \tag{22}$$

Note that this illustrates only the extreme limits of narrow well width (<5 nm) and wide well width (>40 nm). For the more usual cases of intermediate well widths, one must use the appropriate trial wave function of the system (Ossau et al., 1987). The spin components and valence-band effects have been neglected.

At high fields where $\gamma \gg 1$, the magnetic energy dominates and the slope is given by $(1/2)\hbar\omega_C$. For excited exciton states E_B is smaller, which means the linear-like behavior appears at a lower magnetic field. Thus, the upper state in Fig. 21 is associated with a weakly bound exciton, while the lower state is the ground state 1s exciton. In this way, magneto-optics not only provides a measure of the carrier mass but also helps in identifying the $B = 0$ states.

c. *Exciton Localization in a Magnetic Field*

At high magnetic fields and low temperatures, excitons are localized by shallow potential fluctuations in the 2D quantum well. Potential minima arise from monolayer well-width fluctuations, impurities in the quantum well, and remote ionized impurities. At temperatures above about $T = 10$ K, excitons become delocalized via thermal activation from these localized states at potential minima. As the temperature is reduced, exciton transport must proceed by tunneling between localized exciton states. Activated hopping and tunneling mechanisms have been measured by resonant Raman scattering from the LO phonon in undoped GaAs-(Al,Ga)As quantum well systems (Zucker et al., 1988).

Such measurements rely on the variation of the intensity of the inelastic light scattering with respect to the degree of exciton localization. For increasing localization, the exciton dephasing time becomes long. This reduces the homogeneous exciton linewidth, thus increasing the resonant scattering strength. At each temperature and magnetic field, the intensity of the LO phonon peak at 36.6 meV is measured at resonance, i.e., when the laser wavelength has been adjusted to give maximum scattering. Figure 22 shows

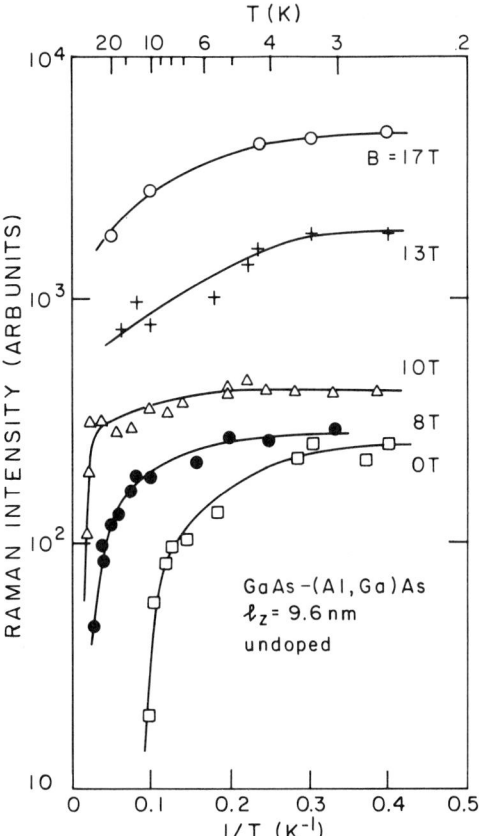

FIG. 22. Intensity of resonant Raman scattering from LO phonons in GaAs quantum wells versus temperature at various magnetic fields. The intensity is a measure of the exciton localization and shows thermal activation above $T \sim 6$ K for $B = 0$, and tunneling between localized states at lower temperature. High magnetic fields push the onset of tunneling to higher temperatures (knee region) by reducing the size of the exciton. (From Zucker et al., 1988.)

the Raman intensity as a function of temperature and magnetic field. In the lower curve at $B = 0$, the intensity decreases rapidly above $T \simeq 6$ K, with an activation energy of 2.4 meV. This represents the energy needed to make a transition from the localized exciton states, having a large scattering strength, to delocalized states, having small scattering strength. Below $T = 2$ K, the slope is reduced to 0.014 meV, where phonon-assisted tunneling between localized minima dominates.

An increasing magnetic field causes shrinkage of the exciton envelope. This leads to two pronounced effects: the overall intensity increases, and the activated regime has an onset at higher temperature. A larger field generates a larger oscillator strength for the excitonic transition, since the oscillator strength of the exciton is proportional to $|\Psi(r=0)|^2$, the propability of finding the electron and hole at zero separation. Ignoring a field-induced change in the exciton–phonon interaction, the resonant Raman intensity is proportional to $|\Psi(r=0)|^4$. Thus, the large intensity increase observed between $B=0$ and $B=17$ T corresponds to a decrease in exciton radius of a factor of 2. The onset of thermal activation has a similar origin. At larger fields the smaller excitons are localized to smaller regions that are deeper. The energy needed to reach an extended state becomes larger, leading to a higher temperature at which thermal activation becomes dominant. This is also reflected in the higher activation energies at higher fields. In a field of $B=8$T, the activation energy increases from the $B=0$ value by 20%, to 3.0 meV.

Although, magnetoresistance is a useful technique to measure the transport properties of either electrons or holes, in general, optical methods allow one to measure the transport of *uncharged* particles (excitons). For excitons, both electrons and holes may contribute to localization, depending on the type of potential fluctuations. For example, well-width fluctuations will affect the carriers inversely proportional to the effective mass and to a lesser amount on the band discontinuity. Although the optical technique may not be generally applicable to all materials, it relies on optically generated carriers and does not need samples to be extrinsically doped.

d. Competition between Strain and Magnetic Field

In a strain-layer superlattice the valence-band levels can be considerably altered by the built-in strain. In this section it is shown how a magnetic field can compensate and overcome the strain-induced valence-band energy offsets in the GaAs–Ga(As,P) strain-layer system.

p-type doped strain-layer quantum wells can have hole masses closer to the light-hole mass, in contrast to unstrained materials with heavy-hole mass values (Osbourn, 1986). Emission spectroscopy to fields of $B=45$ T shows that the mass changes from light to heavy for increasing field (Jones et al., 1987).

Two materials with different bulk lattice constants can be joined without defects provided the layers are thinner than approximately 20 nm. These strain-layer superlattices have an in-plane lattice constant intermediate between the two bulk values. Thus, one material is under compression in the plane of the layers and in dilation in the plane perpendicular to the layers, since the unit cell volume is roughly conserved. The adjoining material has the reverse tetragonal distortion. GaAs–(In,Ga)As and GaAs–Ga(As,P) are examples of strain-layer materials (Osbourn, 1986).

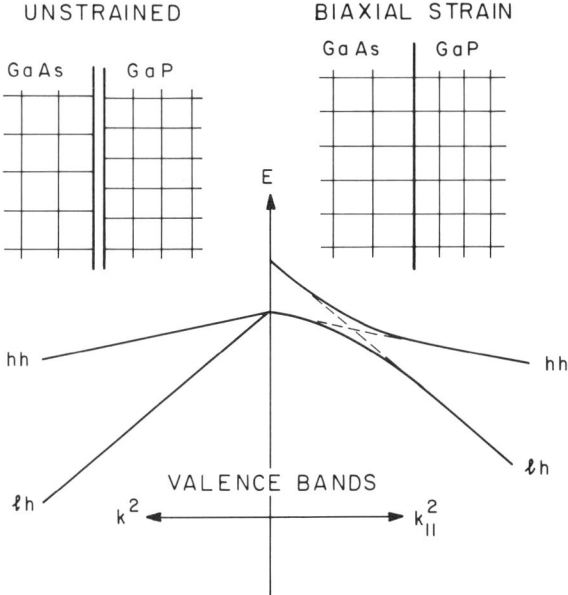

FIG. 23. Schematic representation of strain layer superlattice of GaAs–GaP. The top figures show the lattices before and after biaxial strain relaxation. The lower graph plots the valence-band dispersion showing the strain-induced band mixing for k-vector in the plane of the interface ($k_\parallel$).

Perhaps the most interesting feature of these systems is the substantial change in the valence subbands caused by strain. Figure 23 is a schematic diagram of the valence subband dispersion of the lower-band-gap material, for the inplane motion $k_\parallel$ and the confinement direction $k_\perp$. The solid curves for $k_\parallel$ include coupling of light and heavy holes and lead to large nonparabolicity. Notice that the upper subband for $k_\parallel$ has a light-hole mass, actually $\sim(4/3)m_{lh}$, near $k = 0$; however, the mass increases toward that of the heavy hole for large k. Modulation-doped p-type materials show light-hole masses, and consequent large mobility, which could be useful for complementary logic electronics. The nonparabolic dispersion has been measured using PL spectroscopy on the GaAs–Ga(As,P) strain-layer system (Jones et al., 1987).

Figure 24 shows the energy of the interband recombination emission as a function of applied magnetic field. The slope is large at low fields and makes a transition to a smaller value above $B \approx 8$ T. By matching the slopes to $(1/2)\hbar\omega_c$, Jones and co-workers determined the limiting hole masses as $0.08m_0$ and $0.20m_0$ (Jones et al., 1987). Thus, the striking change in hole mass demonstrates that a large strain energy can be overcome by the confinement energy of a high magnetic field.

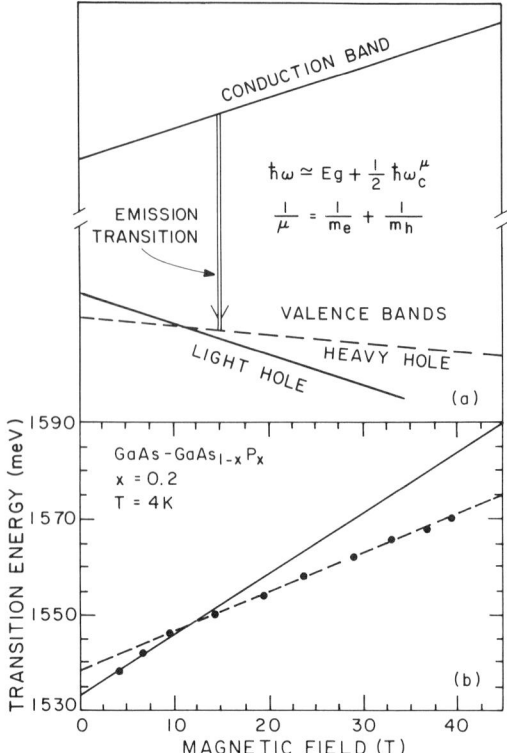

FIG. 24. Magnetic-field-induced light- and heavy-hole transition in a p-type doped strain-layer superlattice. The upper figure is a representation of the emission transition for electrons recombining with heavy and light holes. At lower fields, $B < 10$ T, the holes are in the light-hole band, while for higher fields the lowest energy state is that for heavy holes. The lower plot shows results from photoluminescence measurements (solid points) and expected energies for heavy holes (dashed line) and light holes (solid line).

IV. Magnetic Semiconductors

The focus of this section is on the class of magnetic semiconductors consisting of wide-band-gap ternary alloys of $Cd_{1-x}Mn_xTe$, $Cd_{1-x}Mn_xSe$, and $Cd_{1-x}Fe_xSe$. The order of subjects proceeds from some basic ion-ion (d-d) exchange mechanisms through more complex carrier–ion (sp-d) exchange mechanisms:

(1) spin alignment of the magnetic-ion subsystem to fields of 150 T;
(2) spectroscopy of nearly isolated clusters (pairs and triplets) of nearest-neighbor ions in dilute systems ($x < 0.05$) in fields above $B = 10$ T;

(3) enhanced bound magnetic polaron below $T = 1$ K using Raman spectroscopy;
(4) metal–insulator transition in n-type (Cd,Mn)Se below $T = 1$ K;
(5) spectroscopy of Fe^{2+} ions at high fields.

Reviews of magneto-optics on magnetic semiconductors are contained in Galazka (1987), Brandt and Moshchalkov (1984), and Gaj (1980), and general reviews can be found in Furdyna and Kossut (1988), Aggarwal et al., (1987), and Heiman (1990).

10. Background on Exchange Interactions

There are two important exchange interactions in magnetic semiconductors covered here: the d-d interactions between d-shell magnetic ion spins, and the sp-d interaction between localized magnetic-ion spins and semiconductor carrier spins (s-like conduction electrons and p-like valence holes). The ion–ion exchange is antiferromagnetic and leads to reduced paramagnetism at low concentrations x, spin-glass behavior at moderate x, and defect antiferromagnetism at large x. It is convenient to view the sp-d exchange as a magnetic field amplifier for carrier spins. First, an externally applied magnetic field creates an average spin alignment (magnetization) of the magnetic ions; then the exchange interaction provides a much larger exchange field on the semiconductor carrier spins. This amplification varies as $1/T$ and at liquid helium temperatures is typically 10^2 for electrons and 10^3 for holes. As a result, the Zeeman splitting of the semiconductor bands is huge, $10–10^2$ meV, and leads to large Faraday rotation and magnetically tunable band gaps. In the case of magnetic polarons the external field is replaced by the exchange field created by the carrier spin itself. These will be described in more detail.

a. Magnetic-Ion Subsystem

The primary exchange interaction between the d-shell electron spins, $\mathbf{S}_i$ and $\mathbf{S}_j$, localized on magnetic ions at $\mathbf{R}_i$ and $\mathbf{R}_j$, given by

$$H = -2J_{d-d}(\mathbf{R}_i - \mathbf{R}_j)\bar{S}_i \cdot \bar{S}_j, \qquad (23)$$

is due to the superexchange mechanism (Larsen and Ehrenreich, 1989, 1990). (There is also a smaller exchange of the anisotropic Dzyaloshinski–Moria form $\mathbf{D} \cdot \mathbf{S}_i \times \mathbf{S}_j$; see Larsen and Ehrenreich, 1989.) Since the interaction takes place through the anion wave functions, it is short-ranged and the exchange energy decreases with distance d as $J_{d-d} \propto d^{-7}$ (de Jonge et al., 1987). At distances greater than about third or fourth nearest-neighbor, the dipole-dipole interaction dominates. The magnitude of J_{d-d} depends on the lattice constant, the anion species, and the magnetic-ion species. In general, J_{d-d} increases for lighter anions and cations, and increases in going from Mn to Fe to Co. The nearest-neighbor exchange constant ranges from $J_{NN}/k_B = -6.1$ K for (Cd,Mn)Te (Isaacs et al., 1988b) to $J_{NN}/k_B \sim -40$ K for (Zn,Co)S (Giebultowicz et al., 1990). Accurate values of J_{d-d} can be determined from measurements of field-induced transitions in ion pair clusters described in Section 12. In general, the antiferromagnetic character of this exchange inhibits the field-induced spin alignment, since some ions are usually antialigned and produce an exchange field in opposition to the applied field. Next, it will be shown how the d-d exchange modifies the spin alignment.

The average alignment $\langle S_Z \rangle$ can be measured by a number of techniques. These include direct measurement of the total magnetic moment of the sample via a vibrating-sample magnetometer, and magneto-optic techniques such as reflectivity spectroscopy, Faraday rotation, and spin-flip Raman scattering.

The first complete study of the low-temperature magnetization $M(B)$ of $Cd_{1-x}Mn_xTe$ was measured in magnetic fields up to $B = 15$ T (Gaj et al., 1978). The connection between M (in units of emu/g) and $\langle S_Z \rangle$ is

$$M = \frac{xg\mu_B A}{W_M} \langle S_Z \rangle, \quad (24)$$

in which A is Avogadro's number and W_M is the formula molecular weight. Those results were fit to the empirical relation

$$\langle S_Z \rangle = \tilde{S} B_{5/2} \left[\frac{Sg\mu_B B}{k_B(T + T_0)} \right], \quad (25)$$

where $\tilde{S}$ is an effective spin, $B_{5/2}$ is the Brillouin function, $S = 5/2$, and $g = 2.0$ for Mn^{2+}. The quantities $\tilde{S}$ and T_0 were treated as fitting parameters that vary strongly with x and more weakly with T. It is also convenient to define an analogous effective x (Heiman et al., 1983)

$$\tilde{x} \equiv x \left(\frac{\tilde{S}}{S} \right).$$

Using the measured values of $\tilde{S}(x)$ and $T_0(x)$ for (Cd,Mn)Te, one obtains the empirical relations

$$\tilde{x} = \frac{x}{13x + 1} \quad (26)$$

and

$$T_0 = 38x \text{ K}, \tag{27}$$

for $Cd_{1-x}Mn_xTe$ with $0 \leq x \leq 0.3$. Now, $\tilde{S}(x)$ is less than S due to strong nearest-neighbor antiferromagnetic cancellations between Mn^{2+} ions, and $T_0(x) > 0$ arises from weaker, more-distant-neighbor coupling. For $x < 0.1$ and $T < 5$ K, the ratio $\tilde{x}/x$ as a function of x is generally the same for nearly all magnetic semiconductors and arises from the random statistical distribution of magnetic ions on the lattice sites. This universal function for $\tilde{x}/x$ can be found in Shapira *et al.* (1984).

A convenient way to determine x is from PL measurements of excitons. For (Cd,Mn)Te the PL energies in meV are (Heiman, *et al.*, 1986)

$$E_{L1} = 1588.8 + 1440x \text{ for } 0 < x \leq 0.1$$

$$E_{L2} = 1604.7 + 1397x \text{ for } 0.05 \leq x \leq 0.2 \tag{28}$$

$$E_{L2} = 1575 + 1536x \text{ for } 0.2 \leq x \leq 0.4.$$

b. *sp-d Exchange Interaction*

The sp-d exchange interaction between the carrier spins and the magnetic-ion spins is usually expressed by the Heisenberg Hamiltonian

$$H(\mathbf{r}) = -J\sum[(\mathbf{s} \cdot \mathbf{S}_j)\,\delta(\mathbf{r} - \mathbf{R}_j)], \tag{29}$$

where J is the sp-d exchange constant, $\mathbf{s}$ the spin of the carrier, and $\mathbf{S}_j$ the spin of the jth magnetic ion at position $\mathbf{R}_j$. In the mean-field approximation the Zeeman splitting of the bands becomes

$$E = \pm sx(JN_0)\langle S_z\rangle, \tag{30}$$

using αN_0 and $\beta N_{0/3}$ for the electron and hole exchange energy JN_0, N_0 is the density of cations, and $\langle S_z\rangle$ is the average magnetic-ion spin alignment along the direction of the applied magnetic field. Typically, $\alpha N_0 \sim +1/4$ eV and $\beta N_0 \sim -1$ eV in II-VI type semiconductors. For a dilute system of magnetic ions ($x \ll 1$), $\langle S_z\rangle$ follows a Brillouin function, rising linearly with increasing field and saturating near $S = 5/2$ for Mn^{2+} ions. A more transparent way to appreciate the physical interactions is to write the enhanced g-factor, appropriate in the limit of low fields or moderate temperatures

$(5\mu_B B < kT)$, as

$$\tilde{g}(x, T) \cong g^* + \frac{35}{12}\tilde{x}\left[\frac{JN_0}{k(T + T_0)}\right]g_{Mn}, \qquad (31)$$

where g^* is the bare g-value in the absence of exchange ($g^* \sim 1$) and $g_{Mn} = 2.0$ (Heiman et al., 1983). From the last equation it becomes clear that the enhanced g-value is proportional to the ratio of exchange energy to thermal energy

$$\frac{JN_0}{k(T + T_0)}, \qquad (32)$$

which can be quite large, 10^2–10^3 at liquid helium temperatures. Spin-flip Raman scattering has revealed temperature-dependent g-values of Eq. (31) for donor electrons in $Cd_{1-x}Mn_xSe$ as large as $\bar{g} \sim 200$ at $T \sim 2$ K (Heiman et al., 1983; Peterson et al., 1985) and $\tilde{g} \sim 500$ at $T < 1$ K (Issacs et al., 1990b; Dietl et al., 1991b).

The sp-d exchange interaction is also responsible for a dramatic field-dependent reduction in the band-gap $E_g(B, T)$, due to the Zeeman splitting of both conduction and valence bands. The band-gap tuning is given by

$$E_g(B, T) = E_g(B = 0) - \Delta E_g, \qquad (33)$$

$$\Delta E_g = \frac{1}{2}x(\alpha N_0 - \beta N_0)\langle S_z \rangle, \qquad (34)$$

or

$$\Delta E_g \cong \frac{35}{12}\frac{\tilde{x}(\alpha N_0 - \beta N_0)\mu_B B}{k(T + T_0)} \qquad (35)$$

at low fields or moderate temperatures. This field-dependent band gap led to the demonstration of magnetically tunable lasers with $\Delta E_g \sim 30$–60 meV in bulk (Cd,Mn)Se and (Cd,Mn)Te quantum well structures (Heiman, 1983; Issacs et al., 1986).

c. *Bound Magnetic Polarons*

Bound magnetic polarons (BMP) are a collective state consisting of a loosely localized carrier and a ferromagnetic cloud of aligned magnetic ions (see the reviews: Nagaev, 1983; Wolff, 1988). The mechanism is similar to the RKKY mechanism in metals in which free electrons provide a ferromagnetic exchange between magnetic ions. Here, electrons (holes) are loosely localized in hydrogenic donor (acceptor) orbits. Typical donor radii are 3–6 nm and contain hundreds of magnetic ions, while acceptor radii are 1–1.5 nm and contain tens of magnetic ions. Since the carrier spends a fraction of time near each magnetic ion, the ion feels an exchange field due to the carrier spin. The resulting alignment of ion spins produces a self-consistent exchange field back on the carrier. In this way the carrier lowers its total energy by digging itself a magnetic well. The interaction is always ferromagnetic, independent of the sign of sp-d exchange, since the exchange works twice (carrier → ion, ion → carrier) and the polaron energy is proportional to the square of the sp-d exchange. It is convenient to express the exchange contribution to the BMP by

$$H = -J \sum [(\mathbf{s} \cdot \mathbf{S}_j)|\phi(\mathbf{R}_j)|^2], \tag{36}$$

where a simple hydrogenic wave function $\phi(r) = (\pi a_0^3)^{-1/2} \exp(-r/a_0)$ can sometimes be used for the impurity-bound carrier of radius a_0. After a considerable amount of algebra that takes into account the statistics of the ensemble of magnetic-ion spins, the polaron energy is simply written as (Dietl and Spalek, 1982; Heiman et al., 1983; Ryabchenko and Semenov, 1983)

$$E_{\text{BMP}} \cong \frac{W^2}{4kT}, \tag{37}$$

where W is related to the magnetic susceptability χ by

$$W^2 = W_0^2 \left(\frac{\chi}{\chi_{\text{CURIE}}} \right), \tag{38}$$

and

$$W_0^2 = \frac{35}{96} \frac{\tilde{x}(JN_0)^2}{\pi a_0^3 N_0}. \tag{39}$$

In an experiment the spectral intensity of the Zeeman spin splitting is measured as a function of energy, $I(E)$. At $B = 0$ this is

$$I_0(E) \sim E^2 \exp\left(\frac{-E^2}{2W^2}\right) \exp\left(\frac{E}{2k_B T}\right), \tag{40}$$

which has a spectral peak at

$$E_0 \to W_0\sqrt{2} \qquad (41)$$

in the limit of high temperature. In the general case, the peak position must be determined by numerical methods. In a magnetic field B the spectrum is

$$I_B(E) = \left(\frac{W^2}{\mu_B \tilde{g} BE}\right) \sinh\left(\frac{\mu_B \tilde{g} BE}{W^2}\right) I_0(E). \qquad (42)$$

Since acceptor-bound holes have larger exchange energies and smaller Bohr radii, the BMP energy is typically $10-10^2$ meV for acceptors (Bulgaiski, et al., 1988) and only 1 meV for donors. In the case of donor electrons, the BMP energy becomes sizable below $T = 1$ K because χ becomes large. This type of enhancement was observed in spin-flip Raman scattering down to $T = 0.13$ K and was explained by a model for $\chi(T)$ that included d-d exchange between magnetic-ion neighbors (Isaacs et al., 1988b). Further details are given in Section 13.

11. High-Field Spin Alignment in (Cd,Mn)Te

$Cd_{1-x}Mn_xTe$ is a convenient material for studying the magnetic properties of a dilute, frustrated antiferromagnet. The system exhibits a wide range of magnetic behavior, from paramagnetism at low x, to short-range antiferromagnetism at high x (see Furdyna and Kossut, 1988). The cations form a face-centered cubic lattice having 12 nearest neighbors. This structure gives rise to magnetic order that appears to be type III, where for $x = 1$ a given spin has eight nearest neighbors antiparallel, four parallel, and six next-nearest neighbors antiparallel. The behavior of $\langle S_z \rangle$ for $x \leq 0.1$ and $B < 10$ T can be adequately described by the modified paramagnet model; however, for $x > 0.1$ and $B > 10$ T there are additional contributions. It has been pointed out that at a given B-field the spin alignment can be enhanced by reducing the number of nearest neighbors in: (a) layered materials such as CdTe/CdTe; (b) ordered structures such as famatinite Cd_3MnTe_4 (Heiman, et al., 1987b).

a. Results to 45 T

For low-x concentrations of $Cd_{1-x}Mn_xTe$, $x < 0.05$, $\langle S_z \rangle$ is nearly paramagnetic; it rises linearly with B at small fields and saturates when the magnetic energy overcomes the thermal energy. This paramagnetic-like behavior occurs when the majority of Mn^{2+} ions are isolated singlets and have no other Mn^{2+} spins on nearest-neighbor sites. For increasing values of x, there is an increasing fraction of Mn^{2+} ions in large clusters, which have an antiferromagnetic response that is linear in B.

Measurements of $\langle S_z \rangle$ at $B > 15$ T show large deviations from the modified Brillouin function (Heiman et al., 1987a). Some results of Faraday rotation experiments in pulsed fields to $B = 45$ T are displayed in Fig. 25. These fields were produced by a multiturn copper coil at the MIT Francis Bitter National Magnet Laboratory. A number of significant features are observed in the $\langle S_z \rangle$ versus B plots: (a) increasing x reduces the average spin alignment due to antiferromagnetic cancellations; (b) the paramagnetic contribution below 10 T appears more strongly in the low-x data; and (c) for fields above 15 T there is a linear variation with B. At the lowest concentration, $x = 0.1$, $\langle S_z \rangle$ rises quickly with increasing B, partially saturates by $B \sim 8$ T, then continues to rise linearly with field. This curve, and all others, can be separated empirically into the sum of two components,

$$\langle S_z \rangle = \tilde{S} B_{5/2}(B) + \tilde{S}' B, \qquad (43)$$

where $\tilde{S}'$ is the coefficient of the linear term. Both $\tilde{S}(x)$ and $\tilde{S}'(x)$ decrease for increasing x, with $\tilde{S}$ varying much more rapidly than $\tilde{S}'$. The first coefficient arises mainly from *isolated clusters* that have a total moment even at $B = 0$. For increasing x, $\tilde{S}$ decreases rapidly, since the percentage of spins in isolated

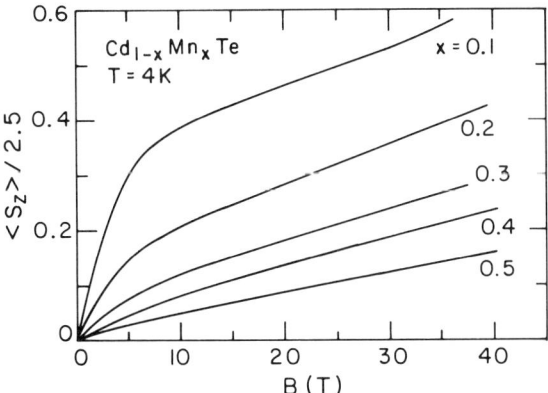

FIG. 25. Average spin alignment $\langle S_z \rangle$ of Mn^{2+} ions ($S = 5/2$) along the applied magnetic field B in the random alloy $Cd_{1-x}Mn_x Te$ for various x-values. Results were obtained from Faraday rotation measurements in a 10-msec pulsed-field magnet. The saturating behavior below $B \sim 10$ T for low x-values is due to the paramagnetic response of nearly isolated singlet ions, while the remaining linear behavior for all x-values arises from large clusters and the infinite network of coupled spins. (From Heiman et al., 1987a.)

clusters decreases. The second coefficient arises from changes in the *internal* magnetic moment of large clusters and the infinite network of antiferromagnetically coupled spins. In the limit $x = 1$, only the linear behavior should remain, as observed in undiluted antiferromagnets.

b. *Results to* 150 T

Eventually, at very high magnetic fields the spin system at all x should saturate. Faraday rotation experiments on $Cd_{1-x}Mn_xTe$ at $T \sim 10$ K were performed up to 150 T (Isaacs et al., 1990a). These fields were generated by an exploding, single-turn coil at the Megagauss Facility of the Institute for Solid State Physics, Tokyo University. For $x \leq 0.2$ the applied field is large enough to overcome completely the internal antiferromagnetic exchange fields. Total spin saturation, $\langle S_z \rangle = 5/2$, is observed in Fig. 26 at $B = 100$ T and 140 T, respectively, for $x = 0.1$ and 0.2. For higher x-values (up to $x = 0.6$) the spin alignment and, hence, the spin saturation become more difficult with increasing x. At large x, there is a large negative exchange field due to

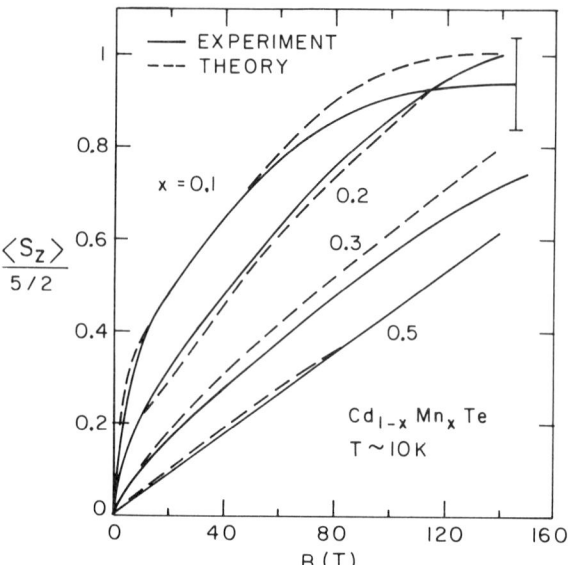

FIG. 26. High-field spin alignment $\langle S_z \rangle$ of Mn^{2+} ions ($S = 5/2$) along the applied magnetic field B in the random allow $Cd_{1-x}Mn_xTe$ for various x-values. Results were obtained from Faraday rotation measurements in a 6-μsec pulsed-field magnet, with the sample temperature $T \sim 10$ K at the beginning of the magnet pulse. Complete saturation, $\langle S_z \rangle = 5/2$, is observed for $x = 0.1$ and 0.2 where the external field overcomes the internal antiferromagnetic exchange field of nearest-neighbor ions. (From Isaacs et al., 1990a.)

nearby ions. A mean-field model that takes into account the random site occupation self-consistently fits both the B- and x-dependence of $\langle S_z \rangle$ remarkably well (Isaacs *et al.*, 1990a). This is shown in Fig. 26 for a variety of x-values. The solid lines represent the experimental data, while the dashed lines were generated from the mean-field theory. There is good agreement between theory and experiment.

It is important to know $\langle S_z \rangle$ at these large fields since the acceptor BMP often has an internal exchange field of this size at low temperatures. To see this, Eq. (36) can be expressed as

$$g\mu_B B_{\text{BMP}} = \frac{-\beta N_0}{\pi a_0^3 N_0} \left(\frac{3}{2} \frac{5}{2}\right) \exp\left(\frac{-2r}{a_0}\right), \tag{44}$$

assuming the temperature is low enough to keep the Mn^{2+} spins aligned with the hole spin. Thus, the exchange field acting on the magnetic ions in the center of the acceptor orbit, produced by the acceptor-bound hole, is

$$B_{\text{BMP}} \sim 200 \text{ T}, \tag{45}$$

using $a_0 = 1.5$ nm and $N_0 = 1.5 \times 10^{22}$ cm^{-3}. The measured $\langle S_z \rangle$ to $B = 150$ T have been used in self-consistent calculations of the BMP energies and nonhydrogenic wave functions (Ram-Mohan and Wolff, 1988). Substantial errors in the BMP energies will result from the use of low-field values of $\langle S_z \rangle$ in calculations, especially for $x > 0.1$.

12. Field-Induced Transitions within Magnetic-Ion Clusters

The internal magnetic structure of small clusters of ions was first demonstrated for Mn-ion pairs (Shapira *et al.*, 1984; Aggarwal *et al.*, 1985). Here, a cluster is a group of Mn^{2+} ions that are connected by nearest-neighbor bonds, but are isolated in the sense that all other Mn ions lie further away than the 12 nearest-neighbor sites. For increasing magnetic field, the total spin alignment of the pair, S_T, increases in integral quantum jumps. This leads to step-like increases in the magnetization and also in the Zeeman splitting. Typically, a step occurs every 10 to 20 T. The fields at which these steps occur give an accurate value ($\pm$ few percent) for the d-d exchange constants. From the magnitude of the step it has been determined that the Mn ions are distributed randomly on the cation sites (Shapira *et al.*, 1984). The following discussion describes magneto-optical studies of ion pairs and triplets. Many studies have been made using nonoptical techniques, such as direct magnetization measurements, but will not be reviewed here (see Shapira, 1990).

a. Magnetic-Ion Pairs

Since the nearest-neighbor d-d exchange constant is of order $J_{NN}/k_B \sim -10$ K, at liquid helium temperatures a pair of ions will be in the lowest-energy, antiferromagnetic $S_T = 0$ configuration. Figure 27 shows that when the magnetic energy $g\mu_B B$ reaches an integer n times twice the exchange energy, $-2J_{NN}$, the system makes a transition to higher S-state plural states, $S_T = 1,2,3,4,$ and 5. Actually, all the transitions are biased to slightly larger

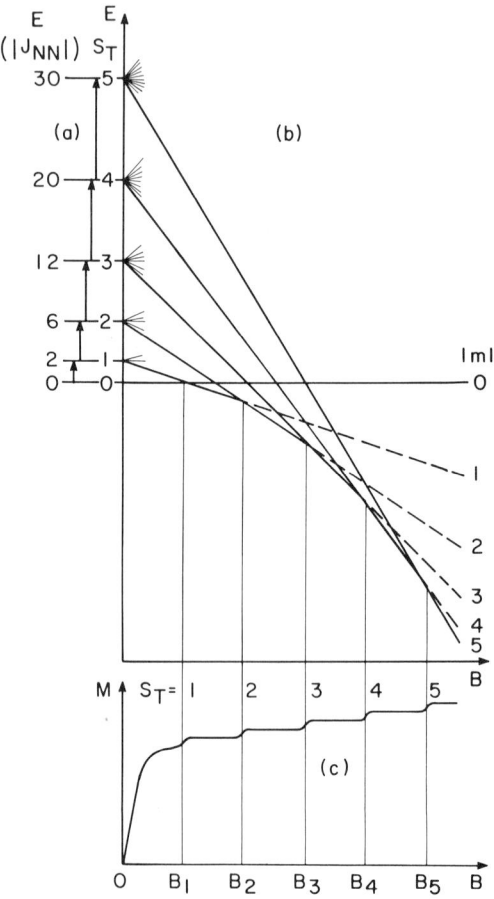

FIG. 27. Schematic diagram showing the total spin S_T of magnetic-ion pair clusters versus applied magnetic field. In the upper left (a), the energy levels at $B = 0$ are shown; the right side (b) plots the lowest level of each state as a function of magnetic field. As each higher level crosses a lower level, at the vertical lines, the pair increases its spin by one unit. The lower curve (c) illustrates the magnetization $M(B)$ for a dilute sample where the contribution of ion pair steps is added to a saturating paramagnetic contribution from isolated singlet ions. (From Foner et al., 1989.)

fields (~1–2 T) by singlet ions at further neighbor sites (Larsen et al., 1986). Thus, to get accurate values for the exchange constant it is better to take the difference between steps,

$$-J_{NN} = \frac{g\mu_B}{2}(B_n - B_{n-1}). \quad (46)$$

Spin-flip Raman scattering to $B = 30$ T at $T = 0.5$–1.3 K has measured these steps in (Cd,Mn)Te and (Cd,Mn)Se (Isaacs et al., 1988b). This is the highest magnetic field ever attained in a Raman scattering experiment. Basically, Raman scattering measures the Zeeman splitting of the donor electrons, which is proportional to the average spin alignment, including that of ion pairs. The Raman Stokes-shift is composed of two terms

$$E_0 = x(\alpha N_0)|\langle S_Z \rangle| + g^*\mu_B B, \quad (47)$$

when the field is larger than a few tesla and magnetic polaron effects can be ignored (Heiman et al., 1983). The first term is due to the s-d exchange of the conduction band and reflects the quantum jumps of S_T through $|\langle S_Z \rangle|$.

The first two steps measured in $Cd_{1-x}Mn_xSe$, $x = 0.05$, are shown in Fig. 28. The difference between the two steps at $B_1 = 12.5$ and $B_2 = 23.7$ T

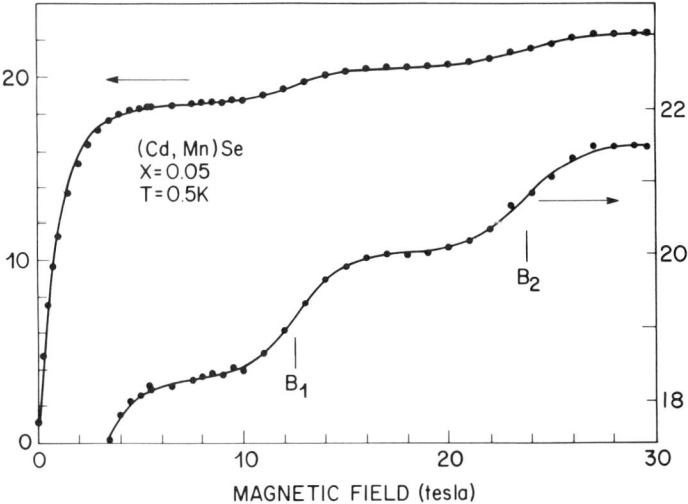

FIG. 28. Magnetic-ion pair steps observed in spin-flip Raman scattering from donor electrons in (Cd,Mn)Se to $B = 30$ T at $T = 0.5$ K. At these high fields the measured Stokes energy E_0 represents the average spin alignment. B_1 and B_2 are the fields at which the total z-component of spin of the coupled ion pairs makes the transitions $S_T = 0 \to 1$ and $S_T = 1 \to 2$. (From Isaacs et al., 1988b.)

yields $J_{NN} = -7.5 \pm 0.3$ K for (Cd,Mn)Se. In (Cd,Mn)Te, the measured value $J_{NN} = -6.1 \pm 0.2$ is somewhat lower, allowing the first three steps to be observed for $x = 0.03$ up to $B = 30$ T. The expected magnitude of the steps, relative to the saturated value of the paramagnetic contribution, is

$$\frac{\delta E}{E_S} = \frac{P_2}{5}\left(\frac{x}{\tilde{x}}\right), \tag{48}$$

where the probability of finding a magnetic ion in a pair is $P_2 = 12x(1-x)^{18}$, assuming a random distribution. Thus, one expects $\delta E/E_S = 0.06$ for $x = 0.03$, and $\delta E/E_S = 0.08$ for $x = 0.05$. The good agreement between these values and the measured values implies that the ratio of pairs to singlets is what one expects for a random distribution of Mn ions. All five steps have been measured in (Cd,Mn)Te by pulsed-field magnetization experiments (Foner et al., 1989).

b. Magnetic-Ion Triplets

There are two types of triplet clusters: open triplets, where a central magnetic ion has two nearest-neighbor ions but those at the ends are not nearest neighbors of each other, and closed triplets, where each of three ions has *two* nearest-neighbor ions in an equilateral triangle. The open triplets are more numerous; 10% of all the magnetic ions are in open triplets for $x = 0.1$.

The first unambiguous demonstration of triplet-ion clusters was reported for $Cd_{1-x}Mn_xTe$, $x = 0.1$, by using Faraday rotation in pulsed fields to $B = 61$ T (Wang et al., 1989). The top part of Fig. 29 shows $M(B)$ at $T = 4.2$ K versus applied field. Compare the deviations of the data points from the solid straight line. These deviations are more clearly shown in the dashed line representing the derivative dM/dB of the experimental data. The sudden increase of slope at $B \sim 30$ T represents the onset of a set of steps due to open triplets. The individual steps cannot be resolved because the combination of thermal broadening and alloy broadening is larger than the step interval. Thus the set of five steps appears as an incline rather than a staircase. To illustrate the individual effects of all clusters, including singlets, pairs, and triplets, the lower part of the figure displays the calculated magnetization of the thermally broadened steps. The arrow at $B = 28$ T marks the onset of the open triplets, and the arrow at $B = 49$ T marks the end of the pairs. The solid curve in the upper part of the figure shows the calculated dM/dB. There is a good coincidence between the measurement and the simple model. A $\Delta B \sim 2$ T shift to higher fields should be made in the calculation to account for the negative bias field of further neighbors, as discussed in the previous section on pairs. Note that this model contains only one adjustable param-

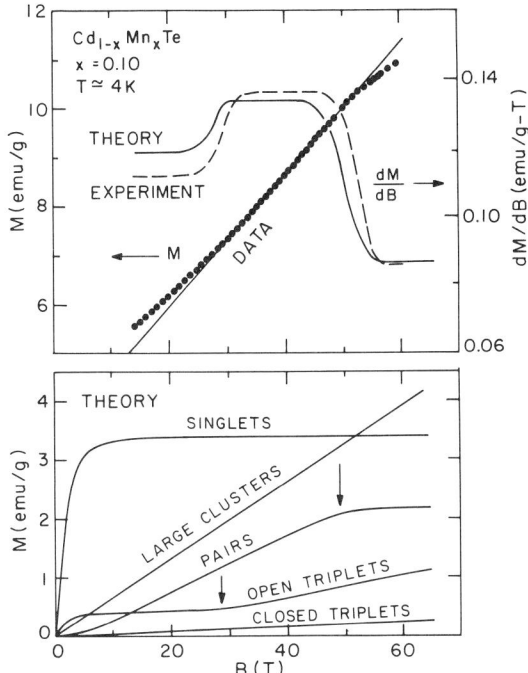

FIG. 29. Magnetization M of (Cd,Mn)Te to $B = 60$ T showing the onset of magnetic-ion triplet clusters at $B = 29$ T and the saturation of ion pair clusters at $B = 52$ T. Results were obtained from Faraday rotation measurements in a 6-msec pulsed-field magnet, at $T \sim 4$ K where individual pair and triplet cluster steps are broadened into a smooth incline. The lower part of the figure plots the contributions from various clusters in a model that considers only nearest-neighbor exchange interactions. The upper part of the figure plots the experimental data as solid points with a straight line drawn through the region where both triplet and pair clusters are active. Also plotted are the measured and calculated derivatives dM/dB, showing the shift due to next-nearest and further-neighbor exchange interactions. (From Wang et al., 1989.)

eter, an additional linear-B term that approximates all the clusters with more than three magnetic ions. These clusters are expected to have a linear B dependence, except for a small paramagnetic-like increase at low fields, which has been neglected.

Further quantitative analysis of data for ion pairs reveals a linear relationship between bias field (average exchange field due to more-distant neighbors) and total magnetization, independent of x for $x \leq 0.1$. This relation leads to a relatively accurate value for the next-nearest-neighbor exchange constant in $Cd_{1-x}Mn_xTe$ of $J_{NNN}/k_B = -1.1 \pm 0.2$ K (Wang, 1991).

There are three methods for determining J_{NN} values. The least accurate method, but perhaps the simplest, is from measurement of the temperature dependence of the magnetic susceptibility for $k_B T \gg |2SJ_{NN}|$, usually near room temperature. From the Curie–Weiss expression

$$\chi = \frac{C}{T - \theta}, \tag{49}$$

$$C = \left(\frac{k_B}{3}\right) S(S + 1) x N_0 g^2 \mu_B^2, \tag{50}$$

one gets

$$J_{NN} = \frac{3 k_B x \theta}{2 z S(S + 1)}, \tag{51}$$

where z is the number of nearest neighbors (see Kittel, 1976). Measurement of C and θ is used not only to determine J_{NN} but to determine the value of S for ions that may not have simple spin-only moments (see Twardowski et al., 1988; Lewicki et al., 1987). A second method uses inelastic neutron scattering at $B = 0$ to measure directly the transition energy of a pair for $S_T = 0 \to 1$ (see Giebultowicz et al., 1990). This technique has the disadvantage that Cd compounds cannot be measured because of the high flux absorption, but it allows measurement of large J_{NN} values that would require unusually large fields, $B > 40$ T, to observe magnetization steps. Present state-of-the-art long-pulsed magnets reach $B \sim 70$ T and thus limit $|J_{NN}|/k_B < 50$ K by the method of magnetization steps.

13. Bound Magnetic Polarons below $T = 1$ K

Bound magnetic polarons contain a localized semiconductor carrier that experiences a local exchange field due to magnetic ions in its vicinity. This local ferromagnetic magnetic bubble can arise from two processes. The more interesting process occurs at low temperature, where the carrier spin is itself responsible for the magnetic-ion alignment. In this "cooperative" regime the carrier spin creates an exchange field for the magnetic ions, and as these ion spins line up their net alignment creates a self-consistent exchange field back on the carrier. Thus, the carrier acts as a ferromagnetic "glue" and digs itself a potential hole in the process of mutual alignment. On the other hand, at high temperatures, in the "spin-fluctuation" regime the net local alignment of ion spins is not caused by the mobile carrier spin. Rather, the alignment results from the finite number of magnetic ions within the carrier vicinity and the resulting stochastic fluctuations in their total moment. The magnetic moment

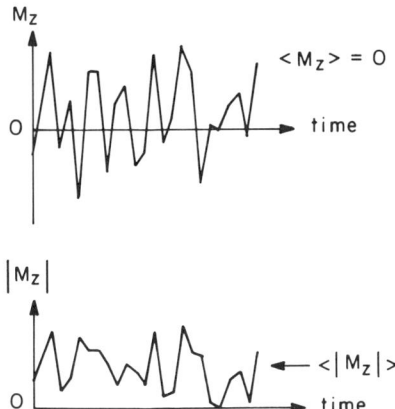

FIG. 30. Illustration of time-varying magnetic moment in a bound magnetic polaron. Isolated magnetic-ion spins within the hydrogenic donor orbit fluctuate stochastically, giving rise to a total M_z that varies in magnitude and direction; thus $\langle M_z \rangle = 0$. However, since at low temperature the electron spin is aligned with M_z, the average exchange field felt by the electron $\langle |M_z| \rangle$ is not zero. (From Heiman, 1990.)

fluctuates in time randomly in both magnitude and direction; however, the time average of the *vector* moment is zero. But since the carrier spin can follow the vector moment, the carrier experiences the *magnitude* of the moment, which has a nonzero average. This is shown schematically in Fig. 30. In both BMP regimes, when the temperature is not so high as to decouple the carrier spin from the exchange field of the magnetic-ion spins, there is a net positive binding energy associated with the BMP.

Although the energy of a donor BMP is a few orders of magnitude smaller than that for acceptors, there are a number of reasons why the donor BMP is an attractive problem for investigating BMPs: spin-flip Raman scattering results are straightforward to interpret; the Stokes shift is a direct measure of the total donor BMP energy (Nawrocki et al., 1981), and the theoretical problem can be exactly solved semiclassically (Dietl and Spalek, 1982; Heiman et al., 1983).

A dramatic increase in the donor BMP energy below $T = 1$ K has been observed with spin-flip Raman scattering in $Cd_{1-x}Mn_xSe$ (Isaacs et al., 1988a). In previous light scattering studies the spin-fluctuation region was found to be dominant above $T \sim 2$ K (Nawrocki et al., 1981; Heiman et al., 1983). There, the Stokes shift E_0 is nearly temperature-independent at $E_0 \sim 1$ meV, even up to $T = 30$ K. These studies found only a small increase in BMP energy between $T = 1.6$ and 2 K. Following this, fiber-optic techniques became available for light scattering spectroscopy at much lower temperatures, $0.35 \leq T \leq 1.8$ K in liquid helium-3, and $T \geq 0.1$ K in a dilution

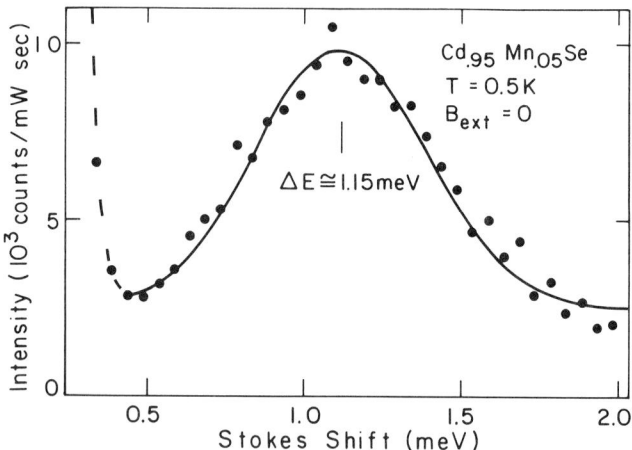

FIG. 31. Spectrum of spin-flip Raman scattering from donor-bound magnetic polarons in (Cd,Mn)Se at $T = 0.5$ K. Solid curve is a fit to the experimental points as described in the text. The curve is Gaussian-like due to the stochastic distribution of magnetic moments within the ensemble of donors. (From Isaacs et al., 1988a.)

refrigerator (Isaacs and Heiman, 1987). Figure 31 shows the light scattering spectrum of a donor BMP at $T = 0.5$ K. This spectrum was taken without an external magnetic field and shows a broad peak centered at $E_0 = 1.15$ meV. The finite shift is the Zeeman splitting of the donor electron spin in the local exchange field of the BMP. The sizable linewidth results from stochastic fluctuation of the local field about the average. If one follows the peak position as a function of temperature, a large increase in energy is observed below $T \sim 1$ K, plotted in Fig. 32. The increase is due to the cooperative BMP mechanism arising from an increasing susceptibility at low temperatures.

The order-of-magnitude increase in the magnetic susceptibility below $T = 1$ K was subsequently explained by using a further-neighbor cluster model (Isaacs et al., 1988a). In this model, the susceptibility increase is attributed to a small population of magnetic ions that have neither nearest-neighbors (NN) nor next-nearest-neighbor (NNN) ions. These isolated ions polarize more easily since they have much weaker antiferromagnetic exchange fields. This can be shown by comparing the exchange energy of the polaron with the further-neighbor exchange energy. The polaron begins to form when the exchange energy $E_{\text{exch}}(r = 0)$ at the center of the Bohr orbit is larger than both $k_B T$ and $2J$. Here,

$$E_{\text{exch}}(r = 0) = \frac{5(\alpha N_0)}{4\pi a_0^3 N_0}. \tag{52}$$

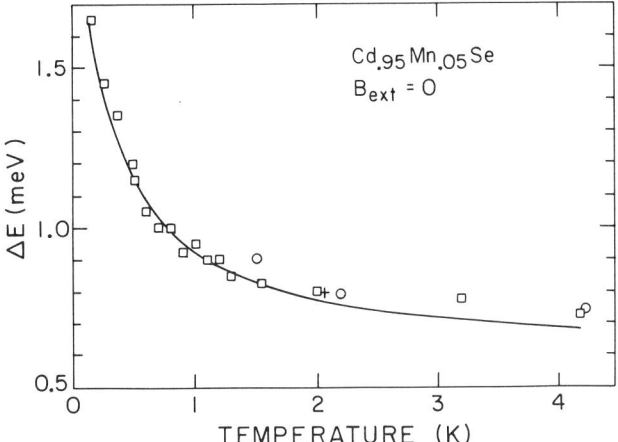

FIG. 32. Bound magnetic polaron energy of donor electrons in (Cd,Mn)Se versus temperature down to $T = 0.13$ K. Spin-flip Raman scattering was used to determine the average Zeeman splitting ΔE from spectral fits like those shown in Fig. 31. The increase in ΔE below $T \sim 1$ K is due to the cooperative polaron in which the spin of the localized donor electron (radius ~ 3.8 nm) polarizes the magnetic ions in its vicinity. (From Isaacs et al., 1988a.)

For $\alpha N_0 = 260$ meV, $a_0 = 3.8$ nm, and $N_0 = 1.8 \times 10^{22}$ cm^{-3}, one finds $E_{\text{BMP}}(r = 0) \approx 0.1$ meV. Since $J_{\text{NN}} = 0.65$ meV and $J_{\text{NNN}} \sim 0.1$ meV,

$$J_{\text{NNNN}} < E_{\text{BMP}} \leq J_{\text{NNN}}, J_{\text{NN}}, \tag{53}$$

thus the carrier is only expected to cause significant alignment of Mn^{2+} ions that are well isolated from other Mn^{2+} ions, greater than second-nearest-neighbor distances.

This work on the donor BMP represents an experimental realization of an important statistical–mechanical model of the BMP. In this model the partition function is exactly calculable in the semiclassical limit, where the number of magnetic ions within the carrier's orbit is large, $\sim 10^2$ (Wolff, 1988). In the case of acceptor BMP the interactions are large enough to distort the wave function from a simple hydrogenic form. This theoretical problem is more difficult and must treat the wave function self-consistently, as has been done using numerical methods (Ram-Mohan and Wolff, 1988).

14. METAL–INSULATOR TRANSITION BELOW $T = 1$ K

The metal–insulator transition (MiT) in (Cd,Mn)Se has a number of interesting problems that can be investigated with spin-flip Raman scattering.

Recent transport measurements on the metallic side of the MiT show an unexpected carrier freeze-out below $T \sim 0.5$ K, which is absent in CdSe without Mn (Sawicki et al., 1986). This suggests that carriers become somewhat localized even within the metallic regime, and might be associated with the large BMP enhancement found in the insulating samples described in the previous section. In the following, it is shown that the diffusion coefficient for electrons can be measured via angle-dependent spin-flip Raman scattering. Before presenting the results on (Cd,Mn)Se (Dietl et al., 1991a,b; Isaacs et al., 1990b) we describe how the method has been applied to nonmagnetic CdS.

The diffusion coefficient D_s for conduction electron spin in CdS was determined from the measured linewidth in spin-flip Raman scattering (Scott et al., 1972). Following this, a study was carried out for a range of carrier concentrations near the critical concentration n_C for the onset of the MiT (Geschwind and Romestain, 1984). The transition is usually estimated from the Mott relation.

$$n_C \cong \left(\frac{0.26}{a_0}\right)^3, \tag{54}$$

corresponding to the density at which the average donor–donor spacing is roughly twice the Bohr diameter. For CdS, $a_0 = 2.5$ nm and $n_C \cong 1.1 \times 10^{18}$ cm^{-3}, while for CdSe, $a_0 = 3.8$ nm and $n_C \cong 3 \times 10^{17}$ cm^{-3}. The Raman study of CdS verified that *spin* diffusion was equivalent to the *charge* diffusion for a range of concentrations from near n_C to well above n_C, by relating the spin-flip linewidth to the measured resistivity.

The spin-flip Stokes scattering intensity appropriate to the case of delocalized electrons is (Scott et al., 1972; Wolff et al., 1976)

$$I(q, E) = -C[n(E) + 1]\mathrm{Im}\,\chi_\perp(E) \tag{55a}$$

$$= \frac{C[n(E) + 1]\chi_0 E\Gamma}{(E - E_0)^2 + \Gamma^2}, \tag{55b}$$

where q is the scattering wave vector, C is a constant, $n(E) = 1/(e^{E/kT} - 1)$, and χ_0 is the static (Pauli) susceptibility of the electron spins. The half-width at half-maximum can be expressed by

$$\Gamma = \frac{1}{T_2} + D_S q^2, \tag{56}$$

and T_2 is the transverse spin relaxation (spin coherence) time. The diffusional damping of the second term arises from collisionally narrowed Doppler broadening. If the electrons did not experience any collisions, then as the electrons moved with velocity **v** they would experience a Doppler-shifted laser frequency $(\mathbf{v} \cdot \mathbf{k}_L)$ and radiate at any angle θ the Raman-shifted frequency $-(\mathbf{v} \cdot \mathbf{k}_S)$. Thus, one would measure a broadened Raman line from the distribution of velocities. However, the electrons undergo collisions that prevent the spins from diffusing. Since $q = |k_L - k_S| \cong 2k_L \sin(\theta/2)$ can be varied by an order of magnitude, between forward and backward scattering, in practice, both T_2 and D_S can be determined separately. For example, using the scattering angles $\theta = 5°$ and $85°$, the ratio of q^2 is 0.004 and then

$$\frac{1}{T_2} \cong \Gamma_{FWD} \tag{57}$$

and

$$D_S \cong \frac{\Gamma_{BKD} - \Gamma_{FWD}}{q_{BKD}^2}. \tag{58}$$

When electron correlation and many-body effects can be neglected, D_S will be the same as the charge diffusion D_C. For a classical gas

$$D_C = \frac{k_B T \mu}{e}, \tag{59}$$

via the Einstein relation, or, for a degenerate electron gas,

$$D_C = \frac{2E_F \mu}{3e} = \frac{2E_F \tau}{3m}, \tag{60}$$

where μ is the electron mobility, τ the scattering time, and E_F the electron-density-dependent Fermi energy.

Spin-flip Raman scattering experiments were carried out on $Cd_{1-x}Mn_xSe$, $x = 0.05$, down to $T = 0.2$ K (Dietl et al., 1991b). The sample was doped with In to a density just above the MiT, with $n = 7 \times 10^{17}$ cm^{-3}. A dilution refrigerator was equipped with three optical fibers: a collection fiber for transmitting the scattered light, and two input fibers for the laser light, one each for forward and backward scattering. In this way, both T_2 and D_S could be determined at each temperature. The critical part of the experiment was to reduce the laser power as low as possible when operating below $T = 0.5$ K, since the sample temperature is limited by laser power and not by the cooling capacity of the refrigerator (0.4 W at $T = 0.1$ K). It was found that only 10 μW of incident light limited the Mn^{2+} spin temperature to $T = 0.2$ K, although the bath could be substantially below $T = 0.1$ K. The spin temperature was determined from the power dependence of the Raman shift by using both previously measured values of $E_0(T)$ (Isaacs et al., 1988a) and results from susceptibility measurements $\chi(T)$. It appears that the absorbed light is well coupled to the Mn^{2+} spins, while the spins are only weakly coupled to the helium bath.

Figure 33 shows spectra taken at $T = 2.0$ and 0.2 K for both forward and backward scattering. A small field $B = 2.5$ kG was applied in order to shift the peak sufficiently away from the strong elastic scattering peak. Notice the broader lines for backward scattering, indicating a significant contribution from the spin diffusion. Also note that the line shapes have sharper peaks and broader Lorentzian-like tails more appropriate to lifetime broadening; in insulating samples (Heiman et al., 1983) the line shape is Gaussian-like and reflects the Gaussian distribution of BMP energies.

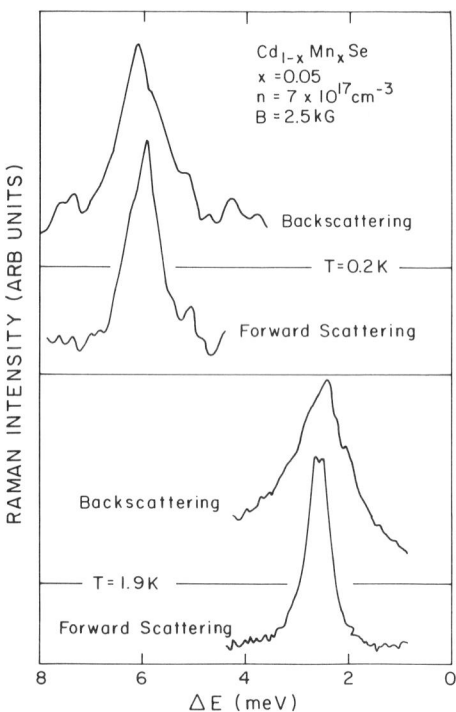

FIG. 33. Spin-flip Raman scattering spectra for electrons at the metal–insulator transition in (Cd,Mn)Se. The top two spectra are for forward scattering and backscattering at $T = 0.2$ K, and the lower spectra were taken at $T = 1.9$ K. Electron diffusion broadens the backscattering spectra as described in the text.

TABLE I

PARAMETERS FROM SPIN-FLIP RAMAN SCATTERING FROM ELECTRONS IN $Cd_{1-x}Mn_xSe$, $x = 0.05$, AND $n = 7 \times 10^{17}$ cm^{-3}

T (K)	Γ_{FWD} (meV)	Γ_{BKD} (meV)	$1/T_2$ (sec^{-1})	D_S (cm^2/sec)
0.2	0.24	0.40	0.35×10^{12}	3.1
2.0	0.20	0.62	0.29×10^{12}	6.3

Table I lists the measured linewidths for both scattering directions, and values for T_2 and D_S determined from these linewidths. Typical uncertainties are $\pm 20\%$ for all quantities except temperature. There is a sizable decrease in D_S for the lowest temperature. This two fold decrease is considerably larger than the 10% decrease observed (Sawicki et al., 1986) in the conductivity σ on a similar sample with $n = 8.6 \times 10^{17}$ cm^{-3}. The decrease in spin diffusion appears to be influenced by BMP formation. There are two arguments supporting this conjecture (Sawicki et al., 1986). First, as the polarons form, the electron wave function might become localized to some extent, which decreases the effective donor radius and hence decreases D_C and, likewise, σ. Second, ferromagnetic regions associated with polarons give rise to efficient spin-disorder scattering, since the localization length (roughly the donor size) is of the order of the Fermi wave vector. The scattering time can be estimated from D_S, assuming $D_S = D_C$, and is $\tau = 1.6 \times 10^{-14}$ sec at $T = 0.2$ K. This value obtained from Raman scattering agrees with $\tau = 2.1 \times 10^{-14}$ sec obtained from conductivity measurements on a similar sample (Sawicki et al., 1986). It is remarkable that such a close correspondence exists for these diverse experimental techniques.

Spin-flip Raman scattering below $T = 1$ K has been shown to be a useful probe of spin diffusion at the metal–insulator transition (Dietl, et al., 1991b). It may also provide a unique opportunity to determine if both localized and delocalized electrons exist on the metallic side of the MiT. The BMP associated with localized carriers should have an additional Raman shift due to the exchange field of the polaron. Thus, the two types of electrons, those that are localized and those that are not localized, might give rise to two Raman peaks. It is important to pursue these studies, since there has been no direct experimental verification of the coexistence of these two types of carriers.

15. SPECTROSCOPY OF Fe AND OTHER IONS

Unlike the fivefold degenerate ground state of Mn^{2+} (3d^6, $L = 0$, $S = 5/2$) that splits in a magnetic field into five levels equally spaced by $g\mu_B B$, the odd

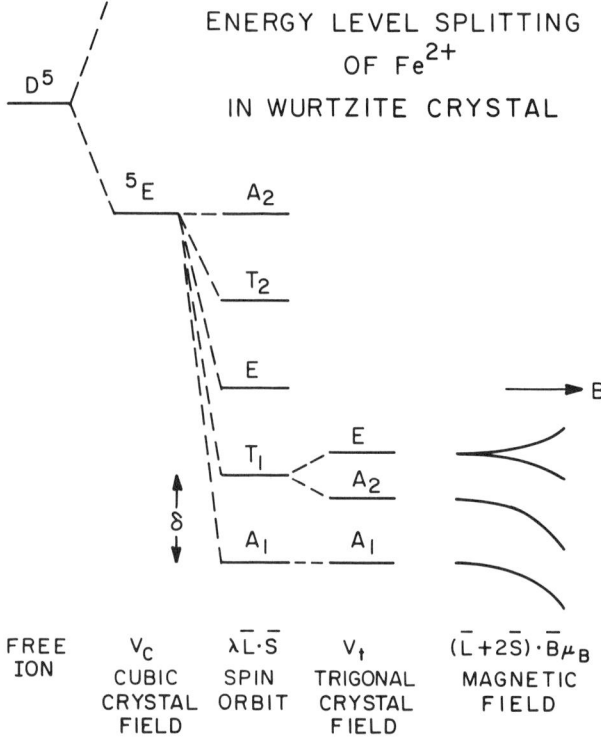

FIG. 34. Diagram of lower-energy levels of Fe^{2+} in a wurtzite crystal structure. Interactions that lead to splitting of the levels are shown at the bottom.

number of 3d electrons (5) and nonzero orbital quantum number in Fe^{2+} ($3d^5$, $L = 2$, $S = 2$) leads to a more complicated situation. The low-energy manifold of states is shown in Fig. 34. Since the ground state is a singlet (A_1), it does not split in a magnetic field and thus does not have a permanent magnetic moment. This property of a Van Vleck ion is expressed as

$$\langle \phi_0 | \mu_Z | \phi_0 \rangle = 0, \qquad (61)$$

where μ_Z is magnetic dipole moment operator and $|\phi_0\rangle$ is the unperturbed ground-state wave function of the collective set of 3d electrons. However, the ground state can acquire a moment *induced* by an applied magnetic field. This field distorts the wave functions and mixes them together. In standard perturbation theory the ground-state wave function becomes

$$|\phi'_0\rangle = |\phi_0\rangle + \sum \frac{B \langle N | \mu_Z | 0 \rangle}{E_N - E_0} |\phi_N\rangle, \qquad (62)$$

where $E_N - E_0$ is the energy difference between the ground and excited states. The induced moment in the ground state is now

$$\langle \phi_0' | \mu_Z | \phi_0' \rangle = \sum \frac{2B |\langle N | \mu_Z | 0 \rangle|^2}{E_N - E_0}. \tag{63}$$

As long as $k_B T \ll E_N - E_0$, the moment is *temperature-independent*.

In practice, Van Vleck ions are usually identified by susceptibility measurements where the usual T^{-1} behavior at high temperatures gives way to a temperature-independent susceptibility at low temperature. A much more striking manifestation of Van Vleck ions has been observed in spin-flip Raman scattering from donor electrons in (Cd,Fe)Se (Heiman et al., 1988a). Figure 35 shows the Zeeman splitting of donor electrons in CdSe with comparable concentrations of Fe and Mn. Notice the large BMP splitting as $B \to 0$ in the Mn alloy compared with the negligible splitting in the Fe alloy. This lack of a large BMP energy is a consequence of the small local moment within the donor orbit caused by the nonmagnetic ground state of Fe^{2+}. In contrast, the finite number ($\sim 10^2$) of Mn^{2+} within a donor orbit can give a sizable local moment due to stochastic fluctuations in the individual moments, as discussed in the section on magnetic polarons.

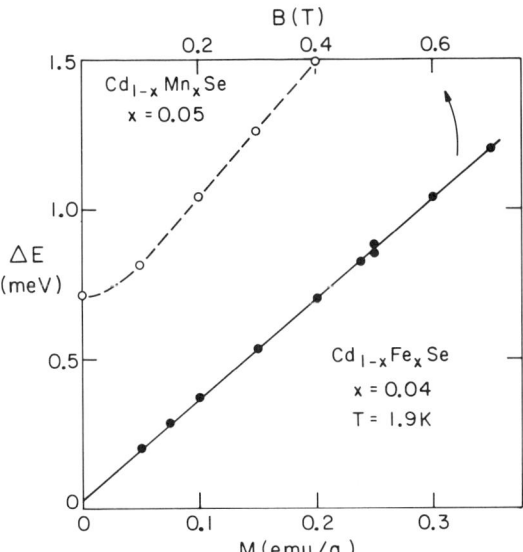

FIG. 35. Comparison of bound magnetic polaron in Mn- and Fe-doped CdSe. Donor electron spin splitting ΔE from spin-flip Raman scattering is plotted versus average magnetization M. The curves were drawn through the data points for visual effect. The larger energy for the Mn material is due to the BMP not found in the Fe material. (From Heiman et al., 1988a.)

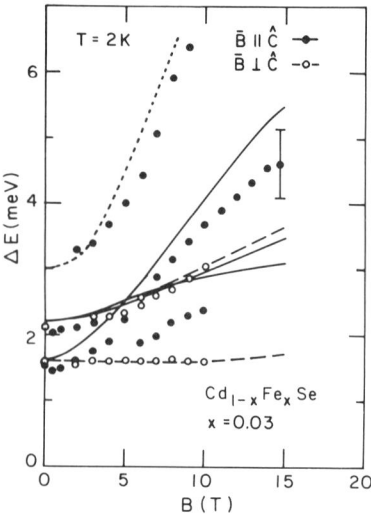

FIG. 36. Transition energies between internal levels of Fe^{2+} in CdSe from Raman scattering at high magnetic fields. The solid and long-dashed curves are from calculations (Scalbert et al., 1989a). The short-dashed curve is plotted at twice the energy of upper solid curve. (From Heiman, 1990.)

In order to calculate the magnetic properties of alloys containing Fe^{2+}, we must know the energy spacing between Fe^{2+} levels. It has been shown that Raman scattering can give accurate values for the energy differences between the ground A_1 state and the lower excited states (Heiman et al., 1989b). Figure 36 shows the results of Raman scattering from (Cd,Fe)Se up to $B = 15$ T. There are three peaks for $B \to 0$. The first peak at 1.5 meV is assigned to the $A_1 \to A_2$ transition, and the second at 2.0 meV to $A_1 \to E_1$. The third peak at 3.0 meV has been assigned to two $A_1 \to E_1$ excitation (Scalbert et al., 1989a). At small B all the lines shift proportional to B^2 as expected for the wave function mixing process.

The selection rules and field dependencies of these transitions will be examined next. First, notice the striking difference in the field-dependencies when applying the field either parallel or perpendicular to the crystal hexagonal c-axis. This behavior results from the selection rules for field-induced mixing of the levels. For B perpendicular to the c-axis the $A_1 \to A_2$ transition is field-independent, reflecting the fact that mixing is forbidden by symmetry. Also in this geometry, only the lower-energy transition of the twofold E_1 level was observed. In the other geometry, B parallel to the c-axis, a crossing is observed for the $A_1 \to A_2$ and $A_1 \to E_1$ transitions; the $A_1 \to A_2$ has a large symmetry-allowed mixing, whereas the $A_1 \to E_1$ is forbidden and relatively field-independent. The linewidth for $A_1 \to A_2$ was found to increase significantly for increasing field owing to the small field-induced splitting of this transition.

A comparison with theory is shown by the curves in Fig. 36. No adjustable parameters were used in the calculations; the input parameters, the crystal-field parameter $10D_q = -325$ meV and spin-orbit coupling parameter $\lambda = -11.8$ meV, were obtained from other measurements (Scalbert et al., 1989a). There is a reasonable agreement between theory and experiment for most of the transitions.

Theory predicts that the third-lowest energy level should be 4.2 meV above the ground state at $B = 0$, which does not agree with the third-lowest transition observed at 3.1 meV at $B = 0$. The dotted curve, representing twice the energy of the highest $A_1 \to A_2$ measured transitions, shows good agreement with the upper data points. Thus, the upper transition is derived from a second-order excitation of the lowest $A_1 \to A_2$ transition. Higher-order excitations up to sixth order have been observed (Scalbert et al., 1990).

An interesting interaction between the spin-flip transition and the internal Fe^{2+} transitions has been found (Scalbert, 1989b). At a field of $B = 2$ T the spin-flip energy is equal to the $A_1 \to A_2$ and $A_1 \to E$ energies, and the transitions exhibit anticrossing. This anticrossing is a measure of the interaction strength between the donor-electron spin and the internal Fe^{2+} spin configuration. In essence, this points to a second-order BMP effect in which the exchange field from the donor-electron spin induces a small magnetic moment in the Fe^{2+} ions. The exchange field applied to the Fe^{2+} ions at the donor radius $r = a_0$ is only 0.1 T. In addition to a small value of the BMP energy, the BMP cannot be observed directly in spin-flip Raman scattering; as the spin flips, the induced moments in the Fe^{2+} ions will also reverse, and thus the spin energy will be the same after the spin-flip process.

Spin-flip Raman scattering has also been applied to CdSe with Co^{2+} (Bartholomew et al., 1989). This is discussed in another chapter devoted to Raman scattering.

These are a few examples how Raman scattering in magnetic fields can be used to obtain information more accurately than other methods, and in some cases it provides unique knowledge not normally accessible. So far, low-temperature magneto-optics has been applied mostly to magnetic semiconductors with Mn^{2+} ions, but new studies are beginning to be devoted to other, more complex magnetic ions such as Fe^{2+} and Co^{2+}. In closing, the ground state energies of the 3d transition metals in II-VI semiconductors are summarized in Fig. 37. On the right side, the conduction and valence band edges of the II-VI materials are shown with their relative band offsets, determined by comparing the transition metal levels in the various semiconductors (Caldas, et al., 1984; Langer, et al., 1988). It is interesting to note that Fe in

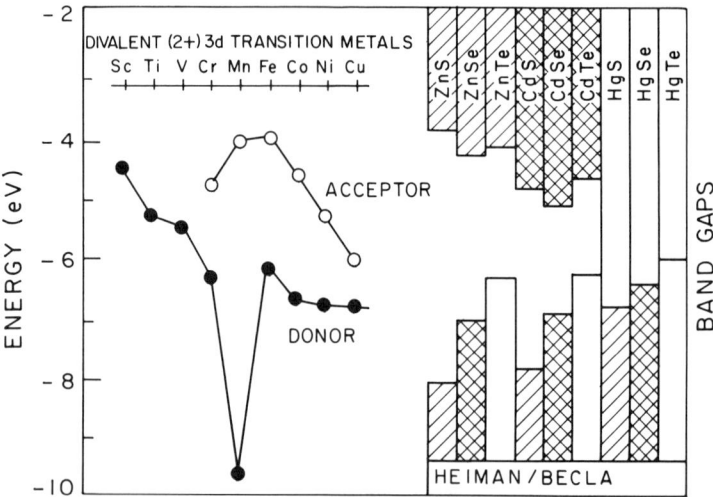

FIG. 37. Electronic energy levels for $3d^{2+}$ transition metal ions in II-VI semiconductors. The solid circles are for ground state donor transitions $(2+ \rightarrow 3+)$ and the open circles are for acceptor transitions $(2+ \rightarrow 1+)$. The bars on the right show the conduction and valence band edges on a relative scale with ± 0.2 eV uncertainty. (From Heiman, et al., 1990.)

HgSe and Sc in CdSe have electronic levels lying *above* the bottom of their conduction bands. This coincidence has led to the observed ionic charge ordering in (He,Fe)Se envisioned by J. Mycielski (Mysielski, 1988).

Acknowledgments

It has been a pleasure to work with a virtual "melting pot" of excellent scientists that have come though the MIT National Magnet Lab. Those involved in the work on magnetic semiconductors were exceptional students Eric Isaacs, Xiaomei Wang, and Marianne Shih; inspirational collaborators Peter Wolff, Yaacov Shapira, Si Foner, Carmen Huber, Mike Graf, Jan Misiewicz, Roshi Aggarwal, and Manfred Dahl at MIT; Athos Petrou, and Tomasz Dietl; Noboru Miura, Shojiro Takeyama, and K. Nakao in Tokyo; and finally, Piotr Becla, Aaron Wold, and Witold Giriat, who supplied many exquisite crystals. Much of the work on 2D GaAs systems over the past six years was inspired by Aron Pinczuk of AT&T Bell Labs. Other fruitful collaborators on this work include Bennet Goldberg, Xiao-Lu Zheng, Lushalan Liao, Gervais Favrot, and Jane Zucker. The Magnet Lab staff, Larry Rubin, Bruce Brandt, Jean Morrison, Margaret O'Meara, and Scott

Hannahs, provided invaluable assistance. Support for much of this work was supplied by the National Science Foundation through grants DMR-8807419, DMR-8807682, and the cooperative agreement DMR-8813164.

References

Abstreiter, G., Merlin, R., and Pinczuk, A. (1986). *IEEE J. Quant. Electron.* **QE-22**, 1771.
Ackermann, H., Jones, E. D., Schirber, J. E., and Overmyer, D. L. (1985). *Cryogenics* **25**, 496.
Aggarwal, R. L., Jasperson, S. N., Shapira, Y., Foner, S., Sakakibara, T., Goto and, T., Miura, N., Dwight, K., and Wold, A. (1985). Proc. 17th Int. Conf. Phys. Semicond., p. 1419 (Chadi, J. D., and Harrison, W. A., eds.). Springer-Verlag, New York.
Aggarwal, R. L., Furdyna, J. K., and von Molnar, S., eds. (1987). Mat. Res. Soc. Symp. 89.
Altarelli, M. (1985). *Festkorperprobleme* **25**, 381.
Anderson, A. C. (1970). *Rev. Sci. Istrum.* **41**, 1446.
Ando, T., and Murayama, Y. (1985). *J. Phys. Soc. Japan* **54**, 1519.
Ando, T., Fowler, A. B., and Stern, F. (1982). *Rev. Mod. Phys.* **54**, 437.
Bangert, E., and Landwehr, G. (1985). *Superlatt. Microstrut.* **1**, 363.
Bartholomew, D. U., Suh, E.-K., Ramdas, A. K., Rodriguez, S., Debska, U., and Furdyna, J. K. (1989). *Phys. Rev. B***39**, 5865.
Bastard, G. (1985). *Festkorperprobleme* **25**, 389.
Bauer, G. E. W., and Ando, T. (1988). *Phys. Rev. B***37**, 3130.
Boebinger, G. S., Stormer, H. L., Tsui, D. C., Chang, A. M., Hwang, J. C. M., Cho, A. Y., Tu, C. W., and Weigmann, G. (1987). *Phys. Rev. B***36**, 7919.
Brandt, N. B., and Moshchalkov, V. V. (1984). *Adv. Phys.* **33**, 193.
Broido, D. A., and Sham, L. J. (1985). *Phys. Rev. B***31**, 888.
Buhmann, H., Joss, W., Klitzing, K. von, Kukushkin, I. V., Martinez, G., Plaut, A. S., Ploog, K., and Timofeev, V. B. (1990). *Phys. Rev. Lett.* **65**, 1056.
Bulgaiski, M., Becla, P., Wolff, P. A., Heiman, D., and Ram-Mohan, L. R. (1988). *Phys. Rev. B***38**, 10512.
Caldas, M. J., Fazzio, A., and Zunger, A. (1984). *Appl. Phys. Lett.* **45**, 671.
Côté, R. and MacDonald, A. H. (1990). *Phys. Rev. Lett.* **65**, 2662.
Dahl, M., Heiman, D., Pinczuk, A., Goldberg, B. B., Pfeiffer, L. N., and West, K. W. (1992). *Phys. Rev. B* (Rapid Communications, to be published).
Das Sarma, S. (1981). *Phys. Rev. B***23**, 4592.
de Jonge, W. J. M., Trardowski, A., and Denissen, C. J. M. (1987). *Mat. Res. Soc. Symp.* **89**, 153.
Dietl. T., and Spalek, J. (1982). *Phys. Rev. Lett.* **48**, 355.
Dietl. T., Sawicki, M., Isaacs, E. D., Heiman, D., Dahl, M., Graf, M. J., Alov D. L., and Gubarev, S. I. (1991a). Proc. 20th Int. Conf. Phys. Semicond. World Scientific, Singapore.
Dietl. T., Sawicki, M., Isaacs, E. D., Dahl, M., Heiman, D., Graf, M. J., Alov D. L., and Gubarev, S. I. (1991b). *Phys. Rev. B*. **43**, 3154.
Duggan, G. (1988). *Phys. Rev. B***37**, 2758.
Foner, S. (1986). *Appl. Phys. Lett.* **49**, 982.
Foner, S., and Kolm, H. H. (1956). *Rev. Sci. Instrum.* **27**, 547.
Foner, S., Shapira, Y., Heiman, D., Becla, P., Kershaw, R., Dwight, K., and Wold, A. (1989). *Phys. Rev. B***39**, 11793.
Furdyna, J. K., and Kossut, J., eds. (1988). "Diluted Magnetic Semiconductors. Academic Press, Boston.
Gaj, J. A. (1980). *J. Phys. Soc. Jpn. Suppl. A***49**, 797.
Gaj, J. A., Planel, R. R., and Fishman, G. (1978). *Sol. State Commun.* **29**, 435.

Galazka, R. R. (1987). Proc. 18th Int. Conf. Phys. Semicond., (Engstrom, O., ed.), p. 127. World Scientific, Singapore.
Geschwind, S., and Romestain, R. (1984). In "Light Scattering in Solids IV" (Cardona, M. and Guntherodt, G., eds.), p. 164. Springer-Verlag, New York.
Giebultowicz, T. M., Rhynn, J. J., and Furdyna, J. K. (1990). *J. Appl. Phys.* **67**, 5096.
Goldberg, B. B., Heiman, D., Pinczuk, A., Tu, C. W., Gossard, A. C., and English, J. H. (1988a). *Surf. Sci.* **196**, 209.
Goldberg, B. B., Heiman, D., Pinczuk, and Gossard (1988b). *Surf. Sci.* **196**, 619.
Goldberg, B. B., Heiman, D., Graf, M. J., Broido, D. A., Pinczuk, A., Tu, C. W., English J. H., and Gossard, A. C. (1988c). *Phys. Rev.* **B38**, 10131.
Goldberg, B. B., Heiman, D., and Pinczuk, A. (1989). *Phys. Rev. Lett.* **63**, 1102.
Goldberg, B. B., Heiman, D., Pinczuk, A., Pfeiffer, L. N., and West, K. W. (1990). *Phys. Rev. Lett.*, **65**, 641.
Goldberg, B. B., Heiman, D., Pinczuk, A., Pfeiffer, L. N. and West, K. W. (1991). *Phys. Rev. B* **44**, 4006.
Golnik, A., Gaj, J. A., Nawrocki, M., Planel, R., and Benoit à la Guillaume, C. (1980). *J. Phys. Soc. Jap.* **49** Suppl. *A*, 819.
"Guinness Book of World Records 1991" (1991). p. 191, Bantam Books, New York.
Heiman, D. (1983). *Appl. Phys. Lett.* **42**, 775.
Heiman, D. (1985). *Rev. Sci. Instrum.* **56**, 684.
Heiman, D. (1987). *Bull. Am. Phys. Soc.* **32**, 577.
Heiman, D. (1990). *Acta. Phys. Polonica A***77**, 411.
Heiman, D., Wolff, P. A., and Warnock, J. (1983). *Phys. Rev. B***27**, 4848.
Heiman, D., Becla, P., Kershaw, R., Ridgley, D., Dwight K., Wold, A., and Galazka, R. R. (1986). *Phys. Rev. B***34**, 3961.
Heiman, D., Isaacs, E. D., Becla, P., and Foner, S. (1987a). *Phys. Rev. B***35**, 3307.
Heiman, D., Isaacs, E. D., Becla, P., and Foner, S. (1987b). *Mat. Res. Soc. Symp.* **89**, 21.
Heiman, D., Pinczuk, A., Gossard, A. C., Fasolino, A., and Altarelli, M. (1987c). Proc. 18th Int. Conf. Phys. Semicond. (Engstrom, O., ed.) p. 617. World Scientific, Singapore.
Heiman, D., Petrou, A., Bloom, S. H., Shapira, Y., Isaacs, E. D., and Giriat, W. (1988a). *Phys. Rev. Lett.* **60**, 1876.
Heiman, D., Goldberg, B. B., Pinczuk, A., Tu, C. W., Gossard, A. C., and English, J. H. (1988b). *Phys. Rev. Lett.* **61**, 605.
Heiman, D., Goldberg, B. B., Pinczuk, A., Tu, C. W., Gossard, A. C., English, J. H., Broido, D. A., Santos, M., and Shayegan, M. (1989a). "Application of High Magnetic Fields in Semiconductor Physics II" (Landwehr, G., ed.), p. 278. Springer-Verlag, Berlin.
Heiman, D., Isaacs, E. D., Becla, P., Petrou, A., Smith, K., Marsella, J., Dwight, K., and Wold, A. (1989b). Proc. 19th Int. Conf. Phys. Semicond. (Zawadski, W.,) ed., p. 1539. Academy of Science, Warsaw.
Heiman, D., Zheng, X. L., Sprunt, S., Goldberg, B. B., and Isaacs, E. D. (1989c). *SPIE* **1055**, 96.
Heiman, D., Dahl, M., Wang, X., Wolff, P. A., Becla, P., Petrou, A., and Mycielski, A. (1990). *Mat. Res. Soc. Symp. Proc.* **161**, 479.
Heiman, D., Pinczuk, A., Dennis, B. S., Pfeiffer, L. N., and West, K. W. (1992). *Phys. Rev. B* (to be published).
Herlach, F., ed. (1985). "Strong and Ultrastrong Magnetic Fields and Their Applications," Springer-Verlag, Berlin.
Herlach, F., and Miura, N. (1987). "High Magnetic Fields in Semiconductor Physics" (Landwehr, G., ed.), p. 536, Springer-Verlag, Berlin.
Isaacs, E. D., and Heiman, D. (1987). *Rev. Sci. Instrum.* **58**, 1672.
Isaacs, E. D., Heiman, D., Zayhowski, J. J., Bicknell, R. N., and Schetzina, J. F. (1986). *Appl. Phys. Lett.* **48**, 275.

Isaacs, E. D., Heiman, D., Graf, M. J., Goldberg, B. B., Kershaw, R., Ridgley, D., Dwight K., Wold, A., Furdyna, J. K., and Brooks, J. S. (1988a). *Phys. Rev. B***37**, 7108.
Isaacs, E. D., Heiman, D., Becla, P., Shapira, Y., Kershaw, R., Dwight, K., and Wold, A. (1988b). *Phys. Rev. B***38**, 8412.
Isaacs, E. D., Heiman, D., Wang, X., Becla, P., Nakao, K., Takeyama, S., and Miura, N. (1990a). *Phys. Rev. B***43**, 3351.
Isaacs, E. D., Dietl, T., Sawicki, M., Heiman, D., Dahl, M., Graf, M. J., Alov, D. L., and Gubarev, S. I. (1990b). Proc. LT-19, *Physica B* **165–166**, 235.
Jones, E. D., Biefeld, R. M., Fritz, I. J., Gourley, P. L., Osbourn, G. C., Schirber, J. E., Heiman, D., and Foner, S. (1987). *Mat. Res. Soc. Symp.* **77**, 467.
Kallin, C., and Halperin, B. I. (1984). *Phys. Rev. B***30**, 5655.
Kallin, C., and Halperin, B. I. (1985). *Phys. Rev. B***31**, 3635.
Katayama, S., and Ando, T. (1989). *Sol. State Commun.* **70**, 97.
Kittel, C. (1976). "Introduction to Solid State Physics." Wiley, New York.
Klitzing, K. von, Dorda, G., and Pepper, M. (1980). *Phys. Rev. Lett.* **45**, 494.
Komarov, A. V., Ryabchenko, S. M., Terletskii, O. V., Zheru, I. I., and Ivanchuk, R. D. (1977). *Sov. Phys. JETP* **46**, 318.
Kukushkin, I. V., and Timofeev, V. B. (1986). *JETP* **43**, 387.
Kukushkin, I. V., and Timofeev, V. B. (1988). *Surf. Sci.* **196**, 196.
Kukushkin, I. V., Klitzing, K. v., and Ploog, A. (1988). *Phys. Rev. B***37**, 8509.
Lam, P. K., and Girvin, S. M. (1984). *Phys. Rev. B***30**, 473; and (1985). *Phys. Rev. B***31**, 613(E).
Langer, J. M., Delerve, C., Lannoo, M., Heinrich, H. (1988). *Phys. Rev B.* **38**, 7723.
Larsen, B. E., and Ehrenreich, H. (1989). *Phys. Rev. B***39**, 1747.
Larsen, B. E., and Ehrenreich, H. (1990). *J. Appl. Phys.* **67**, 5084.
Larsen, B. E., Hass, K. C., and Aggarwal, R. L. (1986). *Phys. Rev. B***33**, 1789.
Laughlin, R. B. (1983). *Phys. Rev. Lett.* **50**, 1395.
Lewicki, A., Spalek, J., and Mycielski, A. (1987). *J. Phys. C: Sol. State* **20**, 2005.
Liao, L. B., Heiman, D., Pinczuk, A., MacDonald, A. H., Pfeiffer, L. N., and West, K. W. (1992) *Phys. Rev.* (to be published).
Lounasmaa, O. V. (1974). "Experimental Principles and Methods Below 1 K." Academic Press, London.
Maan, J. C., Belle, G., Fasolino, A., Altarelli, M., and Ploog, K. (1984). *Phys. Rev. B***30**, 2253.
MacDonald, A. H. (1985). *J. Phys. C: Sol. State Phys.* **18**, 1003.
MacDonald, A. H., Oji, H. C. A., and Girvin, S. M. (1985b) *Phys. Rev. Lett.* **55**, 2208.
MacDonald, A. H., Liu, K. L., Girvin, S. M., and Platzman, P. M. (1986). *Phys. Rev. B***33**, 4014.
Moschalov, V. V., and Brandt, N. B. (1984). *Adv. Phys.* **33**, 193.
Mycielski, A. (1988). *J. Appl. Phys.* **63**, 3279.
Nagaev, E. L. (1983). "Physics of Magnetic Semiconductors." MIR, Moscow.
Nawrocki, M., Planel, R., Fishman, G., and Galazka, R. R. (1981). *Phys. Rev. Lett.* **46**, 735.
Osbourn, G. C. (1986). *IEEE J. Quant. Electron.* **QE-22**, 1677.
Ossau, W., Jakel, B., and Bangert, E. (1987). "High Magnetic Fields in Semiconductor Physics," (Landwehr, G., ed.) p. 213. Springer-Verlag, Berlin.
Perry, C. H., Worlock, J. M., Smith, M. C., and Petrou, A. (1987). "High Magnetic Fields in Semiconductor Physics," (Landwehr, G., ed.), p. 202. Springer-Verlag, Berlin.
Peterson, D. L., Bartholomew, D. U., Debska, U., Ramdas, A. K., and Rodriguez, S. (1985). *Phys. Rev. B***32**, 323.
Petrou, A., and McCombe, B. D. (1991). "Landau Level Spectroscopy" (Landwehr, G., and Rashba, E. I., eds.), Ch. 13. Elsevier, New York.
Pinczuk, A., and Heiman, D. (1987). "High Magnetic Fields in Semiconductor Physics," (Landwehr, G., eds.), p. 227. Springer-Verlag, Berlin.
Pinczuk, A., and Worlock, J. M. (1982). *Surf. Sci.* **113**, 69.

Pinczuk, A., Shah, J., Miller, R. C., Gossard, A. C., and Wiegmann, W. (1984). *Sol. State Commun.* **50**, 735.
Pinczuk, A., Heiman, D., Sooryakumar, R., Gossard, A. C., and Weigmann, W. (1986). *Surf. Sci.* **170**, 573.
Pinczuk, A., Heiman, D., Gossard, A. C., and English, J. H. (1987). Proc. 18th Int. Conf. Phys. Semicond. (Engstron, O., ed.), p. 557. World Scientific, Singapore.
Pinczuk, A., Valladares, J. P., Heiman, D., Gossard, A. C., English, J. H., Tu, C. W., Pfeiffer, L., and West, K. (1988). *Phys. Rev. Lett.* **61**, 2701.
Pinczuk, A., Valladares, J. P., Heiman, D., Pfeiffer, L., and West, K. (1990). *Surf. Sci.*, **229**, 384.
Prang, R. E., and Girvin, S. M. eds. (1987). "The Quantum Hall Effect." Springer-Verlag, New York.
Ram-Mohan, L. R., and Wolff, P. A. (1988). *Phys. Rev.* **B38**, 1330.
Rezayi, E. H., and Haldane, F. D. M. (1985). *Phys. Rev.* **B32**, 6924.
Richardson, R. C., and Smith, E. N. (1988). "Experimental Techniques in Condensed Matter Physics at Low Temperatures." Addison-Wesley, Reading, MA.
Rogers, D. C., Singleton, J., Nicholas, R. J., and Foxon, C. T. (1987). "High Magnetic Fields in Semiconductor Physics," (Landwehr, G., ed.) p. 223. Springer-Verlag, Berlin.
Rose-Innes, A. C. (1973). "Low Temperature Laboratory Techniques." English Universities Press, London.
Ryabchenko, M., and Semenov, Yu. (1983). *Sov. Phys. JETP* **57**, 825.
Sakaki, H., Arakawa, Y., Nishioka, M., Yoshino, J., Okamato, H., and Miura, N. (1985). *Appl. Phys. Lett.* **46**, 492.
Sawicki, M., Dietl, T., Kossut, J., Igalson, J., Wojtowicz, T., and Plesiewicz, W. (1986). *Phys. Rev. Lett.* **56**, 508.
Scalbert, D., Cernogora, J., Mauger, A., Benoit a la Guillaume, C., and Mycielski, A. (1989a). *Sol. State Commun.* **69**, 453.
Scalbert, D., Gaj, J., Mauger, A., Cernogora, J., and Benoit à la Guillaume, C. (1989b). *Phys. Rev. Lett.* **62**, 2865.
Scalbert, D., Gaj, J., Mauger, A., Cernogora, J., and Benoit à la Guillaume, C. (1990). Proc. Int. Conf. on II-IV Compounds, Berlin, 1989.
Schirber, J. E. (1970). *Cryogenics* **10**, 478.
Schlesinger, Z., Allen, S. J., Yafet, Y., Gossard, A. C., and Weigmann, W. (1985). *Phys. Rev.* **B32**, 5231.
Schmitt-Rink, S., Chemla, D. S., and Miller, D. A. B. (1989). *Adv. Phys.* **38**, 89.
Scott, J. F., Damen, T. C., and Fleury, P. A. (1972). *Phys. Rev.* **B6**, 3856.
Sham, L. J. (1987). "High Magnetic Fields in Semiconductor Physics" (Landwehr, G., ed.), p. 288. Springer-Verlag, Berlin.
Shapira, Y. (1990). *J. Appl. Phys.* **67**, 5090.
Shapira, Y., Foner, S., Ridgley, D. H., Dwight, K., and Wold, A. (1984). *Phys. Rev.* **B30**, 4021.
Shapira, Y., Oliveira, N. F., Ridgley, D. H., Kershaw, R., Dwight, K., and Wold, A. (1986). *Phys. Rev.* **B34**, 4187.
Stormer, H. L., Chang, A. M., Tsui, D. C., Huang, J. C. M., Gossard, A. C., and Weigmann, W. (1984). *Phys. Rev. Lett.* **53**, 997.
Tarucha, S., Okamoto, H., Iwasa, Y., and Miura, N. (1984). *Sol. State Commun.* **52**, 815.
Tien, Z. J., Worlock, J. M., Perry, C. H., Pinczuk, A., Aggarwal, R. L., Stormer, H. L., Gossard, A. C., and Weigmann, W. (1982). *Surf. Sci.* **113**, 89.
Tsui, D. C., Stormer, H. L., and Gossard, A. C. (1982). *Phys. Rev. Lett.* **48**, 1559.
Turberfield, A. J., Haynes, S. R., Wright, P. A., Ford, R. A., Clark, R. G., Ryan, J. F., Harris, J. J., and Foxon, C. T. (1990). *Phys. Rev. Lett.* **65**, 637.
Twardowski, A., Lewicki, A., Arciszewska, M., de Jonge, W. J. M., Swagten, H. J. M., and Demianiuk, M. (1988). *Phys. Rev.* **B38**, 10749.

Uenoyama, T., and Sham, L. J. (1989). *Phys. Rev.* **B39**, 11044.
Wang, X., Heiman, D., Foner, S., and Becla, P. (1989). *Phys. Rev.* **B41**, 1135.
Wang, X. (1991). PhD. Thesis. Massachusetts Institute of Technology, Department of Physics.
Weigmann, W. (1985). *Phys. Rev.* **B32**, 5231.
White, G. K. (1979). "Experimental Techniques in Low Temperature Physics." Clarendon Press, London.
Wolff. P. A. (1988). "Diluted Magnetic Semiconductors," Ch. 10. Academic Press, Boston.
Wolff, P. A., Ramos, J. G., Yuen, S. (1976). "Theory of Light Scattering in Condensed Matter" (Bendow, B., Birman, J. L., and Agranovich, V. M., eds.), p. 475. Plenum, New York.
Worlock, J. M., Pinczuk, A., Tien, Z. J., Perry, C. H., Stormer, H. L., Dingle, R., Gossard, A. C., Weigmann, W., and Aggarwal, R. L. (1981). *Sol. State Commun.* **40**, 867.
Worlock, J. M., Maciel, A. C., Perry, C. H., Tien, Z. J., Aggarwal, R. L., Gossard, A. C., and Weigmann, W. (1983). "Application of High Magnetic Fields in Semiconductor Physics" (Landwehr, G., eds.), p. 186. Springer-Verlag, Berlin.
Yang, Y. C., and Sham, L. J. (1987). *Phys. Rev. Lett.* **58**, 2598.
Zhang, F. C., Volovic, V. Z., Guo, Y., and Das Sarma, S. (1985). *Phys. Rev.* **B32**, 6920.
Zheng, X.-L., Heiman, D., Lax, B., and Chambers, F. A. (1988a). *Superlatt. Microstruct.* **4**, 351.
Zheng, X.-L., Heiman, D., Lax, B., Chambers, F. A., and Stair, K. A. (1988b). *Appl. Phys. Lett.* **52**, 984.
Zheng, X.-L., Heiman, D., Lax, B., and Chambers, F. A. (1989). *Phys. Rev.* **B40**, 10523.
Zucker, J. E., Isaacs, E. D., Heiman, D., Pinczuk, A., and Chemla, D. S. (1988). *Surf. Sci.* **196**, 563.

CHAPTER 2

Transient Spectroscopy by Ultrashort Laser Pulse Techniques

Arto V. Nurmikko

DIVISION OF ENGINEERING AND DEPARTMENT OF PHYSICS
BROWN UNIVERSITY, PROVIDENCE, RHODE ISLAND

I. INTRODUCTION 85
II. INSTRUMENTATION FOR TIME-RESOLVED SPECTROSCOPY 88
   1. Sources 88
      a. Solid-State Sources 89
      b. Dye Laser Sources 91
      c. Nonlinear Optical Generation of Ultrashort Pulses 92
   2. Modulators and Ultrafast Switches 93
   3. Detectors 94
III. TIME-RESOLVED SPECTROSCOPY IN SEMICONDUCTORS: APPROACHES 96
   4. Transient Photoluminescence: Spectrum 97
   5. Excite-Probe Techniques 101
IV. PORTFOLIO OF EXPERIMENTAL EXAMPLES IN TIME-RESOLVED SPECTROSCOPY . . . 105
   6. Bulk Semiconductors (Low Excitation Conditions) 106
   7. Quantum Wells and Superlattices: Linear Regime 116
   8. Nonlinear Transient Optical Spectroscopy: Recent Results 126
V. CONCLUDING REMARKS 131
   REFERENCES 133

I. Introduction

In the past decade, a range of advanced laser equipment and associated optoelectronic instrumentation has emerged from specialized research laboratories to the general scientific and commercial marketplace to make possible spectroscopic studies of semiconductors by sophisticated time-resolved methods in the picosecond and femtosecond time scale. As a consequence, a veritable explosion has occurred in the application of such techniques in various areas of physical sciences in general and in semiconductor sciences in particular. On subpicosecond time scales, electron motion in real or momentum space within complex heterostructures can today be tracked for direct microscopic and detailed physical insights. For instance, an electron wave packet with a group velocity of 10^7 cm/sec will advance a distance of 100 Å in 100 fsec, an event well within the direct resolution of modern time-resolved spectroscopic instrumentation. In this chapter I will summarize the main technical features and virtues of these powerful experimental techniques and illustrate them in the context of contemporary semiconductor problems, with emphasis on basic physical (electronic) phenomena in crystalline materials and heterostructures. Of special interest are material microstructures such as quantum wells and superlattices, which, in general, have added considerably to the vitality of contemporary semiconductor science. With such a focus on illustrating the application of transient spectroscopic methods to basic semiconductor physics problems, a large domain of very active semiconductor optoelectronics device work that is driven by application to wide-bandwidth optical communication systems will be unavoidably excluded. Yet ultrafast electron phenomena are clearly of importance in high-speed semiconductor lasers and detectors. Ultrafast laser annealing work and spectroscopy of amorphous semiconductors is also left outside this selective review. Similarly, spectroscopy of surfaces is not treated, although it should be noted that this field is now developing fairly rapidly. The main purpose here is to demonstrate to the practicing scientist the wide range of possibilities, with state-of-the art examples, that exist in condensed-matter experiments regarding dynamical aspects of spectroscopically accessible excitations. One no longer needs to be a laser specialist to exploit these opportunities, though some familiarity is, of course, very helpful.

This chapter is organized along the following lines. We begin with a review of relevant experimental technologies (sources and recording instrumentation) that, while not comprehensive, is aimed at providing concrete examples of the laboratory equipment which are typically employed in a modern ultrafast laser laboratory. The reader should note that the rapid pace of technical progress in this area of modern optoelectronics and lasers leads to a relatively short half-life of such instrumentation before the onset of partial obsolescence. The discussion about instrumentation is followed by discussion of the principal spectroscopic approaches and techniques used in transient optical studies of semiconductors. The methods can be conveniently divided into those where spontaneous emission following an impulse of optical excitation is registered, i.e., transient luminescence, and those where short laser pulses are employed as spectroscopic probes to measure the electronic and/or vibronic states in the photoexcited system, the so-called pump-and-probe techniques. The latter can be viewed as a form of photomodulated spectroscopy, which, in addition to reflectance and transmission measurements, may also include transient Raman probes and other nonlinear optical processes such as four-wave mixing. The choice between these methods often depends on the physical details of the problem at hand, including the spectral and temporal regime of interest, material constraints, and available instrumentation. With many available laser sources, the level of initial excitation can be varied over an enormous range, corresponding to a small perturbation of the material system under study, or to induce very large departures from equilibrium (up to a phase transition or material damage). Most of the balance of the chapter is devoted to illustrations and discussion of examples of actual time-resolved measurements in contemporary semiconductor science. An attempt has been made to cover a range of physical phenomena and material circumstances to provide a wider demonstration for the usefulness of transient spectroscopies on the picosecond and femtosecond time scales. The choices for examples to illustrate these approaches in the context of recent and ongoing work a necessarily subjective; it would by now be somewhat encyclopedic to write an all-encompassing review. Accordingly, I apologize in advance to those colleagues in the field whose work I did not have a chance to refer to.

One question frequently raised concerns the overall usefulness of time-resolved studies when compared with conventional steady-state and continuous-wave optical spectroscopies. There are certainly circumstances, especially in the linear regime of optical response by a semiconductor, where transient spectroscopy does not always provide uniquely new physical information. However, even there it usually offers a powerful quantitative complement to conventional continuous-wave measurements and provides a direct means of characterizing dynamical properties in the system under study, following its initial impulsive departure from equilibrium. Especially in those circumstances where a multitude of energetic relaxation pathways exist in the photoexcited semiconductor, time-resolved spectroscopy can be invaluable in sorting things out. In the case of nonlinear optical response of the semiconductor, short laser pulses can be extraordinarily useful in measurements of the system response, in effect by time-dependent correlation functions in the optical susceptibility. Elsewhere, in the regime of very high excitation (including phase transitions), the desired photoinduced changes may only be realized on ultrashort time scales without material damage or excessive heating effects, and such large departures from equilibrium may simply be unaccessible to steady-state spectroscopies. Examples of specific physical aspects of electronic excitations for which time-resolved methods have recently given unique insight include

- dephasing processes of electrons and excitons,
- electron–phonon inelastic scattering (including intervalley scattering),
- exciton–exciton and exciton-free carrier interaction,
- determination of exciton–polariton group velocity,
- optical Stark effect,
- electron transport in layered structures such as superlattices, and
- spin-related effects in magnetic semiconductors.

Examples of these and other subjects where transient optical techniques have added a new dimension in semiconductor spectroscopy will be given in Section III.

II. Instrumentation for Time-Resolved Spectroscopy

The purpose of this section is to give an overview of the laser and electro-optic instrumentation that is typical of a modern ultrafast spectroscopic facility. One of the important characteristics of the field from a user's point of view has been the rapid technological development that has frequently made yesterday's instrumentation less competitive in terms of performance, if not altogether obsolete. For the user who mainly wishes to benefit from this instrumentation for spectroscopic use and is not conducting laser research per se, this pace of development poses frequent problems in the choice and acquisition of equipment, while attempting to balance what is usually a significant cost factor. Here, an attempt is made to summarize the status of ultrashort pulsed laser and wide-bandwidth electro-optic instrumentation that is most relevant to semiconductor spectroscopy. I would caution the reader, however, that while a certain level of maturity appears to be presently setting in, e.g., for femtosecond dye lasers (in terms of the pulse width reduction), other aspects of laser performance, nonlinear optical techniques and materials, as well as advanced detector and data acquisition equipment are likely to continue their pace of development unabated. One useful and recommended reference source for tracking such developments is the Proceedings of the Biannual Conference on Ultrafast Phenomena published by Springer-Verlag.

1. SOURCES

As the principal spectral regions of interest in a semiconductor or its microstructure are usually related to critical points in the joint density of states, most commonly at the fundamental absorption edge, the availability of a short-pulse laser source over a maximum wavelength range is clearly the ideal circumstance for an aspiring experimentalist for handling a number of different semiconductors. By and large, there is no such ideal source, but instead discrete chunks in the optical spectrum are typically covered by specific sources (see, however, the notes on continuum generation). Generally speaking, much of the visible and near-infrared regions are now quite accessible for picosecond, and to a somewhat lesser extent, femtosecond sources for resonant or near resonant excitation and probing. Short visible and ultraviolet wavelengths (say $\lambda < 3000$ Å) as well as longer infrared wavelengths ($\lambda > 3$ μm) are more cumbersome, though accessible in principle by nonlinear wave-mixing techniques. Work in the infrared is an added challenge due to the considerably more modest detector technology presently available in this spectral region.

a. Solid-State Sources

The main workhorse today in a typical ultrashort pulse laser laboratory is the continuous-wave (cw) mode-locked Nd:YAG laser or its cousin, the Nd:YLF laser, in the 1.06-μm wavelength regime. These sources provide sub-100-psec pulses (50–80 psec) at high repetition rates (~ 100 MHz) and reasonable pulse energies (~ 10 nJ), with average powers reaching 20–30 W. While the use of the Nd:YAG systems is more widespread, the YLF laser has its advantage in a smaller thermal index gradient within the laser rod, a property that leads to improved beam pointing stability. This is an important issue in those applications where the infrared laser beam is tightly focused, e.g., for coupling into single-mode fibers or for generation of high intensities in nonlinear applications. Recently, however, stabilization schemes that specifically address the beam-pointing problem in Nd:YAG lasers have appeared in commercial devices. Generally, given the available peak powers of the mode-locked Nd:YAG or YLF lasers, efficient second-harmonic generation to the green at $\lambda = 0.532$ μm (at conversion efficiencies exceeding 10%) and third-harmonic generation to the near-UV at $\lambda = 0.339$ μm (at efficiencies up to $\approx 5\%$) generation are readily accomplished. For the second-harmonic generation (2ω), the most common nonlinear crystal is KTP (potassium tantalum phosphate), although at this writing LBO as a possibly more advantageous alternative is being introduced commercially. The third-harmonic generation usually proceeds via frequency summation of the YAG and its second harmonic ($\omega + 2\omega = 3\omega$) in β-barium borate (BBO) or LBO.

In an ultrafast laser laboratory for semiconductor spectroscopy, a Nd:YAG or YLF laser often forms the first optical power source in a photonic food-chain chain, which may next be followed by a synchronously pumped dye laser or other short-pulse optically pumped laser device. In the process of synchronous pumping, for example, dynamical pulse compression effects occur, yielding typically dye laser pulse of $\tau \sim 1$–5 psec in duration. The Nd:YAG and YLF lasers are themselves also quite compatible with fiber-optic pulse-compression schemes where nonlinear effects in the fiber (self-phase modulation) have been employed either at 1.06 μm or 0.52 μm to obtain pulses of subpicosecond duration (Johnson *et al.*, 1984). Apart from the use of this radiation directly as photoexcitation of a semiconductor, the compressed pulses have been used to generate subpicosecond dye laser emission in the synchronous pumping arrangement (Johnson and Simpson, 1986).

Pulse, actively or passively mode-locked, low-repetition-rate (~10 Hz) Nd:YAG lasers are occasionally still used in semiconductor spectroscopy, although, apart from high pulse energies, they offer few advantages over the cw sources. The more useful role for pulsed YAG systems in ultrashort applications is in their use as amplifiers for extremely high power (>terawatt) generation.

A recent significant development that promises to augment the short pulse solid state source area in the Ti:sapphire laser (e.g., Sanchez *et al.*, 1988). In the cw mode of operation, the tuning range in an Ar^+-ion laser or the frequency-doubled YAG laser-pumped Ti:sapphire laser extends from about 700 to 900 nm. In cw applications, these lasers have now become developed to the level of performance at which organic dye lasers operate, and are increasingly replacing dye lasers in this spectral range (as well as the range in the blue accessible by second-harmonic generation by the Ti:sapphire source). The Ti:sapphire laser is also a useful candidate for pumping of dye lasers at infrared wavelengths, and it is already seeing use in amplifier applications of short pulses as well. At longer wavelengths, further development of YAG-based sources that involve other rare-earth elements is likely to provide opportunities also for short-pulse spectroscopy beyond the immediate reach of the Nd:YAG or YLF lasers. Following early mode locking studies, there has been a flurry of very recent commercial activity. While this article was being written, femtosecond mode-locked Ti:sapphire lasers were being introduced to the market. These sources offer considerable amplitude stability, typically in 100 fsec pulses (nearly transform limited) at high repetition rates (~100 MHz) and, most important, a wide wavelength tunability. It is already clear that these new sources will present a significant challenge to the near infrared and blue dye lasers and probably surpass them in a variety of time-resolved spectroscopic applications. Furthermore, a high power all solid-state ultrashort pulse system is likely to emerge shortly, with amplification of the femtosecond pulses by a Ti:sapphire amplifier.

The use of the YAG laser in pumping mode-locked color center lasers has given some added spectroscopic opportunities at selected wavelengths up to and slightly beyond $2\mu m$ (Islam *et al.*, 1989). These sources are usually somewhat more complex experimental arrangements from the user's point of view, given, for example, the need for cryogenic cooling of the active medium. However, current progress is yielding pulses as short as 100 fsec from the color center lasers, making them increasingly attractive for spectroscopic use. We also mention in passing ongoing research in the use of optical fibers (and fiber amplifiers such as the Er-doped systems) for ultrashort pulse generation in the infrared, where the use of nonlinear (soliton) effects is made to compress or maintain the pulse width in the picosecond regime and below (Mitschke and Mollenauer, 1987).

Although the overall technical advances made in the past decade with semiconductor lasers probably dwarf all other coherent sources, these compact, low-power lasers have yet to be incorporated into transient spectroscopic studies in a substantial way. However, both gain switching (Nagarajan *et al.*, 1989) and active mode-locking methods (Bowers *et al.*, 1989) have yielded low-jitter picosecond (and subpicosecond) pulses from GaAs- and InGaAs-based lasers, including surface-emitting configurations (Jiang *et al.*, 1991). While the absence of substantial wavelength tunability will continue to place limitations in their versatile spectroscopic use, it should be only a matter of time and some initiative before their application becomes more widespread, especially in those experimental circumstances where portability or limited access to sample space restrict the use of the larger conventional short-pulse sources (e.g., high magnetic fields or high-pressure ambients).

b. *Dye Laser Sources*

By far the most versatile and widely employed sources in picosecond and/or femtosecond laser spectroscopy today are the organic dye lasers whose wide-gain bandwidth offers the combination of ultrashort pulses and wavelength tunability. These two advantages are readily realized in the synchronously pumped configuration, where the laser gain is periodically modulated at the resonator round-trip transit time by the mode-locked Nd:Yag (YLF) or Ar^+-ion pump laser. In the most basic configuration of a linear resonator and a birefringent wavelength tuning element, very stable pulses at 100-MHz repetition rate of typical duration of 1–5 psec are obtained, with average powers typically ~ 100 mW (pulse energies a fraction of 1 nJ, that is, some 10^{10} photons per pluse) and wavelength range from ~ 580 to 900 nm with the choice of readily available commercial dyes. Recently, the synchronous pumping by using the third harmonic of a cw mode-locked YAG laser (at 353 nm) has also permitted the use of such blue dyes as the stillbenes and coumarins in picosecond transient semiconductor spectroscopy.

The one distinct disadvantage of synchronous pumping schemes is that it becomes increasingly cumbersome to generate pulses that are substantially shorter than 1 psec. With a compressed YAG pulse, dye laser pulses on the order of 400–500 fsec are relatively straightforward to obtain, but beyond this both the inherent gain switching dynamics and resonator instabilities become inhibiting factors. (Also recall that the YAG laser has a finite beam-pointing instability problem arising from thermal index gradients in the laser rod; this leads to spatial fluctuations in the photoexcited dye, which in turn becomes a significant source of instability when attempting to reach the femtosecond regime. The YLF source, on the other hand, avoids this problem.)

An elegant source of very short dye laser pulses is the passively mode-locked laser, especially in the colliding pulse (CPM) version (Fork et al., 1981). While limited so far to relatively few fixed wavelengths in the visible and near infrared (e.g., Knox, 1987), it readily produces very stable pulses of ~ 100 fsec in duration at peak power levels comparable with those in synchronous pumping. Typically a cw Ar^+-ion laser is used as a source of excitation, although recently a mode-locked source (frequency-doubled Nd:YAG) has yielded a hybrid operation now exploited in commercially available systems. The hybrid laser is subject to a combination of synchronous pumping and passive mode-locking by a saturable absorber and has some flexibility in wavelength tuning.

The use of subsequent dye laser *amplifiers* permits today the generation of ultrashort pulses of femtosecond scale duration at very high intensities. With subsequent pulse-compression techniques, pulses that only contain a few optical cycles have been generated (Fork et al., 1987); their application in practical spectroscopy is quite challenging, however, since a material medium of finite thickness will result in pulse broadening by group velocity dispersion. Through nonlinear optical interactions in transparent liquid or solid media, it is also possible (and practical) to generate a partial continuum of ultrashort pulse radiation for use as a white light spectroscopic probe in time-resolved experiments (Fork et al., 1983; Knox et al., 1985). The continuum can also be amplified to generate ultrashort pulses at selected wavelengths, such as in the blue green (Schoenlein et al., 1991). Such instrumentation, which is slowly increasing in its utilization among nonspecialists, provides an extraordinarily powerful experimental arrangement in the study of very fast transient optical phenomena. It represents today the cutting edge of the state-of-the-art short-pulse spectroscopies and will undoubtedly be an important element in future applications to semiconductor studies.

c. Nonlinear Optical Generation of Ultrashort Pulses

Nonlinear techniques have traditionally played an important role in extending the wavelength range obtained from fixed-photon energy laser sources. The techniques of harmonic generation have already been referred to. Beyond this, both sum and difference generation have been widely employed through the use of suitable nonlinear crystals to extend the generation and application of ultrashort pulse techniques to the ultraviolet and infrared wavelengths. Practical issues of the problem concern the availability of crystals with a large second-order susceptibility $\chi^{(2)}$ that must be transparent and afford phase velocity matching for the optical frequencies of interest. The usual strategy of obtaining phase matching through the birefringence of an anisotropic ($\chi^{(1)}$) crystal is frequently accompanied by substantial beam walk-off effects; that is, the nonlinearly generated beam has a Poynting vector that is not parallel to that of the pump beam(s) even in otherwise collinear wave vector geometry. This limits the maximum useful length of the crystal, something that is not always of significant detriment for ultrashort pulses, given the need to limit the crystal length for minimizing pulse broadening from dispersive group velocity effects. The transverse beam profile of the nonlinearly generated pulse is often uniaxially distorted, however, and can deviate substantially from the fundamental Gaussian beam model. At short visible wavelengths, the earlier use of KH_2PO_4 (KDP) and $LiIO_3$ is today being partially supplanted by BBO and LBO, while in the infrared (3–12 μm), commercial availability of $AgGaS_2$ and $AgGaSe_2$ is replacing $LiNbO_3$ and $LiIO_3$.

Optical parametric oscillators, devices utilizing $\chi^{(2)}$-type nonlinearities within optical resonator structures, have also recently been developed for use in the picosecond and subpicosecond pulse generation. A very attractive feature of such parametric schemes in general, is the wavelength tuning that they afford in principle over wide spectral ranges. For, example a KTP-based parametric oscillator has generated subpicosecond pulses from 0.7 to 4.5 μm with commensurate changes in the external resonator optics required over this range (Edelstein *et al.*, 1989). While some engineering improvements are probably still necessary before the parametric oscillator schemes become stable user-friendly sources for the nonexpert spectroscopist, these sources are very attractive in principle and promising for future semiconductor research.

2. MODULATORS AND ULTRAFAST SWITCHES

The use of electro-optic or acousto-optic modulation schemes to actively mode-lock cw Ar^+-ion and Nd:YAG lasers is a routine practice. Rapid switching of semiconductor lasers is another application of a fast electrical switching means to generate moderately short optical pulses, as already mentioned. Generally, however, such external modulator schemes are too slow to be useful in the generation of ultrashort optical radiation. On the other hand, standard electro-optic and acousto-optic modulators are widely used as fast "choppers" to impose additional control on a high-repetition-rate ultrashort pulse laser for optical signal processing. For example, in the case of mode-locked dye lasers the considerable low-frequency laser noise (from hydrodynamical fluctuations in the dye jet) often necessitates the use of amplitude modulation (chopping) frequencies in the 1–10 MHz range for improved synchronous detection.

A different application of fast modulator methods, accessing the subpicosecond regime, uses fast *photoconductive switching* (Auston, 1987;

Grischkowsky, 1988) within a high-frequency microwave stripline to generate ultrashort electrical pulses. The stripline is designed onto a semiconductor, or contains a semiconducting section, so that a charged portion of the stripline can be rapidly electrically connected to the uncharged portion (i.e., form a pulse) by using short laser pulses to induce a photoconductive path. The very process of rapid photoconductive switching has been studied in GaAs from the standpoint of hot electron effects in such a stripline configuration with ultrafast time resolution built into this electrical–optical experimental device arrangement (Meyer *et al.*, 1988). A potentially useful source of free space propagating short electrical pulses, that is, terahertz electromagnetic radiation, has been recently discovered in connection with fast photoconductive transients at absorbing semiconductor surfaces (Darrow *et al.*, Zhang *et al.*, 1990). The electric field in the surface depletion region of GaAs or Si, for example, can induce a very fast photoconductive current, initiated by a femtosecond incident laser pulse, thereby creating a dipole-like photoconductive emitting antenna. At this writing, applications of such short pulse electromagnetic radiation in semiconductor spectroscopy is just beginning. For example, as a probe of free carrier response, this novel source may offer an attractive counterpart to more conventional transient transport experiments.

3. DETECTORS

Very high speed *photodiodes* have been developed in recent years in research laboratories and available detectors have speeds of response better than 50 psec, often making them very useful in the characterization of pulsed laser sources in conjunction with commensurate sampling electronics (oscilloscopes). However, due to the small effective areas (< 100 μm diameter), such detectors are not generally useful in spectroscopic applications such as transient luminescence. For higher sensitivity in the subnanosecond time domain, fast *photomultipliers* (PMT) based on microchannel plate technology are now approaching the 100 psec limit at visible wavelengths by such manufacturers as Hamamatsu and ITT. An alternative approach with comparable or slightly better time resolution is through synchronous coincident photon counting methods with fast single-channel PMTs, where fast digital electronic circuitry can be effectively used (Yamazaki *et al.*, 1985). Both of these approaches are relatively inexpensive in comparison with the, by far, most versatile and direct recording of luminescence events on picosecond timescale, that is, by a *streak camera*. The streak camera is a very versatile electron–optical chronograph for which in single-shot operation the 1–2 psec resolution is available in commercial instruments (e.g., by Hamamatsu Inc.

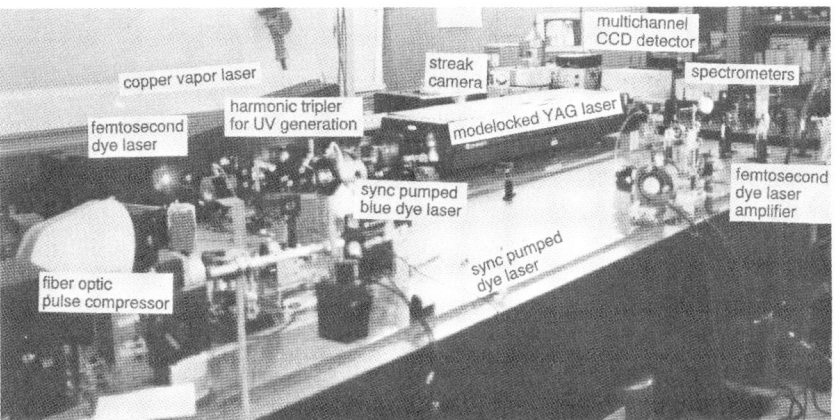

FIG. 1. Photograph showing a partial view of an ultrafast laser facility devoted to semiconductor spectroscopy. The view emphasises the laser sources and associated electro-optic instrumentation.

and Hadland Photonics Inc.) and in high-repetition-rate events, which give the advantage of signal averaging (so-called synchroscan mode), an approximately 10-psec resolution is possible. The time averaging, however, implies that nearly photon counting sensitivity is possible if a good photodetector is employed as a readout of the streak camera output phosphor screen.

In a particularly powerful combination for time-resolved spectroscopy, one takes advantage of the two-dimensional recording ability of the streak camera, by reading out also associated spectral information at the photoanode plane by an array detector such as, e.g., a sensitive charge-coupled device (CCD) detector (See, e.g., Fig. 3). A multichannel plate PMT has also been configured for such fast spectral–temporal readout with coincident photon counting methods, although with more modest time resolution (McMullan *et al.*, 1987).

To conclude this section, Fig. 1 shows an example of a modern university-based ultrafast laser facility, taken from the author's laboratories. The partial available view concentrates on the sources and other relevant electro-optic instrumentation that has been used to study a wide range of semiconductors and their heterostructures. The most recent addition (not shown in the figure) is a modelocked Ti:sapphire laser. Other important considerations in designing such a facility include the access of the spectroscopic probes onto the samples themselves, often within a cryogenic environment. In the facility of Fig. 1, for example, experiments can be conducted to temperatures as low as 0.3 K and in magnetic fields to 13 T either by direct optical or fiber-optic access to the sample region.

III. Time-Resolved Spectroscopy in Semiconductors: Approaches

In this section we discuss the principal spectroscopic approaches to time-resolved spectroscopy of semiconductors. These are the methods of transient luminescence and excite-probe (pump-probe) methods, respectively. Throughout the discussion we focus entirely on the use of short laser pulses to generate the initial impulse of excitation. Short-pulse electron beam methods have also been applied to investigate luminescence transients on subnanosecond time scales, although the lack of energy specific resonant excitation and the difficulty to achieve picosecond electron beam pulses (or shorter) have made this approach a less viable alternative. As already mentioned, fast photoconductive switching can produce extraordinarily short electrical transients that are relevant in ultrafast transport experiments. At terahertz frequencies, far-infrared (FIR) radiation has been recently generated also by such switching in transmission lines coupled to suitable antenna structures (deFonzo and Lutz, 1989).

The general schematic of Fig. 2 summarizes the range of different possible spectroscopic approaches. Following the excitation of a semiconductor sample at the time of arrival t_0 by an incident short-laser pulse of duration τ_p with center optical frequency ω_{ex} (or photon energy $\hbar\omega_{ex}$), several options are, in principle, open for spectroscopy of the nonequilibrium system. *Spontaneous* (radiative) recombination effects give rise to a time-dependent spectrum, labeled by $I_{PL}(\omega, t - t_0)$ in the figure, whose details yield kinetic information about the status of the electronic and other excitations of the material. The kinetic processes may include specific one-electron energy

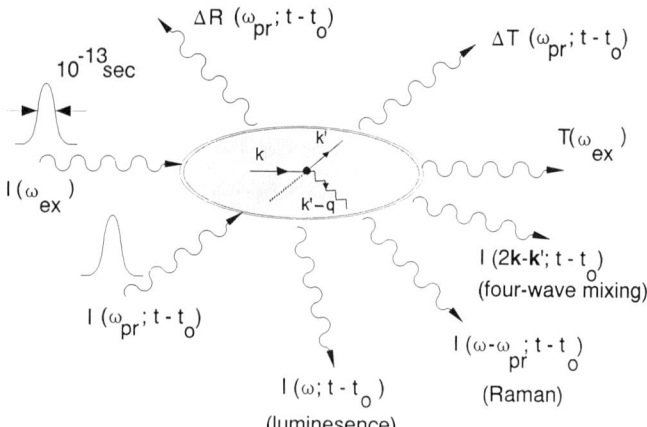

FIG. 2. General configuration schematic for transient spectroscopy showing approaches by luminescence, excite-probe, and light scattering techniques.

relaxation paths (in real or energy–momentum space); examples of other excitations are phonons or Coulombically correlated electron–hole states. The dynamics of these excitations can also be probed by a second (weaker) pulse of laser light that is of short duration on the time scale of events of interest. The transient spectroscopy is now realized experimentally by varying the probe pulse frequency ω_{pr} and time of arrival $t_{pr} - t_0$ at the sample. Hence, the probe beam senses the state of the photoexcited system through *induced* emission (or absorption), a process that can be measured by monitoring changes in the probe beam's transmission or reflection (ΔT and ΔR in Fig. 2). Generally, the probe intensity is kept weak so that the probe-induced emission rates remain small in comparison with other kinetic rates in the physical problem under study.

If the probe beam is used in an inelastic scattering mode, the pump-probe experiment may take the form of a *transient Raman* experiment. In this instance, the spontaneous Raman spectrum of the photoexcited medium as measured by the scattered probe light $I(\omega_{pr} - \omega, t - t_0)$ is the object of spectroscopic measurement (Fig. 2). In experiments to data, the Raman process most commonly used involves nonequilibrium optical phonons. Generally, the wave vector and polarization selection rules that apply in cw Raman spectroscopy require similar attention to geometrical details in transient Raman experiments as well. Another nonlinear method of transient spectroscopy employs the *four-wave mixing* technique in which the excitation is delivered by two pump beams intersecting at the sample. This methods is in many instances equivalent to a transient grating approach where the probe beam is diffracted from the spatially periodic, pump-induced variation in the optical dielectric constant $\delta\varepsilon(r, t)$ (Jarasiunas *et al.*, 1978).

The theoretical treatment of the range of possibilities depicted is most naturally treated by focusing on specific instances. We next consider explicitly a few simple circumstances where a generic description of the problem is possible for simple cases of one-electron (one-hole) excitations in transient luminescence and pump-probe experiments, respectively.

4. Transient Photoluminescence: Spectrum

Transient photoluminescence in its simplest form is designed to yield time-resolved spectra of a bulk semiconductor or a microstructured material, in which the spectral details give quantitative insight to the energy relaxation and recombination processes, most typically for electron–hole pairs excited initially above the band gap in energy (i.e., interband excitations). Let us recall that in the absence of exciton effects, i.e., when the electron–hole Coulomb energy is small in comparison with the carrier kinetic energies, the spectrum for a *thermalized* carrier distribution in steady state is written readily as

$$I(\omega) \sim \int D(E) f_c\left(E_e - \frac{E_F^e}{kT_e}\right) f_v(E_h, E_F^h) \, dE, \tag{1}$$

where $D(E)$ represents the joint conduction–valence band density of states with f_c and f_v their quasi-Fermi distribution functions, respectively. In this expression, the wave vector conservation laws as well as the dependence of the optical matrix elements on energy are not explicitly displayed. The one aspect of interest here is the use of luminescence spectra to identify the distribution functions f_c and f_v in the case of free-electron–hole pair interband excitations. In the case of exciton effects, either bound or free, the spectral energy positions and lines shapes are used as additional indicators to identify these species.

In the application of time-resolved methods, the emphasis is often on the initially *nonthermal* distribution of electrons and holes, and it is their evolution that is usually being studied before and until thermalized distributions are reached (if at all). If the relaxation rates toward thermalized distributions are much faster than, e.g., the electron–hole pair recombination time, this being the most common situation, then the transient methods offer direct quantitative information about hot-carrier energy relaxation processes that are only indirectly inferred from steady-state measurements. If excitonic effects are important, transient luminescence spectra can give direct information about the formation time of such bound states. Examples of these aspects are given later in this chapter. In addition, the measurement of the luminescence decay associated with the thermalized (or nonthermal) carrier (or exciton) population gives direct information about the relevant recombination rates and may be used to identify radiative and nonradiative decay channels, a point of both fundamental interest and relevance for optoelectronic applications, especially in the case of novel materials and their microstructures. Furthermore, in the case of spatially inhomogeneous photoexcitation or material microstructures, time-dependent spectra can be sometimes correlated with transport measurements to obtain such quantities as diffusion constants and Fermi velocities. Finally, if very fast time resolution is available, the study of the time dependence of the polarization of transient luminescence can give insight about electron–hole phase-relaxation mechanisms in the photoexcited system. It should be emphasized that many of the physical processes cited in this paragraph are also accessible through the excite-probe methods reviewed here.

Intraband excitations are, in principle, also compatible with transient luminescence experiments, though very few examples exist in the literature to date. These experiments have the virtue of focusing on unipolar hot-electron (or hot-hole) processes in terms of the initial excitation, so electron-hole interactions are not a complicating issue. On the other hand, detection of intraband luminescence is technically quite difficult because of the generally small matrix elements for these transitions especially in the bulk (phonon-assisted), and a need to employ spectroscopic methods at infrared or far-infrared wavelengths where the short-pulse optoelectronic instrumentation is still rather underdeveloped (note, however, the progress being made with FIR generation referred to in the opening paragraph of Section III). In quantum well structures or with impurity-related transitions, however, the matrix elements associated, e.g., with intersubband resonances are quite competitive with interband transitions.

The specific laboratory techniques applied for time resolving of photoluminescence events depend first and foremost on the time-scale in question and the dynamical range therein, as well as the wavelength domain. For direct detection of the luminescence transients in the visible and near infrared, the principle tools are the streak camera and time-correlated photon counting. These were discussed in Section II in connection with other instrumentation issues. As pointed out, the very small effective areas that are necessary for high-speed photodiodes generally exclude these from luminescence experiments. A particularly effective application of the streak camera is the use of a detector array to read the output phosphor, such as a microchannel plate photomultiplier or a CCD detector array. Such readout detectors, apart from sensitivity, also give the overall instrument the capability for simultaneous recording of the entire spectrum of interest. A schematic of this arrangement is shown in Fig. 3, together with a sample result (Ding *et al.*, unpubl).

One important experimental aspect in recording spectra with high temporal resolution is the right choice for a spectrometer. Any dispersive spectrometer will, for fundamental reasons, broaden the luminescence transients, thereby impairing the time resolution. In terms of practically acceptable compromise, the best choice for a monochromator is a double-grating arrangement in which the gratings are paired in a subtractive mode to minimize the temporal broadening. The eventual spectral–temporal resolution that can be accomplished depends greatly on the sensitivity required (in terms of the slit widths of the spectrometer), but performance at the level of 10 psec–0.5 meV temporal spectral resolution has been routinely achieved in the author's laboratory with close to photon counting limits of (single-channel) detection sensitivity.

One approach that circumvents the limitations by streak cameras (or similar fast photoelectron devices) both in the spectral and temporal domain involves the use of nonlinear optical methods, as illustrated schematically in Fig. 4. In this case, while photoluminescence is again excited by an ultrashort-pulse laser source at center optical frequency ω_{ex}, some of the

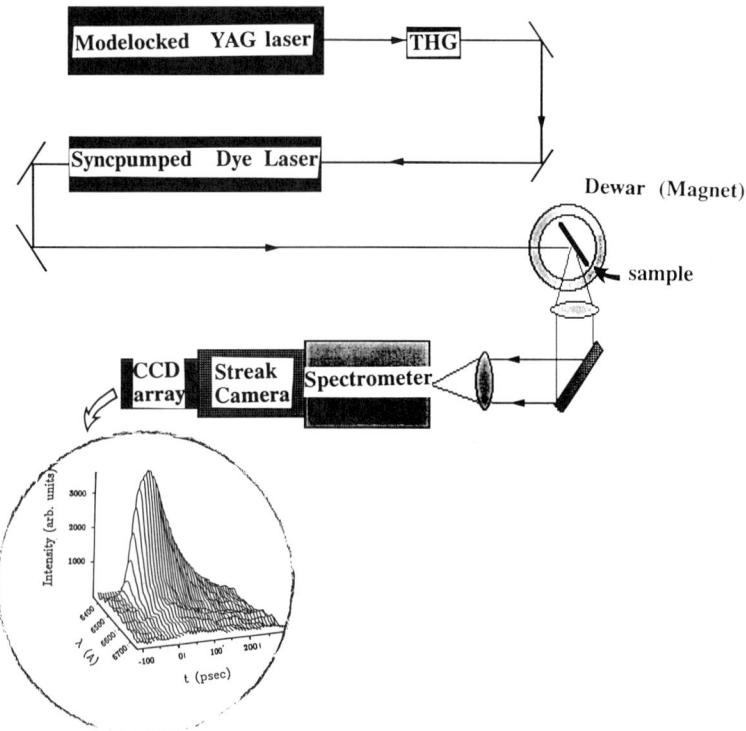

FIG. 3. Laboratory diagram for general time-resolved luminescence studies with streak camera detection. With a two-dimensional multielement detector reading the streak camera output phosphor, both temporal and spectral information can be simultaneously recorded (inset).

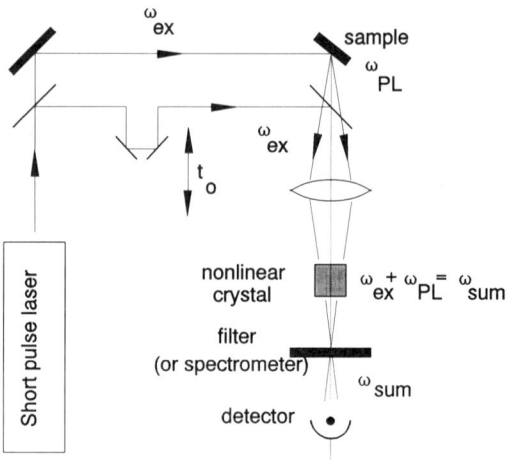

FIG. 4. Schematic illustration of time-resolved luminescence experiments based on nonlinear frequency mixing. The transient luminescence spectrum and excitation pulses are subject to sum frequency generation in the nonlinear crystal, which acts as a fast nonlinear gate in the process.

excitation is now also split off and directed to a nonlinear crystal. The transient luminescence at frequency ω_{PL} is also focused into this crystal, where, through the second-order susceptibility $\chi^{(2)}$, an optical wave at the sum frequency $\omega_{ex} + \omega_{PL}$ is generated, with an amplitude proportional to $\int I_{ex}(\omega_{ex}, t) I_{PL}(\omega_{PL}, t - t_0) dt$. Phase-matching conditions for the pairs of frequencies involved, as well as the transparency range and laser damage threshold of the nonlinear crystal, are the key practical factors in determining (and limiting) the practically achieved upconversion efficiency. By varying the time of arrival of the ultrashort laser pulse at the crystal, the upconversion occurs, in effect, within a fast optical gate (with "open" time determined by the duration of the excitation pulse), which endows the method with a time-resolved spectroscopic performance. Generally, the phase-matched optical bandwidths are relatively narrow for a fixed orientation of the nonlinear crystal so that angular adjustments (for tuning the phase-matching conditions for varying ω_{PL}) are usually necessary during the recording of the luminescence spectrum in semiconductor experiments. At visible and near-infrared wavelengths, Shah and collaborators (1987) have developed this technique for transient luminescence spectroscopy in the subpicosecond time domain with thin crystals of $LiIO_3$ as the nonlinear gate. At longer infrared wavelengths (≈ 5 μm), we mention as an example work of Heyen et al., (1989), where transient luminescence in narrow-gap semiconductors has been studied by using $AgGaS_2$ in the upconversion of infrared luminescence into the visible.

5. *Excite-Probe Techniques:*

The excite-probe (pump-probe) methods represent a form of *photomodulation spectroscopy*, which, as one example of general modulation techniques, is also widely used in continuous-wave spectroscopy. Due to the differential aspect of the technique, one must pay close attention to the connection between experimentally obtained spectral lines shapes and the physical nature of the optical transitions involved. Here we discuss general aspects of the method and defer specific material examples to Section IV. As shown schematically in Fig. 2, excite-probe methods can be used in both transmission and reflection, the choice often made on the basis of the sample transparency. Structures containing quantum wells can often be prepared for transmission experiments in the highly absorbent region above the fundamental edge, while for bulk samples reflectance methods are the usual choice (except for impurity-related transitions for which sufficient sample transparency may be available). In excite-probe experiments configured for transmission, the measured photoinduced fractional change through the sample of the probe beam, $\Delta T(\omega_{pr}, t - t_0)/T(\omega_{pr})$, is simply connected to the excitation-induced chances in the absorption coefficient in the small-signal limit as

$$\frac{\Delta T(\omega_{pr}, t - t_0)}{T(\omega_{pr})} = -\Delta\alpha(\omega_{pr}, t - t_0)L, \qquad (2)$$

where L is the thickness of the photoexcited region of the sample. In the case of free-electron–hole pairs, it is easy to derive expressions for $\Delta\alpha$ by accounting for the excitation-induced changes in the occupancy factors within the bands that are probed. By far, the largest-amplitude signals are obtained if the probe is tuned close to a particular interband resonance. This corresponds, of course, to probing the excess electron–hole pair distribution in the corresponding range of the joint density of states spectrum. For example, in the case of weak injection of density δn electrons into a perfect parabolic conduction band, we have

$$\Delta\alpha(\omega_{pr}) \sim \frac{-(\hbar\omega_{pr} - E_0)^{1/2}[\exp(-m_h/m_e + m_h)(\hbar\omega_{pr} - E_0)/kT_e]\delta n}{2(2\pi m_{de}kT_e)^{3/2}\hbar^{-3}}, \qquad (3)$$

where E_0 is the interband resonance in question and the electron and hole effective masses are identified, together with the joint density-of-states effective mass m_{de} at that critical point. The constant A contains the matrix elements. For photomodulation in semiconductors of zinc blende symmetry, the energy E_0 corresponds usually either to the fundamental edge at E_g or to the spin-orbit split-off transition at $E_g + \Delta_0$ for $k \approx 0$. In case of the latter, the hole component is usually unimportant due to the rapid relaxation of holes to the valence-band top, so in this way the experiment may be made sensitive to the excess electron distribution only.

In the case of a low density of free excitons, the photomodulated line shape can also be fairly easily obtained analytically if the following two approximations hold. First, assume that the pump-generated exciton (or free-carrier) density is well below the exciton–free-electron–hole plasma transition the rough equivalent of semiconductor–metal Mott transition), and, second, assume that polariton effects are negligible. In this limit, simple collisional broadening effects analogous to scattering processes in atomic gases yield the following photomodulated spectrum:

$$\Delta\alpha \approx \frac{-A[\Gamma^2 + 4B(\hbar\omega_{pr} - E_x)\Gamma + (\hbar\omega_{pr} - E_x)^2]}{\pi[(\hbar\omega_{pr} - E_x)^2 + \Gamma^2]^2}. \qquad (4)$$

In this expression, the initial unmodulated exciton resonance has been taken as a Lorentzian line-shape function with a linewidth Γ and an asymmetry parameter B. This description fits well, for example, the prototype excitone semiconductor Cu_2O (Hefetz et al., 1985).

Another instance where the spectral line shapes are usually straightforward to derive corresponds to the case of excitons bound to impurities or disorder-induced fluctuations, under conditions where the free-exciton (or free-carrier) concentrations are low (typically $< 10^{16}$ cm^{-3}). In the lowest approximation, such an exciton transition is analogous to a two-level atomic system and is chiefly subject to pump-induced population saturation effects in the photo-modulated probe spectrum; i.e., a certain fraction of the impurity sites are occupied by pump-generated excitons.

Figure 5 summarizes schematically the various spectral line shapes expected from the photomodulation experiments discussed so far. More concrete examples of specific semiconductors and structures will be given in Section IV. The spectra in Fig. 5 all correspond to the idealized models of excess electron–hole systems existing in a single phase. It is important to keep in mind that during the course of a time-resolved experiment the photo-excited system may in fact undergo an evolution, e.g., from free-carrier to free-exciton state, and further to bound exciton states before the final recombination steps are completed. This presents, in fact, a very attractive application of time-resolved spectroscopy to the study of the kinetics of such phase changes. Effects of finite disorder or lower dimensionality (2D and below) have not been considered here either; examples of their influence are given later when representative experimental results are reviewed. For photomodulated reflectance experiments one can use similar physical arguments to obtain spectral line shapes for $\Delta R/R$. For small deviations from equilibrium (i.e.,

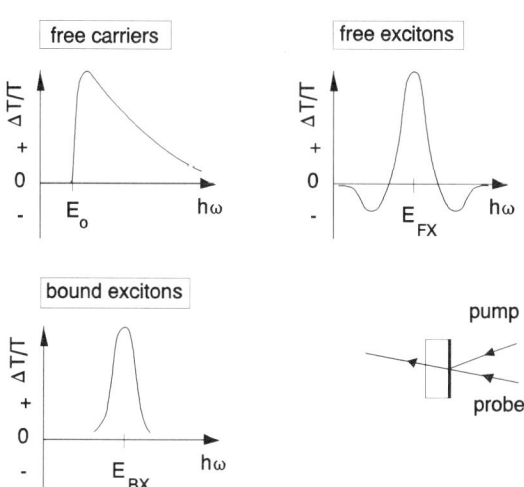

FIG. 5. Schematic illustration for characteristic line shapes expected in photomodulation experiments with free carriers, free excitons, and bound excitons (small-signal limit).

linear optical regime), the use of Kramers–Kronig transforms provides a reasonable approximation for the calculation of index of refraction changes $\Delta\eta$ from the differential changes in the absorption in Fig. 5.

A schematic arrangement of a high-amplitude resolution excite-probe arrangement is shown in Fig. 6. Both the pump and probe sources are (synchronously) mode-locked tunable dye lasers typically at a repetition rate of ~ 100 MHz. The pump laser is chopped (electro- or acousto-optically) at a high frequency ($\omega_m \sim$ MHz) while the probe laser at a lower frequency. As mentioned in Section II, high-frequency chopping is useful in that it places the detector-derived photomodulated signals well outside the dominant spectral region of dye laser jet noise (which persists up to a few hundred kilohertz). For the photodiode, any reasonably performing standard detector will suffice. The photodiode is usually followed by a narrowband electronic amplifier (centered about ω_m) and a high-frequency lock-in amplifier. For an ideal diode, the minimum detectable amplitude changes $\Delta T/T$ or $\Delta R/R$ in the probe power are readily calculated to be in the detector shot noise limit as

$$\frac{\Delta T}{T} \approx \left[\frac{4\hbar\omega_{pr}\Delta v}{P_{pr}\eta}\right]^{1/2} \tag{5}$$

where P_{pr} is the average transmitted probe power (kept below detector saturation). The measurement (integration) times enters inversely to the electronic bandwidth Δv. With the arrangement in Fig. 6, values for $\Delta T/T$ approaching

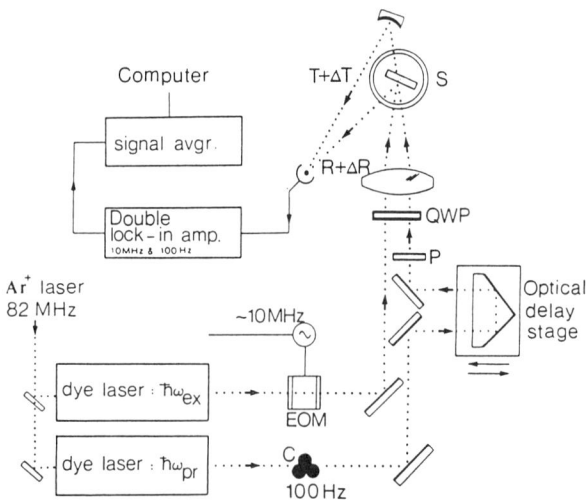

FIG. 6. Excite-probe arrangement for a pair of model-locked continuous-wave dye lasers in a double-amplitude-modulation scheme.

10^{-8} have been achieved. Because of the finite limitations imposed by laser temporal jitter, the time resolution in this highly amplitude sensitive scheme is limited in practice to about 10 psec. For subpicosecond excite-probe spectroscopy, the alternative avenue is to use "white light" continuum probes as mentioned in Section II. However, in this instance, and in spite of the optical multichannel detection, the slower repetition rates (typically a few kilohertz) due to dye laser amplifier constraints results in a reduction by several orders of magnitude in the sensitivity to photoinduced effects.

In the case of Raman excite-probe spectroscopy where the state of the photoexcited system is measured by inelastic scattering from phonons or electronic excitations, recent advances in multielement array detector technology have been at least as important as the progress with laser sources. This new generation of detectors has made an enormous difference in making many experiments possible that were previously considered impractical in the context of Raman scattering in general and its transient analog in particular. In typical transient experiments (of which there are still relatively few examples), one generally looks for small amounts of anti-Stokes signal as a measure of the population associated with nonequilibrium excitations such as optical phonons. As a consequence, the actual photon counting levels of the signal can be very low indeed. However, both microchannel plate photomultipliers and scientific-grade, cooled CCDs are today commercially available with appropriate multichannel readout electronics that permit signal acquisition at light levels well below that of 0.1 photon per second over the spectrum of interest. The actual Raman spectrum itself is dispersed by a spectrometer where the multielement detector array is placed at the spectrometer's exit slit plane (without the slit, of course).

IV. Portfolio of Experimental Examples in Time-Resolved Spectroscopy

In this section we connect the technical discussion in the preceding sections to a range of experimental results from recent technical literature. Within the past few years alone, a plethora of examples exists of the applications of transient spectroscopies in semiconductor science, both basic and applied, so that a subjective choice is evident in the examples used in this section. The increase in the number of investigators who have made contributions to the field is partly a recognition of the scientific power of these techniques and partly the more readily commercially available ultrafast laser systems. We begin with a selective review of representative work in bulk semiconductors and follow this with contemporary examples of studies in semiconductor microstructures, principally quantum wells and superlattices. Following this, specific examples will be given of recent experimental activity where semiconductors and their microstructures are subject to high levels of electronic excitation by ultrashort laser pulses. An attempt has been made below to describe different physical phenomena in the (rough) sequential order that corresponds to the typical time-ordered spectrum of events following the initial photoinjection of hot nonequilibrium electrons and holes in a semiconductor.

6. BULK SEMICONDUCTORS (LOW EXCITATION CONDITIONS)

For the prototypical material, Fig. 7 shows schematically a portion of the energy spectrum in a direct-band-gap tetrahedrally bonded bulk semiconductor (such as GaAs), together with some optical transitions and energy relaxation paths for the photoexcited system (absorption indicated by the vertical upward arrows). In most time-resolved experiments to date, the interband transitions of practical interest lie near the center of the Brillouin zone. In principle, however, higher-lying critical point resonances at E_1 and E_2 can be used, e.g., to probe electron occupation over a much wider range of the energy–momentum space. In the III-V and especially II-VI compound semiconductors, the electron and hole coupling to the polar longitudinal-optic (LO) phonon through the Fröhlich interaction is strong so that for small wave vector changes (intraband energy relaxation) the LO phonon inelastic scattering is the predominant hot-electron (or hole) energy loss mechanism

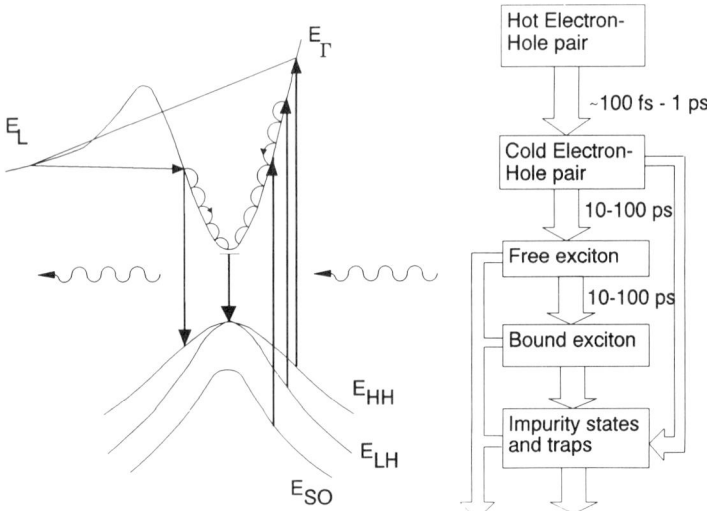

FIG. 7. Simplified energy band diagram for a bulk direct-gap semiconductor showing primary optical and electronic energy relaxation pathways.

under low excitation conditions (i.e., weak or moderate carrier–carrier scattering). In the early application and development of the ultrafast optical techniques, the problem of energy relaxation of hot photocarriers in bulk semiconductors was arguably the prototype experiment. Unfortunately, however, the initial conditions of photoexcitation varied widely between different experiments and researchers in terms of initial excess electron–hole energy and the carrier density. Not surprisingly, this research into *hot-electron* relaxation produced a range of quantitative results that only recently have been usefully connected to theory and transport experiments. We will not review the history of the hot-electron effects in bulk semiconductors (mainly GaAs) here, but we show selected recent samplings only. For further information the reader is referred to, e.g., the two most recent proceedings of the Hot Electron Conferences in 1987 and 1989 (Solid State Electronics) and a somewhat earlier but representative review article (Lyon, 1986).

In terms of the hot-electron (hole) cooling rates, the electron–hole pair density and the prevalent lattice temperature are of great influence, apart from material parameters and band-structural details. Generally, the timescale in question for energy relaxation by the (unscreened) LO phonon path occurs on a picosecond time scale (say ~ 200–300 fsec per one-phonon emission step). When the initial photon energy is such that the hot electrons may also scatter to adjacent conduction-band valleys, the use of time-resolved methods becomes particularly helpful in delineating the energetic relaxation paths. One example of such work is by Shah and co-workers (Shah *et al.*, 1987). Following Fig. 7, consider the specific problem of initial photoexcitation of electrons from the valence band of GaAs into the Γ_6 conduction-band states but at initial excess energies at $t = 0$ that may be comparable to the L-valley energy ($\Delta E_{\Gamma L} \approx 300$ meV). The energy relaxation pathway must then include a finite role for the L-valley, an issue of practical importance in a number of hot-electron (transport) devices. Shah and co-workers employed time-resolved photoluminescence methods by the nonlinear upconversion techniques where the spectrum over an approximately 200-meV range above the direct GaAs band gap was studied. Figure 8 shows their spectra produced by excitation pulses of less than 0.5 psec in duration at a photon energy of $\hbar\omega_{ex} = 2.04$ eV (some 600 meV above the GaAs band gap). The feature that is different from similar experimental data at lower photon energies of excitation (though still many LO phonon energies above the band gap) is the considerably longer delay that is here observed on a picosecond time scale to characterize the electron cooling (hole relaxation is assumed to be very fast). The authors conclude that the main cause for this delay is the additional energy pathway for electrons, namely the detour through the L-valley aided by efficient deformation potential scattering of large wave vector phonons. Another aspect of this work is that comparisons with Monte Carlo simulations provided both a measurement for the Γ–L deformation potential

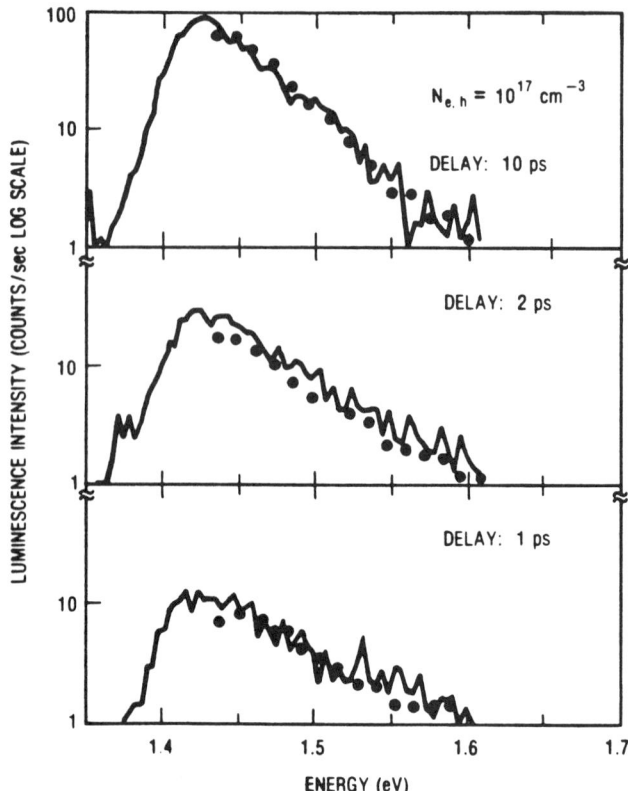

FIG. 8. Transient luminescence spectra mirroring the hot-electron distribution of GaAs at $T = 300$ K excited well above band gap with subpicosecond pulses (at zero time delay). Initial injection density is also indicated and the dots refer to a Monte Carlo calculation. (After Shah et al., 1987.)

($6.5 \pm 1.5 \times 10^8$ eV/cm) as well as an appreciation of the role of carrier–carrier scattering, which, upon higher levels of optical excitation induced a further slowing down of the electron cooling process from the subpicosecond to the picosecond time domain.

Recent variations in studies of photoexcited hot-electron energy relaxation includes work in a doped semiconductor, where additional scattering of hot photoelectrons by the cold, "background" equilibrium electrons (in n-type material) is at issue (e.g., Rieck et al., 1989). In these experiments the equilibriation of a small number of excess electrons and holes within the background bath is investigated with a view toward developing a more complete understanding of carrier–carrier interaction processes. In the context of contemporary work in the field, however, a great deal of attention has been given to the relationship between hot-carrier relaxation and electron–optical phonon coupling in lower-dimensional systems, notably quantum wells. We continue on this point with further examples of the hot-electron spectroscopy later in this section.

A general aspect that needs to be kept in mind in all of the optical measurements purporting to identify and characterize hot-electron relaxation with a connection to electronic transport is the obvious point that interband optical experiments produce both electrons and holes. Due care must therefore be exercised to ensure that the experiment is indeed compatible with a *unipolar* hot-carrier intepretation, in spite of the inherently *bipolar* nature of the optical experiment. There are occasions when the electron component can be sufficiently isolated so that the hot-hole contribution need not be explicitly considered, but, in this author's opinion, there are a surprising number of examples in the literature where insufficient attention has been paid to this issue. In the case of impurity-to-band transitions the unipolar nature can be fairly well approximated. While not reviewing transient spectroscopy of impurities in this chapter, we note in this connection recent investigations of the photorefractive effect in GaAs where the associated free-carrier dynamics have been investigated by the transient grating techniques (Smirl *et al.*, 1988). In the case of hot-electron effects in metals, the situation is dominated by the very large background electron density. At the same time, the absorption processes are usually a superposition of intraband and interband contributions. Subpicosecond methods have been applied recently in pump-probe experiments to deduce from the measured electron relaxation rate values for the electron–phonon coupling (deformation potential constant), including the new high-temperature superconducting compounds (Brorson *et al.*, 1987; Doll, *et al.*, 1989; Brorson *et al.*, 1990).

Since the process of hot-electron (and hole) energy relaxation also results in a *nonequilibrium phonon* population, time-resolved Raman scattering methods have been gainfully employed to examine selected aspects of the buildup and decay of the excess phonons (von der Linde *et al.*, 1980; Kash *et al.*, 1985). The latter work (in bulk GaAs) is contemporary in the sense that the experiments were conducted under a moderately low level of excitation so that the electron–phonon system was not displaced very far from equilibrium; the experiments being made possible by the availability of the sensitivity multichannel detectors discussed in Section II. Figure 9 illustrates the principal result of this work, which the authors have since extended to the alloy $In_xGa_{1-x}As$ where the LO phonon modes exhibit two-mode behavior (Kash *et al.*, 1987).

We mention in passing recent work in which subpicosecond laser excitation has been used to directly initiate terahertz-frequency acoustic phonons in semiconductors and related media (Grahn *et al.*, 1989). Following the absorption of an ultrashort laser pulse in a suitable transducer layer, the electronic

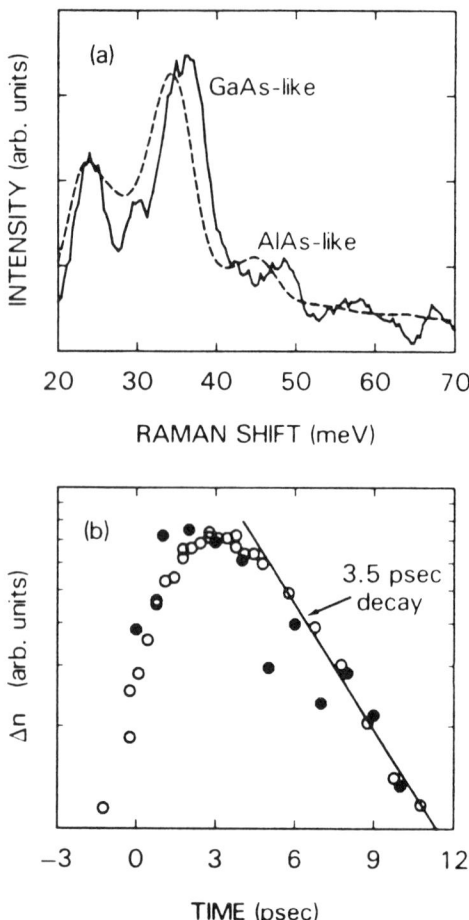

FIG. 9. Raman scattering from bulk $Al_{0.24}Ga_{0.76}As$ at $T = 300$ K; (a) shows the equilibrium two-mode alloy spectrum; (b) displays the transient LO phonon nonequilibrium population in GaAs (open circles) and the GaAs-like LO mode in the alloy. (After Kash et al., 1987.)

impulse of excitation can be rapidly transferred to the acoustic degrees of freedom so that a high-frequency acoustic pulse is launched into the medium of interest. In this way, very precise velocity and attenuation measurements of high-frequency (near zone edge) acoustic phonons have been made in a number of materials. Impulse excitation of coherent optical phonons has also been accomplished on a femtosecond time scale by a nonlinear optical approach, analogous to transient Raman scattering in GaAs (Cho et al., 1990) and in Bi and Sb (Cheng et al., 1990).

Particularly at low lattice temperatures, the thermalized electron–hole pair state before the final recombination event is usually an excitonic one in an undoped semiconductor (this two-particle state schematized in Fig. 7 as a discrete level below the conduction-band edge). Hence, we can raise the question about the kinetics of the additional energetic relaxation step for the *formation of the exciton* from the free-pair continuum through the excited states of the exciton. Generally in III-V and II-VI compound semiconductors, the exciton binding energy E_x is smaller than the longitudinal optical phonon energy $\hbar\omega_{LO}$. Accordingly, the energy loss required of the free pair for exciton formation is taken up predominantly by acoustic phonons (with relaxation times on the order of 100 psec). In addition, except in ultrapure materials, the lowest excitonic energy state is usually not free but associated with an impurity or other point defect so that a further kinetic step is needed, e.g., for the formation of a donor-bound exciton. The relevant time constants for reaching the lowest exciton state from cold free pairs will, of course, vary a great deal with particular materials, impurity background and temperature. While relatively little systematic work exists in this area through time-resolved methods, we cite here a pump-probe measurement in a II-VI bulk semiconductor (Cd,Mn)Se, where the exciton formation for a donor-related impurity-bound state was measured to be about 20 psec from the free-exciton state (Harris and Nurmikko, 1983). Acoustic phonons were the most readily available sources of energy dissipation in this case.

In a semiconductor with a large exciton binding energy, the energy relaxation within the excitonic states themselves can be studied by time-resolved spectroscopic means under initial excitation to an excited exciton state. This permits additional insight to be developed into *exciton–exciton* collisonal interactions. One example of such work are excite-probe experiments in Cu_2O (Hefetz *et al.*, 1985), where transient photomodulation effects within a free-exciton gas were used to make contact with excitonic scattering within several of the nP transitions of the well-known yellow series in this classic exciton semiconductor. In particular, these authors obtained specific quantitative detail about the cross section for the nP-1S exciton scattering process. Figure 10 shows their principal results in the form of a photomodulated spectrum (top trace) and an example of its time dependence (bottom trace). The injection of a moderate density of excitons by the pump pulse creates an exciton gas that relaxes quickly to the 1S band (on the time scale of interest). The probe pulses are tuned to the nP excitonic resonances, which then experience added collisional broadening resulting from the interaction of the probe-induced excitons with the 1S excess (pump-induced) population; a close analogy to collision broadening in atomic gases (the line shape is somewhat modified from that in the schematic of Fig. 5 because of asymmetry effects of the nP resonances). The photomodulated spectrum is in excellent agreement with theory, and the dependence of the time constants on the photon energy of excitation gives further information about the scattering

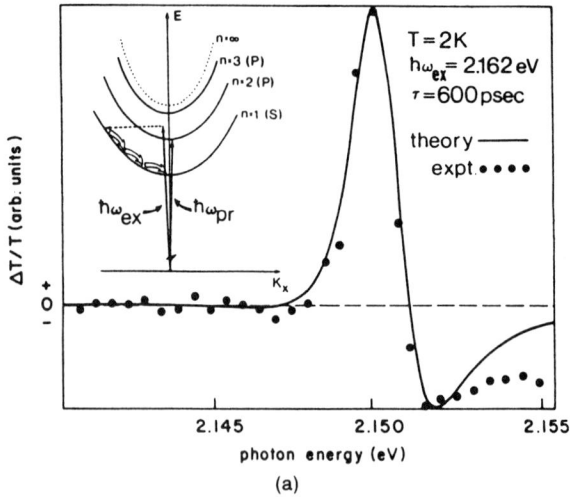

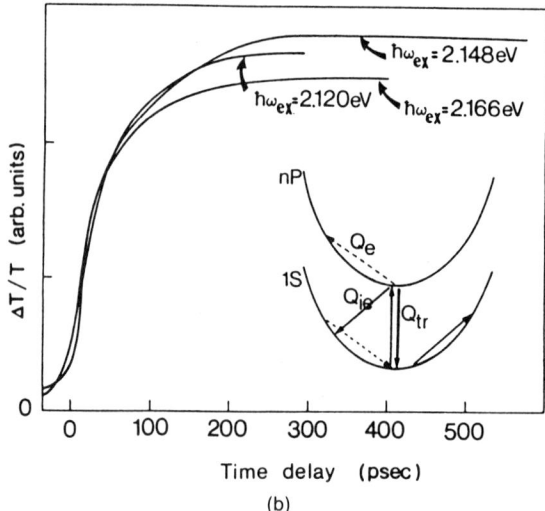

Fig. 10. (a) Photomodulated spectrum near the 2P exciton resonance in Cu_2O at $T = 2$ K, 600 psec following the excitation. Solid line refers to a calculated line shape with no adjustable parameters. The inset shows the relevant energy-level scheme. (b) Time evolution of the probe signal for three different photon energies of excitation. The inset shows the principal exciton scattering channels. (After Hefetz et al., 1985.)

schematic. We will discuss the issues associated with higher-density exciton systems later in this section in the context of lower-dimensional systems.

Let us return for a moment to the early stages in the life of the optically produced transient electronic excitations, while considering the use of time-resolved spectroscopies to identify the loss of electron phase coherence. In a typical crystalline semiconductor the eigenstates involved in an interband transition have a well-defined parity and symmetry; hence electron–hole pair excitations are created selectively in a well-defined initial state of polarization (momentum) by incident polarized light. In real time this initial phase coherent state decays, that is, loses its polarization memory, by elastic or inelastic scattering events. This question of *phase relaxation* has been studied in bulk semiconductors by several workers. In GaAs, for example, Oudar and co-workers (Oudar *et al.*, 1984, 1985) employed subpicosecond, time-resolved polarization-sensitive techniques at room temperature. In their experiments the authors took advantage of the fact that interband optical matrix elements for GaAs (and other zinc blende semiconductors) are anisotropic in momentum space due to the symmetry associated with the conduction- and (heavy hole versus light hole) valence-band wave functions. Under ultrashort-pulse excitation from a linearly polarized source at time t_0, the initial electron–hole momentum has a degree of anisotropy given by $|\hat{\mathbf{a}} \cdot \mathbf{p}_{cv}(\mathbf{k})|^2 = 1 + \lambda P_2(\cos \theta)$, where θ is the angle between the unit vector of polarization $\hat{\mathbf{a}}$ and the electronic momentum $\mathbf{k}$, and P_2 is the second-order Legendre polynomial. The symmetry difference for the heavy- and light-hole bands gives them an anisotropy parameter of opposite sign $\lambda = \pm 1$, respectively. Figure 11 illustrates the results of Oudar *et al.*, by showing the incremental changes in the reflected probe polarization, measured under conditions where the initial electron–hole pair energy is less than 1 LO phonon energy. The phase loss time under these experimental conditions at modest injection conditions was extracted to be approximately 190 fsec and assigned by the authors to carrier–carrier scattering (within the photoexcited gas) since phonon-intermediated memory loss mechanisms were expected to be considerably slower.

The question of phase relaxation of resonantly created free excitons has also been studied in bulk GaAs by transient degenerate four-wave mixing methods. In the experiment, one in effect prepares an optically induced phase grating from which time-dependent diffraction (with appropriate wave vector conservation) can give, in principle, a direct measure of the phase-relaxation rates. Dephasing times of the order of a few picoseconds have been measured by Kuhl and co-workers at low lattice temperatures (Schultheis *et al.*, 1986).

In the presence of intentionally incorporated magnetic impurities, the phase relaxation involves explicitly *spin-exchange* processes between the photoexcited carriers and the magnetic ions. These processes are of prime

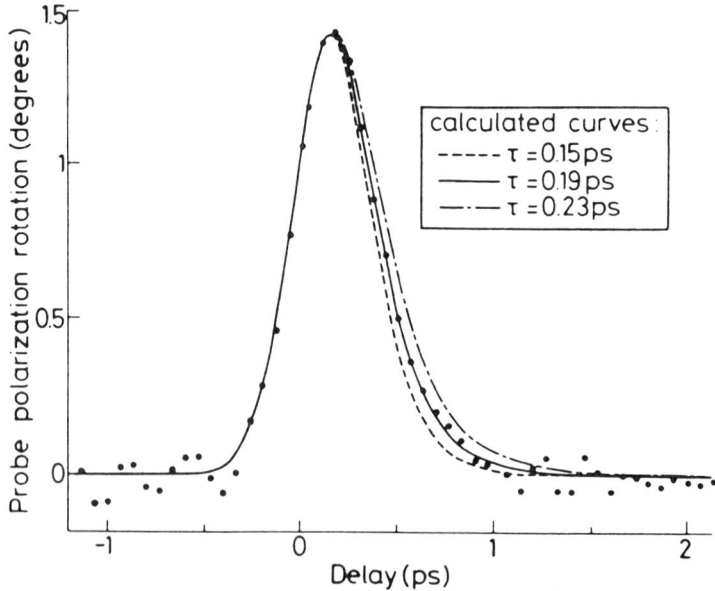

FIG. 11. Results from a measurement of electron phase relaxation by transient polarization spectroscopy in bulk GaAs. Calculated curves are derived from a model based on momentum matrix anisotropy. (After Oudar et al., 1984.)

interest in the so-called diluted magnetic semiconductors (DMS), typically II-VI compound semiconductors incorporating transition-metal elements such as Mn. In the case of hot electrons and holes, the exchange scattering accelerates the phase memory loss rates substantially (Krenn et al., 1989). A very different situation can ensue in the case of resonant optical injection of "cold" excitons in a DMS crystal at low temperatures. Now a collective effect can develop where the exchange coupling leads to the buildup of local spin-aligned quasiferromagnetic domains about a localized exciton, the so-called bound magnetic polaron effect. In bulk II-VI DMS alloys such as $Cd_{1-x}Mn_xSe$, the kinetics of the bound magnetic polaron formation have been studied on a picosecond time scale by both pump-probe and time-resolved photoluminescence techniques (Harris and Nurmikko, 1983; Awschalom et al., 1987; Zayhowski et al., 1985). As an example of the effect, Fig. 12 shows the transient photomodulated transmission spectrum in the region of the neutral donor-bound exciton in $Cd_{1-x}Mn_xSe$. In contrast with normal-bound exciton spectra, which is stationary in energy, distinct spectral red shifts are seen. These were ascribed to effective Zeeman shifts within the bound exciton complex as the energy lowering magnetic complex (magnetic polaron) is forming by dynamical alignment of the Mn-ion spins within the

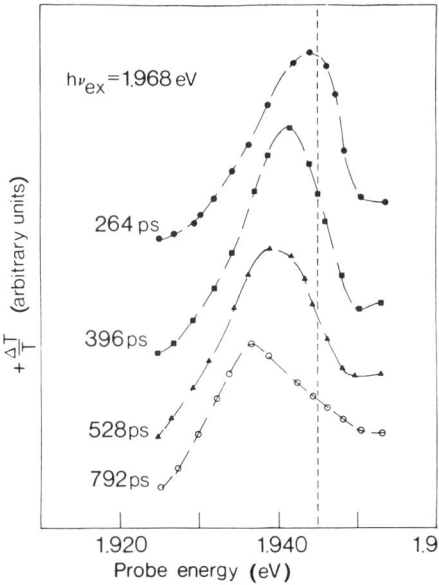

FIG. 12. Transient luminescence in n-$Cd_{0.90}Mn_{0.10}Se$ at $T = 2$ K, originating from a donor bound exciton. The spectral diffusion indicates the formation of the bound magnetic polaron complex. (After Harris and Nurmikko, 1983.)

exciton orbit. More recently Awschalom and co-workers (1989) have extended such work on spin dynamics to $CdTe/Cd_{1-x}Mn_xTe$ quantum wells.

Our final topic in this series of examples of transient spectroscopy in bulk semiconductors touches briefly on the question of disorder. As is well known, the presence of disorder projects a finite density of states into the band gap of a crystalline semiconductor. This leads to the concept of a mobility edge, which, while being of fundamental importance in amorphous semiconductors, can also be used to describe the energy demarcation between localized and extended electronics states in weakly disordered crystalline materials. In this context, problems concerning the dynamics in the *exciton localization* process in doped semiconductors and undoped alloys have been studied through ultrafast time-resolved method, with particular emphasis on the energy relaxation into and within these states. Ultrashort-pulse spectroscopies have also been usefully employed in the past decade with amorphous semiconductors in this context as well. For crystalline semiconductor alloys, compositional disorder is responsible for the localized states. The details of electronic energy transfer within such an energetically inhomogeneous and spatially randomly distributed range of states have been studied, for example, in II-VI alloys such as $CdS_{1-x}Se_x$ by Cohen and co-workers (Cohen, 1985) and in $Cd_{1-x}Mn_xTe$ (Zhang and Nurmikko, 1985), as well as in connection with the nitrogen isoelectronic trap in the III-V ternary $GaAs_{1-x}P_x$ (Kash, 1984). In such experiments, the principal signature of energy transfer and relaxation between localized states is obtained from time-resolved spectral diffusion, frequently under direct initial resonant excitation into the localized states. From the dependence of the spectral diffusion rates on temperature, photon energy of excitation, and other parameters, the experimental input is used in physical models where real-space energy transfer in localized states occurs via mechanisms such as phonon-assisted tunneling or multiple trapping, to mention two key processes.

Apart from the physics of electronic energy relaxation within localized states in a weakly disordered semiconductor, the question of phase relaxation can also be raised, especially under conditions of direct resonant optical excitation into these states. Recently, for example, Knoll *et al.* have employed picosecond degenerate four-wave mixing techniques to observe photon echoes from localized excitons in $CdS_{1-x}Se_x$ and concluded that phase-relaxation times as long as few hundred picoseconds (!) are possible in this system (Knoll *et al.*, 1990).

7. Quantum Wells and Superlattices: Linear Regime

Perhaps no other single development in the past decade has so rejuvenated the field of basic and applied semiconductor science as the appearance of artificial heterostructures such as quantum wells and superlattices. A brief look at the preceding volumes in this series alone shows what a large sector of today's research encompasses the study of lower-dimensional and related phenomena in such structures. Underpinning this development are the sophisticated epitaxial growth methods that in many elemental and compound semiconductors can now produce buried heterostructures that contain ultrathin layers at close to atomic-level precision in the definition of their individual thickness. We will not review the basic optical properties of quantum wells and superlattices, but refer the reader to relevant recent material on the subject (e.g., Weisbuch, 1988; Chemla and Pinczuk, 1986; Coleman, 1988). This section is devoted to illustration of time-resolved spectroscopic methods as they have been applied to the study of a range of dynamical phenomena involving primarily electronic excitations in quantum wells and superlattices. There is a rich literature on the subject that has emerged in the past few years and that is still in the process of growth; hence only a selective sampling is attempted here. Since much of the basic semiconductor physics in heterostructures involves questions about electronic confinement and its impact on quasi-two-dimensional excitations, the use of time-resolved spectroscopies has played an important role in providing direct kinetic information and insight concerning such issues.

Let us begin by considering a schematic illustration of the kinds of electronic pathways that are typically of interest in a semiconductor heterostructure, as shown in the real-space potential energy-band diagram of a superlattice or multiple quantum well (MQW) in Fig. 13. The schematic shows the periodic potential due to multilayer growth in the z-direction for either conduction electrons or valence holes, most typically near the Γ-valley in a zinc blende material. Following the initial excitation by an ultrashort laser pulse at a photon energy $\hbar\omega_1$ across the barrier band gap, the initial electron state in the figure is indicated by placing the real-space wave function $\Psi_e(\mathbf{r})$ in the barrier layers (e.g., $Al_{1-x}Ga_xAs$ in a $GaAs-Al_{1-x}Ga_xAs$ heterostructure). In the event of thick barrier layers, we may focus on the behavior of a single quantum well that is simply multiply repeated; in the opposite limit the wave function extension over several quantum well periods requires a true superlattice approach (description in terms of electronic states that extend throughout the structure in the z-direction). In the QW limit, apart from the question of initial hot-electron and -hole energy relaxation to the band edges of the barrier E_{cb}, the key additional kinetic steps now include their capture into the quantum well, energy relaxation within the well itself through the quantized subbands (including possible localization at interfaces, and exciton formation steps), and the recombination processes within the well. Since there are two translational degrees of freedom in an ideal quantum well, lateral transport in the layer (xy) plane is possible and can also be studied spectroscopically.

In the superlattice limit, the dynamical aspect of the extended state character in the z-direction involves the question of "vertical transport" from the

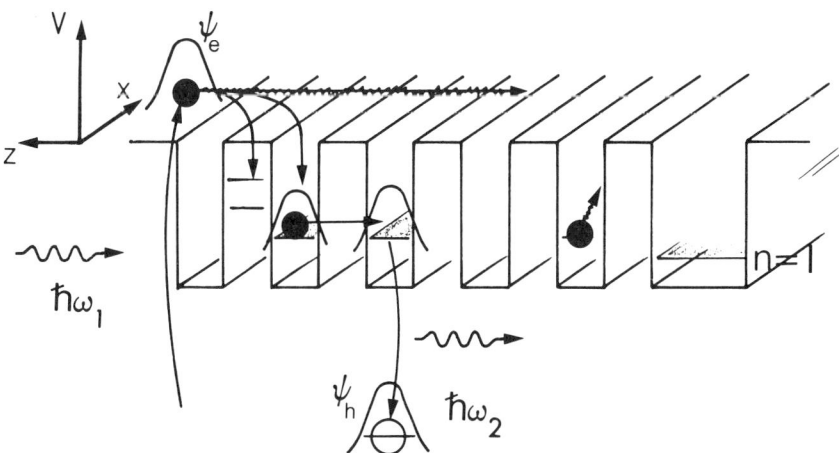

FIG. 13. Schematic (in real space) of a conduction band of a superlattice showing some of the energetic pathways that have been studied by time-resolved techniques.

standpoint of tunneling. Specific thin single-barrier or coupled double quantum well structures are most commonly employed to study the specifics of the tunneling phenomena in detail. We show now examples of these kinetic processes as investigated by time-resolved methods while noting that apart from intrinsic interest, such information is also very valuable as a diagnostic tool, especially in the development of new and incompletey characterized semiconductor heterostructures.

Considering the importance of efficient collection of electrons and holes into a quantum well and the need to understand the kinetic details that determine the effective cross section of the *carrier capture* (e.g., as a function of QW thickness L_w and the carrier energy), it is somewhat surprising that capture cross sections are still quantitatively rather poorly known. One of the first experiments to show by time-resolved methods that electron transfer from barrier to QW layers could be spectroscopically studied was that by Harris and co-workers in the case of a wide-well GaAs–$Ga_xIn_{1-x}P$ double heterojunction (Harris et al., 1982). However, only recently has there been some experimental success in attempts to establish the actual rate of probability of carrier capture in the GaAs–$Al_{1-x}Ga_xAs$ quantum wells, that is, to go beyond simple classical arguments that are based on carrier diffusion across the barrier layer to the quantum well. The capture problem has been considered theoretically from the standpoint of inelastic electron scattering by optical phonons in the range of virtual and real bound states of the quantum well with respect to the continuum states of the barrier (Brum and Bastard, 1986). The theory leads, among other things, to predictions of the capture cross section that should exhibit an oscillatory variation as a function of the well width L_w. Such details have not yet been discovered even qualitatively in the laboratory. Rather, experiments by Oberli et al. (1989, 1989(a)) on strained $Ga_{1-x}In_xAs$–GaAs quantum wells by time-resolved photoluminescence show qualitatively two different regimes of behavior: in the thin-well limit ($L_w \leq 100$ Å), where the kinetics of the carrier capture is dominated by the hot-carrier cooling (with respect to the band gap of the barrier layer), and the "classical" thick-well limit, where diffusion determines the effective capture rate (actual trapping probability into the quantum well being approximately unity). In these experiments, the dynamically evolving luminescence spectrum was measured with subpicosecond time resolution to give an energy "map" of the nonthermal carrier distribution, with particular emphasis on the photon energy range corresponding to interband transitions near (and below) the barrier band gap. A present discrepancy concerns the quantitative discrepancy with theory, the latter predicting considerably longer capture times than implied by experiments.

As in the case of bulk material, the question of *hot-electron relaxation* is also of key importance in quantum well (following capture), both for fundamental reasons and device design. Early on, questions about the effects of lower dimensionality were raised, especially in terms of modified electron–phonon interaction and changes in the scattering rates in quasi-2D conditions. First a score of spectroscopic and transport experiments were reported in undoped quantum wells, to be followed later by experiments in modulation-doped quantum wells as well as several theoretical reports on this subject. The use of time-resolved optical spectroscopies has brought considerable additional insight to the interplay between hot carriers and hot phonons, although difficulties in the interpretation of experimental results have also occurred. These difficulties have frequently been the result of poorly defined experimental conditions, specifically with respect to the level of excitation intensity and the absence of flexible wavelength tunability in the excitation (and probe) source(s). Some of the early work in the field was carried out by Ryan and co-workers by transient luminescence techniques where the carrier cooling in GaAs–$Al_{1-x}Ga_xAs$ quantum wells was studied, including well-width dependence, excitation level, and the influence of a magnetic field (Ryan et al., 1985). Subsequently, Leo et al., for example, have systematically investigated the issue of hot-electron relaxation by drawing on comparison of data obtained through both transient and steady-state luminescence, including the effects of a background cold-carrier gas (Leo et al., 1988). Figure 14 shows an example of their work, where the left panel displays the transient spectra for an undoped quantum well (at three delay times following the short-pulse excitation at $t = 0$ psec). The right panel shows results of spectral analysis from which the evolution of the effective carrier temperature (assumed to be defined by rapid carrier–carrier scattering) is deduced. The carrier (electron) cooling is seen to be distinctly dependent on the degree of doping and the level of initial optical excitation. One particular conclusion from this and other recent work is that the basic carrier–phonon scattering rates for bulk GaAs and its quantum wells are, in fact, quite comparable, as also predicted theoretically (das Sarma and Mason, 1985). Allowing for carrier–carrier scattering and heating of the phonon system by the relaxing electron–hole gas, one finds that the actual intrinsic carrier cooling rates between bulk and quantum wells of GaAs are also rather similar.

The current understanding of the hot-carrier energy relaxation in GaAs and related III-V quantum wells is that the apparently slow cooling rates encountered in several quantum wells studied are not a consequence of a lower-dimensional effect on the electron–LO phonon interaction (including screening) but rather the fact that a sufficient nonequilibrium phonon population is typically produced in the experiments that shows the overall cooling process (i.e., the final channeling of excess energy into the long wavelength acoustic phonons). As far as differences between electrons and holes go, the hole energy relaxation rates have been measured to be considerably faster in GaAs (Hopfel et al., 1986), a circumstance that is qualitatively similar in most zinc blende cubic semiconductors primarily due to the large density

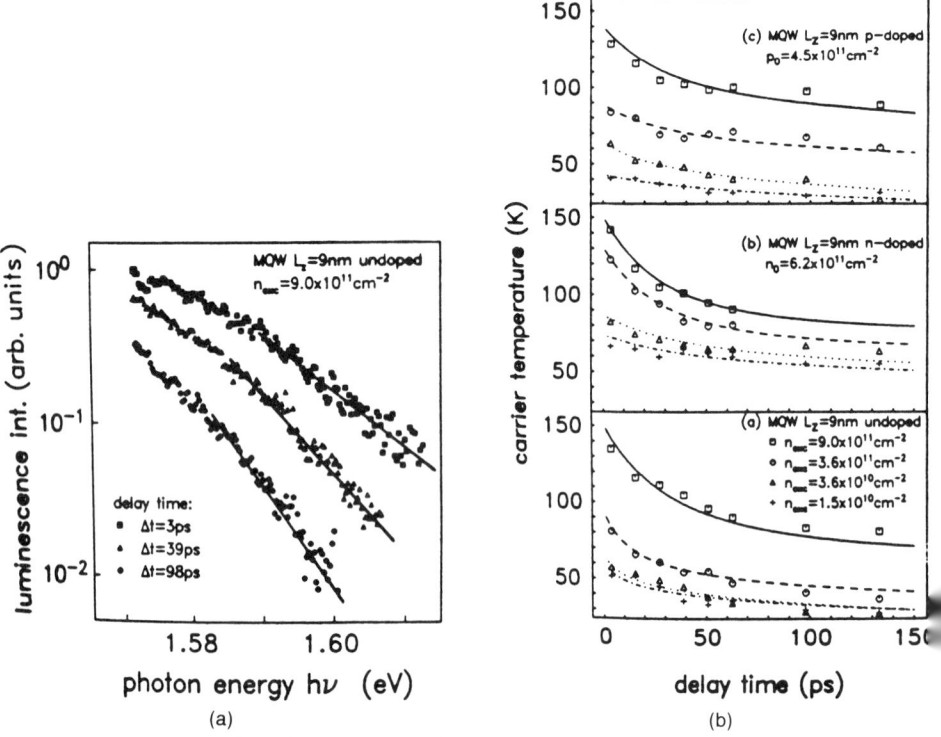

FIG. 14. Illustration of hot-electron relaxation in a GaAs–(Ga,Al)As MQW structure ($L_w = 90$ Å): (a) High-energy portion of transient luminescence spectrum under high injection level. (b) Experimentally determined transient cooling rates for undoped and modulation-doped wells. The lines are calculated fits to the model. (After Leo et al., 1988.)

of states at the valence-band top. As mentioned in the preceding section dealing with bulk materials, inelastic light scattering experiments have been used to measure the decay time of the hot phonons. Finally, transient electronic Raman scattering has been employed also to measure the intersubband relaxation time in GaAs quantum wells (Oberli et al., 1987).

Recombination kinetics in quantum wells have been extensively studied in many semiconductor heterostructures, both from the standpoint of isolating the intrinsic processes as well as providing a diagnostic technique in attempts to separate the contributions of radiative processes from nonradiative ones (the latter usually of extrinsic, defect-related origin). In undoped quantum wells at low lattice temperatures, the excitonic component frequently dominates the radiative recombination process in quantum wells. Transient luminescence was employed by Feldmann et al. to argue that the well-known increase in the exciton oscillator strength seen in optical absorption in quasi-

2D systems was also responsible for the specific observed dependence of the radiative lifetime as a function of quantum well thickness and temperature (Feldmann et al., 1987). Figure 15 shows a summary of their lifetime measurements for the GaAs–$Al_{1-x}Ga_xAs$ case. The decay time was obtained from an approximately exponential transient decrease of the exciton luminescence in the energy range overlapping the $n = 1$ heavy-hole exciton. The general feature of an *increase* of the radiative lifetime with increasing temperature has also been seen in the strongly excitonic ZnSe quantum wells (Hefetz et al., 1986). For completely free excitons, the reduction in radiative recombination rate with increasing temperature rests physically on the idea of larger exciton phase-space occupation (away from $K \approx 0$, where K is the exciton center-of-mass wave vector). In the presence of a significant background free-carrier density in the quantum well, excitons are no longer energetically stable and the radiative recombination can be viewed simply in terms of free carriers (e.g., photoholes with equilibrium electrons). It can be readily shown that in this case the well thickness dependence of the radiative matrix element vanishes, and a different prediction about the temperature (and carrier density) dependence of the radiative lifetime can be made. This has been experimentally shown through transient luminescence in GaAs–$Al_{1-x}Ga_xAs$ modulation-doped quantum wells (Matsusue and Sakaki, 1987) and in narrow-gap PbTe QWs where excitonic effects are always negligible (Heyen et al., 1989).

The structural extension of a single quantum well to a situation with two quantum wells separated by a thin barrier layer gives the opportunity to

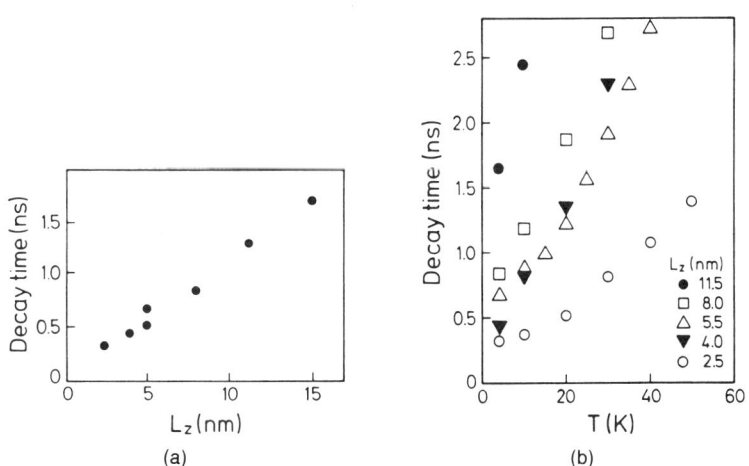

FIG. 15. (a) Exciton luminescence decay time versus quantum well thickness for a GaAs–(Ga,Al)As MQW sample at $T = 5$ K. (b) Temperature dependence of the decay time. (After Feldmann et al., 1987.)

examine optically questions related to *resonant tunneling*, a subject of considerable microwave device interest. Here, too, time-resolved spectroscopic methods have by now been employed by several groups in microstructures of varying complexity to characterize the tunneling rates, including the presence of applied electric fields. One fundamental challenge in resonant tunneling studies concerns the experimental identification of the "coherent" (one-step) and "sequential" (two-step) tunneling processes as individually distinct. (Tsuchiya *et al.*, (1987) were among the first to employ transient photoluminescence methods to examine the temporal details of resonant tunneling. Their aim was to measure the tunneling escape rate from a single quantum well with two thin barrier layers under initial resonant excitation selectively to the $n = 1$ quantum well resonance. Figure 16 shows a result that expresses the decay time of the luminescence signal from the quantum well (equated by the authors to the tunneling rate) as a function of barrier thickness. The experiment demands that the absorption (and emission) from the large amount of bulk GaAs constituting the cap and buffer layers does not spectrally obscure the excitonic resonance that identifies the quantum well. Furthermore, the finite Stokes shift between absorption and emission at the $n = 1$ exciton resonance (from exciton localization effects) is necessary to prevent elastically scattered laser light from saturating the optical detection.

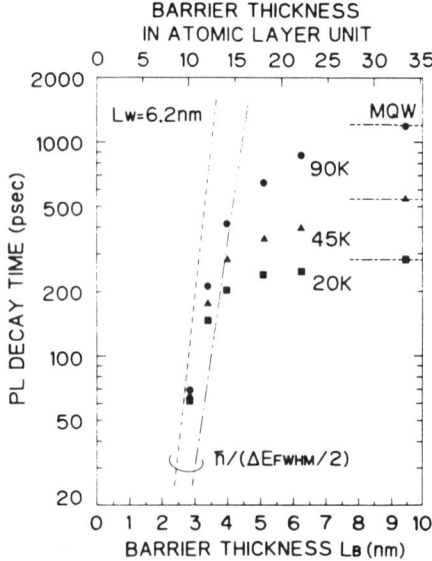

FIG. 16. Photoluminescence decay time from a single GaAs–AlAs quantum well with thin barriers in the tunneling dominated regime. The lines are a theoretical fit for two different barrier heights. (After Tsuchiya *et al.*, 1987.)

The effects of applied electric fields on tunneling rates have been reported by Norris et al., 1989) by similar methods. Furthermore, coupled quantum well structures composed of a pair of adjacent wells ($L_{w1} \neq L_{w2}$) that are separated by a thin barrier layer have been studied by Oberli et al. (1989b) and Matsusue et al. (1989). In such structures the confined particle states in the quantum wells can be tuned into and out of a tunneling resonance with simultaneous spectroscopic observation of the luminescence decay time in each. Figure 17 illustrates the results of Oberli et al., where the abrupt decrease in the luminescence decay time at the $n = 1$ heavy-hole resonance of the "emitting" quantum well is interpreted as the signature of the resonant tunneling event. Recently, these authors have also shown evidence of optical phonon intermediated tunneling in the asymmetric coupled quantum well structures (Oberli et al., 1990). In an elegant application of transient four-wave mixing and pump-probe techniques to a tunneling-related problem Leo et al. (1991) have observed coherent oscillations of the electron wavepacket in a double quantum well structure, the oscillations originating from

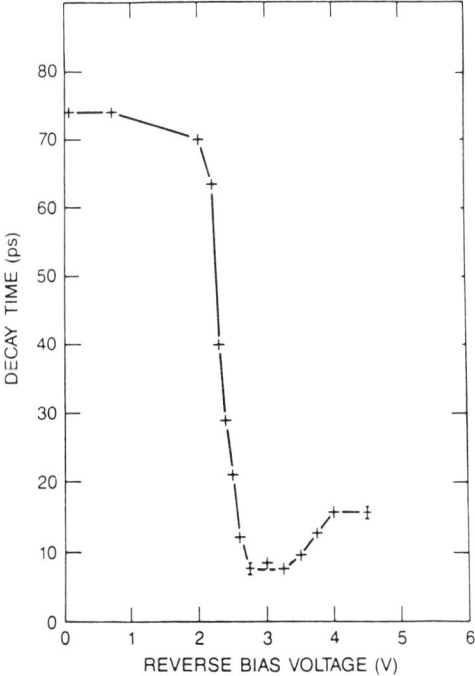

FIG. 17. Luminescence decay time versus bias voltage from one of the GaAs quantum wells for an asymmetric coupled GaAs–(Ga,Al)As well structure. The sharp reduction with field occurs as resonant tunneling conditions are obtained. (After Oberli et al., 1989.)

transitions between the symmetric and antisymmetric conduction electron states inherent in such a structure.

With continuing improvements in recent years in the structural quality of semiconductor superlattices in general, and in the interfacial quality in the GaAs–Al$_{1-x}$Ga$_x$As heterostructures in particular, investigation of *vertical transport* (z-direction in Fig. 13) with emphasis on intrinsic effects due to true superlattice dispersion has now become possible both through transport and optical measurements. Some of the seminal work in this area by time-resolved and other spectroscopic methods is due to the French group of B. Deveaud and collaborators (e.g., Chomette *et al.*, 1989; Lambert *et al.*, 1989). Vertical transport over superlattice thicknesses exceeding 1 μm has been observed in time-of-flight studies by transient photoluminescence, designed to investigate the electron and hole drift, diffusion, and localization within the superlattice minibands. The structures of choice in these experiments usually included a single, separate enlarged quantum well as a spatially specific optical marker either in a normal superlattice or in a periodic structure grown as stepwise graded (in Al composition) so as to provide an additional potential energy gradient for driving the diffusive process. The latter structures were particularly useful in that they show dramatic slowing in the vertical transport when the supercell lattice constant is increased to the point where the transport no longer occurs simply through the extended miniband states, as shown in Fig. 18 (Deveaud *et al.*, 1988). The spectra show the carrier distribution in the structure where the feature labeled EW identifies the single-collector well that is being supplied by carrier flow from the rest of the superlattice. In the bottom trace, where the superlattice period has increased to 60 Å, the very slow accumulation into the collector well is interpreted as a breakdown of the extended superlattice (Bloch) state picture and the important role of localized states at miniband edges (hopping transport).

Of other recent applications of ultrashort-pulse techniques to quantum wells and superlattices in terms of characterization of basic physical properties by weak optical probes, the issues of electron phase relaxation should also be mentioned. A recent observation of quantum beats in GaAs MQWs by a self-diffracted transient grating technique represents a particularly striking observation of excitonic coherence effects (Göbel *et al.*, 1990). Previously four-wave mixing methods have been used to investigate dephasing effects of such quasi-2D excitons (Kuhl *et al.*, 1989). Within the past two years, considerable progress has been made in the characterization of electron-hole coherence in quantum well systems through the application of ultrashort-pulse techniques. Among the first of such experiments, four-wave mixing techniques were used to investigate dephasing of quasi-2D excitons (Kuhl *et al.*, 1989). An observation of transient quantum beats in GaAs MQWs by a self-diffracted grating technique represents a particularly striking observation of excitonic coherence (Göbel *et al.*, 1990). An elegant series of experi-

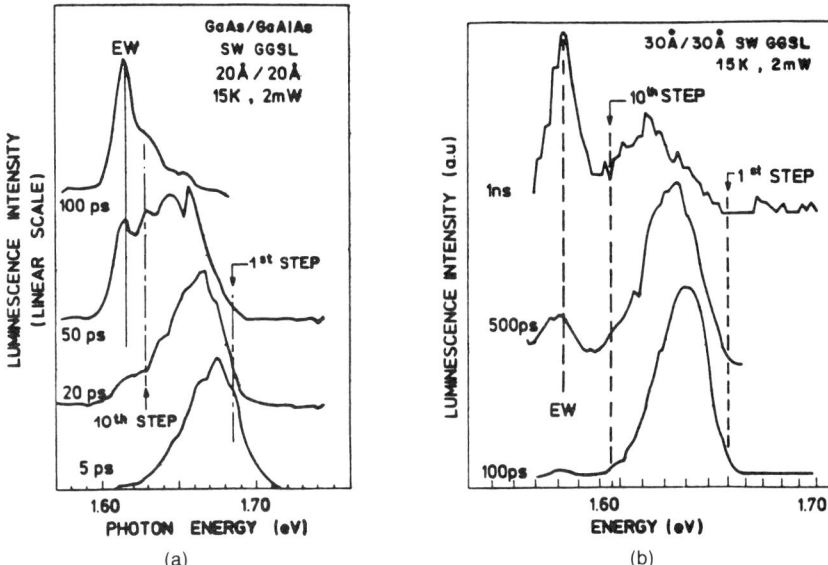

FIG. 18. Illustration of the use of time-resolved luminescence to study vertical transport in a graded GaAs–(Ga,Al)As superlattice at $T = 15$ K (signals obtained from a collector well). Electron–hole pair diffusion in the minibands occurs in (a) but is inhibited for the wider-well case in (b). The arrival at the collector well (EW) is indicated. (After Deveaud et al., © 1988 IEEE.)

ments by Webb et al. (1991) have shown new insight into the exciton phase relaxation in GaAs MQWs by observation of stimulated photon echoes and free polarization decay. The experimental technique was a three-pulse picosecond four-wave mixing in a situation where the homogeneously broadened component of the exciton resonance contributes a prompt signal (the free induction decay), whereas the inhomogeneous component yields a delayed signal (the stimulated two-photon echo). In another set of contemporary experiments, coherent oscillations associated with exciton states in a double GaAs quantum well (DQW) have been observed (Leo et al., 1991). Both the four-wave mixing and excite-probe techniques were employed. The coupled quantum wells, one wide, the other narrower, exhibit symmetric and asymmetric states whose energy separation can be fine-tuned by an applied electric field. In a somewhat related recent experiment, such coherent exciton states have been observed to emit electromagnetic radiation in the terahertz frequency range from the oscillating exciton dipole created by an incident resonant ultrashort optical pulse (Roskos et al., 1992). In yet another facet of the exciton quantum beat phenomena, transient oscillatory behavior has been observed in magneto-absorption in GaAs QWs (Bar-Ad and Bar-Joseph;

1991). These authors observe the beat between Zeeman states of the ground state exciton on a picosecond timescale, and draw quantitative information about the effective g-factors in this system.

Finally, we note that short pulse spectroscopy has also been applied to examine the role of Coulomb interaction for hot photoexcited electron-hole pairs from the standpoint of hot exciton processes, where the transition from Raman scattering (phase coherent limit) to thermalized limit (fully incoherent) has been studied in ZnTe QWs (Pelekanos *et al.*, 1991; Stanley and Hegarty, 1990).

8. Nonlinear Transient Optical Spectroscopy: Recent Results

The use of high-intensity ultrashort laser pulses to induce nonlinear optical effects in semiconductors is subject to considerable current activity. It is also a field where a substantial literature has accumulated over the past two decades. Traditionally, nonlinear optics as a field has heavily relied on the high optical intensities that ultrashort-pulse laser sources can provide. Two-photon absorption and stimulated Raman scattering are just two examples of the many phenomena that have been usefully investigated by short-pulse techniques, primarily at photon energies below the band gap. With intense ultrashort laser pulses, it is also relatively easy to obtain avalanche breakdown and damage in semiconductors excited below the band gap. Furthermore, ultrafast melting and rapid thermal annealing are also subjects of substantial continuing interest. We refer the reader, for example, again to the recent Proceedings of the Conferences on Ultrafast Phenomena (Springer), as well as a specific earlier compilation on the subject (Pilkuhn, 1985). In this section we briefly review some of the more recent experimental developments that specifically relate to highly excited electronic states in semiconductor microstructures in the context of above or near band gap interband excitation.

Under intense excitation at above band-gap photon energies, high-density electron–hole populations can be produced so that many-body Coulombic effects and the exclusion principle play an important role. On the other hand, large nonlinear coherent polarizations can be induced if the excitation energy is kept below (but near) the fundamental edge. The optical properties of such interacting many-electron–hole systems, in both 3D and 2D, have been reviewed by a number of authors in recent literature. One excellent reference source is Schmitt-Rink *et al.* (1989).

Of the already wide range of phenomena discovered and studied to date, we touch first on the subject of excitons and their interaction with excitons and free-electron–hole pairs, respectively, in quantum wells. Experimental investigations in this area have been greatly facilitated by sophisticated advances in the excite-probe techniques on subpicosecond time scales, notably by the developments made by Knox and collaborators (Knox et al., 1985). In particular, the use of white light continuum pulses in conjunction with multichannel spectroscopy has opened avenues of research where steady-state methods could only hint at possibilities earlier.

We consider first an example where the exciton resonance in a quantum well is probed in absorption or reflection, during and after the incident excitation produces a dense free-electron–hole pair population at $t_0 = 0$. The physical phenomena that are responsible for the pump-induced changes in the absorption coefficient and the index of refraction of the sample typically originate from *screening, exclusion principle*, and *band-gap renormalization* effects. The band-gap renormalization, for example, is a mechanism for inducing the exciton to electron–hole plasma (EHP) transition (insulator–metal transition) both in 3D and 2D semiconductors. This follows from the remarkably near constancy of the exciton (interband) energy as a function of pair density so that at some critical density the EHP energy represents the lowest state. (For example, the scaling in a 2D system for the gap normalization with electron density N is obtained from random-phase approximation (RPA) theory approximately as $\Delta E_g \sim (Na_0^2)^{1/3}$, with a_0 the exciton Bohr radius.) The gap renormalization phenomena has been investigated spectroscopically under a range of experimental circumstances, steady-state and transient, although some controversy exists about the interpretation of results in highly photoexcited, hot semiconductors (as compared with cold modulation-doped quantum wells (Deveaud, 1990; Cingolani et al. 1990).

An illustrative experimental result, highlighting the benefits of an ultrafast spectroscopic technique, is shown in Fig. 19, which displays the room temperature differential transmission spectrum $\Delta T(\omega_{pr}, t - t_0)/T(\omega_{pr})$ of aGaAs–Al$_{1-x}$Ga$_x$As MQW structure at room temperature near the main exciton resonances (Knox et al., 1985). The initial excitation of some 10^{10} cm^{-2} pairs at $t = t_0$ occurs in the continuum between the $n = 1$ and $n = 2$ QW exciton states so that the $n = 1$ heavy- and light-hole states are not immediately affected by occupancy effects (Pauli principle) by the excess carriers. Screening by the direct Coulomb interaction, on the other hand, would be effectively instantaneous even on the 100-fsec time scale. While such a relatively high excess free-pair density in 3D (appropriately normalized) would practically obliterate the exciton, the survival of the exciton absorption peaks here gives tangible evidence of the weakening of screening in 2D conditions. With time, as the excess carriers equilibrate and relax toward the $n = 1$ exciton ground state, the differential transmission increases, showing the reduction of the $n = 1$ exciton oscillator strength as a consequence of exclusion principle effects. This corresponds, to a good approximation, to filling of the exciton

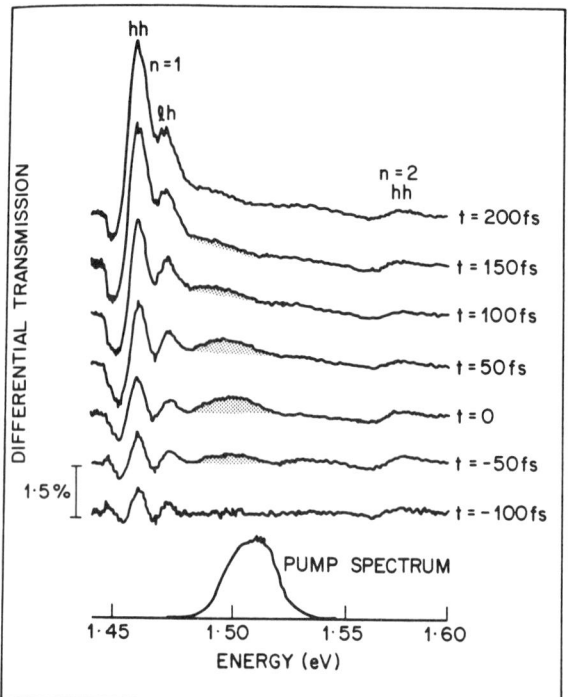

FIG. 19. Ultrashort-pulse spectroscopy with continuum probe pulses: Differential transmission through GaAs–(Ga,Al)As MQW structure at $T = 300$ K ($L_w \approx 100$ Å). The excitation pulse (spectrum shown) initially burns a population hole in the continuum pair states above the $n = 1$ exciton; subsequent thermalization bleaches the exciton due to exclusion principle effects. (After Knox et al., 1985.)

phase space, which in k-space spans the volume $\sim (a_0)^{-3}$ and in real space the quantum well volume commensurate for hard disk excitons of 2D radius $a_0/2$ (Schmitt-Rink, 1989).

The phase-space filling effects are further accentuated if the initial excitation occurs directly into the $n = 1$ resonance as shown in the differential transmission traces of Fig. 20 for a GaAs–$Al_{1-x}Ga_xAs$ MQW (Knox et al., 1986). Following strong initial bleaching of the $n = 1$ exciton absorption by such filling effects, its partial recovery is seen to have set in by about 0.3 psec in this room temperature experiment, since the excitons are ionized to free-particle states by LO phonon (Frohlich) interaction (where their contribution is primarily one of screening). This time constant is in good agreement from measurements of the homogeneous exciton linewidth in linear absorption at room temperature, which is dominated by the broadening from the exciton dissociation rate in GaAs quantum wells. Closer examination of the spectral

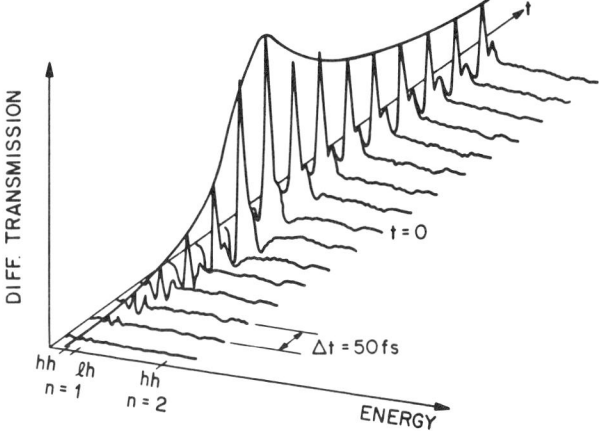

FIG. 20. Differential transmission of the sample shown in Fig. 18 but now under condition of resonant $n = 1$ excitation: the exclusion principle effects are very large in initially weakening the exciton state (as probed by the weaker wideband probe pulses); exciton LO phonon ionization step in ~ 0.3 psec leads to partial recovery in $\delta\alpha$. (After Knox et al., 1986.)

position of the exciton resonance during such high-density resonant excitation shows finite blue shifts (Peyghambarian et al., 1984), which have been shown to originate from short-range hard-core repulsive interactions in a 2D exciton gas where the cancellation by the van der Waal's like exciton–exciton attractive potential is not quite as complete as it is in three dimensions.

An example of the coherent nonlinear effects where virtual exciton interactions play a role is the *optical Stark effect* recently investigated in quantum well structures. While virtual, the cross section for the processes is enhanced considerably by tuning the incident short-pulse laser immediately below the excitonic band gap. Although the effect could, in principle, be observed by continuous-wave excitation, other effects, including thermal and two-photon excitation of real particles, make it experimentally quite difficult to isolate in such circumstances. With ultrashort pulses one not only takes advantage of the concomitant high intensities (electromagnetic fields) but has the benefit of studying transient aspects of the coherent electronic polarizations through the optical Stark effect.

An illustration of the first demonstration of this effect is shown in Fig. 21, obtained for a GaAs–$Al_{1-x}Ga_xAs$ MQW structure as a transient red shift of the exciton absorption edge in a pump-probe experiment (Mysyrowicz et al., 1986; von Lehmen et al., 1986). The theoretical description of the exciton Stark shift at optical frequencies in the adiabatic regime have been put forth both in terms of a "dressed two-level atom model" (Mysyrowicz et al., 1986) and, more rigorously, by considering the interactions in the coherently

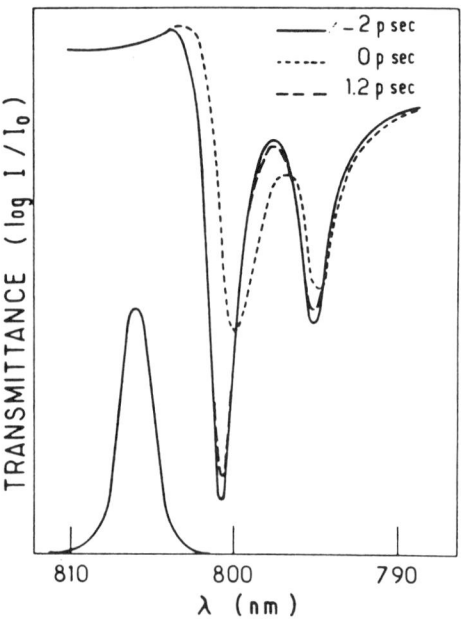

FIG. 21. Optical Stark effect in a GaAs–(Ga,Al)As MQW sample as "seen" in transmission as an instantaneous spectral blue shift by an ultrashort broadband probe pulse. The excitation at $t_0 = 0$ psec occurs slightly below the $n = 1$ exciton absorption edge (pump spectrum shown). (After Mysyrowicz et al., 1986.)

driven virtual exciton gas (Schmitt-Rink et al., 1989). Recent developments in this area include the observation of spectral and temporal oscillations in the pump-probe experiments on femtosecond time scales, prior to the steady-state Stark effect regime. These ultrashort transients have been analyzed in terms of coherent response of the virtual exciton gas in the absence of polarization dephasing collisions (Joffre et al. 1989). An example of these results is shown for a high-quality epitaxial layer of GaAs in Fig. 22 with a very narrow 1S exciton absorption line (hence, long phase-relaxation time). The figure shows very clearly the spectral coherent "beats" indicative of strong coherence in the interband polarizations. The study of such transient coherent effects in semiconductors is still in its early stage (note examples in the preceding section), but it is also one of the more dramatic illustrations of the power of ultrafast transient spectroscopy.

Very recently, the transient nonlinear effects with quasi-2D excitons in GaAs quantum wells have been studied in strong magnetic fields. Stark and co-workers have shown in femtosecond pump-probe experiments up to fields of 12 Tesla how the exciton-exciton Coulomb interactions are strongly modified by the "quasi-zero-dimensional" confinement in per-

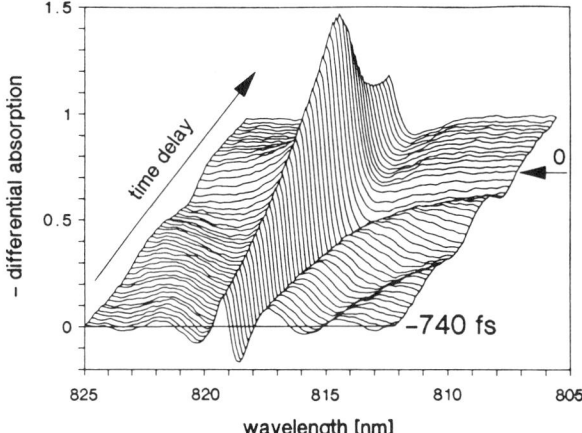

FIG. 22. Ultrafast differential excite–probe experiment in bulk GaAs in an optical Stark effect experiment. Differential absorption spectra are shown in 20-fsec steps; the distinct spectral oscillations are related to coherent electron–hole pair states. (After Joffre et al., © 1989 IEEE.)

pendicular fields (Stark et al., 1990). As a function of the magnetic field, the exciton spectrum and associate eigenstates are change from the regime where the Coulomb interaction dominates to that where the Landau quantization becomes important. The 1s exciton state remains strongly Coulomb-correlated; this leads to a fundamental question about the (lack of) interaction between such "point-like" electron-hole pair composites. A rather different circumstance can be encountered in modulation-doped quantum wells, especially when the electron density corresponds to a Fermi-level proximity to an energetically nearby empty conduction subband. This leads to a strong singularity effect (peak) in the luminescence which is physically composed of a Coulombically interacting many-electron/one-hole exciton-like state. The kinetics of formation of such a many-body complex in magnetic fields has been recently studied through picosecond time-resolved photoluminescence techniques (Chen et al., 1992). In undoped GaAs epitaxial layers, there is evidence of a nonequilibrium edge singularity state which has been identified from femtosecond pump-probe experiments (Foing et al., 1992).

V. Concluding Remarks

In this chapter I have provided a subjective review of the field of ultrafast spectroscopy of semiconductors from a contemporary point of view. The main purpose, as stated in the introduction, has been to give a general reader a rounded picture about the possibilities that this fascinating technical activity offers. Undoubtedly, in the years to come, technical developments with both ultrafast sources and detectors will continue to open new avenues and fuel further innovation. In terms of sources, femtosecond lasers with widewavelength tunability would represent an important next milestone. With ultrathin material structures, the problems of group velocity dispersion may be overcome so that full advantage can be taken of the shortest laser pulses available today (~ 10 fsec). Equally, the extension of ultrafast techniques to the vacuum ultraviolet and the soft x-ray regions, on one hand, and to the mid-infrared, on the other, would considerably enlarge the domain of semiconductors (and other materials) for study. The role of synchrotron-based souces is likely to be important in such future endeavors for source development.

In terms of the semiconductor physics itself, identification of a number of important research directions can be made in the context of transient spectroscopy of the near future. Clearly, new materials, especially in hetero- and nanostructure form, will offer ample opportunities for new discovery and useful characterization. Quantum wires and dots and other forms of architecture will probably call increasingly on optical studies for understanding the electronic and vibronic degrees of freedom. For example, with the long mean-free paths (tens of micrometers) available in high-mobility GaAS 2D electron systems, it should be possible to design direct optical measurements of ballistic electron transport over macroscopic distances. The mixing of optical and optoelectronic techniques on an ultrashort time scale presents a particularly attractive possibility in this context, as shown in early work on "excitonic optoelectronics" (Knox *et al.*, 1989).

The study of surface phenomena by optical methods (both in terms of steady-state and transient spectroscopy) is a field that is likely to see increasing and important activity. While left out of this chapter, a number of rather advanced experiments where ultrashort-pulse laser techniques have been used to both induce and characterize surface phenomena have been reported within the past year or two. With more compact and semiportable laser and optoelectronic instrumentation being developed, opportunities will exist for integrating ultrafast spectroscopic equipment, e.g., within advanced epitaxial facilities for in situ experiments and optical processing during the layered growth.

Finally, apart from studies of laser melting, there has been little use made of ultrafast optical techniques in the study of phase-transition phenomena. Structural, magnetic, and superconducting–normal transitions are instances where selective resonance excitation from a femtosecond source to induce such a transition should permit the identification of the dominant real-time processes and energy pathways. In systems with long-range order, macroscopic quantum phenomena may also be optically accessible.

References

Auston, D. H. (1987). "Ultrashort Optical Pulses" (Kaiser, W., ed.), Ch. 5. Springer-Verlag, New York.
Awschalom, D., Warnock, J., and von Molnar, S. (1987), *Phys. Rev. Lett.* **58**, 812.
Awschalom, D., (1989), *Phys. Rev. Lett.* **62**, 199.
Bar-Ad, S., and Bar-Joseph, I. (1991). *Phys. Rev. Lett.* **66**, 2491.
Bowers, J. E., Morton, P. A., Mar, A., and Corzine, S. (1989). *IEEE J. Quant. Elect.* **QE-25**, 1426.
Brorson, S. D., Fujimoto, J. G., and Ippen, E. P. (1987). *Phys. Rev. Lett.* **59**, 1962.
Brorson, S. D., Kazeroonian, A., Moodera, J. S., Face, D. W., Cheng, T. K., Ippen, E. P., Dresselhaus, M. S., and Dresselhaus G. (1990). *Phys. Rev. Lett.* **64**, 2172.
Brum, J. A., and Bastard, G. (1986). *Phys. Rev.* **B33**, 1420.
Chemla, D., and Pinczuk, A., eds. (1986). Special Issue of *IEEE J. Quant. Elect.* **QE-22**, 1579–1798.
Chen, W., Fritze, M., Nurmikko, A. V., Ackley, D., Hong, J. M., and Chang, L. L. (1992). *Phys. Rev. B* (in press).
Cheng. T. K., Brorson, S. D., Kazeroonian, A. S., Moodera, J. S., Dresselhaus, G., Dresselhaus, M. S., and Ippen, E. P. (1990). *Appl. Phys. Lett.* **57**, 1004.
Cho, G. C., Kütt, W., and Kurz, H. (1990). *Phys. Rev. Lett.* **65**, 764.
Chomette, A., Deveaud, B., Clerot, F., Lambert, B., and Regreny, A. (1989). *J. Luminescence* **44**, 265.
Cohen, E. (1985). Proc. 17th Int. Conf. on Physics of Semiconductors (Chadi, J. and Harrison, W. eds.). Springer-Verlag.
Coleman, J. J., ed. (1988). Special Issue of *IEEE J. Quant. Elect.* **QE-24**, 1609–1798.
Darrow, J. T., Hu, B. B., Zhang, X.-C., Auston, D. H., and Smith, P. R. (1990). *Appl. Phys. Lett.* **56**, 886.
Das Sarma, S., and Mason, B. A. (1985). *Ann. Phys.* **163**, 78.
deFonzo, A. P., and Lutz, C. R. (1989). *Appl. Phys. Lett.* **54**, 2186.
Deveaud, B. (1990), Proc. 20th Int. Conf. on the Physics of Semiconductors, Thessaloniki.
Deveaud, B., Shah, J., Damen, T. C., Lambert, B., Chomette, A., Regreny, A. (1988). *IEEE J. Quant. Elect.* **QE24**, 1641.
Ding, J. D., and Nurmikko, A. V. (1989), unpublished.
Doll, G. L., Esley, G. L., Brorson, S. D., Dresselhaus, M. S., Dresselhaus, G., Cassanho, A., Jenssen, H. P., and Gabbe, D. R., (1989). *Appl. Phys. Lett.* **55**, 402.
Edelstein, D. C., Wachman, E. S., and Tang, C. L. (1989). *Appl. Phys. Lett.* **54**, 1728.
Feldmann, J., Peter, G., Göbel, E. O., Dawson, P., Moore, K., Foxon, C., and Elliott, R. J. (1987). *Phys. Rev. Lett.* **59**, 2337.
Foing, J.-P., Hulin, D., Joffre, M., Jackson, M., Oudar, J.-L., Tanguy, C., and Combescot, M. (1992). *Phys. Rev. Lett.* (in press).
Fork, R. L., Greene, B. I., and Shank, C. V. (1981). *Appl. Phys. Lett.* **38**, 671.
Fork. R. L. Shank, C. V., Hirlimann, C., and Yen, R. (1983). *Opt. Lett.* **8**, 1.
Fork, R. L., Brito Cruz, C. H., Becker, P. C., and Shank, C. V. (1987). *Appl. Phys. Lett.* **52**, 483.
Grahn, H., Maris, H., and Tauc, J. (1989). IEEE *J. Quant. Elect.* **QE-25**, 2562.
Grischkowsky, D. R. (1988). *IEEE J. Quant. Elect.* **QE-24**, 221.
Göbel, E. O., Leo, K., Damen, T. C., Shah, J., Schmitt-Rink, S., Schäfer, W., Müller, J., and Köhler, K. (1990). *Phys. Rev. Lett.* **64**, 1801.
Harris, J., and Nurmikko, A. V. (1983). *Phys. Rev.* **B28**, 1181.
Harris, J., and Nurmikko, A. V. (1985). *Phys. Rev. Lett.* **5128**, 1472.
Harris, J., Sugai, S., and Nurmikko, A. V. (1982). *Appl. Phys. Lett.* **40**, 885.
Hefetz, Y., Zhang, X.-C., and Nurmikko, A. V. (1985). *Phys. Rev.* **B31**, 5371.
Hefetz, Y., Goltsos, W. C., Lee, D., Nurmikko, A. V., Kolodziejski, L. A., and Gunshor, R. L. (1986). *Superlattices and Microstructures* **2**, 455.

Heyen, E. T., Hagerott, M., Nurmikko, A. V., and Partin, D. L. (1989). *Appl. Phys. Lett.* **54**, 653.
Hopfel, R. A., Shah, J., and Gossard, A. C. (1986). *Phys. Rev. Lett.* **56**, 765.
Islam, M., Sunderman, E., Soccolich, C., Bar-Joseph, I., Sauer, N., Chang, T.-Y., and Miller, B. (1989). *IEEE J. Quant. Elect.* **QE-25**, 2454.
Jarasiunas, K., Hoffman, C., Gerritsen, H., and Nurmikko, A. V. (1978). *Appl. Phys. Lett.* **33**, 536.
Jiang, W. B., Friberg, S. R., Iwamura, H., and Yamamoto, Y. (1991). *Appl. Phys. Lett.* **58**, 807.
Joffre, M., Hulin, D., Foing, J.-P., Chambaret, J.-P., Migus, A., and Antonetti, A. (1989). *IEEE J. Quant. Elect.* **QE-25**, 2505.
Johnson, A. M., and Simpson, W. M. (1986). *IEEE J. Quant. Elect.* **QE-22**, 133.
Johnson A. M., Stolen, R. H., and Simpson, W. M. (1984). *Appl. Phys. Lett.* **44**, 729.
Kash, J. (1984). *Phys. Rev.* **B29**, 7069.
Kash, J. A., Tsang, J. C., and Hvam, J. M. (1985). *Phys. Rev. Lett.* **54**, 2151.
Kash, J. A., Jha, S. S., and Tsang, J. C. (1987). *Phys. Rev. Lett.* **58**, 1869.
Kuhl, J., Schultheis, L., and Honold, A. (1989). "Festkörperprobleme: Advances in Solid State Physics," vol. 29, p. 157. Springer-Verlag, New York.
Knoll, G., Siegner U., Shevel, S. G., and Göbel, E. O. (1990). *Phys. Rev. Lett.* **64**, 792.
Knox, W. H. (1987). *J. Opt. Soc. A.* **B4**, 1771.
Knox, W., Fork, R., Downer, M., Miller, D. A. B., Chemla, D. S., Shank, C. V., Gossard, A., and Wiegmann, W. (1985). *Phys. Rev. Lett.* **54**, 1306.
Knox, W., Hirliman, C., Miller, D. A. B, Shah, J., Chemla, D. S., Shank, C. V., Gossard, A., and Wiegmann, W. (1986). *Phys. Rev. Lett.* **56**, 1191.
Knox, W., Henry, J., Goossen, K., Li, K., Tell, B., Miller, D. A. B., Chemla, D. S., Gossard, A. C., English, J., and Schmitt-Rink, S. (1989), *IEEE J. Quant. El.* **QE-25**, 2586.
Krenn, H., Kaltenegger, K., Dietl, T., Spalek, J., and Bauer, G. (1986). *Phys. Rev.* **B39**, 10918.
Lambert, B., Clerot, F., Chomette, A., Deveaud, B., Talalaeff, G., and Regreny, A. (1989). *J. Luminescence* **44**, 277.
Leo, K., Rühle, W. W., and Ploog, K. (1988). *Phys. Rev.* **B38**, 1947.
Leo, K., Shah, J., Gobel, E. O., Damen, T. C., Schmitt-Rink, S., Shafer, W., and Kohler, K. (1991). *Phys. Rev. Lett.* **66**, 201., *Phys. Rev.* **B44**, 5726.
Lyon, S. A. (1986). *J. Luminescence* **35**, 121.
Matsusue, T., and Sakaki, H. (1987). *Appl. Phys. Lett.* **50**, 1429.
Matsusue, T., Tsuchiya, M., and Sakaki, H. (1989). Proc. OSA Topical Meeting on Quantum Wells in Optoelectronics, Salt Lake City, OSA Technical Digest Series 10.
McMullan, W. G., Charbonneau, S., and Thewalt, M. L. W. (1987). *Rev. Sci. Instr.* **58**, 1626.
Meyer, K. Pessot, M., and Mourou, G., (1988). *Appl. Phys. Lett.* **53**, 2254.
Mitschke, F. M., and Mollenauer, L. F. (1987). *Opt. Lett.* **12**, 407.
Mysyrowicz, A., Hulin, D., Antoneeti, A., Migus, A., Masselink, W. T., and Morkoç, H. (1986). *Phys. Rev. Lett.* **56**, 2748.
Nagarajan, R., Kamiya, T., Kasukawa, A., and Okamoto, H. (1989). *Appl. Phys. Lett.* **55**, 1273.
Norris, T. B., Song, X. J., Schaff, W. J., Eastman, L. F., Wicks, G., and Mourou, G. A. (1989). *Appl. Phys. Lett.* **54**, 60.
Oberli, D., Wake, D., Klein, M. V., Klem. J., Henderson, T., and Morkoç, H. (1987). *Phys. Rev. Lett.* **59**, 696.
Oberli, D., Shah, J., Jewell, J. Damen, T. C., and Chand, N. (1989a). *Appl. Phys. Lett.* **54**, 1028.
Oberli, D., Shah, J., Damen, T. C., Tu, C. W., and Miller, D. A. B. (1989b). Proc. OSA Topical Meeting on Quantum Wells in Optoelectronics, Salt Lake City, OSA Technical Digest Series 10.
Oberli, D., Shah, J., Damen, T. C., Kuo, J. M., Henry, J. E., Lary, J., and Goodnick, S. M. (1990). *Appl. Phys. Lett.* **56**, 1239.
Oudar, J. L., Hulin, D., Migus, A., Antonetti, A., and Alexandre, F. (1984). *Phys. Rev. Lett.* **53**, 384.
Oudar, J. L., Migus, A., Hulin, D., Grillon, G., Etchepare, J., and Antonetti, A. (1985). *Phys. Rev. Lett.* **55**, 2074.

Pelekanos, N., Ding, J., Nurmikko, A., Kobayashi, M., and Gunshor, R. L. (1991). *Phys. Rev.* B**44**, 1111.
Peyghambarian, N., Gibbs, H., Jewell, J., Antonetti, A., Migus, A., Hulin, D., and Mysyrowicz, A. (1984). *Phys. Rev. Lett.* **53**, 2433.
Pilkuhn, M. H., ed. (1985), "High Excitation and Short Pulse Phenomena." North-Holland, Amsterdam.
Rieck, B., Goldstein, M., Roskos, H., Seilmeier, A., and Kaiser, W. (1989). *Solid State Elect.* **32**, 1405.
Roskos, H., Nuss, M. C., Shah, J., Tell, B., and Cunningham, J., (1991) Proc. Conf. on Quantum Optoelectronics, Salt Lake City, OSA Technical Digest Series 7.
Roskos, H., Nuss, M., Shah, J., Leo, K., Miller, D. A. B., Schmitt-Rink, S., and Köhler, K. (1992). *Phys. Rev. Lett.* (in press).
Ryan, J. F., Taylor, R. A., Turberfield, A. J., Maciel, A. Worlock, J. M., Gossard, A. C., and Wiegmann, W. (1985). *Phys. Rev. Lett.* **53**, 1841.
Ryan, J. F., Taylor, R. A., Turberfield, A. J., and Worlock, J. M. (1985b). *Physica* **134B**, 318, 403.
Sanchez, A., Strauss, A. J., Aggarwal, R. L., and Fahey, R. E. (1988), *J. Quant. Elect.* **QE24**, 995.
Schmitt-Rink, S., Chemla, D. S., and Miller, D. A. B. (1989). *Adv. Phy.* **38**, 89–188.
Schoenlein, R. W., Bigot, J.-Y., Portella, M. T., and Shank, C. V. (1991). *Appl. Phys. Lett.* **58**, 801.
Schultheis, L., Kuhl, J., Honold, A., and Tu, C. W. (1986). *Phys. Rev. Lett.* **57** (1635).
Shah, J., Damen, T. C., Deveaud, B., and Block, D. (1987). *Appl. Phys. Lett.* **50**, 1307.
Shah, J., Damen, T. C., Tsang, W. T., Gossard, A. C., Lugli, P. (1987), *Phys. Rev. Lett.* **59**, 2222.
Smirl, A. L., Valley, G. C., Bohnert, K. M., and Boggess, T. F. (1988). *J. Quant. Elect.* **QE-24**, 289.
Solid State Electronics: Special issues for the Fifth and Sixth International Conferences on Hot Carriers (1988) and (1990), **31**, 319–820; **32**, 1049–1648.
Springer: series of volumes for the Proceedings of Conference on Ultrafast Phenomena, published in Series of Chemical Physics by Springer-Verlag (Berlin).
Stanley, J., Hegarty, J., (1990), Proc. Int. Conf. Quantum Electr., Anaheim, CA.
Stark, J. B., Knox, W., Chemla, D. S., Schäfer, W., Schmitt-Rink, S., and Stafford, C. (1990). *Phys. Rev. Lett.* **65**, 3033.
Tsuchiya, M., Matsusue, T., and Sakaki, H. (1987). *Phys. Rev. Lett.* **59**, 2356.
Turko, B. T., Nairn, J. A., and Sauer, K. (1983). *Rev. Sci. Instr.* **54**, 118.
von der Linde, D., Kuhl, J., and Klingenburg, H. (1980). *Phys. Rev. Lett.* **44**, 1505.
von Lehmen, A., Chemla, D. S., Zucker, J., and Heritage, J. P. (1986). *Opt. Lett.* **11**, 609.
Webb, M. D., Cundiff, S. T., and Steel, D. G. (1991). *Phys. Rev. Lett.* **66**, 934.
Weisbuch, C. (1988), "Semiconductors and Semimetals" (A. Beer and R. Willardson, eds.), Academic Press, New York. Vol. 24, pp. 1–125.
Yamazaki, I., Tamai, N., Kume, H., Tsuchiya, H., and Oba, K. (1985). *Rev. Sci. Instr.* **56**, 1187.
Zayhowski, J. J., Jagannath, Kershaw, R., Ridgley, D., Dwight, K., and Wold, A. (1985). *Solid State. Comm.* **55**, 941.
Zhang, X.-C., Nurmikko, A. V. (1985). Proc. 17th Int. Conf. on Physics of Semiconductors (Chadi, J. and Harrison, W. eds.). Springer-Verlag, New York.
Zhang, X.-C., Hu, B. B., Darrow, J. T., and Auston, D. H. (1990). *Appl. Phys. Lett.* **56**, 1011.

CHAPTER 3

Piezospectroscopy of Semiconductors

A. K. Ramdas and S. Rodriguez

DEPARTMENT OF PHYSICS, PURDUE UNIVERSITY
WEST LAFAYETTE, INDIANA

I.	INTRODUCTION	137
II.	ELASTICITY, SYMMETRY, LATENT ANISOTROPY, AND DEFORMATION POTENTIALS	139
	1. *Elasticity*	139
	2. *Symmetry*	142
	3. *Latent Anisotropy*	143
	4. *Deformation Potential Theory*	144
III.	EXPERIMENTAL TECHNIQUES	145
IV.	VALENCE AND CONDUCTION BANDS OF ELEMENTAL AND COMPOUND SEMICONDUCTORS UNDER EXTERNAL STRESS	150
	5. *Deformation Potentials of Band Extrema*	151
	6. *Relative Intensities of Stress-Induced Components of an Optical Transition*	161
V.	INTERBAND TRANSITIONS AND ASSOCIATED EXCITONS	166
	7. *Direct Transitions*	166
	8. *Indirect Transitions*	172
	9. *Effects in Heterostructures*	175
VI.	LYMAN SPECTRA OF SHALLOW DONORS AND ACCEPTORS	179
VII.	ZONE-CENTER OPTICAL PHONONS	197
VIII.	OPTICAL PHONONS AND ELECTRONIC STATES	206
	10. *Coupled LO Phonon–Plasmon Modes in n-GaSb*	207
	11. *Zone-Center Optical Phonons in Si in the Presence of Free Carriers*	210
IX.	CONCLUDING REMARKS	213
	ACKNOWLEDGMENTS	215
	REFERENCES	215

I. Introduction

A comprehensive and detailed view of matter in all states of aggregation emerges from spectroscopic investigations. All the important features of atomic structure, based on the positive nucleus and the extranuclear electrons bound to it by Coulomb attraction, were discovered from a study of atomic spectra: electronic energy levels; electron and nuclear spin; selection rules governing the electric dipole, quadrupole, and magnetic dipole transitions;

and so on. This list should convey the comprehensive nature of spectroscopy. The impact of spectroscopy on condensed matter physics can be illustrated similarly with advances in semiconductor physics through the applications of spectroscopy (Ramdas, 1990). The absorption edge and the spectrum of the recombination radiation have been crucial in establishing the nature of the valence-band maximum and the conduction-band minimum. The electronic levels of shallow-donor-bound electrons and acceptor-bound holes have been discovered and delineated in an exceptionally thorough manner from a study of the infrared excitation spectra (Ramdas and Rodriguez, 1981). Raman, Brillouin, and infrared spectroscopy have provided a detailed knowledge of the frequency and symmetry of the collective and localized vibrations (Hayes and Loudon, 1978; Möller and Rothschild, 1971). When performed under an external perturbation, spectroscopic studies provide special microscopic insights into the quantum numbers that characterize the energy levels of the system under study. For example, Zeeman and Stark effects reveal their degeneracies associated with the orbital angular momentum as well as electronic and nuclear spins. In the case of solids, yet another perturbation technique has been fruitfully exploited namely spectroscopic observations carried out with the system under an external uniaxial stress. Spectroscopy thus performed is appropriately labeled *piezospectroscopy* (Kaplyanskii, 1964a). It is unique to solids because only they can sustain shear. The elastic anisotropy of matter in the crystalline state permits the imposition of a rich variety of stress-induced perturbations.

Brewster (1815, 1816) reported that an isotropic medium becomes birefringent when subjected to uniaxial stress. (See also Coker and Filon, 1957.) Indeed this is the first reported observation of an optical effect under an external perturbation. The photoelastic effect discovered by Brewster is, of course, ultimately traced to the stress-induced changes—splittings and shifts—in the optical transitions that underlie the refractive index of the medium. The situation is analogous to the circular birefringence (Faraday effect) and the linear birefringence (Voigt effect) exhibited by matter in the presence of an external magnetic field, on the one hand, and the Zeeman effect, on the other. Surprisingly the first example of piezospectroscopy is very recent, namely the modification of the Raman spectrum of α-quartz under uniaxial compression reported by Marieé and Mathieu (1946). The power of piezospectroscopy was recognized by a number of investigators in the early 1960s (see, for example, Feofilov and Kaplyanskii, 1962; Aggarwal and Ramdas, 1964, 1965a; Fitchen, 1968) and has become an important tool in the hands of solid-state spectroscopists. Interband and excitonic transitions, electronic transitions of shallow donors and acceptors and of defects, and optical phonons as well as vibrational modes of defects are examples of phenomena in semiconductors that have been fruitfully studied with piezospectroscopy. Absorption and reflectivity in the visible, near and far infrared, photoluminescence, and Raman and four-wave mixing spectroscopy are the spectroscopic techniques that, in conjunction with uniaxial stress, have been ingeniously employed in these studies, thereby providing significant insights into the nature of collective and localized excitations.

While the theoretical analysis of piezospectroscopy can be carried out by standard perturbation techniques, symmetry arguments are especially powerful in the interpretation of the experimental results. A quantitative theoretical formulation of the shifts and splittings of spectroscopic features, the relative intensities of the stress-induced components, dichroism and polarization effects observed in absorption and emission spectra, and the polarization features of the stress-induced components of Raman lines are most effectively addressed with symmetry arguments, for which group theory provides the most appropriate mathematical language.

In this chapter the focus is on the applications of piezospectroscopy in semiconductor physics in the context of localized and collective excitations, electronic and vibrational in character. The theoretical tools and the general principles are discussed in Section II, the experimental techniques in Section III. Applications of piezospectroscopy in semiconductor physics are presented in the rest of the chapter with a selection of examples that illustrate the power and scope of the technique.

II. Elasticity, Symmetry, Latent Anisotropy, and Deformation Potentials

In order to specify the reduction in the symmetry of a crystal when subjected to a uniaxial stress, one has to deduce the strain tensor using the theory of elasticity. From a knowledge of the reduced symmetry, one can draw *qualitative* conclusions about the removal of level degeneracies and the selection rules governing the stress-induced components of the optical transitions or the Raman lines. The quantitative description of piezospectroscopy can be formulated in terms of a phenomenological theory—the so-called deformation potential theory; group theory can be effectively exploited in the development of the theory. The phenomenological constants that appear in the deformation potential theory are deduced from experiments but can be calculated, in principle, in terms of microscopic models.

1. Elasticity

Under the influence of an external mechanical force, a body undergoes a deformation characterized by a displacement field $\mathbf{u}(x_1, x_2, x_3)$, x_1, x_2, and x_3

being the coordinates of a point referred to an arbitrary Cartesian coordinate system. We consider a homogeneous strain within the elastic limit as defined by the components of a symmetric second-rank tensor $\{\varepsilon_{ij}\}$:

$$\varepsilon_{ij} = \frac{1}{2}\left(\frac{\partial u_i}{\partial x_j} + \frac{\partial u_j}{\partial x_i}\right); \quad i, j = 1, 2, 3. \tag{1}$$

The strain ellipsoid is defined by the quadratic form $\sum_{i,j} \varepsilon_{ij} x_i x_j$. The strains result from externally applied forces, which in turn give rise to stresses between adjacent volume elements of the body. The stress tensor $\{\sigma_{ij}\}$ is defined by its components σ_{ij}, the force per unit area on a surface element normal to the i-axis projected on the j-axis. We note that $\{\sigma_{ij}\}$ is symmetric and defines a stress quadric $\sum_{i,j} \sigma_{ij} x_i x_j$. The stress–strain relations are represented by the generalized Hooke's law

$$\varepsilon_{ij} = s_{ijkl}\sigma_{kl}, \tag{2a}$$

where summation over k, l is implied and s_{ijkl} are the *elastic compliance constants*. Equivalently,

$$\sigma_{ij} = c_{ijkl}\varepsilon_{kl}, \tag{2b}$$

c_{ijkl} being the *elastic stiffness constants* or *elastic moduli*. The following one-suffix notation, widely used in the theory of elasticity, is shown in terms of correspondence with the two-suffix notation:

$$\begin{array}{cccccccccccc}
\sigma_{11} & \sigma_{22} & \sigma_{33} & \sigma_{23} & \sigma_{31} & \sigma_{12} & \varepsilon_{11} & \varepsilon_{12} & \varepsilon_{33} & 2\varepsilon_{23} & 2\varepsilon_{31} & 2\varepsilon_{12} \\
\downarrow & \downarrow & \downarrow & \downarrow & \downarrow & \downarrow & \downarrow & \downarrow & \downarrow & \downarrow & \downarrow & \downarrow \\
\sigma_1 & \sigma_2 & \sigma_3 & \sigma_4 & \sigma_5 & \sigma_6 & e_1 & e_2 & e_3 & e_4 & e_5 & e_6
\end{array} \tag{3}$$

and Eqs. (2a) and (2b) are rewritten as

$$e_i = s_{ij}\sigma_j \tag{4a}$$

and

$$\sigma_i = c_{ij}e_j, \tag{4b}$$

where $i, j = 1, 2, \ldots, 6$. The following relations should be noted: $s_{ijkl} = s_{mn}$, $m, n = 1, 2,$ or 3; $2s_{ijkl} = s_{nm}$, when $m = 1, 2,$ or 3 and $n = 4, 5,$ or 6, or vice versa; $4s_{ijkl} = s_{mn}$ when both m and $n = 4, 5,$ or 6. For c_{ijkl}, $c_{ijkl} = c_{mn}$ for all permissible values of the suffixes. The work dW, done when the strain components change from a state of strain given by ε_{ij} to that of $\varepsilon_{ij} + d\varepsilon_{ij}$, is given by

$$dW = \sigma_{ij}d\varepsilon_{ij} = c_{ijkl}\varepsilon_{kl}d\varepsilon_{ij}, \tag{5}$$

so that

$$c_{ijkl} = \frac{\partial^2 W}{\partial \varepsilon_{ij}\partial \varepsilon_{kl}} = \frac{\partial^2 W}{\partial \varepsilon_{kl}\partial \varepsilon_{ij}} = c_{klij}. \tag{6}$$

Thus, recalling the symmetric nature of the stress and strain tensors, together with Eqs. (6), one can conclude

$$c_{ijkl} = c_{jikl} = c_{ijlk} = c_{jilk} = c_{klij} = c_{lkij} = c_{klji} = c_{lkji}. \tag{7}$$

These equalities dictate that the number of independent elastic moduli is 21. As a consequence of symmetry, the number of independent constants reduces from 21 for a triclinic to 3 for a cubic crystal. The group-theoretical proof for this result is given in Bhagavantam (1966).

We refer the reader to Nye (1957), Bhagavantam (1966), and Musgrave (1970) for pedagogic presentations of the theory of elasticity and its application to crystalline media. Cady (1964) has a convenient collection of the arrays of c_{ij}'s or s_{ij}'s for the 32 crystallographic point groups. The semiconductors selected to illustrate the applications of piezospectroscopy belong to the tetrahedrally coordinated zinc blende III-V or II-VI compound semiconductors, such as GaAs or CdTe of T_d symmetry, or the group IV elemental semiconductors like Si or Ge with O_h symmetry. Thus, for convenience, we give the stress–strain relations applicable to the T_d and O_h point groups:

$$\begin{bmatrix} \varepsilon_{xx} \\ \varepsilon_{yy} \\ \varepsilon_{zz} \\ 2\varepsilon_{yz} \\ 2\varepsilon_{zx} \\ 2\varepsilon_{xy} \end{bmatrix} = \begin{bmatrix} s_{11} & s_{12} & s_{12} & 0 & 0 & 0 \\ s_{12} & s_{11} & s_{12} & 0 & 0 & 0 \\ s_{12} & s_{12} & s_{11} & 0 & 0 & 0 \\ 0 & 0 & 0 & s_{44} & 0 & 0 \\ 0 & 0 & 0 & 0 & s_{44} & 0 \\ 0 & 0 & 0 & 0 & 0 & s_{44} \end{bmatrix} \begin{bmatrix} \sigma_{xx} \\ \sigma_{yy} \\ \sigma_{zz} \\ \sigma_{yz} \\ \sigma_{zx} \\ \sigma_{xy} \end{bmatrix}. \tag{8}$$

Here x, y, and z are along the cubic axes.

A force **F** applied along the unit vector $\hat{\mathbf{n}}$ gives rise to a stress tensor

$$\boldsymbol{\sigma} = T\hat{\mathbf{n}}\hat{\mathbf{n}}, \tag{9}$$

where T, the force per unit area, is defined positive for tension and negative for compression. The components of $\boldsymbol{\sigma}$ with respect to a Cartesian coordinate system $\hat{x}_1, \hat{x}_2, \hat{x}_3$ are $\sigma_{ij} = Tn_i n_j$ ($n_i = \hat{\mathbf{n}} \cdot \hat{x}_i$). A hydrostatic pressure p corresponds to $\boldsymbol{\sigma} = -p\mathbf{1}$, where $\mathbf{1}$ is the unit tensor. An arbitrary stress can be regarded as a superposition of hydrostatic and shear stresses. The former corresponds to the trace of $\boldsymbol{\sigma}$. A stress tensor with zero trace can be regarded as three shear tensors originating from couples of forces in three orthogonal planes. For example, the stress defined by $-\sigma_{11} = \sigma_{22} = \sigma$, all other components being zero, is equivalent to a pure shear as can be seen by rotating the coordinate axes by 45° about $\hat{x}_3$. This transformation, defined by $\hat{n}_1 = (\hat{x}_1 + \hat{x}_2)/\sqrt{2}$ and $\hat{n}_2 = (-\hat{x}_1 + \hat{x}_2)/\sqrt{2}$, $\hat{n}_3 = \hat{x}_3$ yields $\sigma'_{ij} = \hat{n}_i \cdot \boldsymbol{\sigma} \cdot \hat{n}_j$ with $\sigma'_{12} = \sigma'_{21} = \sigma$ while all other components are zero.

2. Symmetry

In general, under an external perturbation, crystal symmetry is lowered. The fundamental principle used to deduce this lower symmetry is called the *Curie principle* (Curie, 1894; Nye, 1957), which states that only those symmetry operations are allowed that are common to both the unperturbed system and to the perturbation itself. This condition restricts the new symmetry to a subgroup common to the original group and that of the perturbation.

When a homogeneous unaxial compression is applied to a crystal, the strain tensor can be represented by a triaxial ellipsoid that has at least D_{2h} point group symmetry $[D_{2h}(2/m\ 2/m\ 2/m)]$ (see Elliot and Dawber, 1979). If two of its major axes are equal, the ellipsoid acquires rotational symmetry about the third, and the point group symmetry of the strain ellipsoid is $D_{\infty h}$, whereas if all three axes are equal it becomes a sphere characterized by the three-dimensional rotation–reflection group $O(3)$. In order to determine the symmetry operations of the strained crystal, we must know the orientation of the strain ellipsoid relative to the crystallographic axes of the unstrained crystal. In principle, this orientation can be determined for a given direction of compressive force by using the stress–strain relationship together with a principal axes transformation, which yields the magnitude and direction of the principal axes of the strain ellipsoid. The twofold axes of the D_{2h} point group symmetry are then the three principal axes. The symmetry elements common to the strain ellipsoid and the point group of the strained crystal constitute the new point group of the strained crystal (Peiser *et al.*, 1963; Wachtman and Peiser, 1965).

An alternative and more convenient procedure is to treat the stress itself as the imposed condition and find the symmetry elements common to the unstrained crystal and to the stress tensor. The equivalence of these two procedures follows from the invariance of the elastic stiffness tensor $\mathbf{c}$ under the operations of the point group of the undeformed solid. In fact, if R is such a symmetry operation, $P_R \boldsymbol{\sigma} = P_R(\mathbf{c} \cdot \boldsymbol{\varepsilon}) = (P_R \mathbf{c} P_R^{-1}) \cdot (P_R \boldsymbol{\varepsilon}) = \mathbf{c} \cdot (P_R \boldsymbol{\varepsilon})$. Thus, if R leaves $\boldsymbol{\varepsilon}$ invariant, it also leaves $\boldsymbol{\sigma}$ invariant.

TABLE I

REDUCTION OF T_d SYMMETRY UNDER UNIAXIAL STRESS

Direct of applied force, F	Surviving symmetry operations	Reduced symmetry
$F \parallel [100]$	$2S_4 \parallel [100]$; $C_2 \parallel [100]$; $2C'_2 \parallel [010], [001]$;	D_{2d} (Tetragonal)
$F \parallel [110]$	$\sigma_d \parallel (110), (1\bar{1}0)$; $C_2 \parallel [001]$	C_{2v} (Orthorhombic)
$F \parallel [111]$	$C_3 \parallel [111]$; $\sigma_d \parallel (1\bar{1}0), (01\bar{1})$	C_{3v} (Trigonal)

An alternative procedure, which is particularly simple when the external perturbation is a uniaxial compression, consists of using the symmetry properties of the stress tensor. In this case the stress ellipsoid has $D_{\infty h}$ point group symmetry and can be conveniently represented by a right circular cylinder with its center coinciding with the center of the crystal and its axis of revolution along the direction of the force. The symmetry operations common to the unstrained crystal and to the cylinder representing the stress can then be easily determined by inspection.

As an illustrative example, we consider the point group T_d, the site symmetry of substitutional donors and acceptors in Si and Ge. The symmetry operations of T_d are E (identity), eight C_3 (threefold rotation along $<111>$), three C_2 (twofold rotation about x, y, or z, the three coordinate axes parallel to [100], [010], and [001], respectively), six S_4 (fourfold rotation–reflection about x, y, or z), and six σ_d (diagonal reflection planes $\{110\}$). In Table I the reduction of T_d symmetry under uniaxial stress $F \parallel [100]$, [110] or [111] is displayed.

Once the new, reduced symmetry of the crystal in the presence of the external perturbation is determined, the correlation between the irreducible representations of the two groups can be obtained. From such a correlation, the removal of the degeneracy of a particular energy level can be readily deduced The polarization characteristics of the stress-induced components of an optical transition (in absorption, emission, or inelastic light scattering) can also be deduced directly since the energy levels involved in the transition must conform to the irreducible representations of the new symmetry group. Polarization studies in absorption (emission) and in Raman scattering thus are useful in determining unambiguously the irreducible representations characterizing the initial and final states of the stress-induced components.

3. LATENT ANISOTROPY

Consider a cubic crystal like $NaClO_3$, which is optically isotropic. Each individual ClO_3^- ion has C_{3v} symmetry and is clearly anisotropic; the overall isotropy demanded by the cubic symmetry is ensured by the "orientational degeneracy" of the four ClO_3^- ions per unit cell with their C_3 axes along the four body diagonals of the cube. The trigonal symmetry and the "uniaxial" anisotropy of the ClO_3^- ion thus remain "latent." Consider the lowest conduction-band (C.B.) minima of silicon, i.e., the "absolute" conduction-band minima. They lie along $\mathbf{k} \parallel \langle 100 \rangle$ about 80% of the magnitude of the distance in reciprocal space from the center of the Brillouin zone to its boundary; these are the so-called Δ minima with C_{4v} symmetry. (For labeling band extrema of the electronic structure of a semiconductor we use the notation in Bouckaert et al., 1936 with modifications in Koster et al., 1966. A convenient table correlating the different notations in the literature appears in Table X of Kane, 1969.) However, optical isotropy is ensured by the energy equivalence of the C.B. minima along the six cubic directions, so the anisotropy of a single (or individual) C.B. minimum remains latent. Anisotropic local centers may be produced in cubic crystals whose anisotropy remains latent because of their orientational degeneracy, ensured by the overall random distribution of such centers in the body of the crystal, e.g., substitutional Ni in Ge, which displaces along $\langle 100 \rangle$ as a result of a Jahn–Teller distortion (Ludwig and Woodbury, 1959).

The orientational degeneracy can be deduced in a straightforward fashion (Feofilov and Kaplyanskii, 1962; Kaplyanskii, 1964b). The point group of the constituent, e.g., the site symmetry of ClO_3^- in the $NaClO_3$ crystal, is a subgroup of the crystallographic point group. The orientational degeneracy (R) is equal to the ratio of the orders of the symmetry group of the crystal (G) and that of the site symmetry (g). For a cubic crystal of symmetry O_h and site symmetry D_{4h}, C_{4v}, C_4, C_3, and C_2, $R = 3, 6, 12, 16$, and 24, respectively.

Under ordinary conditions all possible positions of defects are populated with equal probability. In a crystal the centers form R differently oriented groups with statistically the same number of centers in each group. Similar remarks apply to multivalley semiconductors and crystals that have anisotropic constituents.

The latent anisotropy can be exposed by the application of external fields to the crystals. The external field may affect the different subsets and the associated energy levels differently. For example, a stress with compressive force along a specific cubic direction, say [100], will affect the [100], [$\bar{1}$00] conduction-band minima of Si differently from those that are along [010], [0$\bar{1}$0], [001], and [00$\bar{1}$]. Piezospectroscopic effects associated with the removal of latent anisotropy are of considerable physical interest (Kaplyanskii, 1964b).

4. DEFORMATION POTENTIAL THEORY

While the symmetry arguments outlined provide qualitative predictions, they do not quantitatively characterize the shifts and splittings in terms of parameters that can be experimentally determined nor do they correlate the positions and relative intensities of the stress induced components. A deeper insight can be obtained by the use of perturbation theory following Bardeen and Shockley (1950) who developed a phenomenological theory in connection with mobilities of charge carriers in nonpolar crystals (semiconductors and insulators). In order to calculate the mean-free path of carriers, they postulated that shifts in energy bands of the crystal are produced by thermal vibrations; these shifts "may be used as varying potentials in calculating the behaviors of electrons and holes" and the potentials are referred to as *deformation potentials* (see also Rodriguez and Kartheuser, 1985).

Let $\psi_k(\Gamma_\alpha)$ denote the energy states of a crystal. Here α labels the irreducible representations of the group of the crystal and k designates the different rows of Γ_α. The splittings and shifts of a given energy level for an arbitrary applied stress are obtained by solving the eigenvalue problem for the crystal in the presence of the perturbing potential D. To terms linear in strain, D is given by

$$D = \sum_{ij} D_{ij}\varepsilon_{ij}, \qquad i,j = x, y, z, \qquad (10)$$

where the ε_{ij}'s are components of the strain tensor and the D_{ij}'s are the deformation potentials. Both $\{D_{ij}\}$ and $\{\varepsilon_{ij}\}$ are symmetric second-rank tensors. We assume that the first-order perturbation theory is sufficient, and this assumption as well as the linearity of Eq. (10) with respect to strain is to be justified in the context of experimental results. The diagonalization of the matrix whose elements are $\langle \psi_{k'}(\Gamma_\alpha)|D|\psi_k(\Gamma_\alpha)\rangle$, with $k', k = 1, \ldots, l_\alpha$ yields the splittings and shifts of the energy level belonging to Γ_α of dimension l_α. The number of independent elements of this matrix is equal to the number of times the totally symmetric representation appears in the reduction of $[\Gamma_\alpha \times \Gamma_\alpha] \times \Gamma_D$, where $[\Gamma_\alpha \times \Gamma_\alpha]$ is the symmetric direct product of the reduction Γ_α with itself and Γ_D is the representation generated by $\{D_{ij}\}$. When latent anisotropy is present, this approach is applied to each subset separately, the net result being a superposition of the shifts and splittings of the energy levels associated with each subset. The special aspects of the deformation potential theory as they apply to the specific phenomena (for example, electronic excitations, band to band or localized; local modes or long-wavelength (zone-center) lattice vibrations, etc.) will be presented in the appropriate sections.

III. Experimental Techniques

In the literature there appear several descriptions of apparatus used for subjecting samples to uniaxial stress. Since samples are often cooled to sufficiently low temperatures (down to the boiling point of liquid nitrogen or liquid helium), a versatile uniaxial stress apparatus is incorporated in a suitable optical cryostat used in typical spectroscopic measurements. Cuevas and Fritzsche (1965) designed a stress frame, piston, pull rod, and lever arrangement that was adapted by Feldman (1966), Balslev (1966), Bhargava and Nathan (1967), and Pollak and Cardona (1968) for spectroscopic measurements. (See, for example, Fig. 1 in Feldman's paper.) A calibrated spring and a linear variable differential transformer were employed for the application of a known stress. Stresses up to 10^{10} dyne/cm^2 could be achieved under experimental conditions. Schawlow et al. (1961) and Thomas (1961) describe simple arrangements involving a loading platform and a push rod, the entire arrangement being immersed in liquid nitrogen or cooled helium gas. The design employed by Seiler and Addington (1972) involves a rotating cam to transmit the stress to the sample and allows the force to be applied in different directions with respect to a specific external field, e.g., a magnetic field. Vogelmann and Fjeldly (1974) have designed a uniaxial stress apparatus with analog pressure readout. In the remainder of this section we give a description of a glass cryostat (Fig. 1) and the uniaxial stress insert used for a number of years in our laboratory (Fisher et al., 1963; Tekippe et al., 1972).

As can be seen from this diagram the cryostat consists of two parts. One unit consists of the liquid coolant reservoir and sample block and will be called the centerpiece. The remainder we shall call the outer piece. The sample is clamped over one of the two identical apertures in a copper block. Either the sample or the reference aperture can be interposed in the light beam by rotating the centerpiece at the 29/42 joint. In Fig. 1, the top 35/20 ball can be used to connect the volume containing the coolant to a high-speed vacuum pump so that one can reduce the vapor pressure of the gas above the liquid and thus lower the bath temperature and still allow rotation about 29/42. The outer piece consists of three coaxial tubes. The outermost tube forms the vacuum wall of the cryostat; the inner two constitute the liquid nitrogen radiation shield and are joined to the outer tube at the top in two ring seals and thence to the 71/60 cone. Thus, the liquid nitrogen is open to the atmosphere only at the 24/40 cone, which is used as the filling port. The inner and outer vacuum spaces are interconnected via the apertures through the two walls of the nitrogen jacket. The sizes of these apertures are chosen to just accommodate the light beam and so minimize the undesired room temperature radiation reaching the sample. Since the two walls of the nitrogen shield are rigidly connected at the top and the bottom, bellows are provided to permit independent expansion and contraction of these two tubes. The optical windows (alkali halide, sapphire, CaF_2, crystalline quartz, or Mylar) are mounted onto brass plates, which in turn are bolted to brass bosses waxed to the outside surface of the outer piece.

When uniaxial stress experiments are carried out, a stainless steel centerpiece, shown in Fig. 2, is inserted in place of the glass centerpiece. The

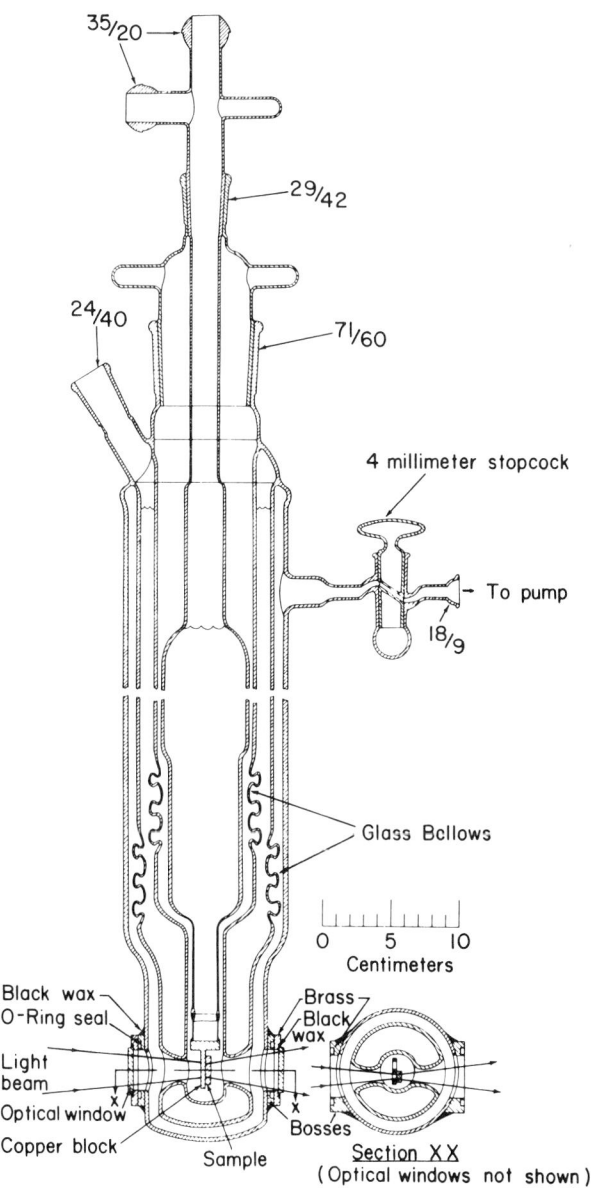

FIG. 1. Glass optical cryostat with glass centerpiece. (After Fisher et al., 1963.)

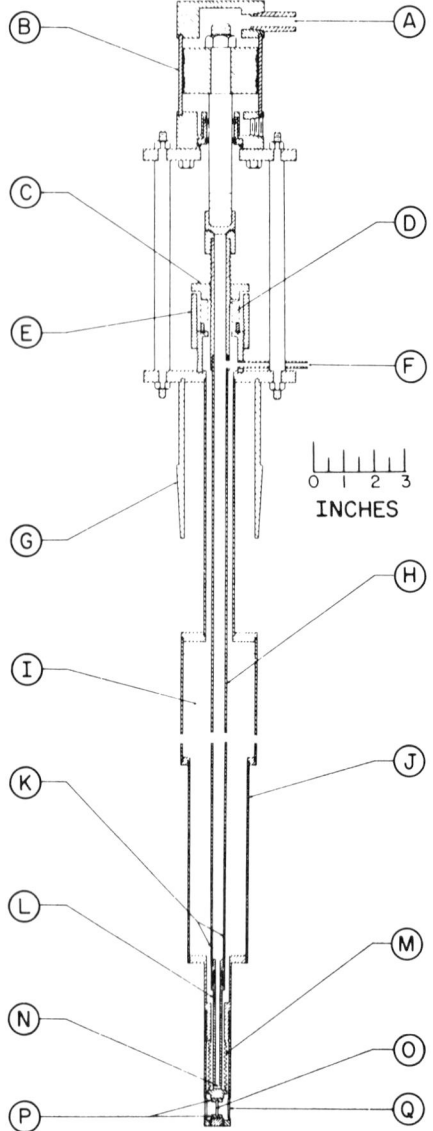

FIG. 2. Stainless steel-stress centerpiece. This replaces the glass centerpiece shown in Fig. 1: A—gas pressure inlet; B—pressure head; C—height adjustment collar; D,M—bellows; E—safety collar; F—coolant outlet; G—stainless steel cone joint; H—hollow push rod; I—coolant chamber; J—walls of the coolant reservoir; K,N—small holes allowing the coolant to flow into I from H; L—copper portion of the push rod; O—sample; P—copper cups; Q—copper tailpiece. (After Tekippe et al. 1972.)

centerpiece is mated to the optical cryostat by means of the stainless steel cone joint G. The pressure head B is pressurized with nitrogen gas through the port A, and the resultant force produced by the piston is transmitted by a hollow push rod H that passes through the coolant I and makes contact with the sample O. The sample is compressed against the bottom of a hollow copper cylinder Q that is threaded to the coolant reservoir J.

The cryostat is designed for either transmission or photoconductivity measurements. For luminescence or light scattering experiments a third window whose axis is at right angles with respect to that of the other two is provided. The design features described allow a ready adaption into an immersion cryostat. The pressure insert can be modified to fit the commercial stainless cryostats in which either liquid helium or the evaporating helium vapor cools the sample and, along with a heater, enables measurements from, say, 1.8 to 300 K.

As remarked in the introduction, a variety of *modulation techniques* have been exploited in the study of optical phenomena exhibited by semiconductors. These have focussed on the free and bound excitonic features near the absorption edge; onsets in phonon-assisted "indirect" transitions; maxima or discontinuities in the reflectivity spectra at the onset of direct and other interband transitions between states with large joint density of states. The modulation is achieved by an alternating applied perturbation (e.g., stress or electric field), which in turn modulates the optical constants (refractive index n and extinction coefficients k, or both) associated with specific transitions in an especially pronounced manner. The modulated signal—say in reflectivity—studied as a function of photon energy and detected with a lock-in amplifier produces a sharp signature at the position of one of the optical transitions listed while ignoring the generally featureless background; i.e., instead of weak maxima, steps, or discontinuities superimposed on a strong background, one observes easily recognized signatures—with a "derivative" character typical of the perturbation technique used. In addition to these modulation techniques that exploit the modulation of a physical property, there exists yet another that is an "external modulation": the wavelength of the incident radiation from a monochromator is oscillated by some device, in turn producing in the optical spectrum "derivative" signatures of the "onset", "step", or the "discontinuity," associated with one or the other optical process already mentioned. We refer the readers to the treatise by Cardona (1969) and to the review article by Aggarwal (1972), where modulation techniques have been thoroughly discussed. Piezospectroscopy in combination with one or the other modulation technique has been applied to the study of excitons, direct and indirect transitions and features associated with the spin-orbit split valence band. For example, uniaxial stress in combination with piezomodulation has been employed by Balslev (1967a,b); in combination with electromodulation by Pollak and Cardona (1968), Laude *et al.*

(1970), and Chandrasekhar and Pollak (1977); and in combination with wavelength modulation by Balslev (1966), Laude et al. (1970, 1971), Pollak and Aggarwal (1971), and Mathieu et al. (1979).

IV. Valence and Conduction Bands of Elemental and Compound Semiconductors under External Stress

Many significant optical processes in elemental and compound semiconductors are associated with electronic transitions between the valence-band maximum and the conduction-band minimum: direct and indirect transitions; free and bound excitons; interband and excitonic recombination radiation; shallow impurity states with effective-mass characteristics that are associated with the conduction-band minimum or minima (donors) or the valence-band maximum (acceptors).

In order to visualize conveniently the specific optical transitions in the context of the nomenclature used to designate band extrema, we show in Fig. 3 part of the band structure of Ge adapted from Cohen and Chelikowsky (1988). The "direct" transitions in Ge originate at the Γ_8^+ valence-band maximum and terminate at the Γ_7^- conduction-band minimum, both extrema

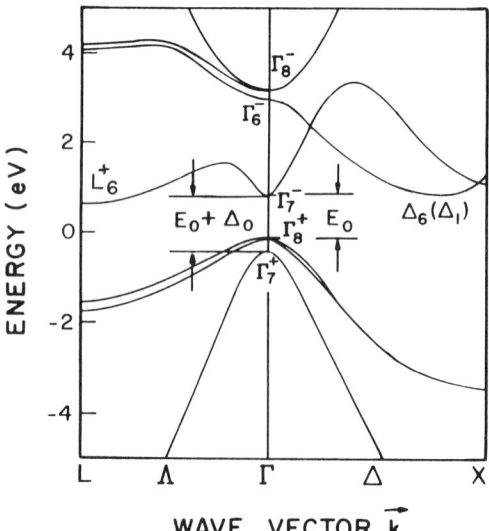

FIG. 3. Electronic band structure of Ge. E_0 and $E_0 + \Delta_0$ correspond to the direct transitions $\Gamma_8^+ \to \Gamma_7^-$ and $\Gamma_7^+ \to \Gamma_7^-$, respectively; $\Gamma_8^+ \to L_6^+$ is the indirect transition. In Si the absolute conduction-band minimum occurs along Δ at the point labeled Δ_6, Δ_1 being the designation ignoring spin. In addition, the lowest conduction-band minimum of Si at the Γ-point is Γ_6^-. (Adapted from Cohen and Chelikowsky, 1988.)

being at the center of the Brillouin zone; these are labeled E_0 transitions. "Indirect" transitions, assisted by the absorption or the emission of phonons, originate at Γ_8^+ and terminate at L_6^+ conduction-band minima at the zone boundary along $\langle 111 \rangle$. For Si, the conduction-band minimum at the zone center is Γ_6^-, and the direct transitions correspond to $\Gamma_8^+ \to \Gamma_6^-$, whereas six Δ_1 minima along $\langle 100 \rangle$ constitute the lowest conduction-band extrema, and correspondingly $\Gamma_8^+ \to \Delta_1$ are the indirect transitions. In Si as well as in Ge, the Γ_7^+ valence-band maximum is split off from Γ_8^+ by Δ_0 as a consequence of spin-orbit interaction; the corresponding direct transitions to the zone-center conduction-band minimum, i.e., to Γ_7^- (Ge) and Γ_6^- (Si), are labeled $E_0 + \Delta_0$ transitions. (In the III-V and II-VI semiconductors, the absence of a center of inversion causes features in the valence-band maximum such as terms linear in k and could, in principle, lead to differences in experimental results as compared to those in Si and Ge.)

5. Deformation Potentials of Band Extrema

We first consider the energy shifts of orbitally nondegenerate states of tetrahedrally coordinated elemental and compound semiconductors as a function of stress. We recall that, by virtue of Kramers' theorem (Kramers, 1930), the Bloch states of electrons in a crystal always possess an even degeneracy. The main consequence of this theorem is that a state at wave vector **k** is degenerate with one at $-\mathbf{k}$, their spin states being orthogonal. However, if the crystal has a center of inversion, there is a spin degeneracy for each value of **k**. Thus, in Si and Ge the Bloch states have twofold degeneracy at each point in the Brillouin zone: note that at the zone center (Γ-point) this degeneracy can be two- and fourfold. Application of stress cannot remove the two-fold degeneracy but will, in general, split the quadruplet levels into two doublets.

We now discuss the energy shifts of the electronic levels at the extrema of the conduction band of the elemental group IV semiconductors. The absolute minima of the conduction band occur along Δ in Si and at the L-point in Ge. There are local minima at Γ whose classification is Γ_6^- in Si and Γ_7^- in Ge. In both cases we rewrite the deformation potential in the form

$$\sum_{ij} D_{ij}\varepsilon_{ij} = \frac{1}{3}\left(\sum_i D_{ii}\right)\left(\sum_j \varepsilon_{jj}\right)$$
$$+ \frac{1}{6}[(2D_{zz} - D_{xx} - D_{yy})(2\varepsilon_{zz} - \varepsilon_{xx} - \varepsilon_{yy}) + 3(D_{xx} - D_{yy})(\varepsilon_{xx} - \varepsilon_{yy})]$$
$$+ 2\sum_{i<j} D_{ij}\varepsilon_{ij}, \qquad (11)$$

where $i, j = x, y, z$ denote the cubic axes. The quantity $\sum_i D_{ii}$ belongs to the totally symmetric representation Γ_1^+ of O_h, $2D_{zz} - D_{xx} - D_{yy}$ and $\sqrt{3}(D_{xx} - D_{yy})$ generate Γ_3^+, and D_{yz}, D_{zx}, D_{xy} generate Γ_5^+. Thus, as expected, the Γ_6^- and Γ_7^- levels shift by

$$\Delta E(\Gamma_{6,7}) = \frac{1}{3}\left\langle \Gamma_{6,7}\left|\sum_i D_{ii}\right|\Gamma_{6,7}\right\rangle \sum_j \varepsilon_{jj}, \tag{12}$$

i.e., proportional to the dilatation.

The conduction-band edge of Si is a Δ_1 state. The group of Δ, often designated by this symbol, is C_{4v}, and the decomposition of the deformation potential operator appropriate for this case is

$$\sum_{ij} D_{ij}\varepsilon_{ij} = \frac{1}{2}(D_{xx} + D_{yy})(\varepsilon_{xx} + \varepsilon_{yy}) + D_{zz}\varepsilon_{zz} + \frac{1}{2}(D_{xx} - D_{yy})(\varepsilon_{xx} - \varepsilon_{yy})$$
$$+ 2D_{yz}\varepsilon_{yz} + 2D_{zx}\varepsilon_{zx} + 2D_{xy}\varepsilon_{xy}. \tag{13}$$

Here $D_{xx} + D_{yy}$ and D_{zz} belong to Δ_1, D_{xy} to Δ_2, $D_{xx} - D_{yy}$ to Δ_2', and D_{yz} and D_{zx} to Δ_5. (Note that $\Delta_1, \Delta_1', \Delta_2, \Delta_2'$, and Δ_5 correspond to $\Gamma_1, \Gamma_2, \Gamma_4, \Gamma_3$, and Γ_5, respectively in the notation of Koster et al., 1968.) We immediately conclude that the energy shift of a specific Δ_1 state is given by

$$\Delta E = \left\langle \Delta_1 \left|\sum_{ij} D_{ij}\varepsilon_{ij}\right|\Delta_1\right\rangle = \Xi_d(\varepsilon_{xx} + \varepsilon_{yy}) + (\Xi_d + \Xi_u)\varepsilon_{zz}, \tag{14}$$

where

$$\Xi_d = \frac{1}{2}\langle\Delta_1|D_{xx} + D_{yy}|\Delta_1\rangle \tag{15}$$

and

$$\Xi_u = \frac{1}{2}\langle\Delta_1|2D_{zz} - D_{xx} - D_{yy}|\Delta_1\rangle \tag{16}$$

are called the dilatational and shear deformation potential constants, respectively.

A similar treatment can be given for the L_6^+ minimum of the lowest conduction band of germanium. Here, the deformation potential is best expressed in terms of invariants in which the $\hat{\zeta}$ axis is the axis of highest rotational symmetry of the group of L, namely [111]. It is sometimes convenient to select $\hat{\xi}$ and $\hat{\eta}$ parallel to $[11\bar{2}]$ and $[\bar{1}10]$, respectively, as we do here. Then

$$\sum_{ij} D_{ij}\varepsilon_{ij} = \frac{1}{2}(D_{\xi\xi} + D_{\eta\eta})(\varepsilon_{\xi\xi} + \varepsilon_{\eta\eta}) + D_{\zeta\zeta}\varepsilon_{\zeta\zeta}$$

$$+ \frac{1}{2}(D_{\xi\xi} - D_{\eta\eta})(\varepsilon_{\xi\xi} - \varepsilon_{\eta\eta}) + 2D_{\xi\eta}\varepsilon_{\xi\eta}$$

$$+ 2D_{\zeta\xi}\varepsilon_{\zeta\xi} + 2D_{\zeta\eta}\varepsilon_{\zeta\eta}. \quad (17)$$

Now $D_{\xi\xi} + D_{\eta\eta}$ and $D_{\zeta\zeta}$ belong to L_1, and the pairs $(D_{\xi\xi} - D_{\eta\eta})$, $D_{\xi\eta}$ and $(D_{\zeta\xi}, D_{\zeta\eta})$ generate L_3. Using the orthogonality theorem of the theory of group representations and time reversal symmetry, we find the energy shift of a given L_6^+-valley

$$\Delta E = \left\langle L_6^+ \left| \sum_{ij} D_{ij}\varepsilon_{ij} \right| L_6^+ \right\rangle = \Xi_d(\varepsilon_{\xi\xi} + \varepsilon_{\eta\eta}) + (\Xi_d + \Xi_u)\varepsilon_{\zeta\zeta}$$

$$= \left(\Xi_d + \frac{1}{3}\Xi_u\right)(\varepsilon_{xx} + \varepsilon_{yy} + \varepsilon_{zz}) + \frac{2}{3}\Xi_u(\varepsilon_{yz} + \varepsilon_{zx} + \varepsilon_{xy}). \quad (18)$$

Here

$$\Xi_d = \frac{1}{2}\langle L_6^+ | D_{\xi\xi} + D_{\eta\eta} | L_6^+ \rangle \quad (19)$$

and

$$\Xi_u = \frac{1}{2}\langle L_6^+ | 2D_{\zeta\zeta} - D_{\xi\xi} - D_{\eta\eta} | L_6^+ \rangle. \quad (20)$$

Equations (14) and (18) can be represented compactly as

$$\Delta E(\mathbf{k}^{(n)}) = \sum_{ij}(\Xi_d \delta_{ij} + \Xi_u \hat{k}_i^{(n)}\hat{k}_j^{(n)})\varepsilon_{ij}, \quad (21)$$

where $\hat{k}^{(n)}$ is a unit vector parallel to $\mathbf{k}^{(n)}$, the position of the nth conduction-band minimum in $\mathbf{k}$-space.

In order to compare experimental results with theory it is useful to note that the shift of the "center of gravity" of the valleys is given by

$$\langle \Delta E^{(n)} \rangle = \left(\Xi_d + \frac{1}{3}\Xi_u \right)(\varepsilon_{xx} + \varepsilon_{yy} + \varepsilon_{zz}). \tag{22}$$

The shift of a given valley n with respect to the center of gravity is then given by

$$\delta E^{(n)} = \pm \Xi_u T(s_{11} - s_{12})\left[(\hat{k}^{(n)} \cdot \hat{F})^2 - \frac{1}{3} \right] \tag{23}$$

for Si. Similarly for Ge,

$$\delta E^{(n)} = \pm \frac{1}{2}\Xi_u T s_{44}\left[(\hat{k}^{(n)} \cdot \hat{F})^2 - \frac{1}{3} \right]. \tag{24}$$

Here + and − signs correspond to tension and compression, respectively. Labeling the six Δ_1 valleys in Si as $1,\ldots,6$, along [100], [$\bar{1}$00], [010], [0$\bar{1}$0], [001], and [00$\bar{1}$], respectively, for a general direction of **F** (compressive force) with θ and ϕ defined in Fig. 4, Eq. (23) leads to

$$\delta E^{(1,2)} = -T\Xi_u(s_{11} - s_{12})\left(\cos^2\theta \sin^2\phi - \frac{1}{3} \right),$$

$$\delta E^{(3,4)} = -T\Xi_u(s_{11} - s_{12})\left(\sin^2\theta \sin^2\phi - \frac{1}{3} \right), \tag{25}$$

$$\delta E^{(5,6)} = -T\Xi_u(s_{11} - s_{12})\left(\cos^2\phi - \frac{1}{3} \right).$$

Figure 5 depicts $\delta E^{(n)}$ as a function of θ and ϕ for a given compressive force as given by Eqs. (25) (Tekippe et al., 1972). Similarly, labeling the four L_6^+-valleys of Ge along [111], [1$\bar{1}\bar{1}$], [$\bar{1}\bar{1}$1], and [$\bar{1}$1$\bar{1}$], respectively, as 1, 2, 3, and 4, Eq. (24) leads to

$$\delta E^{(1)} = -A - B - C,$$

$$\delta E^{(2)} = A - B + C, \tag{26}$$

$$\delta E^{(3)} = -A + B + C,$$

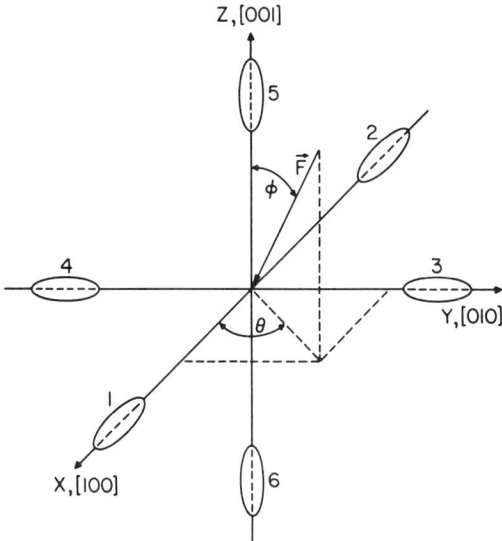

FIG. 4. Constant energy ellipsoids of the Δ_1 conduction-band minima of Si along $\langle 100 \rangle$. Also shown, for convenience, are the coordinate axes, the direction of the applied compressive force **F** and its polar coordinates. (After Tekippe et al., 1972.)

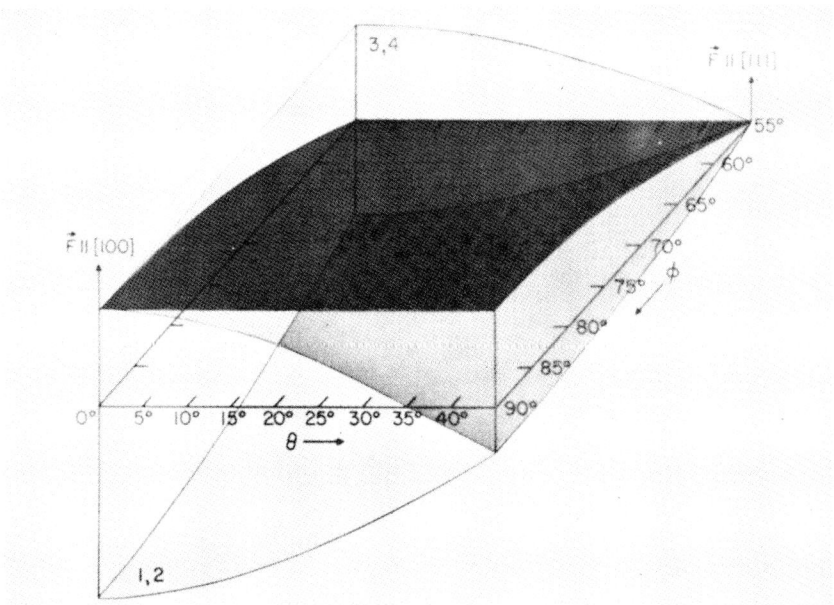

FIG. 5. Energy shifts of the Δ_1 conduction-band valleys of Si from the center of gravity as a function of θ and ϕ defined in Fig. 4 [Eq. (25)]. The horizontal plane corresponds to the center of gravity of the valleys. The energy sheet labeled 1,2 corresponds to valleys along [100] and [$\bar{1}$00], the sheet labeled 3,4 to those along [010] and [0$\bar{1}$0], and the third sheet corresponds to valleys along [001] and [00$\bar{1}$]. (After Tekippe et al., 1972.)

and

$$\delta E^{(4)} = A + B - C.$$

where

$$A = \frac{1}{3} \Xi_u T s_{44} \sin\theta \cos\phi \sin^2\phi,$$

$$B = \frac{1}{3} \Xi_u T s_{44} \sin\theta \cos\phi \sin\phi,$$

$$C = \frac{1}{3} \Xi_u T s_{44} \cos\theta \sin\phi \cos\phi.$$

Here θ and ϕ are defined as in Fig. 4. Figure 6 displays $\delta E^{(n)}$ as a function of θ and ϕ as given by Eqs. (26) (Tekippe, 1973; Martin et al., 1981).

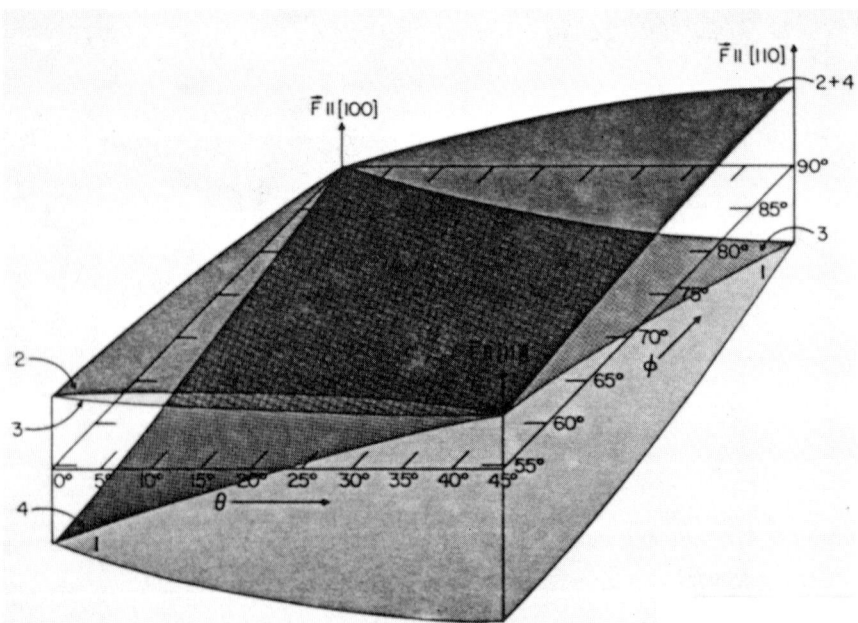

FIG. 6. Energy shifts of the L_6^+ conduction-band valleys of Ge from the center of gravity as a function of θ and ϕ, defined in Fig. 4, for a given compressive force [Eq. (26)]. Energy sheets labeled 1, 2, 3, and 4 correspond to the valleys along [111], [1$\bar{1}\bar{1}$], [$\bar{1}\bar{1}$1], [$\bar{1}$1$\bar{1}$], respectively. (After Tekippe, 1973; Martin et al., 1981.)

As an example in which the levels are orbitally degenerate, we consider here the deformation potential theory for the Γ_5^+ (Γ_{25}') states at the top of the valence band of elemental semiconductors having the diamond structure. We disregard the spin of the electron for the time being and designate the three degenerate states associated with the top of the valence band by the symbols $\varepsilon_1, \varepsilon_2, \varepsilon_3$, which transform under the operations of O_h as yz, zx, and xy, where x, y, z are the components of a vector referred to the cubic axes. The changes in energy upon deformation are obtained by diagonalizing the 3×3 matrix

$$\sum_{ij} \varepsilon_{ij} \langle \varepsilon_\mu | D_{ij} | \varepsilon_\nu \rangle,$$

$\mu, \nu = 1, 2, 3$. We can rewrite the quantity $\sum_{ij} \varepsilon_{ij} D_{ij}$ as a linear combination of

$$(\varepsilon_{xx} + \varepsilon_{yy} + \varepsilon_{zz})(D_{xx} + D_{yy} + D_{zz}),$$

$$(2\varepsilon_{zz} - \varepsilon_{xx} - \varepsilon_{yy})(2D_{zz} - D_{xx} - D_{yy}) + 3(\varepsilon_{xx} - \varepsilon_{yy})(D_{xx} - D_{yy}),$$

and

$$\varepsilon_{yz} D_{yz} + \varepsilon_{zx} D_{zx} + \varepsilon_{xy} D_{xy}. \tag{27}$$

We recall that $D_{xx} + D_{yy} + D_{zz}$ belongs to the totally symmetric representation Γ_1^+ of the cubic group O_h, that $2D_{zz} - D_{xx} - D_{yy}$ and $\sqrt{3}(D_{xx} - D_{yy})$ belong to Γ_3^+ (Γ_{12}) and D_{yz}, D_{zx}, D_{xy} to Γ_5^+ (Γ_{25}'). $D_{xx} + D_{yy} + D_{zz}$ has only diagonal matrix elements in the basis $\{\varepsilon_\mu\}$ ($\mu = 1, 2, 3$), and all its elements are equal. The matrices associated with $2D_{zz} - D_{xx} - D_{yy}$ and $\sqrt{3}(D_{xx} - D_{yy})$ are proportional to $2I_z^2 - I_x^2 - I_y^2$ and $\sqrt{3}(I_x^2 - I_y^2)$, respectively, where

$$I_x = \begin{pmatrix} 0 & 0 & 0 \\ 0 & 0 & -i \\ 0 & i & 0 \end{pmatrix}, \quad I_y = \begin{pmatrix} 0 & 0 & i \\ 0 & 0 & 0 \\ -i & 0 & 0 \end{pmatrix}, \quad I_z = \begin{pmatrix} 0 & -i & 0 \\ i & 0 & 0 \\ 0 & 0 & 0 \end{pmatrix}$$

are the components of the angular momentum operator for $I = 1$ along the cubic axes. In the same way the matrix elements of D_{yz}, D_{zx}, and D_{xy} in the same basis are proportional to $\{I_y, I_z\} = I_y I_z + I_z I_y$, $\{I_z, I_x\}$, and $\{I_x, I_y\}$, respectively; thus, the operator $\sum_{ij} \varepsilon_{ij} D_{ij}$ in the basis generating the Γ_5^+ (Γ_{25}') states has the form of the matrix

$$-a \sum_i \varepsilon_{ii} - 3b \sum_i \varepsilon_{ii} \left(I_i^2 - \frac{1}{3} I^2 \right) - \sqrt{3} d \sum_{i<j} \varepsilon_{ij} \{I_i, I_j\}, \tag{28}$$

where a, b, and c are deformation potential constants; a is referred to as the hydrostatic deformation potential constant, whereas b and c are the shear deformation potential constants that appear for force **F** along [001] and [111], respectively.

We consider now the effect of the spin-orbit interaction. The deformation potentials D_{ij} depend, of course, on the spin of the electron. However, in most circumstances this correction is negligible, so we can take as the deformation potential contribution to the effective second-order Hamiltonian the matrix (28) multiplied (in the sense of the direct product of matrices) by the unit 2×2 matrix. However, the spin-orbit coupling splits the sixfold degeneracy of the $\varepsilon_\mu \alpha, \varepsilon_\mu \beta$ states ($\mu = 1, 2, 3$) into a Γ_8^+ quadruplet and a Γ_7^+ doublet (α and β designate the two orthogonal spin states, taken, to fix the ideas, so that σ_z is diagonal, $\hat{z}$ being one of the cubic axes, and $\boldsymbol{\sigma}$ is the Pauli spin matrix).

The matrix elements of the spin-orbit interaction in the representation $\varepsilon_\mu \chi_\pm$ ($\mu = 1, 2, 3$), where $\chi_+ = \alpha$ and $\chi_- = \beta$, are

$$\frac{\hbar}{4m^2c^2} \langle \varepsilon_{\mu'} | \nabla U \times \mathbf{p} | \varepsilon_\mu \rangle \cdot \langle \chi_{\sigma'} | \boldsymbol{\sigma} | \chi_\sigma \rangle,$$

where $U(\mathbf{x})$ is the crystal potential. Now the components of the pseudovector $\nabla U \times \mathbf{p}$ generate the Γ_4^+ (Γ_{15}') representation of O_h, so its matrix representation is simply a constant times **I**. Thus, we write the spin-orbit interaction in the form

$$\frac{1}{3} \Delta \mathbf{I} \cdot \boldsymbol{\sigma}.$$

The Hamiltonian matrix for the six $\Gamma_5^+ \times \Gamma_6^+$ states for **k** near the center of the Brillouin zone and in the presence of strain in the diamond structure is of the form

$$H = Ak^2 - a \sum_i \varepsilon_{ii} - 3 \sum_i (Bk_i^2 + b\varepsilon_{ii})\left(I_i^2 - \frac{1}{3}I^2\right)$$

$$- \sqrt{3} \sum_{i<j} (Dk_ik_j + d\varepsilon_{ij})\{I_i, I_j\} + \frac{1}{3} \Delta \mathbf{I} \cdot \boldsymbol{\sigma}. \tag{29}$$

The calculation of the eigenvalues of H is simplified if we notice that H is invariant under the time-reversal operator, $\mathbf{I} \to -\mathbf{I}$, and $\boldsymbol{\sigma} \to -\boldsymbol{\sigma}$. Thus, the eigenvalues of H in Eq. (29) are doubly degenerate (Kramers, 1930).

The eigenvalues of this matrix to first order in ε and second order in $\mathbf{k}$ are

$$-\frac{2}{3}\Delta + Ak^2 - a\sum_i \varepsilon_{ii} \tag{30}$$

and

$$E_\pm(\mathbf{k}, \varepsilon) = Ak^2 - a\sum_i \varepsilon_{ii} + \frac{1}{3}\Delta \pm (R_k + R_{k\varepsilon} + R_\varepsilon)^{1/2} \tag{31}$$

with

$$R_k = B^2 k^4 + (D^2 - 3B^2)\sum_{i<j} k_i^2 k_j^2, \tag{32}$$

$$R_{k\varepsilon} = -Bbk^2 \sum_i \varepsilon_{ii} + 3Bb\sum_i k_i^2 \varepsilon_{ii} + 2Dd\sum_{i<j} k_i k_j \varepsilon_{ij}, \tag{33}$$

and

$$R_\varepsilon = b^2 \sum_i \varepsilon_{ii}^2 + \sum_{i<j}(d^2 \varepsilon_{ij}^2 - b^2 \varepsilon_{ii}\varepsilon_{jj}). \tag{34}$$

In the elemental semiconductors Si and Ge, the spin-orbit splitting of the $\Gamma'_{25} \times \Gamma_6$ valence band is positive so that, in the absence of stress, the top of the valence band is a quadruplet Γ_8^+. The split-off band Γ_7^+ lies at an energy Δ_0 below the fourfold level. In the presence of stress the Γ_7^+ shifts by an amount proportional to the dilatation. A similar shift occurs for the Γ_8^+ level, but, in addition, it splits into two doublets under shear stress. If the energy separation between Γ_8^+ and Γ_7^+, i.e., Δ_0 is large compared to the strain energy, the deformation potential operator of the Γ_8^+ level is

$$D = -a\sum_i \varepsilon_{ii} - b\sum_i \varepsilon_{ii}\left(J_i^2 - \frac{1}{3}J^2\right) \frac{d}{\sqrt{3}} \sum_{i<j} \varepsilon_{ij}\{J_i, J_j\}. \tag{35}$$

(This approximation is more accurate for Ge than for Si.) Simple examples for uniaxial stress along three high-symmetry directions are given here. The components of the strain tensor for $\mathbf{F} \parallel [001]$ are

$$\varepsilon_{xx} = \varepsilon_{yy} = s_{12}T, \quad \varepsilon_{zz} = s_{11}T, \quad \varepsilon_{yz} = \varepsilon_{zx} = \varepsilon_{xy} = 0.$$

The dilatation produces a shift as stated earlier, but the associated shear splits the levels associated with $J_z = \pm 1/2$ from those for which $J_z = \pm 3/2$ by

$$\Delta E_{100} = 2b(s_{11} - s_{12})T, \tag{36}$$

the $J_z = \pm 1/2$ level lying above $J_z = \pm 3/2$ if $bT(s_{11} - s_{12}) > 0$. For $\mathbf{F} \parallel [111]$,

$$\varepsilon_{xx} = \varepsilon_{yy} = \varepsilon_{zz} = \frac{(s_{11} + 2s_{12})T}{3}$$

and

$$\varepsilon_{yz} = \varepsilon_{zx} = \varepsilon_{xy} = \frac{Ts_{44}}{6},$$

the level corresponding to $J_n = \pm 1/2$ lying above $J_n = \pm 3/2$ by

$$\Delta E_{111} = \frac{d}{\sqrt{3}} s_{44} T \tag{37}$$

if $Tds_{44} > 0$. Here J_n is the projection of $\mathbf{J}$ in the [111] direction.

In general, the splitting of the Γ_8^+ level under uniaxial stress applied along an arbitrary direction $\hat{e}$ is given by

$$(\Delta E_e)^2 = (\Delta E_{100})^2 + 3(\Delta E_{111}^2 - \Delta E_{100}^2) f(\hat{e}), \tag{38}$$

where

$$f(\hat{e}) = e_y^2 e_z^2 + e_z^2 e_x^2 + e_x^2 e_y^2. \tag{39}$$

The quantity $f(\hat{e})$ has extremal values of 1/3, 1/4, and 0 along $\langle 111 \rangle$, $\langle 110 \rangle$, and $\langle 100 \rangle$, respectively, the first being an absolute maximum and the third an absolute minimum.

These considerations are applicable not only to the band states and excitons but also to the localized levels of electrons or holes in semiconductors, i.e., electronic impurity states. Under such conditions the energy levels are classified according to the irreducible representations of the symmetry group of the site of the localized state. The results described hold, for example, for donor and acceptor states except that there is no question of a wave vector dependence. The acceptor-bound or hole states in elemental semiconductors are classified according to the irreducible representations of the double group $\bar{T}_d$, but the results are the same as in Eq. (35). Of course, the parameters a, b, and d, in principle, can be different for different Γ_8 levels. However, for shallow acceptors, these parameters can be calculated in terms of those of the band levels. For the donor-bound shallow electronic levels, the deformation potential constants of corresponding conduction-band extrema characterize their piezospectroscopic behavior.

6. Relative Intensities of Stress-Induced Components of an Optical Transition

Spectroscopic investigations provide information on the symmetry assignments of the quantum mechanical states associated with the initial and final states for given optical transitions. The energies of the transitions allow us to obtain band parameters and, in the presence of strain, values for the deformation potential constants.

The magnitude and sign of the deformation potentials are obtained experimentally in conjunction with theoretical studies of the symmetry characteristics of the initial and final states of the stress-induced lines associated with specific optical transitions carried out with incident radiation of varied polarization. But this is often insufficient to obtain unambiguous results. In such cases it is useful to compare the measured relative intensities of the stress-induced components of the absorption lines with the predictions of a symmetry analysis. The relative intensities of these components can be deduced, in the absence of significant hybridization with adjacent levels, by calculating the matrix elements of the electric dipole moment **d** between initial and final states. These calculations provide both the selection rules as well as expressions for the relative intensities, either directly or as functions of at most two adjustable parameters.

We recall that d_x, d_y, d_z, the components of **d** along the cubic axes, generate the Γ_5 representation of T_d and the Γ_4^- representation of O_h. For transitions between Γ_8 states, d_x, d_y, d_z are 4×4 matrices. We consider first transitions between localized states of levels where the symmetry is $\bar{T}_d$. The matrix elements of d_x, d_y, d_z for a transition between two Γ_8 levels are proportional to suitable combinations of products of the components of **J**, the angular momentum operator matrix for $j = 3/2$, which transform in the same way as d_x, d_y, d_z under the operations of T_d. The two sets of matrices

$$U_x = \{J_y, J_z\}, \qquad U_y = \{J_z, J_x\}, \qquad U_z = \{J_x, J_y\} \qquad (40)$$

and

$$V_x = \{J_x, J_y^2 - J_z^2\}, \qquad V_y = \{J_y, J_z^2 - J_x^2\}, \qquad V_z = \{J_z, J_x^2 - J_y^2\} \qquad (41)$$

transform as Γ_5. Thus, the components of **d** are linear combinations of **U** and **V** with suitable, complex-valued, coefficients. It is important to recognize that the shorthand notation **V** for the operators V_x, V_y, V_z does not imply that **V** is a vector. For transitions between Γ_8 and Γ_6 or Γ_7 levels the matrix elements of **d** are

$$\begin{pmatrix} \hat{x} - i\hat{y} & 0 & \sqrt{3}(\hat{x} + i\hat{y}) & 2\hat{z} \\ 2\hat{z} & -\sqrt{3}(\hat{x} - i\hat{y}) & 0 & -(\hat{x} + i\hat{y}) \end{pmatrix} \qquad (42)$$

for $\Gamma_8 \to \Gamma_7$ transitions and

$$\begin{pmatrix} -\sqrt{3}(\hat{x}+i\hat{y}) & 2\hat{z} & \hat{x}-i\hat{y} & 0 \\ 0 & -(\hat{x}+i\hat{y}) & 2\hat{z} & \sqrt{3}(\hat{x}-i\hat{y}) \end{pmatrix} \quad (43)$$

for $\Gamma_8 \to \Gamma_6$ transitions.

These matrix elements are expressed between states that are eigenvectors of J_z corresponding to the values $3/2, 1/2, -1/2, -3/2$ for Γ_8 and $1/2, -1/2$ for Γ_6 and Γ_7, taken in this order.

For transitions between states whose symmetry is O_h, a slightly different situation occurs. Here d_x, d_y, d_z transform as Γ_4^-. For transitions from a $\Gamma_8^\pm$ level to a $\Gamma_8^\mp$ level, the matrix elements of d_i ($i = x, y, z$) are proportional to linear combinations of J_i and J_i^3, because the sets (J_x, J_y, J_z) and (J_x^3, J_y^3, J_z^3) transform as Γ_4^+. Note that for systems having a center of symmetry, dipole transitions can occur only between states of different parity. The matrix elements for electric dipole transitions from $\Gamma_8^\pm$ to $\Gamma_6^\mp$ are proportional to those in expression (42), while those for $\Gamma_8^\pm \to \Gamma_7^\mp$ are proportional to the quantities appearing in (41). For transitions between Γ_6 and Γ_7 states the matrices associated with electric dipole transitions are proportional to the components of the Pauli spin matrix σ.

We emphasize that to obtain the relative intensities for a given set of stress-induced components of an absorption line, it is first required to obtain the energy eigenfunctions under stress and carry out the unitary transformation between the eigenstates of J_z and the actual eigenfunctions. These transformations are trivial to carry out for uniaxial stresses along high-symmetry directions. Applications to the acceptor states will be given in Section VI.

The energy eigenvalues as a function of **k** for states near the top of the valence band of the elemental semiconductors having the diamond structure has been given in Eq. (31). If Δ is large compared with the kinetic energy terms, these levels are the eigenvalues of

$$H = Ak^2 - B\sum_i k_i^2\left(J_i^2 - \frac{1}{3}J^2\right) - \frac{D}{\sqrt{3}}\sum_{i<j} k_i k_j \{J_i, J_j\},$$

to which we add the deformation potential (35) in the presence of strain. For the zinc blende semiconductors, lacking inversion symmetry, there is an additional term H_A, linear in **k**, namely

$$H_A = \frac{C}{\sqrt{3}}\sum_i k_i V_i,$$

where V_x, V_y, V_z are given by Eqs. (41). However, the constant C is usually so small that, for most purposes, H_A can be neglected. It originates in second-order perturbation theory as a term proportional to the product of the perturbation $(\hbar/m)\mathbf{k} \cdot \mathbf{p}$ with the spin-orbit interaction so that is is expected to be small.

In general, application of uniaxial stress splits the Γ_8 (Γ_8^+) level at the top of the valence band of a zinc blende (diamond) semiconductor into two Kramers doublets. The same occurs for the Γ_8 levels of an acceptor in tetrahedrally bonded semiconductors. For a force $\mathbf{F}$ applied along one of the cubic axes, a Γ_8 level splits into Γ_6 and Γ_7 levels of $\bar{D}_{2d}$, the group D_{2d} characterizing the site symmetry in the presence of the stress. If the force $\mathbf{F}$ is applied along the [001] axis, and the deformation potential is expressed in the basis generated by the eigenvectors of J_z, namely ϕ_μ ($\mu = 3/2, 1/2, -1/2, -3/2$), the functions $\phi_{\pm 3/2}$ belong to $\Gamma_6(\bar{D}_{2d})$ while $\phi_{\pm 1/2}$ generate $\Gamma_7(\bar{D}_{2d})$. Under uniaxial stress resulting from an applied force directed along [111], the symmetry of the site becomes C_{3v} and the Γ_8 level splits into $\Gamma_4(\bar{C}_{3v})$, corresponding to $\hat{\zeta} \cdot \mathbf{J} = J_\zeta = \pm 1/2$, where $\hat{\zeta}$ is a unit vector parallel to the [111] direction and $(\Gamma_5 + \Gamma_6)(\bar{C}_{3v})$ is associated with $J_\zeta = \pm 3/2$. Levels Γ_5 and Γ_6 are one-dimensional irreducible representations that are complex conjugates of each other and, hence, correspond to states of equal energy eigenvalues. We represent this level by Γ_{5+6}.

Studies of optical absorption in semiconductors under uniaxial stress are useful in the identification of the properties of the excited states of semiconductors: excitons, impurity states, etc. It is often insufficient to make unambiguous assignments of quantum numbers on the basis of selection rules for optical transitions caused by incident radiation of varied polarization characteristics in the presence of uniaxial stress. An analysis of the relative intensities of these transitions has permitted the elucidation of those characteristics not accessible by other means.

For transitions between Γ_8 levels, the matrix elements of the electric dipole operator are a linear combination of the components of the quantities U_i and V_i ($i = x, y, z$, the cubic axes) defined in Eqs. (40) and (41):

$$d_i = \frac{i}{\sqrt{3}} d_1 U_i - \frac{2}{\sqrt{3}} d_2 V_i. \tag{44}$$

We find the relative intensities of the stress-induced absorption lines by reading off the matrix elements and adding the squares of the magnitudes of matrix elements of d_i between degenerate levels. The form of the matrices U_i and V_i permits the direct determination of the relative intensities for $\mathbf{F} \parallel [001]$. The results of this analysis are summarized in Fig. 7a and Table II for incident radiation polarized parallel or perpendicular to $\mathbf{F}$. These results are accurate as long as hybridization with neighboring levels can be neglected. Otherwise the possibility of a transfer of intensity from one set of lines to another is possible and can also be incorporated into the formalism. Results are also

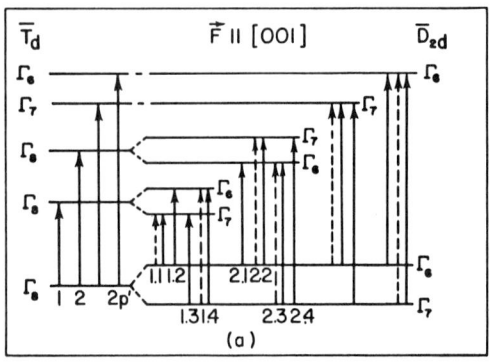

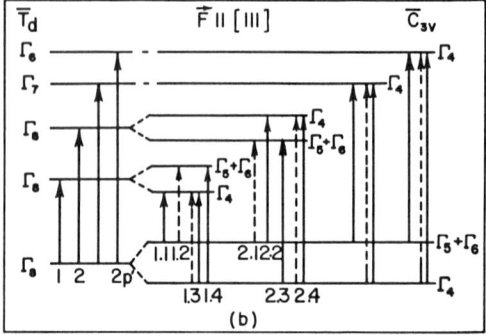

FIG. 7. The allowed transitions from a Γ_8 ground state to Γ_6, Γ_7, and Γ_8 excited states of the double group $\bar{T}_d$ with compressive force **F** along [111] or [001]. The site symmetry is C_{3v} and D_{2d} for the [111] and [001] compressions, respectively; the designations next to the levels denote the irreducible representations of the appropriate double point group following the notation of Koster et al. (1966). The dashed arrows are for electric vector $\mathbf{E} \parallel \mathbf{F}$ while the solid arrows are for $\mathbf{E} \perp \mathbf{F}$.

displayed for transitions from a Γ_8 level to Γ_6 and Γ_7 states by using the rectangular matrices in Eqs. (42) and (43). We note that the relative intensities can be expressed in terms of d_1/d_2, the ratio of two complex numbers, and thus depend on two real parameters u and v defined by

$$u = \left[1 + \left|\frac{d_1}{2d_2}\right|^2\right]^{-1}, \quad v = -\frac{1}{4}\left(\frac{d_1}{d_2} + \frac{d_1^*}{d_2^*}\right)u. \tag{45}$$

In order to investigate the relative intensities of the stress-induced lines for $\mathbf{F} \parallel [111]$, we transform U_i and V_i so that the matrix elements of **d** are referred to a coordinate system $\hat{\xi}, \hat{\eta}, \hat{\zeta}$ in which $\hat{\zeta}$ is directed along [111]. We select $\hat{\xi}$ parallel to $[11\bar{2}]$ and $\hat{\eta}$ along $[\bar{1}10]$ and transform the Γ_8 states by

TABLE II

RELATIVE INTENSITIES FOR STRESS-INDUCED COMPONENTS OF ACCEPTORS IN TETRAHEDRALLY BONDED SEMICONDUCTORS WITH **F** ∥ [001] AND THE ELECTRIC VECTOR, **E**, OF THE INCIDENT RADIATION POLARIZED PARALLEL OR PERPENDICULAR TO **F**

Zero stress transition	Stress-induced components	Sum of the squared amplitudes for the stress-induced components		Relative intensities	
$\bar{T}_d$	$\bar{D}_{2d}$	$\mathbf{E} \perp (\hat{x}, \hat{y})$	$\mathbf{E} \parallel \hat{z}$	$\mathbf{E} \perp (\hat{x}, \hat{y})$	$\mathbf{E} \parallel \hat{z}$
$\Gamma_8 \to \Gamma_8$	$\Gamma_7(\pm\frac{1}{2}) \to \Gamma_7(\pm\frac{1}{2})$	$6|d_2|^2$	0	$\frac{3u}{8}$	0
	$\Gamma_7(\pm\frac{1}{2}) \to \Gamma_6(\pm\frac{3}{2})$	$2|d_1 + d_2|^2$	$2|d_1 - 2d_2|^2$	$\frac{1}{2} - \frac{3u}{8} - \frac{v}{2}$	$\frac{1}{2} + v$
	$\Gamma_6(\pm\frac{3}{2}) \to \Gamma_7(\pm\frac{1}{2})$	$2|d_1 - d_2|^2$	$2|d_1 + 2d_2|^2$	$\frac{1}{2} - \frac{3u}{8} + \frac{v}{2}$	$\frac{1}{2} - v$
	$\Gamma_6(\pm\frac{3}{2}) \to \Gamma_6(\pm\frac{3}{2})$	$6|d_2|^2$	0	$\frac{3u}{8}$	0
$\Gamma_8 \to \Gamma_6$	$\Gamma_7(\pm\frac{1}{2}) \to \Gamma_6$	$2|d_0|^2$	$8|d_0|^2$	1	4
	$\Gamma_6(\pm\frac{3}{2}) \to \Gamma_6$	$6|d_0|^2$	0	3	0
$\Gamma_8 \to \Gamma_7$	$\Gamma_7(\pm\frac{1}{2}) \to \Gamma_7$	$6|d_0'|^2$	0	3	0
	$\Gamma_6(\pm\frac{3}{2}) \to \Gamma_7$	$2|d_0'|^2$	$8|d_0'|^2$	1	4

* The parameters d_1 and d_2 are defined in Eq. (44); d_0 and d_0' are the coefficients of the matrices Eqs. (42) and (43) for $\Gamma_8 \to \Gamma_7$ and $\Gamma_8 \to \Gamma_6$ transitions, respectively; u and v are defined in Eq. (45).

TABLE III

RELATIVE INTENSITIES FOR STRESS-INDUCED COMPONENTS OF ACCEPTORS IN TETRAHEDRALLY BONDED SEMICONDUCTORS WITH **F** ∥ [111] AND THE ELECTRIC VECTOR, **E**, OF THE INCIDENT RADIATION POLARIZED PARALLEL OR PERPENDICULAR TO **F**

Zero stress transition	Stress-induced components	Sum of the squared amplitudes for the stress-induced components		Relative intensities	
$\bar{T}_d$	$\bar{C}_{3v}$	$\mathbf{E} \perp (\hat{\xi}, \hat{\eta})$	$\mathbf{E} \parallel \hat{\zeta}$	$\mathbf{E} \perp (\hat{\xi}, \hat{\eta})$	$\mathbf{E} \parallel \hat{\zeta}$
$\Gamma_8 \to \Gamma_8$	$\Gamma_4(\pm\frac{1}{2}) \to \Gamma_4(\pm\frac{1}{2})$	$8|d_2|^2$	$2|d_1|^2$	$\frac{1}{2}u$	$\frac{1}{2} - \frac{1}{2}u$
	$\Gamma_4(\pm\frac{1}{2}) \to \Gamma_{5+6}(\pm\frac{3}{2})$	$2|d_1|^2 + 4|d_2|^2$	0	$\frac{1}{2} - \frac{1}{4}u$	0
	$\Gamma_{5,6}(\pm\frac{3}{2}) \to \Gamma_4(\pm\frac{1}{2})$	$2|d_1|^2 + 4|d_2|^2$	0	$\frac{1}{2} - \frac{1}{4}u$	0
	$\Gamma_{5,6}(\pm\frac{3}{2}) \to \Gamma_{5,6}(\pm\frac{3}{2})$	0	$2|d_1|^2 + 16|d_2|^2$	0	$\frac{1}{2} + \frac{1}{2}u$
$\Gamma_8 \to \Gamma_6$	$\Gamma_4(\pm\frac{1}{2}) \to \Gamma_4$	$2|d_0|^2$	$8|d_0|^2$	1	4
	$\Gamma_{5,6}(\pm\frac{3}{2}) \to \Gamma_4$	$6|d_0|^2$	0	3	0
$\Gamma_8 \to \Gamma_7$	$\Gamma_4(\pm\frac{1}{2}) \to \Gamma_4$	$2|d_0'|^2$	$8|d_0'|^2$	1	4
	$\Gamma_{5+6}(\pm\frac{3}{2}) \to \Gamma_4$	$6|d_0'|^2$	0	3	0

* The parameters d_1 and d_2 are defined in Eq. (44); d_0 and d_0' are the coefficients of the matrices Eqs. (42) and (43) for $\Gamma_8 \to \Gamma_7$ and $\Gamma_8 \to \Gamma_6$ transitions, respectively; u and v are defined in Eq. (45).

means of the rule

$$\psi_\mu = \sum_{\mu'} \phi_{\mu'} \cdot D^{(3/2)}_{\mu'\mu}(\alpha, \beta, \gamma),$$

where α, β, and γ are the Euler angles of the rotation from $\hat{x}, \hat{y}, \hat{z}$ to $\hat{\xi}, \hat{\eta}, \hat{\zeta}$ ($\alpha = \pi/4$, $\cos\beta = 3^{-1/2}$, and $\gamma = 0$), and $D^{(3/2)}(\alpha, \beta, \gamma)$ is the rotation operator for angular momentum 3/2. The transformations are straightforward, and the results are displayed in Fig. 7b and in Table III. We note that in this case the relative intensities depend only on the parameter u already defined.

V. Interband Transitions and Associated Excitons

In this section we focus on the experimental observations on the piezo-spectroscopy of interband and associated excitonic transitions in the tetrahedrally coordinated group IV, III-V, and II-VI semiconductors using several illustrative examples.

7. Direct Transitions

As an illustrative example of direct transitions, consider the case of GaAs. The E_0 and $E_0 + \Delta_0$ transitions in GaAs at 77 K occur at 1.49 and 1.83 eV, respectively. Chandrasekhar and Pollak (1977) have investigated these transitions as signatures in electromodulated reflectivity, the electromodulation being achieved with Schottky-barrier electroreflectance (SBER). Piezospectroscopy is carried out with specimens simultaneously subjected to uniaxial stress.

The effect of a compressive force $\mathbf{F} \parallel [001]$ on the E_0 and $E_0 + \Delta_0$ transitions in GaAs are shown schematically in Fig. 8. Chandrasekhar and Pollak (1977) discuss the relative contributions of the "unresolved" excitonic and direct interband transitions at E_0 and $E_0 + \Delta_0$ and argue that, at low modulation voltages, excitonic contributions dominate. Figure 9 shows the E_0 exciton thus observed in GaAs at zero stress and for $\mathbf{F} \parallel [001]$ with a stress $T = 8.8 \times 10^9$ dyn/cm^2. The stress dependence of the stress-induced components (observed under conditions favoring interband transitions) is displayed in Fig. 10. The splitting of the E_0 line into $E_0(1)$ and $E_0(2)$ follows from the splitting of the $\Gamma_8(\bar{T}_d)$ quadruplet into $\Gamma_7(\bar{D}_{2d})$ and $\Gamma_6(\bar{D}_{2d})$ doublets. The observation of the low-energy component $E_0(1)$ for both $\mathbf{E} \parallel \mathbf{F}$ and $\mathbf{E} \perp \mathbf{F}$, whereas that of the high-energy component only in $\mathbf{E} \parallel \mathbf{F}$, $\mathbf{E}$ being the electric

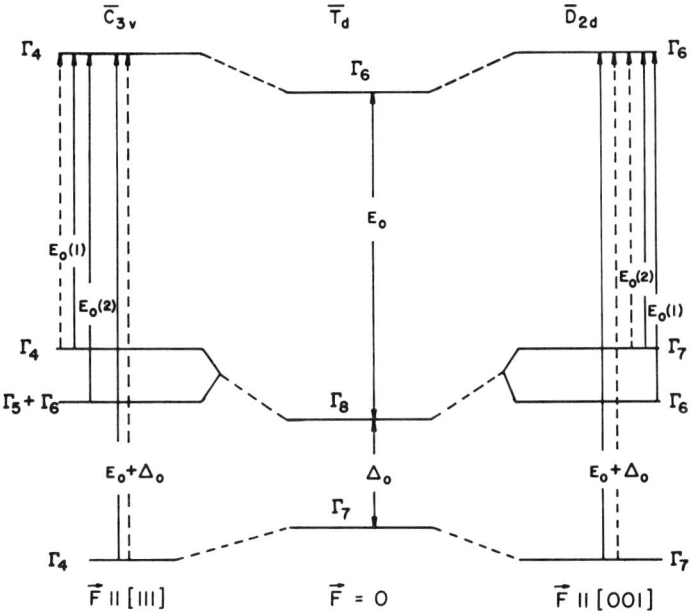

FIG. 8. Schematic representation of the E_0 and the $E_0 + \Delta_0$ direct transitions in the direct-gap III-V and II-VI zinc blende semiconductors that have T_d point group symmetry: (- - - -) Electric dipole transitions allowed for $\mathbf{E} \parallel \mathbf{F}$, where $\mathbf{E}$ is the electric vector and $\mathbf{F}$, the compressive force; (—) electric dipole transitions allowed for $\mathbf{E} \perp \mathbf{F}$.

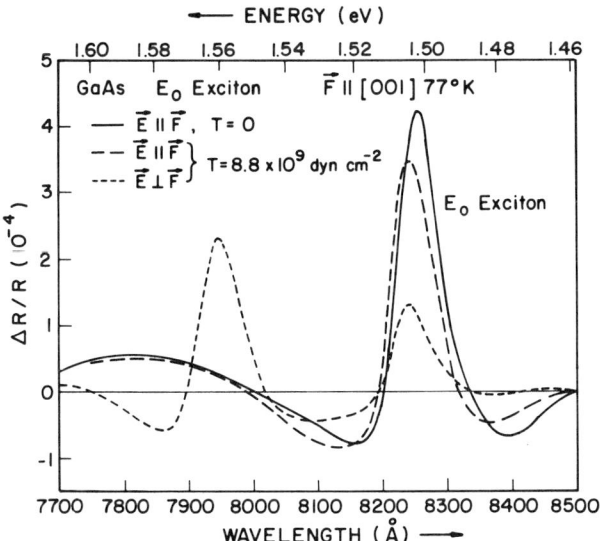

FIG. 9. Electroreflectance spectrum of the exciton associated with the E_0 direct gap of GaAs at zero stress and a stress of 8.8×10^9 dyn/cm² with $\mathbf{F} \parallel [001]$ and spectra recorded with $\mathbf{E} \parallel \mathbf{F}$ (− −) and $\mathbf{E} \perp \mathbf{F}$ (- - - -). The sample was biased so that excitonic transitions are an order of magnitude more intense than the interband transitions. (From Chandrasekhar and Pollak, 1977.)

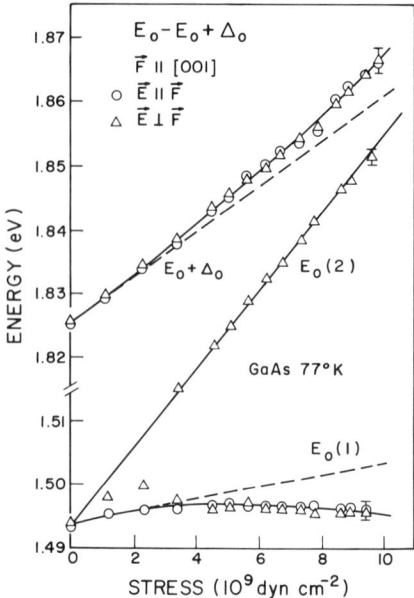

FIG. 10. Stress dependence of the energies of the $E_0 + \Delta_0$ and the $E_0(1)$ and the $E_0(2)$ components of E_0 for $\mathbf{F} \parallel [001]$ with conditions favoring interband transitions. (After Chandrasekhar and Pollak, 1977.)

vector of the incident light, is consistent with the level ordering as depicted in Fig. 8. The $E_0 + \Delta_0$ transition is observed for both $\mathbf{E} \parallel \mathbf{F}$ and $\mathbf{E} \perp \mathbf{F}$ and merely shifts with stress. Stress dependence of the stress-induced components of the E_0 and the $E_0 + \Delta_0$ interband transitions for $\mathbf{F} \parallel [111]$ is displayed in Fig. 11.

The nonlinear stress dependence of the $E_0 + \Delta_0$ and $E_0(1)$ components, clearly noticed in Fig. 10 for $\mathbf{F} \parallel [001]$ and in Fig. 11 for $\mathbf{F} \parallel [111]$, can be traced to the stress-dependent spin-orbit interaction. Such effects are treated in many papers dealing with the piezospectroscopy of direct or excitonic transitions as well as the bound states of acceptors. (See, for example, Pollak and Cardona, 1968; Chandrasekhar et al., 1975.)

According to the predictions based on Tables II and III, the relative intensities of $E_0(1)$ observed in $\mathbf{E} \parallel \mathbf{F}$ and $\mathbf{E} \perp \mathbf{F}$ should be 4:1, whereas for $E_0(1)$ and $E_0(2)$, observed in $\mathbf{E} \perp \mathbf{F}$, it should be 1:3. The experimental results displayed in Fig. 9 are in qualitative agreement with the predictions.

Another illustrative example of piezospectroscopy of excitons in direct-band-gap semiconductors is provided by the investigation of Pollak and Aggarwal (1971) on free and bound excitons in GaSb. Figures 12 and 13

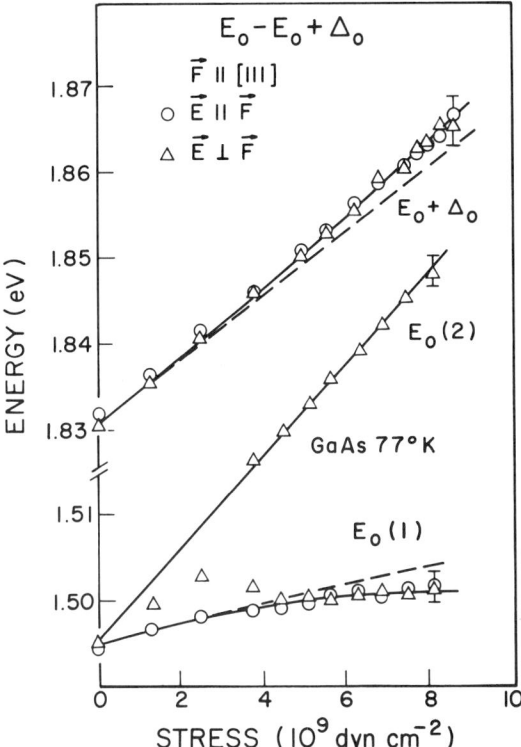

FIG. 11. Stress dependence of the energies of the $E_0 + \Delta_0$ and the $E_0(1)$ and the $E_0(2)$ components of E_0 for $\mathbf{F} \parallel [111]$ with conditions favoring interband transitions. (After Chandrasekhar and Pollak, 1977.)

demonstrate clearly the splitting of the free exciton (α) into the (α_1, α_2) doublet with the expected polarization features and, in the low-stress limit, linear stress dependence of the energy of α_1 and α_2. The bound exciton (γ), on the other hand, does not split; the interpretation of γ has not yet appeared in the literature.

As a convenient reference in the context of the piezospectroscopy of direct interband and excitonic transitions in Si and Ge (symmetry $\bar{O}_h$), we show schematically the relevant diagram in Fig. 14. A comparison of Fig. 14 with Fig. 8 shows that the stress-induced splittings and the polarization features of the components of the E_0 and the $E_0 + \Delta_0$ transitions are qualitatively identical in $\bar{T}_d$ and $\bar{O}_h$. Chandrasekhar and Pollak (1977) have investigated the direct interband transition in Ge, and their results are in agreement with the predictions of Fig. 14.

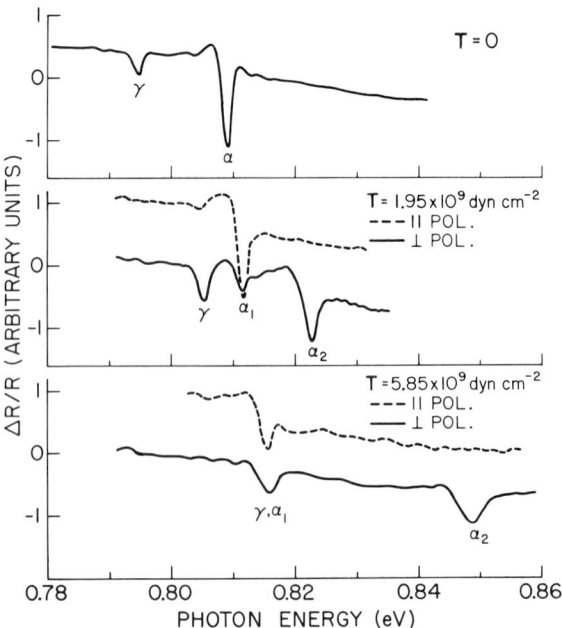

FIG. 12. Wavelength-modulated reflectivity spectra of the free (α) and bound (γ) excitons of GaSb at 1.7 K and at zero stress ($T = 0$), $T = 1.95 \times 10^9$ and 5.85×10^9 dyn/cm². $\mathbf{F} \parallel [111]$, $\mathbf{E} \parallel \mathbf{F}$, and $\mathbf{E} \perp \mathbf{F}$. (After Pollak and Aggarwal, 1971.)

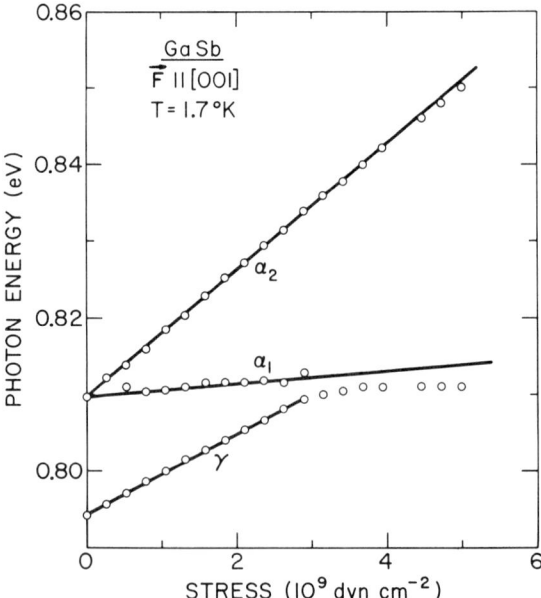

FIG. 13. Energy of the γ-exciton and of the α_1 and α_2 components of the α-exciton as a function of stress for $\mathbf{F} \parallel [001]$. (After Pollak and Aggarwal, 1971.)

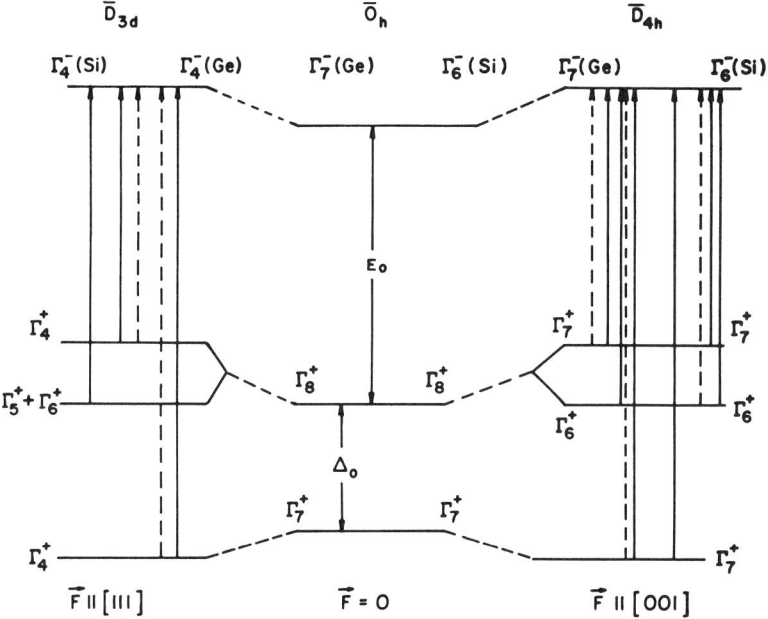

FIG. 14. The E_0 and $E_0 + \Delta_0$ excitons in Si and Ge shown schematically for $\mathbf{F} = 0$, $\mathbf{F} \parallel [111]$, and $\mathbf{F} \parallel [001]$.

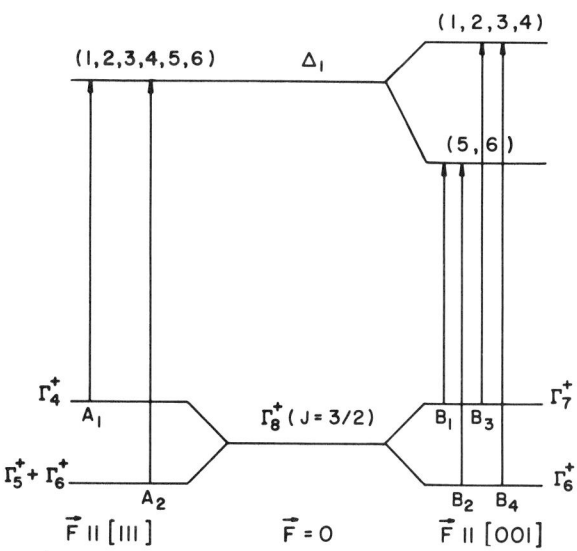

FIG. 15. The $\Gamma_8^+ \rightarrow \Delta_1$ indirect transition in Si for $\mathbf{F} = 0$, $\mathbf{F} \parallel [111]$, and $\mathbf{F} \parallel [001]$ shown schematically. The six conduction-band minima along [100], [$\bar{1}$00]; [010], [0$\bar{1}$0]; [001], [00$\bar{1}$] are labeled 1, 2, 3, 4, 5 and 6, respectively. These six conduction-band minima remain energetically degenerate for $\mathbf{F} \parallel [111]$, whereas for $\mathbf{F} \parallel [001]$, the minima 5 and 6 shift twice as much below their center of gravity as 1, 2, 3, and 4 shift above it (see also Fig. 5).

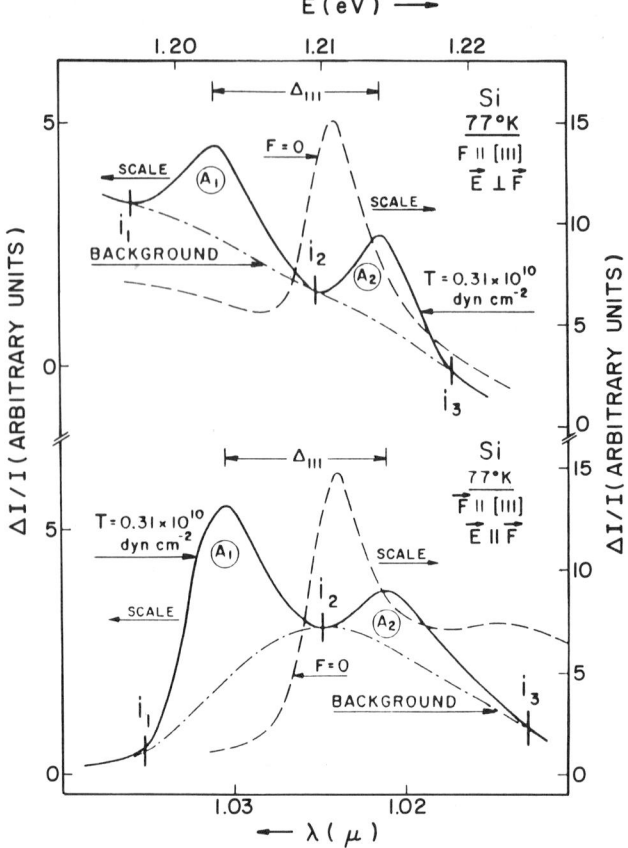

FIG. 16. The $\Gamma_8^+ \to \Delta_1$ indirect transition of Si observed in wavelength-modulated reflectivity and $\mathbf{F} \parallel [111]$. The A_1 and A_2 components shown in Fig. 15 are labeled here accordingly. (From Laude et al., 1971.)

8. Indirect Transitions

As a typical example of indirect transitions observed under uniaxial stress, we consider the TO phonon-assisted $\Gamma_8^+ \to \Delta_1$ transitions in Si. Here we follow closely the investigation of Laude et al. (1971), who have exploited a wavelength modulation spectrometer. Figure 15 shows schematically the expected stress-induced splitting of the indirect transition. The splitting results from the regrouping of the six Δ_1 conduction-band minima into subgroups that are energetically equivalent under the stress [Fig. 5 and Eqs. (21)], on the one hand, and the the splitting of the Γ_8^+, $p_{3/2}$, valence-band maximum, on the other [Eq. (35)]. As expected, the experimental results on the fractional change in transmission resulting from wavelength modulation, i.e., $\Delta I/I$ for $\mathbf{F} \parallel [111]$, shown in Figs. 16 and 17, indeed show the expected A_1 and A_2 components; the separation of A_1 and A_2 yields the deformation potential constant d characterizing the Γ_8^+ valence-band maximum. The stress dependence of the four (B_1, B_2, B_3, B_4) TO phonon-assisted indirect transitions for $\mathbf{F} \parallel [001]$ is shown in Fig. 18. Again the number of components observed conforms to the splitting shown in Fig. 15. The spacings $B_1 - B_2$ and $B_3 - B_4$ allow a determination of the shear deformation potential constant, b, for Γ_8^+, whereas the spacings $B_1 - B_3$ and $B_2 - B_4$ yield Ξ_u, the shear deformation potential constant of the Δ_1 conduction-band minima.

The indirect transitions are governed by selection rules that involve the matrix elements for (1) a vertical optical transition between two electronic states (one of which is the intermediate state) and (2) a transition between

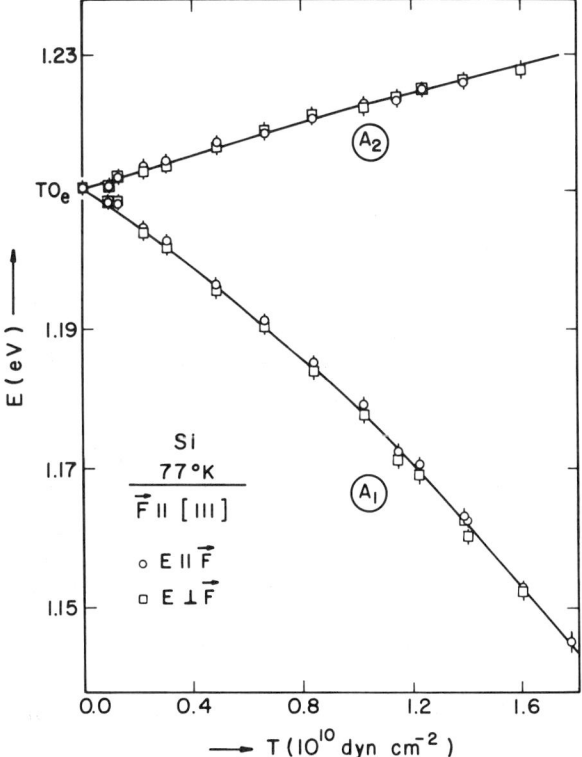

FIG. 17. The stress dependence of the energies of the A_1 and A_2 stress-induced components of the indirect exciton of Si for $\mathbf{F} \parallel [111]$. (After Laude et al., 1971.)

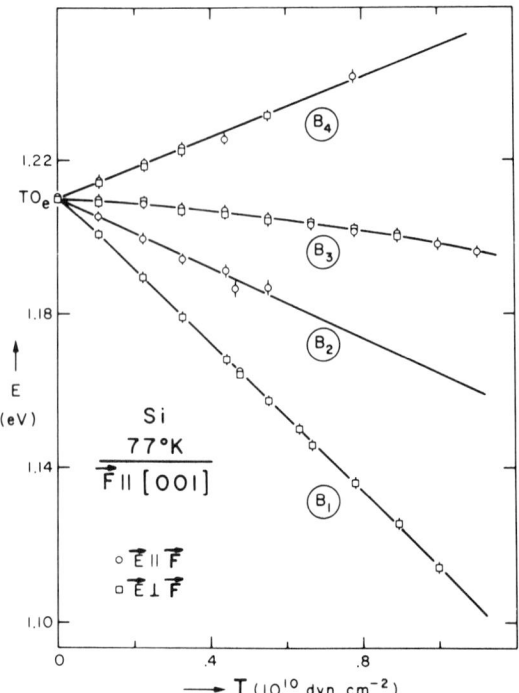

FIG. 18. The stress dependence of the B_1, B_2, B_3, and B_4 stress-induced components of the indirect exciton of Si for $\mathbf{F} \parallel [001]$. (After Laude et al., 1971.)

two electronic states not with the same wave vector and hence requiring the absorption or emission of a phonon. Elliot and Loudon (1960), Lax and Hopfield (1961), and Folland and Bassani (1968) have given the group-theoretical procedures for deriving the selection rules for TO, LO, TA, and LA phonons involving different intermediate states. In the absence of stress, the crystal is optically isotropic and the incident optical transitions conceal the latent anisotropy; under uniaxial stress the anisotropy will be exposed. Erlbach (1966) has calculated precisely these effects for $\mathbf{F} \parallel [001]$, $[110]$, and $[111]$. Laude et al. (1971) have compared such theoretical estimates, explicitly involving either the Γ_6^- (Γ_{15}) (conduction-band minimum) or the Δ_5 (valence-band maximum) as the intermediate state, with the intensities of the stress-induced components observed with $\mathbf{E} \parallel \mathbf{F}$ and $\mathbf{E} \perp \mathbf{F}$.

Mathieu et al. (1979) have deduced the deformation potential constants of the Γ_8 valence-band maximum and the X-conduction-band minima of GaP, employing wavelength-modulated transmission in the investigation of indirect transitions, whereas they studied the E_0 and the $E_0 + \Delta_0$ transitions in electromodulation. The direct and the indirect absorption edges of AlSb have been investigated by Laude et al. (1970). Wardzynski and Sufczynski (1972) and Thomas (1961) studied the direct free exciton in ZnTe and CdTe, respectively. Langer et al. (1970) report results on cubic ZnS and ZnSe as well as on uniaxial ZnO, CdS, and CdSe.

9. Effects in Heterostructures

In the context of interband and excitonic transitions in semiconductors subjected to uniaxial stress, it is of interest to consider novel effects that might arise in the piezospectroscopy of electronic levels in quantum wells fabricated, for example, by molecular beam epitaxy (MBE). Jagannath et al. (1986) and Lee et al. (1988a) have investigated the effect of uniaxial stress on the excitons spatially confined in the GaAs quantum wells in GaAs–Al$_{0.3}$Ga$_{0.7}$As multiple quantum well structures. Using photoluminescence excitation (PLE) spectroscopy, they observe the heavy-hole (E1H) and light-hole (E1L) excitonic transitions in a multiple quantum well grown along [001] (Fig. 19). Note that

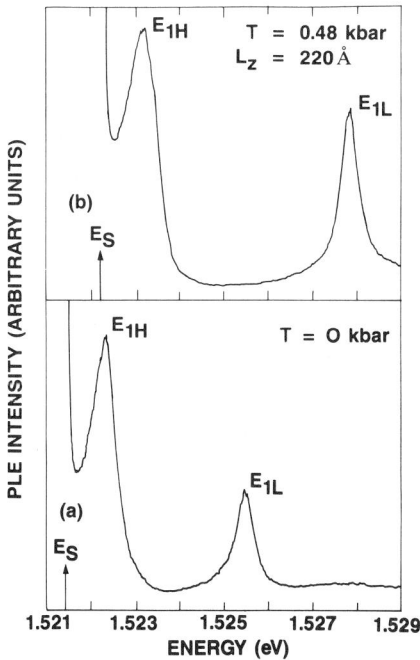

FIG. 19. Photoluminescence excitation spectra of excitons in 220-Å-wide GaAs quantum wells in GaAs–Al$_{0.3}$Ga$_{0.7}$As multiple quantum well structure. E_s labels the energy of the fixed spectrometer setting and E_{1H} and E_{1L} label peaks due to heavy- and light-hole excitons with uniaxial stress $T = 0$ kbar (a) and $T = 0.48$ kbar (b). (After Jagannath et al., 1986.)

in PLE, a specific photoluminescence peak at E_S is monitored while the exciting photon energy is scanned. Under a compressive force along [100], they obtain easily observed shifts of the excitonic transitions (Fig. 19). Here E1H refers to the excitonic recombination radiation associated with the lowest quantum confined electron level in the conduction band and the first heavy-hole level in the valence band of the GaAs well, whereas E1L refers to the corresponding excitonic transition involving the light-hole level. These are also referred to in the literature as 11H and 11L transitions. The experimental observations are analyzed in terms of hydrostatic and shear components of the strain; while the former are characterized by a deformation potential constant close to that of bulk GaAs, the latter clearly involve nonlinear stress-dependent terms that are a function of the overlap integral of the light-hole and heavy-hole wave functions as influenced by the well width.

Jagannath *et al.* (1987) have observed the shifts in the excitonic transition associated with a GaAs single quantum well of a GaAs–AlGaAs heterostructure grown on GaAs–Si as compared to that grown on GaAs–GaAs, the growth being by MBE. The transitions observed in photoluminescence are displayed in Fig. 20. Quantum wells grown on GaAs–GaAs are stress-free, and in the photoluminescence at 4.2 K only the lowest, E1H, quantum confined excitonic transitions are observed, and are identified in Fig. 20a as *A*, *B*, and *C* corresponding to well widths of 180, 72, and 45 Å, respectively. On an identical set of quantum wells but grown on GaAs–Si, the three transitions labeled *A*, *B*, and *C* in Fig. 20b have shifted to lower photon energies. Differential contraction between GaAs and Si from the growth temperature to 4.2 K, the temperature at which the photoluminescence is measured, generates a uniform *biaxial stress* (estimated to be ≈ 2.5 kbar). We note the E1H and the E1L separation increases with decreasing well width; with the increasing biaxial tension, both E1H and E1L shift to lower energies, the latter faster than the former; and for a sufficiently large tension, the E1L and E1H can coincide, and beyond this tension the E1L could actually lie below the E1H transition. Indeed, the transition *A* in Fig. 20b for the 180-Å well corresponds to E1L as a result of this stress tuning.

We conclude this section with yet another example of "built-in" anisotropic strain in a ZnSe–GaAs heterostructure grown by MBE (Lee *et al.*, 1988b). In the pseudomorphic thin (~ 1000 Å) epilayer of ZnSe grown coherently on a (100) GaAs substrate, the lattice mismatch between these constituents results in a contraction parallel to the substrate surface and an extension normal to the plane. As a consequence, the $p_{3/2}(\Gamma_8)$ valence band of ZnSe undergoes a splitting into $|3/2, \pm 3/2\rangle$ and $|3/2, \pm 1/2\rangle$ states. The splitting of the excitonic signature observed in the piezomodulated reflectively is clear seen in Fig. 21; the signature close to the bulk value of unstrained ZnSe observed with a *thick* ZnSe epilayer for which the built-in strain is relieved is shown in Fig. 22.

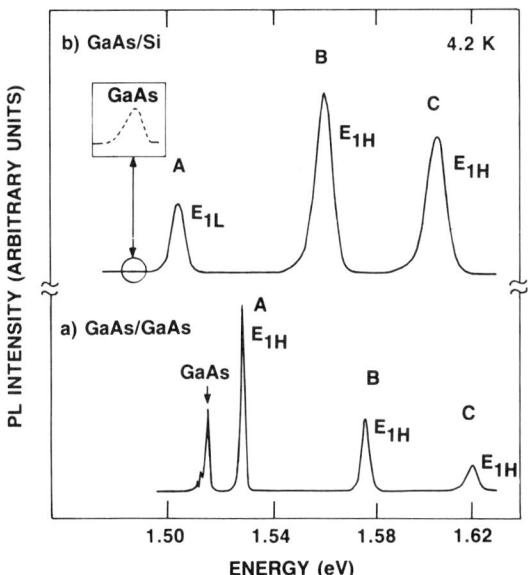

FIG. 20. Low-temperature (4.2 K) photoluminescence (PL) spectra of single quantum wells fabricated on (a) GaAs and (b) GaAs–Si. The features labeled A, B, and C correspond to well widths 180, 72, and 45 Å, respectively. The features labeled GaAs in (a) and (b) refer to the free-exciton signature of the GaAs buffer. (After Jagannath et al., 1987.)

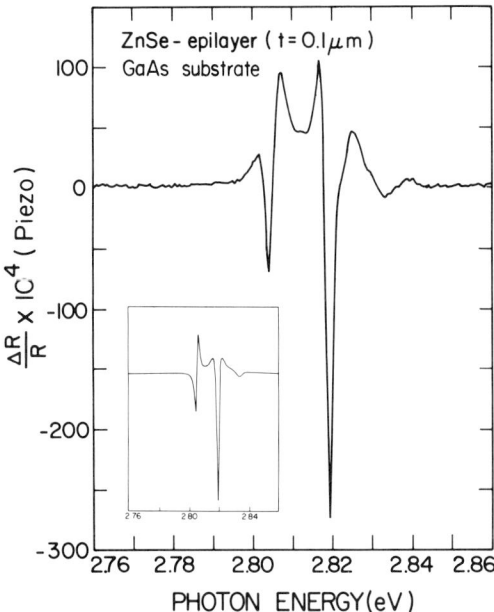

FIG. 21. Piezomodulated reflectivity of ZnSe–GaAs epilayer ($t = 0.1$ μm) for the spectral range in the vicinity of the free exciton of ZnSe. Also shown in the inset of the figure is a theoretical line shape fitting using the first derivative of a Lorentzian function. (After Lee et al., 1988b.)

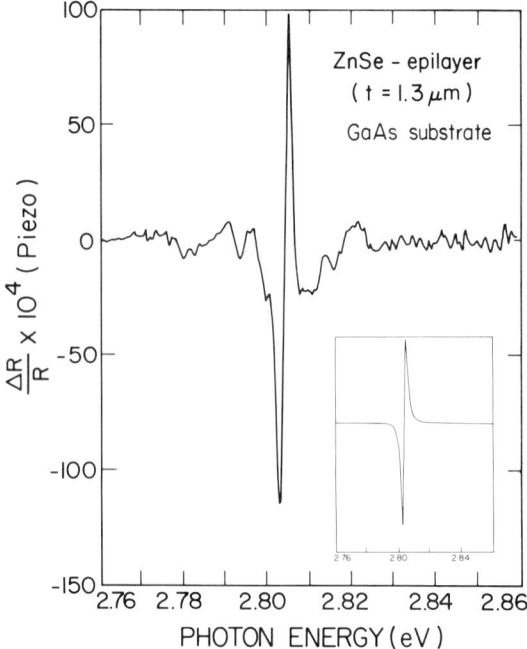

FIG. 22. Piezomodulated reflectivity of ZnSe–GaAs epilayer ($t = 1.3$ μm) for the spectral range in the vicinity of the free exciton of ZnSe. The inset shows a theoretical line shape fitting using the first derivative of a Lorentzian function. (After Lee et al., 1988b.)

The inserts in Figs. 21 and 22 are theoretical line shape fittings derived by using the first derivative of a Lorentzian function (see the appendix in Lee et al., 1988b for the procedure adopted). The deformation potential constants of ZnSe have been deduced from the splittings and the shifts of the signatures in Fig. 21 with respect to the exciton energy of a stress-free bulk crystal, along with the strain calculated from the lattice mismatch. Lee et al. (1988b) obtain a' (the net hydrostatic deformation potential constant characterizing the relative shift of the Γ_6 conduction-band minimum from the Γ_8 valence-band maximum) $= -5.26$ eV and b (the shear deformation potential for the Γ_8 valence-band maximum for $\mathbf{F} \parallel [001]) = -1.27$ eV. These values are in excellent agreement with $a' = -5.4$ and $b = -1.2$ eV obtained for bulk ZnSe (Langer et al., 1970) from the piezospectroscopy of the free exciton observed in the reflectivity spectrum. Ohkawa et al. (1988) have also investigated the biaxial strain in heteroepitaxial ZnSe layers on GaAs substrates; they report splittings in the bound excitons associated with donors and acceptors in the ZnSe layer. Rockwell et al. (1991) have demonstrated how the biaxial strain in the pseudomorphic ZnSe film can be "tuned" with externally applied hydrostatic stress.

VI. Lyman Spectra of Shallow Donors and Acceptors

Foreign atoms in substitutional or interstitial positions, complexes of such atoms, vacancies or interstitials of the host atoms, and nonstoichiometry in compound semiconductors are examples of imperfections that have received a great amount of attention in semiconductor physics (Flynn, 1972). In this section the focus of the discussion is on the piezospectroscopy of the electronic levels associated with shallow donors and acceptors in the elemental and compound semiconductors. The energy levels of these "solid-state analogs" of the H atom and of the He atom can be analyzed in terms of the characteristics of the conduction-band minima (donors) or the valence-band maxima (acceptors). Hence, the theoretical considerations in Section IV are readily adapted to the analysis of the effects of uniaxial stress observed in their excitation spectra (Ramdas and Rodriguez, 1981).

Consider, for example, the group V impurity phosphorus in silicon. On the basis of a variety of evidence (Pearson and Bardeen, 1949), it can be proved that it enters the host in a site normally occupied by a silicon atom, i.e., substitutionally. The saturated (covalent) bonds with its four nearest neighbors are formed by the four of the five of its outermost electrons in the $3s^2 3p^3$ states; the fifth electron not incorporated in this bonding scheme is *donated* to the conduction band. However, it remains bound to the P^+ ion by the Coulomb attraction. It is clearly of interest to find out how tightly the "donor" electron is bound and to establish its energy-level scheme. The potential energy of the donor electron must take into account the adjustment of the charge density of the host in the field of the positively charged donor. The evaluation of this potential is a many-body problem. However, for distances r large compared to the lattice spacing a, this adjustment can be viewed as a polarization of the host described by its static dielectric constant κ. In this limit the potential is

$$U_i(r) = -\frac{e^2}{\kappa r}. \tag{46}$$

Closer to the impurity $U_i(r)$ is more attractive, the dielectric screening being less effective, approaching $-e^2/r$ as r becomes comparable to the size of the P^+ ion. In a crystal the electrons behave under an external field as particles with an effective mass m^* different from the free-electron mass m, and often much smaller. Under these assumptions, the donor electron will have hydrogen-like bound states given by

$$E_n = -\frac{m^* e^4}{2\hbar^2 \kappa^2 n^2}, \tag{47}$$

where $n = 1, 2, 3, \ldots$ For example, in GaAs, $m^* = 0.06650m$ and $\kappa = 12.58$ yield an ionization energy $E_I = -E_1 = 5.72$ meV and the corresponding Bohr radius $a^* = \hbar^2\kappa/m^*e^2 = 100$ Å (see Stillman et al., 1971). Since $a^* \gg a$, the use of an effective kinetic energy $(p^2/2m^*)$ is justified, thereby providing validity for Eq. (47). This simple model (Mott and Gurney, 1948; Bethe, 1942) needs to be modified for actual semiconductors in two important respects: (1) To the extent Eq. (46) fails to describe the true potential, the binding energies of *different* group V impurities in Si or Ge, for example, are not the same. (2) The effective mass m^* for many semiconductors is a *tensor rather than a scalar*, reflecting the nature of the conduction band. In an analogous fashion one can develop a model for substitutional group III impurities in Si or Ge that bind the hole created in the valence band resulting from the formation of the covalent bonds with its nearest neighbors. These impurities that have *accepted* an electron from the valence band to complete the bonding scheme with its four nearest neighbors are called *acceptors*. The details of the bound states of the acceptor reflect the characteristics of the valence-band maximum of the host. (For a convenient summary of the symmetries and mass parameters of the band extrema of the semiconductors discussed in this Chapter, we refer the readers to Appendix C of Long, 1968.)

The concepts of donor and acceptor impurities are easily extended to compound semiconductors, a group VI impurity like Te in a substitutional site on the lattice of the group V atoms and a group II impurity like Zn replacing a group III atom acting as a donor and acceptor, respectively. A group IV impurity in a III-V semiconductor is a donor or an acceptor, depending on whether it substitutes a group III or a group V host atom (Whelan, 1960).

We have thus seen that the donors and acceptors are solid-state analogs of the hydrogen atom. Indeed the energy levels represented in Eq. (47) are simply a "scaled-down" version of the energy levels of the hydrogen atom. In a similar fashion substitutional group VI (II) impurities in Si or Ge that bind two electrons (holes) constitute solid-state analogs of neutral helium.

It is clear from Eq. (47) that donors and acceptors in semiconductors should exhibit optical spectra similar to the well-known Lyman, Balmer,... series of the hydrogen atom. In view of the small m^* and large κ, typical of semiconductors, such spectra will occur in the near to far infrared. Besides, the samples have to be cooled to cryogenic temperatures in order to de-ionize the donors or the acceptors that have small ionization energies. Pioneering experiments by Burstein and co-workers (Burstein et al., 1953, 1956; Picus et al., 1956) on the Lyman spectra of donors and acceptors in silicon showed that at sufficient dilution discrete ground and excited states do exist. Since these early observations, impurity states have been experimentally investigated by using (1) absorption and photoconductivity spectroscopy in the near and far infrared, (2) Raman spectroscopy, and (3) luminescence. Such spectra have been reported for a large number of impurity species in several semiconductors and have been studied under the influence of elastic strain or magnetic fields. The understanding of the electronic-level structure of donors and acceptors has led to the development of some of the most sensitive infrared detectors based on the photoconductivity produced by the photoionization of impurity centers (Putley, 1964; Bratt, 1977). For a comprehensive review of the field of the shallow donors and acceptors explored with spectroscopy, we cite Ramdas and Rodriguez (1981).

The excitation spectrum of phosphorus donors in Si, i.e., of Si(P) is displayed in Fig. 23. The phosphorus was introduced in pure Si by exposing it to a slow neutron flux. It is known that the ^{30}Si isotope occurs with a natural abundance of 3.05%; with neutron capture it converts to ^{31}Si, which is unstable and transmutes to ^{31}P by β^- decay according to $^{30}\text{Si}(n,\gamma)^{31}\text{Si} \xrightarrow{2.6h} {}^{31}\text{P} + \beta^-$. The sample is annealed at 800°C to remove the radiation damage during the neutron irradiation and the nuclear transmutation. The phosphorus is substitutional and hence a donor. The absorption spectrum in Fig. 23 was recorded with a Fourier transform spectrometer, the sample being cooled to

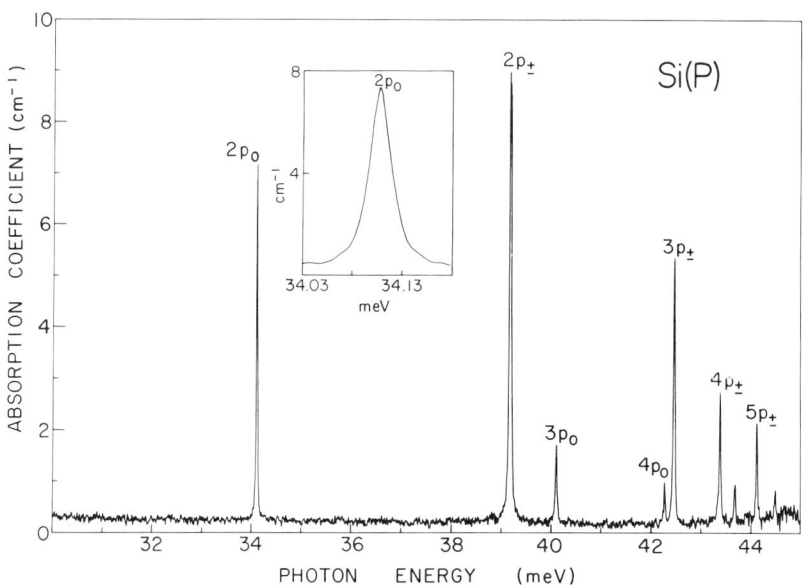

FIG. 23. Excitation spectrum of phosphorus donors in neutron-transmutation-doped silicon. Liquid helium is used as a coolant. Phosphorus donor concentration, $n \sim 1.2 \times 10^{14}$ cm^{-3}. The instrumental resolution with and without apodization is 0.09 and 0.06 cm^{-1}, respectively. The $2p_0$ line with an expanded horizontal scale is shown in the inset. The $2p_\pm$ line has been truncated because, with the thickness of the sample used, the transmission approaches zero at the peak. (After Jagannath *et al.*, 1981.)

liquid helium temperature (Jagannath et al., 1979, 1981). As can be seen, a series of sharp lines is observed in the 32- to 45-meV range. The occurrence of the spectrum—the Lyman spectrum—clearly demonstrates the existence of the bound states expected from a model for the donor that takes into account the multivalley nature of the conduction band of Si, namely the six Δ conduction-band minima that are characterized by the relevant effective-mass parameters.

The theory of the electronic energy levels of donors in Si—their binding energies and their symmetries—has been formulated in the most complete fashion by Kohn and Luttinger (see Kohn, 1957; see also Faulkner, 1969; Pantelides, 1978; Wintgen et al., 1987; and Broecks et al., 1986 for a discussion and the relevant literature). This theory is based on the assumption that the perturbation introduced by the foreign atom, that is the Coulomb potential introduced by the donor, $U_i(\mathbf{r})$ given by Eq. (46), varies slowly within distances of the order of the lattice spacing. It is referred to as the effective-mass theory in which the energy eigenfunctions can be accurately described by a product of the Bloch function at the position of the conduction-band minimum multiplied by an envelope function satisfying a Schrödinger equation of the form

$$[E_v(\mathbf{k}_0 - i\nabla) + U_i(\mathbf{r})]F_v(\mathbf{r}) = EF_v(\mathbf{r}), \qquad (48)$$

where $E_v(\mathbf{k})$ is the energy eigenvalue of the Bloch function for $\mathbf{k}$, near $\mathbf{k}_0$ corresponding to the conduction-band minimum. For Si or Ge,

$$E(\mathbf{k}) = E(\mathbf{k}_0) + \frac{\hbar^2}{2m_t}[(k_x - k_{0x})^2 + (k_y - k_{0y})^2] + \frac{\hbar^2}{2m_l}[(k_z - k_{0z})^2], \qquad (49)$$

$\hat{z}$ being along a $\langle 100 \rangle$ for Si and a $\langle 111 \rangle$ for Ge. Kohn and Luttinger (1955) have shown that $F_v(\mathbf{r})$, the so called enveloped wave function, satisfies

$$-\frac{\hbar^2}{2m_t}\left(\frac{\partial^2 F_v}{\partial x^2} + \frac{\partial^2 F_v}{\partial y^2}\right) - \frac{\hbar^2}{2m_l}\frac{\partial^2 F_v}{\partial z^2} + U_i F_v = (E - E_v(\mathbf{k}_0))F_v. \qquad (50)$$

Equation (50) has axial symmetry about the z-axis rather than spherical symmetry. Hence the effective Hamiltonian commutes with the z-component of the angular momentum, L_z, but not with the other components. Thus the eigenfunctions of Eq. (50) are of the form $R_{km}(r, \theta)e^{im\phi}$, where r, θ, ϕ are the polar coordinates of the electron. Time-reversal symmetry dictates that the states with magnetic quantum numbers m and $-m$ be degenerate. Thus, for a p state, the levels, which are degenerate if $m_t = m_l$, are now split into a singlet p_0 with $m = 0$ and a double $p_\pm$ with $m = \pm 1$.

In Si (Ge), as already stated, there are six (four) equivalent positions in the Brillouin zone for which $E_v(\mathbf{k})$ is a minimum. We designate these by $\mathbf{k}_j$ ($|\mathbf{k}_j| = k_0$). For each of these positions we must solve Eq. (50), bearing in mind that while the structure of the differential equation remains the same, the z-direction is different for each minimum. Arising from these considerations the donor states exhibit a sixfold (fourfold) degeneracy for Si (Ge). In principle, this degeneracy is removed by the crystal potential. The wave functions can be written as

$$\phi_i(\mathbf{r}) = \sum_{j=1}^{N} \alpha_{ij} F_j(\mathbf{r}) \psi_{vk_j}(\mathbf{r}), \tag{51}$$

where $N = 6$ or 4, depending on whether we are considering Si or Ge. Irrespective of the nature of the correction to the impurity potential used in Eq. (50), it must conform to the T_d site symmetry of the impurity. It is thus necessary to characterize the energy states of the donor in terms of the irreducible representations of T_d. These are displayed in Table IV.

From Table IV it can be seen that the N-fold degeneracy of the effective-mass states is, in principle, lifted. This effect, called the chemical splitting, need be considered only for the s states. The ground state of group V donors in Si consists of a singlet $1s(A_1)$, a doublet $1s(E)$, and a triplet $1s(T_2)$, whereas for Ge a singlet $1s(A_1)$ and a triplet $1s(T_2)$ characterizes it. Here we use the group-theoretical notation in Wilson, Decius, and Cross (1955) with $T_1 \equiv F_1$ and $T_2 \equiv F_2$. (In the paper of Kohn and Luttinger, 1955, T_1 and T_2 are interchanged.) The reduction of the N-dimensional representation generated by the states given by Eq. (51) for the 1s ground-state multiplet is

TABLE IV

IRREDUCIBLE REPRESENTATIONS* OF s AND p-LIKE STATES OF GROUP V DONORS IN SILICON AND GERMANIUM

Crystal	State	m	$\Gamma(T_d)$
Si	s-like	0	$A_1 + E + T_2$
	p-like	0	$A_1 + E + T_2$
		±1	$2T_1 + 2T_2$
Ge	s-like	0	$A_1 + T_2$
	p-like	0	$A_1 + T_2$
		±1	$E + T_1 + T_2$

* The nomenclature of column 4 is that of Wilson et al. (1955).

accomplished for Si with the following coefficients:

$$\alpha_{A_1,j} = \frac{1}{\sqrt{6}}(1,1,1,1,1,1), \tag{52}$$

$$\alpha_{E,j} = \begin{cases} \dfrac{1}{\sqrt{12}}(-1,-1,-1,-1,2,2), \\ \dfrac{1}{2}(1,1,-1,-1,0,0); \end{cases} \tag{53}$$

$$\alpha_{T_2,j} = \begin{cases} \dfrac{1}{\sqrt{2}}(1,-1,0,0,0,0), \\ \dfrac{1}{\sqrt{2}}(0,0,1,-1,0,0), \\ \dfrac{1}{\sqrt{2}}(0,0,0,0,1,-1). \end{cases} \tag{54}$$

For Ge the corresponding reduction is achieved by

$$\alpha_{A_1,j} = \frac{1}{2}(1,1,1,1), \tag{55}$$

$$\alpha_{T_2,j} = \begin{cases} \dfrac{1}{2}(1,1,-1,-1), \\ \dfrac{1}{2}(1,-1,-1,1), \\ \dfrac{1}{2}(1,-1,1,-1). \end{cases} \tag{56}$$

From the nature of the linear combinations of the wave functions it is clear (Kohn, 1957) that the $1s(A_1)$ state in both silicon and germanium has the largest amplitude at the donor site and hence should be most affected by species-dependent effects. The p-like states must, in principle, also split in a manner shown in Table IV. However, such states give a small probability density near the nucleus of the donor, and consequently the splittings are too small to be observed experimentally except for Mg^+ (Ho and Ramdas, 1972). The energy separations of the different components of the 1s manifold are called the "valley-orbit" splittings. Figure 24 displays schematically the energy-level scheme for a donor-bound electron in Si with labels reflecting the nature of the envelope function and the group-theoretical symmetry.

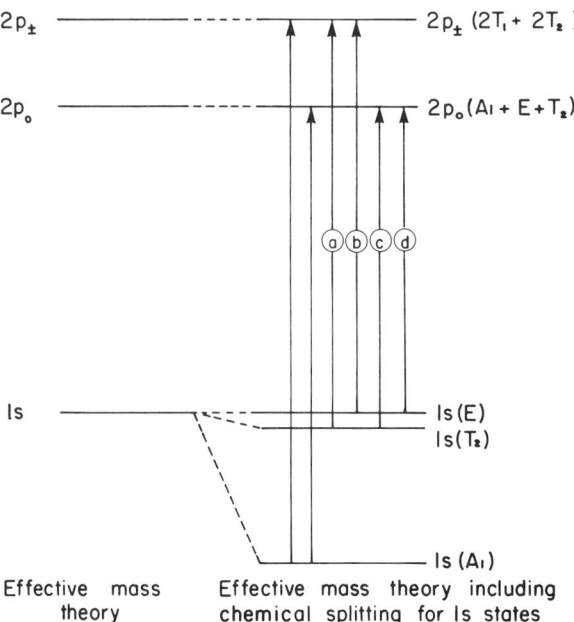

FIG. 24. Energy-level scheme (not to scale) of Group V impurities in silicon for transitions from the 1s states to the $2p_0$, $2p_\pm$ states. The letters next to a level indicate the irreducible representations of T_d; the energy separation between $1s(E)$ and $1s(A_1)$ is usually designated by $6\Delta_c$.

In the effective-mass approximation, the strain dependence of the donor levels is also given by Eq. (21). If it is assumed that both the dielectric constant and the effective masses characterizing the conduction-band minima are unaltered by a small strain, then the energy-level scheme of a group V donor in Si or Ge will be unaffected by the stress since Eq. (50) will be unaltered. However, the energy-level schemes bearing *different* valley labels will be *shifted* relative to one another by amounts given by Eq. (21). Implicit in this deduction is the assumption that Ξ_d and Ξ_u are the same for all the donor levels as well as the conduction-band extremum. Experimentally, the stress dependence of the p states is well described with this assumption. The chemical splitting of $1s(A_1 + E + T_2)$ and $1s(A_1 + T_2)$ ground states of donors in Si and Ge, respectively, requires the use of a Hamiltonian that includes both chemical splitting and strain; such a calculation has been given by Price (1956) for Ge and by Wilson and Feher (1961) for Si.

In order to compare experimental results with the theory, it is useful to recall that the shift of the "center of gravity" of the valleys is given by Eq. (22). From Eq. (25) and Fig. 5 one can thus deduce that, for $\mathbf{F} \parallel [100]$, the excited p states split into two sublevels, the one of lower energy being shifted twice as much below the center of gravity as the other is shifted above it. The shift of the $1s(A_1)$ level from its zero-stress position is given by

$$\Delta E_{gs} = \Delta_c \left[3 + \frac{1}{2}x - \frac{3}{2}\left(x^2 + \frac{4}{3}x + 4\right)^{1/2} \right], \tag{57}$$

where $6\Delta_c$ is the separation between the $1s(A_1)$ and $1s(E)$ states and

$$x = -\frac{\Xi_u(s_{11} - s_{12})T}{3\Delta_c} \tag{58}$$

for compression (Wilson and Feher, 1961). The energy shifts and splittings of the 1s multiplet and an np state for $\mathbf{F} \parallel [100]$, together with their symmetry classification are displayed in Fig. 25 (Aggarwal and Ramdas, 1965b).

In Fig. 26 we show the results of Aggarwal and Ramdas (1965c) for the $1s(A_1) \to n$p transitions in Si(As) subjected to uniaxial compression along [100]; polarized light was used with the electric vector $\mathbf{E}$ either parallel or perpendicular to $\mathbf{F}$. As can be seen, each excitation line splits into two components, np$(+)$ and np$(-)$; here $+$ and $-$ signs denote the high- and low-energy components with respect to the zero-stress position. In Fig. 27 we show the position of the $1s(A_1) \to 2p_\pm(+)$ and $1s(A_1) \to 2p_\pm(-)$ components of Si(P) as a function of stress (Tekippe et al., 1972). The solid straight lines are energy shifts of the two components of the $2p_\pm$ line from the zero-stress position calculated from their energy difference divided in the ratio 2:1. The curved solid lines are due to the shift of the $1s(A_1)$ ground state with stress. The solid points are computed by taking the difference between the straight lines and the data points. The splitting of the $2p_\pm$ level, i.e., $\Delta E_{2p_\pm} = E_{2p_\pm}^{(+)} - E_{2p_\pm}^{(-)}$, as a function of stress for Si(Sb), Si(As), Si(P), and Si(Mg) is displayed in Fig. 28. The linear dependence of $\Delta E_{2p_\pm}$ with stress has the same slope for all the impurities and yields $\Xi_u = (8.77 \pm 0.07)$ eV. A similar study for Si(Bi) by Butler et al. (1975) gave $\Xi_u = (8.77 \pm 0.14)$ eV. The value for Ξ_u determined in this manner from the piezospectroscopy of donors equals within experimental error that of the conduction-band minimum (Tekippe et al., 1972). Jagannath and Ramdas (1981) show that the Ξ_u characterizing the energy levels of interstitial lithium donors in Si is equal to that of the conduction band. Tekippe et al. (1972) found that the Ξ_u for the $1s(A_1)$ state is significantly lower than that for the np states, decreasing with increasing $6\Delta_c$. Jagannath and Ramdas (1981) report that Ξ_u for the $1s(E + T_2)$ state of Si(Li) is identical to that of the np states.

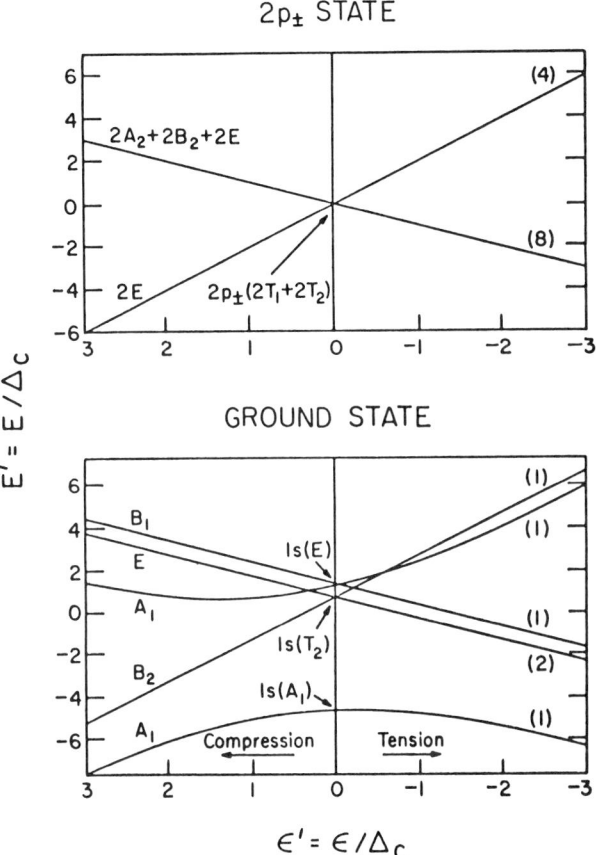

FIG. 25. Splitting of the 1s ground-state multiplet and the $2p_\pm$ state of a group V donor in Si for $\mathbf{F} \parallel [100]$. The numbers in parentheses indicate the degeneracies of the various states while the letters denote the irreducible representations of D_{2d}, the site symmetry of the donor for $\mathbf{F} \parallel [100]$. $E' = E/\Delta_c$ and $\varepsilon' = \varepsilon/\Delta_c$, where E is the energy of a given state, $\Delta_c = 1/6$ [spacing between $1s(E)$ and $1s(A_1)$] and 3ε is the energy difference between the $\langle 100 \rangle$ conduction-band minima. (After Aggarwal and Ramdas, 1965b.)

The striking polarization features of the stress induced components shown in Fig. 26 follow from the electric dipole selection rules shown in Fig. 29. These have been derived within the framework of the effective-mass approximation. In Table V we show the symmetry classification of the stress induced sublevels for donors in Si and Ge with $\mathbf{F} \parallel [100]$, and $[111]$, respectively. The procedure used is similar to that followed to construct Table IV (Aggarwal and Ramdas, 1965a).

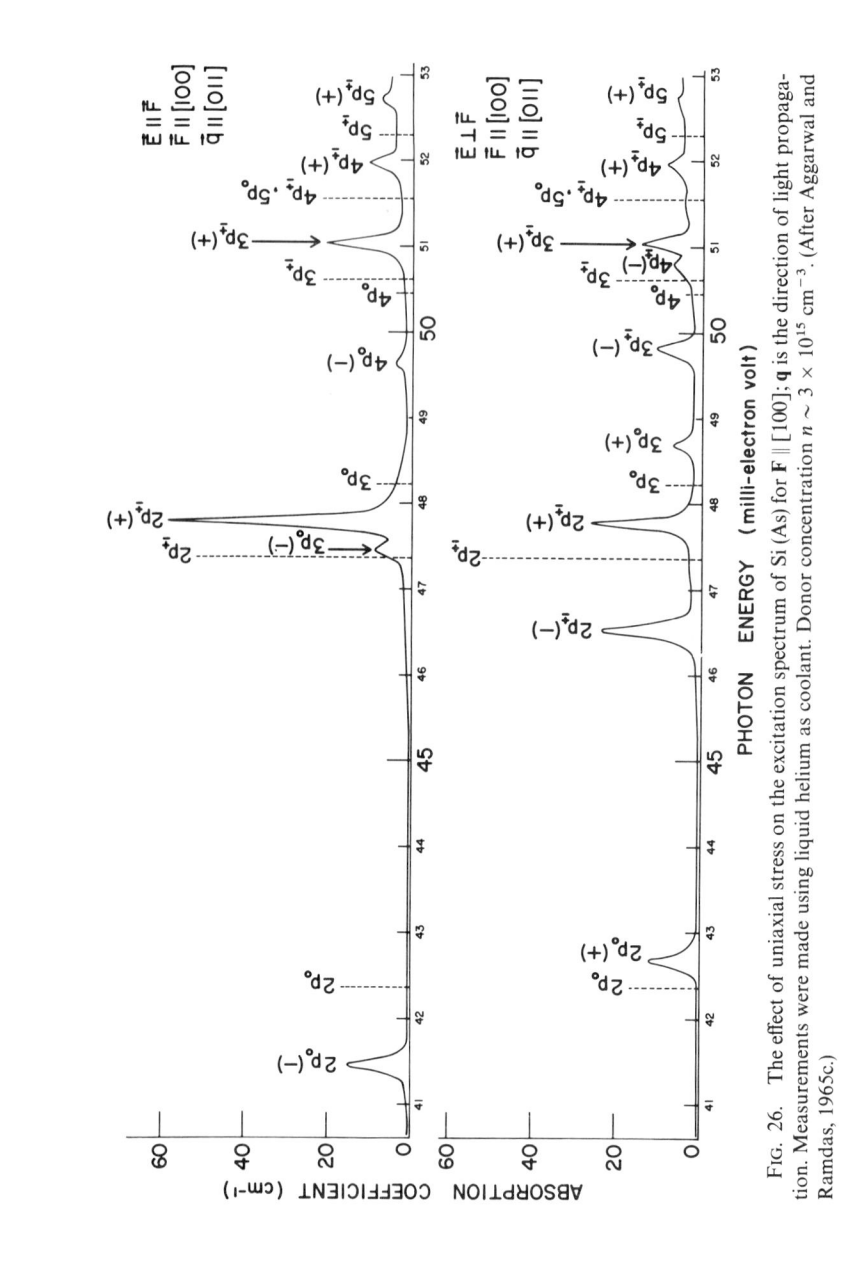

FIG. 26. The effect of uniaxial stress on the excitation spectrum of Si(As) for **F** ∥ [100]; **q** is the direction of light propagation. Measurements were made using liquid helium as coolant. Donor concentration $n \sim 3 \times 10^{15}$ cm^{-3}. (After Aggarwal and Ramdas, 1965c.)

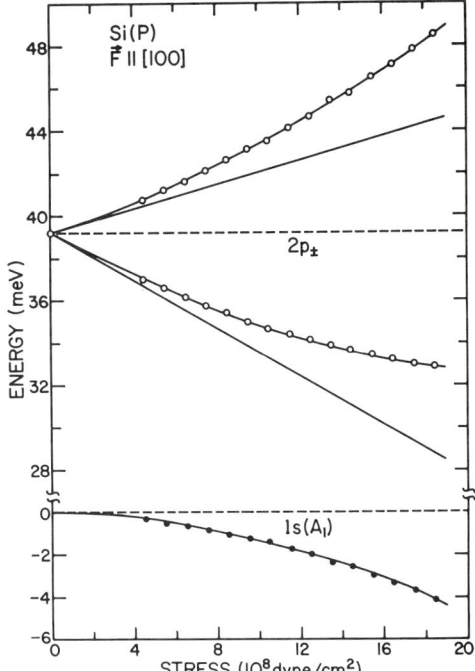

FIG. 27. Stress dependence of the energies of the two components of the $1s(A_1) \to 2p_\pm$ line of Si(P) for $\mathbf{F} \parallel [100]$. (After Tekippe et al., 1972.)

The piezospectroscopy of the $1s(A_1)$, $1s(T_2) \to 2p_0$, $2p_\pm$ Lyman lines of Ge(As) and Ge(Sb) has been investigated by Reuszer and Fisher (1965, 1968). The polarization features of the stress-induced components are once again fully accounted for in a manner similar to that for the donor spectra of Si. The differences in Si and Ge are related to the Δ-minima in the former and the L-minima in the latter and their behavior under uniaxial stress as shown in Figs. 5 and 6. Martin et al. (1979, 1981) have carried out a quantitative piezospectroscopy of group V donors in Ge and deduce $\Xi_u = 16.4 \pm 0.2$ eV from their measurements on the splitting of the p-like excited states. Gorman and Solin (1977) investigated the $1s(A_1) \to 1s(T_2)$ Raman transition of Ge(As) under uniaxial stress. The $1s(T_2)$ state is expected to be characterized by the same Ξ_u that the p-like states as well as the L-minima posses and the higher value of 17.8 ± 0.5 eV obtained by them is somewhat inconsistent with the lower value of 164 ± 0.2 eV reported by Martin et al. (1981). The extensive studies on double donors in Si are reviewed by Grossmann et al. (1987).

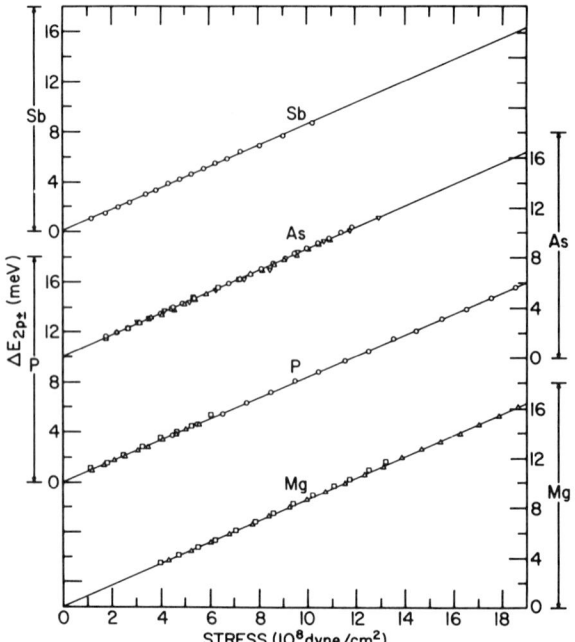

FIG. 28. Splitting of the $1s(A_1) \to 2p_\pm$ line with $\mathbf{F} \parallel [100]$ as a function of stress for Sb, As, P, and Mg donors in Si. The solid lines represent least-squares fits to all the data points for a given impurity. (After Tekippe et al., 1972.)

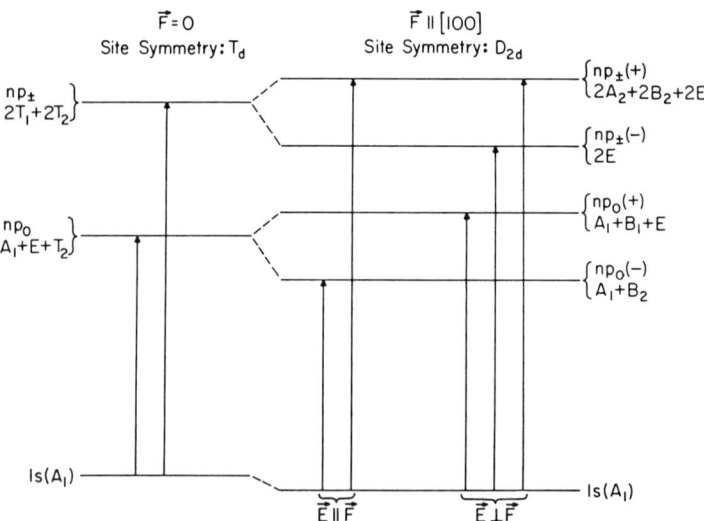

FIG. 29. Energy levels of donors in Si (not to scale) for $\mathbf{F} = 0$ and $\mathbf{F} \parallel [100]$. The arrows indicate the allowed transitions with $1s(A_1)$ as the ground state. The letters next to a level denote the irreducible representations of the appropriate site symmetry. (After Aggarwal and Ramdas, 1965a.)

TABLE V

CLASSIFICATION OF p-STATES OF DONORS IN SILICON AND GERMANIUM UNDER UNIAXIAL COMPRESSION

Crystal	Direction of compression	Symmetry of impurity	m	Level scheme $\mathbf{F} = 0$	Level scheme $\mathbf{F} \neq 0$	
Si	[100]	D_{2d}	0	$A_1 + E + T_2$	$A_1 + B_1 + E$	(3, 4, 5, 6)
					$A_1 + B_2$	(1, 2)
			± 1	$2T_1 + 2T_2$	$2A_2 + 2B_2 + 2E$	(3, 4, 5, 6)
					$2E$	(1, 2)
Ge	[111]	C_{3v}	0	$A_1 + T_2$	$A_1 + E$	(2, 3, 4)
					A_1	(1)
			± 1	$E + T_1 + T_2$	$A_1 + A_2 + 2E$	(2, 3, 4)
					E	(1)

The shallow group III acceptor states in Si and Ge are described in terms of the band parameters and wave functions of the $\Gamma_8^+(p_{3/2})$ valence-band maximum and those of the $\Gamma_7^+(p_{1/2})$ valence-band maximum split-off by spin-orbit interaction. The effective-mass theory for group III acceptor states has been developed by Schechter (1962), Baldereschi and Lipari (1973), Lipari and Baldereschi (1978), Binggelli and Baldereschi (1989), Binggelli et al. (1988), and Buczko and Bassani (1989). The bound states of the hole are shown schematically in Fig. 30, where the $p_{3/2}$ and the $p_{1/2}$ series are associated with the $p_{3/2}$ and $p_{1/2}$ valence bands, respectively. The excitation lines of Si(B) shown in Figs. 31 and 32 (Jagannath et al. 1981) belong to the $p_{3/2}$ series.

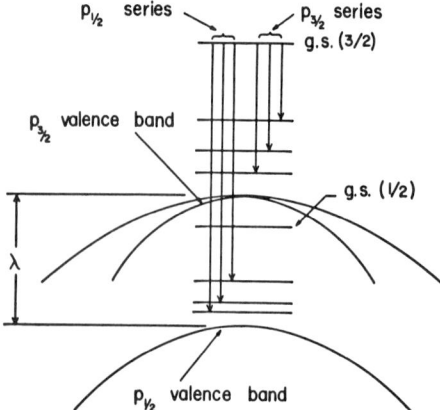

FIG. 30. The spin-orbit split valence band of silicon with associated acceptor states. Here g.s. is the ground state and λ is the spin-orbit splitting at $\mathbf{k} = 0$, labeled Δ_0 in Figs. 3, 8, and 14.

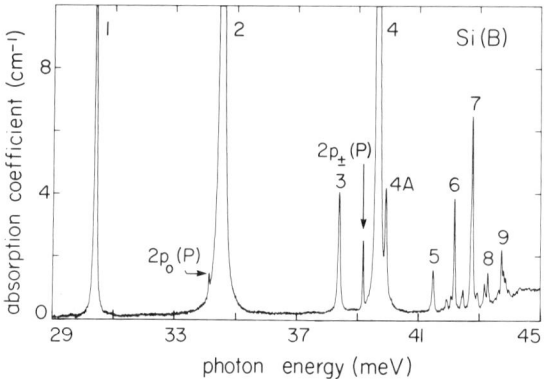

FIG. 31. Excitation spectrum of Si(B), $p(300\ \text{K}) = 8.5 \times 10^{14}\ \text{cm}^{-3}$. The Si(P) lines are labeled P in parentheses. Liquid helium used as coolant. The lines belong to the $p_{3/2}$ series. (After Jagannath et al., 1981.)

The energy states of an acceptor-bound hole can be characterized according to the symmetry deduced from the transformation properties of the products $\Psi_v(\mathbf{r}) = F_v(\mathbf{r})\Psi_{v0}(\mathbf{r})$ of Eq. (48). Both F_v and Ψ_{v0} form bases for the irreducible representations of $\bar{O}_h$. Hence, the symmetry properties of the $\Psi_v(\mathbf{r})$ are obtained by forming the direct product of the irreducible representations of $\bar{O}_h$ to which F_v and Ψ_{v0} belong. In the effective-mass theory, the envelope

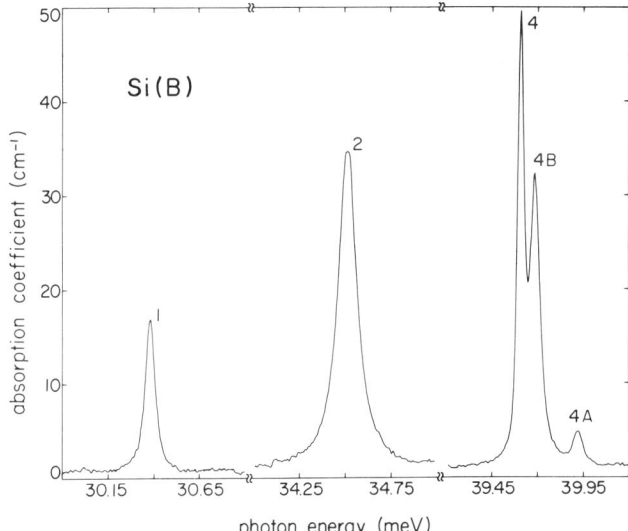

FIG. 32. Lines 1, 2, 4, 4A of Si(B) on an expanded horizontal scale. The sample used is the same as that in Fig. 31 but with reduced thickness. Liquid helium used as coolant. (After Jagannath et al., 1981.)

functions F_v may be approximated by the s, p, d, ... hydrogen functions, which belong to the irreducible representations $D_0^+, D_1^-, D_2^+, \ldots$, respectively, of the orthogonal group in three dimensions; under $\bar{O}_h$, these reduce to Γ_1^+, Γ_4^-, $\Gamma_3^+ + \Gamma_5^+, \ldots$, respectively. Hence, the symmetries of the $p_{3/2}$ impurity states are given by direct products of the irreducible representation of the envelope function and Γ_8^+, the representation of Bloch states at the top of the valence band. They are $\Gamma_1^+ \times \Gamma_8^+, \Gamma_4^- \times \Gamma_8^+, (\Gamma_3^+ + \Gamma_5^+) \times \Gamma_8^+, \ldots$ for the hydrogenic s, p, d, ... states, respectively. This classification scheme was given by Schechter (1962). However, the impurity potential is not in reality spherically symmetric, but must possess the symmetry of the impurity site T_d. Thus, the acceptor wave functions must transform according to the operations of the double group $\bar{T}_d$. The reduction of the representations of $\bar{O}_h$ to those of $\bar{T}_d$ yields additional splittings due to the deviations of the impurity potential from $e^2/\kappa r$. For example, the p-like hydrogenic states that, in the effective-mass theory, yield states of symmetry $\Gamma_4^- \times \Gamma_8^+$ reduce to four levels of symmetry $\Gamma_8, \Gamma_8, \Gamma_7, \Gamma_6$ as shown in Table VI. For further details of the effective-mass theory of acceptors in Si and Ge, we refer the reader to Ramdas and Rodriguez (1981). The excitation lines appearing in the spectra displayed in Figs. 31 and 32 correspond to the $p_{3/2}$ series shown schematically in Fig. 30.

TABLE VI

SYMMETRY OF GROUP III ACCEPTOR STATES ASSOCIATED WITH THE $p_{3/2}$ AND $p_{1/2}$ VALENCE BAND EDGES OF SILICON AND GERMANIUM

Band	Hydrogenic state	Acceptor state[b]	Symmetry under $\bar{T}_d$
$p_{3/2}$	S	$S_{3/2}$	Γ_8
	P	$P_{1/2}$	Γ_6
		$P_{3/2}$	Γ_8
		$P_{5/2}$	$\Gamma_7 + \Gamma_8$
$p_{1/2}$	S	$S_{1/2}$	Γ_7
	P	$P_{1/2}$	Γ_6
		$P_{3/2}$	Γ_8

[a] The symmetries of the Bloch functions, ψ_j, under $\bar{O}_h$ are $\Gamma_8^+(p_{3/2})$ and $\Gamma_7^+(p_{1/2})$.

[b] S, P,... refer to the hydrogenic state and the subindex to the quantum number F defined by equation (3.71) in Ramdas and Rodriguez (1981).

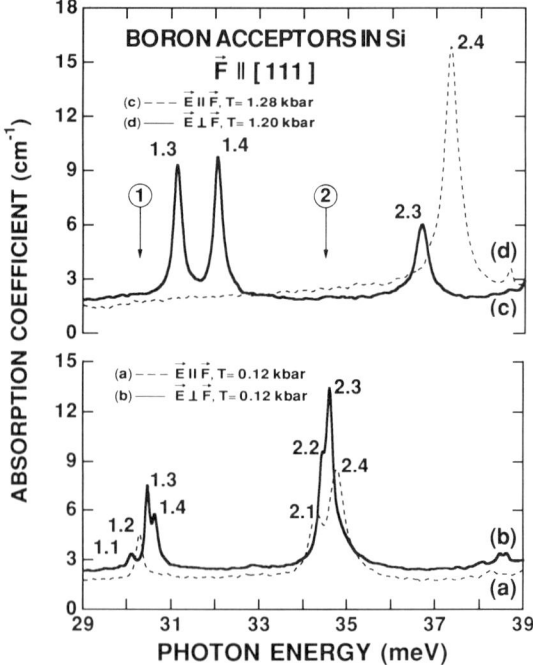

FIG. 33. Effect of uniaxial stress on lines 1 and 2 ($p_{3/2}$ series) of boron acceptors for $\mathbf{F} \parallel [111]$ and (a) $\mathbf{E} \parallel \mathbf{F}$, $T = 0.12$ kbar, (b) $\mathbf{E} \perp \mathbf{F}$, $T = 0.12$ kbar, (c) $\mathbf{E} \parallel \mathbf{F}$, $T = 1.28$ kbar, and (d) $\mathbf{E} \perp \mathbf{F}$, $T = 1.20$ kbar. Measurements were performed with liquid helium as coolant. (From Udo et al. 1991.)

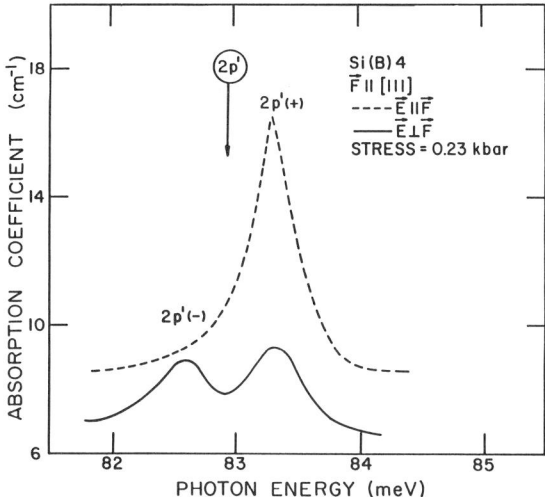

FIG. 34. Splitting of the 2p' line ($p_{1/2}$ series) of Si(B) into 2p'(+) and 2p'(−) components for **F** ∥ [111]. (From Chandrasekhar et al., 1973.)

In Fig. 33 the excitation spectrum of Si(B) under **F** ∥ [111] is shown (Udo et al., 1991). As can be seen, lines 1 and 2 split into four components expected for a $\Gamma_8 \rightarrow \Gamma_8$ transition. The 2p' line of the $p_{1/2}$ series, on the other hand, splits into two components, 2p'(+) and 2p'(−), consistent with a $\Gamma_8 \rightarrow \Gamma_6$ transition as shown in Fig. 34 (Chandrasekhar et al., 1973). The ordering of the ground-state sublevels has been deduced on the basis of the polarization and thermal depopulation effects shown in Figs. 33 and 34 and expected from Fig. 7a. This, in turn, permitted the ordering of the sublevels of the excited states for lines 1 and 2. The positions of the stress-induced components of lines 1 and 2 as a function of stress are shown in Fig. 35 and those of the 2p' line in Fig. 36. The ground-state splitting between the energies of 2p'(+) and 2p'(−) denoted by Δ is plotted in Fig. 37 as a function of stress. The deformation potential constants d_0 and b_0 characterizing the splitting of the Γ_8 ground state can be deduced from these results as well as those for lines 1 and 2. Udo et al. (1991) have obtained $d_0 = -4.41 \pm 0.04$ eV and $b_0 = -1.41 \pm 0.06$ eV from their recent study of lines 1 and 2. The nonlinear stress dependence exhibited in particular by lines 2.1, 2.3, and 1.4 can be explained on the basis of the hybridization of the $\Gamma_5 + \Gamma_6$ sublevels of the excited states of lines 1 and 2. This hybridization results in the unusual stress dependence of the relative intensities of the components. These aspects of the problem are discussed in detail by Chandrasekhar et al. (1973). An analogous situation in the excitation spectrum of Ge(Zn⁻) has been discovered and analyzed by Butler and Fisher (1976) and Duff et al. (1980).

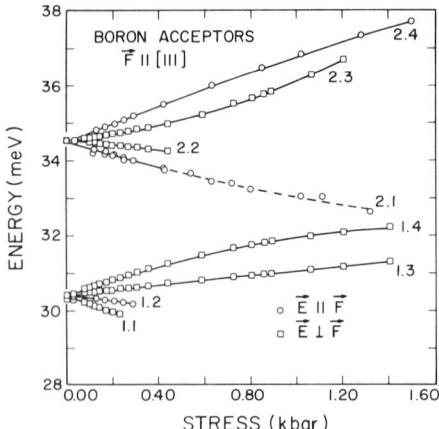

FIG. 35. Stress dependence of the energies of the components of lines 1 and 2 of boron acceptors in silicon for **F** ∥ [111]. (From Udo et al., 1991.)

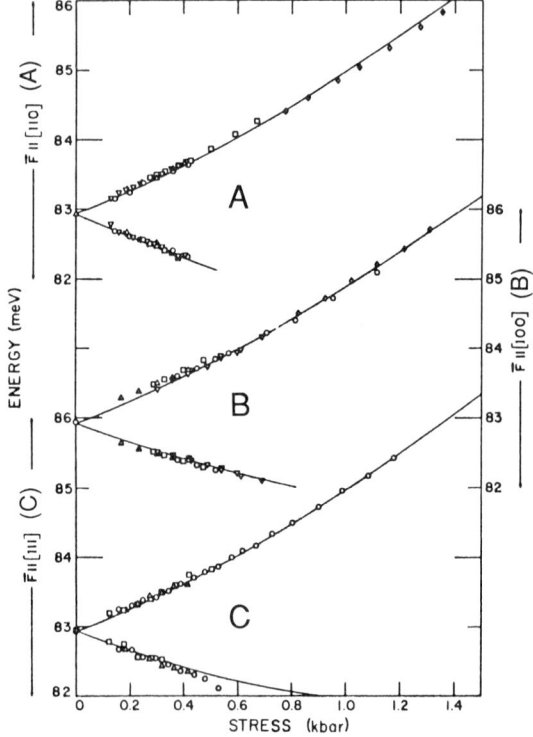

FIG. 36. Positions of the $2p'(+)$ and $2p'(-)$ components of Si(B) for **F** ∥ [111], **F** ∥ [001], and **F** ∥ [110]. (From Chandrasekhar et al., 1973.)

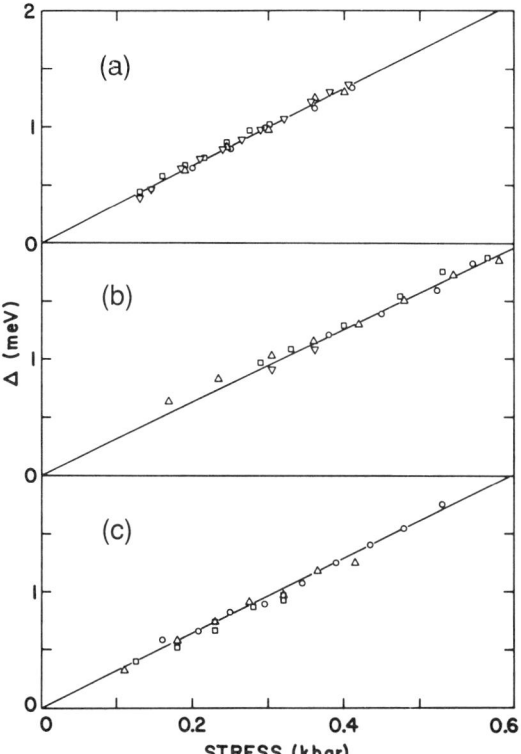

FIG. 37. Spacing Δ between 2p'(+) and 2p'(−) components of Si(B) for **F** ∥ [110] (a), **F** ∥ [001] (b), and **F** ∥ [111] (c). The spacing characterizes the splitting of the Γ_8 ground state. (From Chandrasekhar et al., 1973.)

VII. Zone-Center Optical Phonons

The tetrahedrally coordinated elemental semiconductors (C, Si, Ge, and α-Sn) are characterized by a single, long-wavelength, zero wave vector ($q \sim 0$), i.e., zone-center optical phonon. The zone-center optical phonon is a triply degenerate vibration in which the two interpenetrating face-centered cubic Bravais lattices of a crystal with O_h^7 space group rigidly vibrate against each other. Being symmetric with respect to the center of inversion, it is labeled F_{2g}, the irreducible representation of the O_h point group generated by the normal coordinates (see, for example, Venkatarayudu, 1938; Bhagavantam and Venkatarayudu, 1969). It is Raman active and forbidden in the infrared.

The corresponding zone-center vibration in the tetrahedrally coordinated III-V (e.g., GaAs) and II-VI (e.g., CdTe) semiconductors having T_d^2 space group has F_2 symmetry of the T_d point group; this zone-center vibration undergoes an LO–TO splitting due to the long-range polarization field associated with the partially ionic character of the III–V and II–VI semiconductors. The absence of a center of inversion allows them to appear in the Raman spectrum (Couture-Mathieu and Mathieu, 1953) and to manifest themselves in the reststrahlen reflectivity as well (Haas and Mathieu, 1954). Semiconductors of the wurtzite symmetry (space group C_{6v}^4), also based on the tetrahedral coordination (e.g., CdS, CdSe,...), have a more complex first-order Raman and infrared spectrum, thanks to their larger unit cell, uniaxial structure, and a partially ionic character (Damen *et al.*, 1966; Mitsuishi *et al.*, 1965).

The piezospectroscopy of the zone-center optical phonons, first explored by Marieé and Mathieu (1946) in α-quartz with the limitations of the prelaser Raman spectroscopy, has been investigated since 1970 in a large number of semiconductors and insulators with the well-known advantages associated with the use of lasers, double-grating monochromators, and photoelectric detection. Anastassakis *et al.* (1970) reported for the first time the effect of uniaxial stress on the zone-center optical phonon of a semiconductor, namely the F_{2g} mode of Si. The first-order Raman spectrum of α-quartz studied at liquid helium temperature under uniaxial stress (Tekippe and Ramdas, 1971; Tekippe *et al.*, 1973; Briggs and Ramdas, 1977) and a similar study of the F_{2g} zone-center mode of CaF_2 and BaF_2 as well as the rich first-order spectrum of $B_{12}GeO_{20}$ (Venugopalan and Ramdas, 1973) are examples of piezospectroscopy applied to zone-center optical phonons. Cerdeira *et al.* (1972) enlarged the list of the tetrahedrally coordinated semiconductors thus investigated by studying Ge, GaAs, GaSb, and ZnSe. For an extensive bibliography on such uniaxial stress investigations, we refer the readers to Anastassakis (1980, 1981) and Weinstein and Zallen (1984). We also cite here Grimsditch *et al.* (1978) for work on diamond; Anastassakis and Cardona (1987) on AlSb; Anastassakis *et al.* (1988) on InP; Balslev (1974) on GaP; Anastassakis *et al.* (1990) on Si; Wickboldt *et al.* (1987) and Sood *et al.* (1985) on GaAs. Weinstein and Cardona (1972) investigated the effect of uniaxial stress on the reststrahlen spectrum of GaAs.

In order to illustrate the general theoretical procedure for analyzing the piezospectroscopic effects observed for zone-center optical phonons, we follow the method in Tekippe *et al.* (1973) and Venugopalan and Ramdas (1973). We reproduce the analysis applied to an F_{2g} zone-center optical mode of a crystal with O_h point group as developed in the latter.

Table VII summarizes the decomposition of the F_{2g} representation of the point group O_h in the presence of a compressive force along [001], [111], or [110]. Since Raman lines originate from transitions between the totally symmetric ground state and the excited state having the irreducible representation of the phonon, the representation for the polarizability tensor must

TABLE VII

DECOMPOSITION OF THE IRREDUCIBLE REPRESENTATION F_{2g} OF O_h INTO THE IRREDUCIBLE REPRESENTATIONS OF THE NEW POINT GROUP IN THE PRESENCE OF A UNIAXIAL STRESS. THE DEGENERACIES OF THE REPRESENTATIONS ARE SHOWN IN PARENTHESES

Direction of compression	New point group symmetry	$F = 0$	$F \neq 0$
[001]	D_{4h} (Tetragonal)	$F_{2g}(3)$	$B_{2g}(1)$, $E_g(2)$
[111]	D_{3d} (Trigonal)	$F_{2g}(3)$	$A_{1g}(1)$, $E_g(2)$
[110]	D_{2h} (Orthorhombic)	$F_{2g}(3)$	$A_g(1)$, $B_{2g}(1)$, $E_g(2)$

contain that of the phonon. Thus, the polarizability tensor elements associated with the stress-induced components are those that transform according to the appropriate representation of the new point group for a given component; these are summarized in Table VIII. A perturbation calculation using a phenomenological Hamiltonian linear in strain has been exploited in the piezospectroscopy of the electronic levels as illustrated in the earlier sections. In this approach, the strain-induced potential is expressed in such a manner that each part of the potential transforms according to a particular irreducible representation of the point group of the crystal. Using basis functions appropriate to the irreducible representation of the energy level under consideration, one can obtain the secular equation in first-order perturbation theory. The potential and the basis functions have definite symmetry properties, and thus the group-theoretical matrix element theorem (Tinkham, 1964) can be exploited for determining the nonzero matrix elements. In addition, the independent matrix elements can be determined from the requirement that they remain invariant under the symmetry operations of the point group. Thus, the secular equation can be simplified and expressed in terms of the components of the strain tensor and the independent matrix element or the "deformation potential constant." The secular equation can

TABLE VIII

Polarizability Tensor Components of the Sublevels of the F_{2g} Mode of O_h in the Presence of a Uniaxial Stress: $\hat{x} \parallel [100]$, $\hat{y} \parallel [010]$, $\hat{z} \parallel [001]$; $\hat{x}' \parallel [\bar{1}10]$, $\hat{y}' \parallel [001]$, $\hat{z}' \parallel [110]$; $\hat{x}'' \parallel [11\bar{2}]$, $\hat{y}'' \parallel [\bar{1}10]$, $\hat{z}'' \parallel [111]$

New point group symmetry	Irreducible representation	Polarizability tensor components
D_{4h}	B_{2g}	α_{xy}
	E_g	α_{yz}, α_{zx}
D_{3d}	A_{1g}	$\alpha_{x''x''} + \alpha_{y''y''}, \alpha_{z''z''}$
	E_g	$(\alpha_{x''x''} - \alpha_{y''y''}, \alpha_{x''y''})(\alpha_{y''z''}, \alpha_{z''x''})$
D_{2h}	A_g	$\alpha_{x'x'}, \alpha_{y'y'}, \alpha_{z'z'}$
	B_{2g}	$\alpha_{y'x'}$
	B_{3g}	$\alpha_{y'z'}$

be diagonalized for a given direction of uniaxial stress to yield the eigenvalues and eigenvectors. The eigenvalues give the stress-induced splittings in the energies of the phonons. The eigenvectors in turn can be used to determine the linear combinations of the zero-stress polarizability tensors that are appropriate for describing the polarization characteristics of the stress-induced components. Within the framework of this phenomenological theory, the splittings for different crystallographic orientations of the applied force can be correlated in terms of the deformation potential constants, whereas the intensities of the stress-induced components can be calculated in terms of the zero-stress polarizability tensor components.

In the linear approximation, the strain Hamiltonian can be expressed as in Eq. (27) for the F_{2g} representation of group O_h, which can be written as

$$D = (1/3)(D_{xx} + D_{yy} + D_{zz})(\varepsilon_{xx} + \varepsilon_{yy} + \varepsilon_{zz})$$

$$+ (1/6)(2D_{zz} - D_{xx} - D_{yy})(2\varepsilon_{zz} - \varepsilon_{xx} - \varepsilon_{yy})$$

$$+ (1/2)(D_{xx} - D_{yy})(\varepsilon_{xx} - \varepsilon_{yy})$$

$$+ 2D_{xy}\varepsilon_{xy} + 2D_{yz}\varepsilon_{yz} + 2D_{zx}\varepsilon_{zx}. \tag{59}$$

In Eq. (59) the terms in the potential are regrouped such that $D_{xx} + D_{yy} + D_{zz}$ transforms as A_{1g}, $2D_{zz} - D_{xx} - D_{yy}$ and $\sqrt{3}(D_{xx} - D_{yy})$ as E_g, and D_{xy}, D_{yz}, and D_{zx} as F_{2g}. The basis functions appropriate for the representations A_{1g}, E_g, and F_{2g} in O_h are listed in Table IX. For calculating the matrix elements of D, we need the decomposition of products of the form $f_i f_j$ into the basis functions of O_h, $i, j = \xi, \eta, \zeta$. The decomposition of such products is given in Table X. Using the matrix element theorem and symmetry arguments, the

TABLE IX

BASIS FUNCTIONS FOR A_{1g}, E_g AND F_{2g} REPRESENTATIONS OF THE POINT GROUP O_h

Irreducible representation	Basis function	
A_{1g}	$f_0 = X^2 + Y^2 + Z^2$;	$f'_0 = X^2Y^2 + Y^2Z^2 + Z^2X^2$
E_g	$f_1 = 2Z^2 - X^2 - Y^2$;	$f'_1 = 2X^2Y^2 - Y^2Z^2 - Z^2X^2$
	$f_2 = \sqrt{3}(X^2 - Y^2)$;	$f'_2 = \sqrt{3}(Y^2Z^2 - Z^2X^2)$
F_{2g}	$f_\xi = YZ$;	$f'_\xi = X^2YZ$
	$f_\eta = ZX$;	$f'_\eta = Y^2ZX$
	$f_\zeta = XY$;	$f'_\zeta = Z^2XY$

TABLE X

DECOMPOSITION OF THE PRODUCTS $f_i f_j$ INTO THE BASIS FUNCTIONS OF O_h; $i, j = \xi, \eta, \zeta$

f_i \ f_j	f_ξ	f_η	f_ζ
f_ξ	$\dfrac{f'_0}{3} - \dfrac{f'_1}{6} + \dfrac{f'_2}{2\sqrt{3}}$	f'_ζ	f'_η
f_η	f'_ζ	$\dfrac{f'_0}{3} - \dfrac{f'_1}{6} - \dfrac{f'_2}{2\sqrt{3}}$	f'_ξ
f_ζ	f'_η	f'_ξ	$\dfrac{f'_0}{3} + \dfrac{f'_1}{3}$

secular equation can be expressed in terms of three deformation potential constants a, b, and c, as follows (Venugopalan and Ramdas, 1973):

$$\begin{vmatrix} a\varepsilon_{xx} + b(\varepsilon_{yy} + \varepsilon_{zz}) - \lambda & 2c\varepsilon_{xy} & 2c\varepsilon_{zx} \\ 2c\varepsilon_{xy} & a\varepsilon_{yy} + b(\varepsilon_{zz} + \varepsilon_{xx}) - \lambda & 2c\varepsilon_{yz} \\ 2c\varepsilon_{zx} & 2c\varepsilon_{yz} & a\varepsilon_{zz} + b(\varepsilon_{yy} + \varepsilon_{xx}) - \lambda \end{vmatrix} = 0, \quad (60)$$

where

$$a = (1/9)\{\langle f'_0 | D_{xx} + D_{yy} + D_{zz}\rangle + \langle f'_1 | 2D_{zz} - D_{xx} - D_{yy}\rangle\},$$

$$b = (1/18)\{2\langle f'_0 | D_{xx} + D_{yy} + D_{zz}\rangle - \langle f'_1 | 2D_{zz} - D_{xx} - D_{yy}\rangle\}, \quad (61)$$

$$c = \langle f'_\zeta | D_{xy}\rangle = \langle f'_\eta | D_{zx}\rangle = \langle f'_\xi | D_{yz}\rangle.$$

In cubic crystals, the components of the strain tensor ε_{ij} are related to the components of the stress tensor σ_{ij} through the compliance constants s_{ij} as given in Eq. (8) referred to the cubic axes. Thus, for a given direction of uniaxial stress, the associated strain components can be determined and substituted in Eq. (60) and the eigenvalues and eigenvectors obtained by diagonalizing it.

From Eqs. (60) and (8), we can deduce the expressions for the energy shifts, which can be compared with experimental results for the applied force **F** along [001], [111], or [110]. The eigenvalues (λ) and eigenvectors (Λ) for these three cases are

$$\mathbf{F} \parallel [001]: \lambda_1 = \lambda_2 = T[(a + b)s_{12} + bs_{11}] = \lambda_D,$$

$$\lambda_3 = T[as_{11} + 2bs_{12}] = \lambda_S,$$

$$\Lambda_D = \begin{bmatrix} 1 \\ 0 \\ 0 \end{bmatrix}, \begin{bmatrix} 0 \\ 1 \\ 0 \end{bmatrix}; \quad \Lambda_S = \begin{bmatrix} 0 \\ 0 \\ 1 \end{bmatrix}. \tag{62}$$

$$\mathbf{F} \parallel [111]: \lambda_1 = \lambda_2 = \frac{T}{3}[(a + 2b)(s_{11} + 2s_{12}) - cs_{44}] = \lambda_D,$$

$$\lambda_3 = \frac{T}{3}[(a + 2b)(s_{11} + 2s_{12}) + 2cs_{44}] = \lambda_S,$$

$$\Lambda_D = \frac{1}{\sqrt{6}}\begin{bmatrix} 1 \\ 1 \\ \bar{2} \end{bmatrix}, \frac{1}{\sqrt{2}}\begin{bmatrix} \bar{1} \\ 1 \\ 0 \end{bmatrix}; \quad \Lambda_S = \frac{1}{\sqrt{3}}\begin{bmatrix} 1 \\ 1 \\ 1 \end{bmatrix}; \tag{63}$$

$$\mathbf{F} \parallel [110]: \lambda_1 = T[(a + b)s_{12} + bs_{11}],$$

$$\lambda_2 = \frac{T}{2}[a(s_{11} + s_{12}) + b(s_{11} + 3s_{12}) + cs_{44}],$$

$$\lambda_3 = \frac{T}{2}[a(s_{11} + s_{12}) + b(s_{11} + 3s_{12}) - cs_{44}],$$

$$\Lambda_1 = \begin{bmatrix} 0 \\ 0 \\ 1 \end{bmatrix}, \quad \Lambda_2 = \frac{1}{\sqrt{2}}\begin{bmatrix} 1 \\ 1 \\ 0 \end{bmatrix}, \quad \text{and} \quad \Lambda_3 = \frac{1}{\sqrt{2}}\begin{bmatrix} \bar{1} \\ 1 \\ 0 \end{bmatrix}. \tag{64}$$

In these equations, T is the applied force per unit area and defined to be negative for compression; the subscripts D and S denote doublet and singlet, respectively.

Since any uniaxial stress can be expressed as the sum of a hydrostatic part and shear components of the stress, the effect of a uniaxial stress on a degenerate level can be regarded as a hydrostatic shift of the energy level followed by a splitting due to the shear components of the stress. The eigenvalues in Eqs. (62)–(64) can thus be rewritten as

$$\mathbf{F} \parallel [001]: \lambda_S = \lambda_H + \frac{2}{3}\lambda_{[001]}, \tag{65}$$

$$\lambda_D = \lambda_H - \frac{1}{3}\lambda_{[001]},$$

where $\lambda_H = (T/3)(a + 2b)(s_{11} + 2s_{12})$ and $\lambda_{[001]} = T(a - b)(s_{11} - s_{12})$;

$$\mathbf{F} \parallel [111]: \lambda_S = \lambda_H + \frac{2}{3}\lambda_{[111]}, \tag{66}$$

$$\lambda_D = \lambda_H - \frac{1}{3}\lambda_{[111]},$$

where $\lambda_{[111]} = Tcs_{44}$;

$$\mathbf{F} \parallel [110]: \lambda_1 = \lambda_H - \frac{1}{3}\lambda_{[001]},$$

$$\lambda_2 = \lambda_H + \frac{1}{6}\lambda_{[001]} + \frac{1}{2}\lambda_{[111]}, \tag{67}$$

$$\lambda_3 = \lambda_H + \frac{1}{6}\lambda_{[001]} - \frac{1}{2}\lambda_{[111]}.$$

The polarizability tensors referred to the cubic axes for the F_{2g} mode in the unstressed crystal are

$$\begin{bmatrix} 0 & 0 & 0 \\ 0 & 0 & d \\ 0 & d & 0 \end{bmatrix}, \begin{bmatrix} 0 & 0 & d \\ 0 & 0 & 0 \\ d & 0 & 0 \end{bmatrix}, \begin{bmatrix} 0 & d & 0 \\ d & 0 & 0 \\ 0 & 0 & 0 \end{bmatrix}. \tag{68}$$

TABLE XI

Polarizability Tensors of the Sublevels of the F_{2g} Mode of O_h for $\mathbf{F} \parallel [001]$, [110] or [111]. The Tensors are Referred to the Coordinate Axes Indicated in Each Case. The Eigenvector Corresponding to each Tensor is also Given.

$\mathbf{F} \parallel [001]$ $\hat{x} \parallel [100], \hat{y} \parallel [010], \hat{z} \parallel [001]$		$\mathbf{F} \parallel [110]$ $\hat{x}' \parallel [\bar{1}10], \hat{y}' \parallel [001], \hat{z}' \parallel [110]$		$\mathbf{F} \parallel [111]$ $\hat{x}'' \parallel [11\bar{2}], \hat{y}'' \parallel [\bar{1}10], \hat{z}'' \parallel [111]$	
Eigenvector	Polarizability tensor	Eigenvector	Polarizability tensor	Eigenvector	Polarizability tensor
$\begin{bmatrix} 1 \\ 0 \\ 0 \end{bmatrix}$	$\begin{bmatrix} 0 & 0 & 0 \\ 0 & 0 & d \\ 0 & d & 0 \end{bmatrix}$	$\dfrac{1}{\sqrt{2}}\begin{bmatrix} 1 \\ 1 \\ 0 \end{bmatrix}$	$\begin{bmatrix} 0 & -d & 0 \\ -d & 0 & 0 \\ 0 & 0 & 0 \end{bmatrix}$	$\dfrac{1}{\sqrt{6}}\begin{bmatrix} 1 \\ 1 \\ \bar{2} \end{bmatrix}$	$\begin{bmatrix} \dfrac{-2d}{\sqrt{6}} & 0 & \dfrac{-d}{\sqrt{3}} \\ 0 & \dfrac{2d}{\sqrt{6}} & 0 \\ \dfrac{-d}{\sqrt{3}} & 0 & 0 \end{bmatrix}$
$\begin{bmatrix} 0 \\ 1 \\ 0 \end{bmatrix}$	$\begin{bmatrix} 0 & 0 & d \\ 0 & 0 & 0 \\ d & 0 & 0 \end{bmatrix}$	$\begin{bmatrix} 0 \\ 0 \\ 1 \end{bmatrix}$	$\begin{bmatrix} -d & 0 & 0 \\ 0 & 0 & 0 \\ 0 & 0 & d \end{bmatrix}$	$\dfrac{1}{\sqrt{2}}\begin{bmatrix} 1 \\ 1 \\ 0 \end{bmatrix}$	$\begin{bmatrix} 0 & \dfrac{2d}{\sqrt{6}} & 0 \\ \dfrac{2d}{\sqrt{6}} & 0 & \dfrac{-d}{\sqrt{3}} \\ 0 & \dfrac{-d}{\sqrt{3}} & 0 \end{bmatrix}$
$\begin{bmatrix} 0 \\ 0 \\ 1 \end{bmatrix}$	$\begin{bmatrix} 0 & d & 0 \\ d & 0 & 0 \\ 0 & 0 & 0 \end{bmatrix}$	$\dfrac{1}{\sqrt{2}}\begin{bmatrix} 1 \\ 1 \\ 0 \end{bmatrix}$	$\begin{bmatrix} 0 & 0 & 0 \\ 0 & 0 & d \\ 0 & d & 0 \end{bmatrix}$	$\dfrac{1}{\sqrt{3}}\begin{bmatrix} 1 \\ 1 \\ 1 \end{bmatrix}$	$\begin{bmatrix} \dfrac{-d}{\sqrt{3}} & 0 & 0 \\ 0 & \dfrac{d}{\sqrt{3}} & 0 \\ 0 & 0 & \dfrac{2d}{\sqrt{3}} \end{bmatrix}$

For uniaxial stress along a specified crystallographic direction, the eigenvectors of Eq. (60) can be used to determine the new linear combinations of the zero-stress tensors given earlier. However, these linear combinations are still referred to the cubic axes, and they must be transformed to the laboratory axes. The appropriate linear combinations transformed to the laboratory axes are given in Table XI for **F** ∥ [001], [111], or [110]. The system of axes as well as the eigenvector corresponding to each tensor are also given.

The effect of **F** ∥ [111] on the F_{2g} zone-center optical phonon of Si is displayed in Fig. 38, where the Raman spectrum has been recorded in the right-angle scattering geometry (Anastassakis et al., 1990). The splitting of the triply degenerate mode into a singlet and a doublet is clearly demonstrated. Figure 39 shows the stress dependence of the two components expected and observed for **F** ∥ [111]. The data reveal the linearity with stress of the shifts of the singlet and the doublet, superposed on the hydrostatic shift, as predicted by Eq. (66). The singlet and the doublet character of the stress-induced components are deduced from their distinct polarization behavior deduced from Table XI. Anastassakis et al. (1990) deduce the following values for the deformation potential constants: $(p/\omega_0^2) = -1.85 \pm 0.06$, $(q/\omega_0^2) = 2.31 \pm 0.07$, and $(r/\omega_0^2) = -0.71 \pm 0.02$; here p, q, and r are equivalent to $2\omega_0$ times a, b, and c, respectively, in Eq. (61) and ω_0 = the zero-stress Raman shift = 520 cm^{-1}. It is of interest to note here that Novak et al. (1978) have studied the zone-center optical phonon of Si under a biaxial stress generated by the differential thermal contraction produced in a constrained specimen; stresses ~60 kbar could be reached with this technique.

The "built-in" strain associated with semiconductor heterostructures manifests itself in the positions of the optical phonons observed in quantum

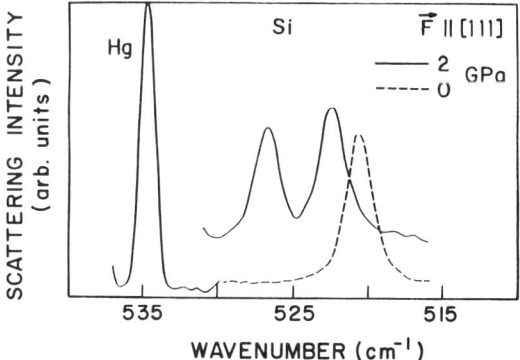

FIG. 38. The effect of [111] compression on the F_{2g} zone-center optical phonon of Si observed in right-angle Raman scattering geometry using the 1.06-μm Nd-YAG line. The zero-stress line occurs at 520 cm^{-1} and splits into a singlet (s) and a doublet (d) as predicted by Eq. (66). 1 GPa = 10 kbar. (After Anastassakis et al., 1990.)

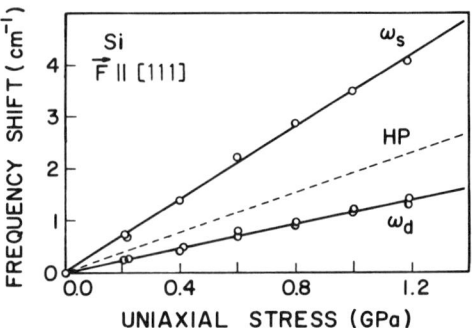

FIG. 39. Frequency shifts of the singlet and doublet components of the zone-center Raman line of Si measured for **F** ∥ [111]. Solid lines are linear least-squares fits to the experimental points. The dashed line shows the calculated shift corresponding to the hydrostatic component (HP). (After Anastassakis et al., 1990.)

well structures and superlattices. Since the lattice mismatch between the constituent layers in many of the heterostructures can be significant, the corresponding strain effects on the positions of the optical phonons have to be appropriately taken into account in discovering and delineating the zone-center optical phonons observed in the Raman spectrum as modified by the special circumstances specific to the structure; the nature of the "propagating," "interface," and "confined" optical phonons may be modified by such strains. See, for example, Cerdeira et al. (1984), who investigated the Raman spectrum of Ge_xSi_{1-x}/Si strained-layer superlattices, and Olego et al. (1988), who studied $ZnSe-ZnS_xSe_{1-x}$ superlattices. We refer the readers to recent review articles by Anastassakis (1991) and Pollak (1990) for further details and discussion.

VIII. Optical Phonons and Electronic States

Interaction between electronic and vibrational excitations plays a significant role in a host of physical phenomena and can be investigated in a particularly controlled manner in semiconductors. Scattering of free carriers by acoustic and optical phonons; the coupling between long-wavelength polar phonons and free carriers characterized by the so-called Fröhlich interaction; the resonant interaction of a discrete electronic transition with phonons resulting in coupled phonon–electron modes; coupled LO phonon–plasmon–

modes, etc. This partial list amply highlights the ubiquitous presence of electron–phonon interaction. These phenomena can be addressed fruitfully by using absorption and Raman spectroscopy, and uniaxial stress can be employed once more as a powerful tool in order to produce a perturbation that yields special insights. In this section we briefly discuss two examples.

10. COUPLED LO PHONON–PLASMON MODES IN n-GaSb

In a partially ionic semiconductor, the zone-center LO phonon can couple to the plasmon of the free carriers via their longitudinal electric fields (Yokata, 1961; Varga, 1965; Singwi and Tosi, 1966). The resulting coupled modes L_- and L_+ have been observed in a number of semiconductors, either directly as Raman lines (Mooradian and Wright, 1966; Mooradian and McWhorter, 1967) or more indirectly from an analysis of the infrared reflectivity spectrum (Perkowitz and Breecher, 1973; Chandrasekhar and Ramdas, 1980). The interest in studying the coupled modes in n-GaSb lies in its electronic band structure (Trommer and Ramdas, 1979), namely within about 90 meV above the absolute conduction-band minimum at the Γ-point of the Brillouin zone occur its subsidiary L-minima (Becker et al., 1961; Auvergne et al., 1974; Noack, 1977). It is thus possible to populate both the Γ- and L-minima with electrons provided by a sufficiently high donor concentration or by applying, at lower doping, uniaxial or hydrostatic stress that decreases the energy difference between them (Noack, 1977); thus, the effect of both types of carriers on the plasmon and, hence, on the coupled modes becomes accessible to study.

In order to observe the bulk properties of a doped semiconductor in Raman scattering, the penetration depth of the incident and the scattered light has to exceed the thickness of the carrier free-surface layer present in most semiconductors. This condition is fulfilled in n-GaSb in the visible red, e.g., for the 6471-Å radiation from a Kr$^+$ laser and for carrier concentrations $\sim 10^{18}$ cm^{-3} or larger. In the Raman experiments performed by Trommer and Ramdas in the standard backscattering geometry from (001) surfaces only the LO modes are allowed.

A typical Raman spectrum of n-GaSb exhibiting the coupled modes is shown in Fig. 40. In the high concentration range the low-frequency coupled mode L_- is almost a totally screened LO phonon and thus at the TO frequency, while the high-frequency mode, L_+, is almost a pure plasmon. In addition, the unscreened LO from the carrier free-surface region as well as some two-phonon features are observed.

In Fig. 41 the dependence of the $L_\pm$ frequencies ($\omega_\pm$) on the total carrier

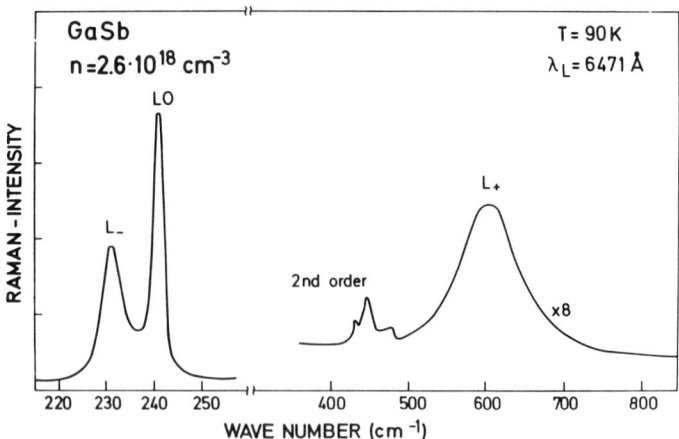

FIG. 40. Coupled plasmon–phonon modes L_+ and L_- of n-GaSb observed in the Raman spectrum. (After Trommer and Ramdas, 1979.)

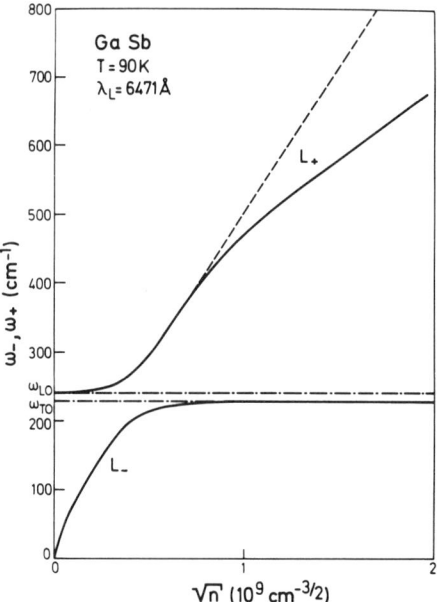

FIG. 41. Variation of the coupled mode frequencies with (electron concentration)$^{1/2}$ in GaSb calculated for $E_c(L) - E_c(\Gamma) = \infty$ (dashed line) and 90 meV (full line). (After Trommer and Ramdas, 1979.)

concentration n as calculated from

$$\omega_\pm^2 = \frac{\omega_0^2 + \omega_{LO}^2}{2} \pm \left[\frac{(\omega_p^2 + \omega_{LO}^2)^2}{4} - \omega_p^2 \omega_{TO}^2\right]^{1/2}, \qquad (69)$$

is shown. Here, ω_p, ω_{LO}, and ω_{TO} are the frequencies of the plasmon, the LO, and the TO phonons, respectively. In the high-frequency limit, i.e., for $\omega_p \gg qv_{F0}, qv_{F1}$, where q is the wave vector of the coupled mode and v_{F0} and v_{F1} are the Fermi velocities of the Γ and L electrons, the plasma frequency is an average of the contributions of both types of carriers and

$$\omega_p^2 = \frac{4\pi e^2}{\varepsilon_\infty}\left[\frac{n_1}{m_0^*} + \frac{n_1}{m_1^*} + \frac{3}{5}q^2(v_{F0}^2 + v_{F1}^2)\right]. \qquad (70)$$

(See Platzman and Wolff, 1973). Hence $\varepsilon_\infty =$ high-frequency dielectric constant; n_0 and n_1 are the carrier concentrations in Γ-, L-valleys, respectively; m_0^* and m_1^* are the corresponding masses for motion parallel to q. The figure is based on $\varepsilon_\infty = 14.4$ (Hass and Henvis, 1962), $m_0^* = 0.0405 m_e$ (Bimberg and Rühle, 1974), $m_{1\perp}^* = 0.079 m_e$ (Albert et al., 1976), and $m_{1\parallel} = 1.59 m_e$ (equal to that of Ge). Since the appropriate mass component of carriers in L-valleys is about three times larger than that of the carriers in the Γ-valley and the ratio of density of states is about 13, the slope of the L_+ frequency in Fig. 41 becomes smaller as soon as the L-minima begin to be occupied.

As discussed in Section IV, a uniaxial stress regroups the energetically equivalent L-minima into subgroups that differ in energy while the hydrostatic component of the stress changes the difference in the band-edge energy, $E_c(L) - E_c(\Gamma)$. The deformation potential constants Ξ_d and Ξ_u of the L-valleys and the hydrostatic deformation potential constant of the Γ-valley defined by Eq. (12), either experimentally known (Noack, 1977) or estimated, provide a basis for interpreting the experimental results on the effect of uniaxial stress on the L_+ and L_- modes. For a compressive force F along [100], the degeneracy of the L-minima is not lifted, but $E_c(L) - E_c(\Gamma)$ decreases. Figure 42 shows the L_+ frequencies as a function of stress T. A good fit to the [100] stress data is obtained by using $E_c(L) - E_c(\Gamma)$ in milli electron Volts $= (90 - 3 \cdot 2)T$ kbar (Noack, 1977). For $F \parallel [110]$, in addition to the hydrostatic shift of the "center of gravity" of the L-valleys with respect to the Γ-valley, the $[\bar{1}\bar{1}1]$, $[11\bar{1}]$ minima regroup and separate from the $[\bar{1}11]$, $[1\bar{1}1]$ minima. The shear deformation potential $\Xi_u(L)$ is deduced from a fit to the data shown in the lower part of Fig. 42. For the two lower concentration samples $\Xi_u = 15 \pm 0.5$ eV gives a good fit (Noack, 1977), whereas for the high-concentration samples a satisfactory fit could be obtained only for $\Xi_u = 4 \pm 0.5$ eV. The origin of this discrepancy has not been established.

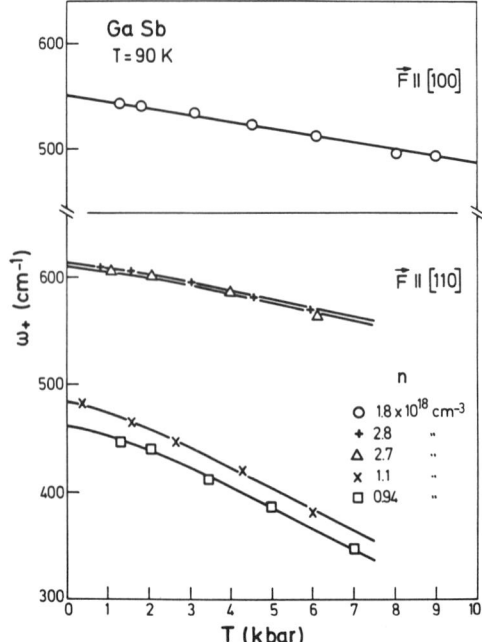

FIG. 42. Variation of the L_+ frequency with uniaxial compressive stress, for five samples with different carrier concentrations n. The solid lines are based on calculations explained in the text. The carrier concentrations were deduced from Fig. 41. (After Trommer and Ramdas, 1979.)

11. Zone-Center Optical Phonons in Si in the Presence of Free Carriers

A number of physical phenomena in semiconductors occur when different types of excitations have nearly the same energy. Strong interactions leading to coupled modes as well as broadening and unusual line shapes are exhibited in the optical spectra. In Si and Ge the zone-center optical phonon of F_{2g} symmetry (alternatively designated as the $\Gamma_{25'}$ phonon) can occur in the Raman spectrum with a frequency in the electronic continuum; this overlap can result in pronounced frequency shifts and line shape effects (Cerdeira and Cardona, 1972; Cerdeira et al., 1973, 1974; Fjeldly et al., 1973) explained in terms of Fano–Breit–Wigner theory. Chandrasekhar et al. (1977, 1978, 1980) have exploited piezospectroscopy to control the free-carrier continuum–$\Gamma_{25'}$ phonon interaction in both n-type and p-type Si. As a particularly striking illustration, we discuss the $\Gamma_{25'}$ phonon in the presence of an electron–hole-liquid (EHL) created by photoexcitation (Guidotti et al., 1979).

Guidotti et al. (1979) studied the Raman spectrum of intrinsic Si illuminated with photons of energies larger than the energy gap by using an Ar^+

pump later. The Si was maintained at low temperatures ($T \leq 15$ K) and subjected to a nonuniform stress that creates a potential energy minimum for excitons in the interior of the bulk. The electron–hole pairs generated near the surface form excitons and are collected into the interior by the potential well where they form an electron–hole–liquid. Raman scattering is studied with the Si thus prepared with a steady-state illumination from the Ar^+ pump laser and with the 10,796-Å line from a Nd-doped $YAlO_3$ laser or the 10,644-Å line from a Nd-doped yttrium aluminum garnet. We note here that Si at low temperatures is transparent to these probe radiations, and hence Raman scattering is observed from the bulk in right-angle scattering geometry.

Figure 43 displays the Raman spectrum and the TA phonon-assisted EHL recombination radiation observed in Si with the YAG and the $YAlO_3$ probe lasers in combination with the Ar^+ pump laser (traces labeled 2 and 3). The EHL luminescence is observed by itself in the absence of the probe laser (trace 1). The shifts A, B, and Γ_{25} track the frequency of the probe laser and hence are identified as Raman lines. The Γ_{25} feature is the F_{2g} triply degenerate zone-center optical phonon and its intensity is independent of the number of electron–hole pairs in the strain well; in contrast, the intensities of A and B are proportional to it, establishing their electronic origin. The frequencies of A and B shift linearly with stress as shown in Fig. 44. They have been identified with electronic transitions between the stress-split $p_{3/2}$ valence band and the stress-shifted $p_{1/2}$ valence band; A corresponds to a transition from the light hole (V_2) to the empty heavy hole (V_1), whereas B is associated

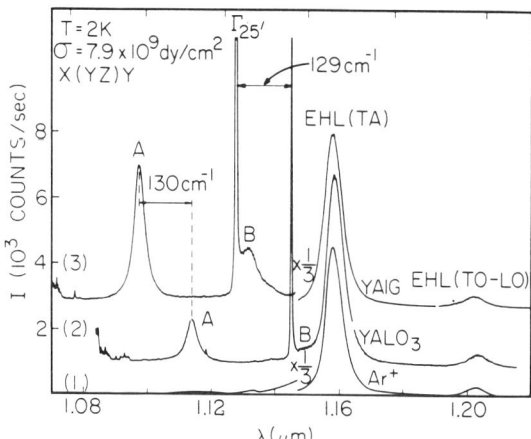

FIG. 43. Overall Raman and luminescence spectrum from a strain well in Si illuminated. (1) with argon-ion pump laser only, (2) with $YalO_3$ probe laser plus the Ar^+ laser, and (3) with yttrium aluminum garnet probe laser plus the Ar^+ laser. The energy difference between the two probe lasers is 132 cm^{-1}. (After Guidotti et al., 1979.)

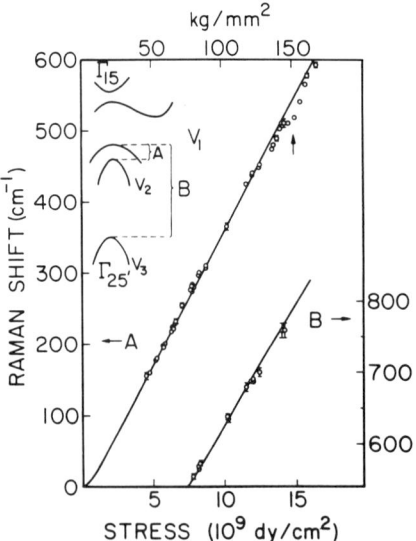

FIG. 44. Stress dependence of the peak position of A and B transitions in Si shown in Fig. 43. The solid lines are shifts calculated from the splitting of the $p_{3/2}$ and the shift of the $p_{1/2}$ valence bands of Si using the deformation potential theory (Section IV). The inset indicates band splittings under stress, and the vertical arrow indicates stress at which A overlaps with the phonon. (After Guidotti et al., 1979.)

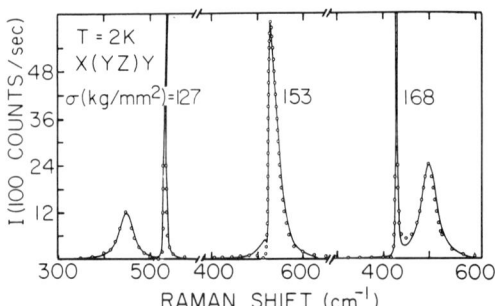

FIG. 45. Stress tuning of antiresonance. The solid lines are experimental traces normalized for system's response. Circles represent the line shape calculated from a Fano analysis (Klein, 1975). Note the marked asymmetry at maximum overlap of the A and $\Gamma_{25'}$ features. (After Guidotti et al., 1979.)

with the transition from the spin-orbit V_3 to V_1. The interpretation is consistent with the selection rules and yields valence-band deformation potential constants that are in reasonable agreement with those obtained from other techniques. The large stress tuning of the energies of A and B allows them to come in close proximity of the $\Gamma_{25'}$ phonon. The symmetry of the Raman tensors for A and $\Gamma_{25'}$ is identical and allows a coupling and a Fano-type interference with marked broadening and asymmetry associated with the antiresonance effects illustrated in Fig. 45. Guidotti et al. analyze the antiresonance effects in the light of the theory expounded by Klein (1975).

IX. Concluding Remarks

Piezospectroscopy has proved to be a powerful tool in the hands of the solid-state spectroscopist and has been exploited with great ingenuity in semiconductor physics as well as in other areas of condensed-matter physics (see, for example, Fitchen, 1968). The symmetries of the ground and excited states involved in a specific optical transition, the quantitative characterization of the stress dependence of their energies through deformation potential constants, and the theoretical analyses providing microscopic origins of the stress effects (Bell, 1972; Cerdeira et al., 1972; Nielsen and Martin, 1985; Rodriguez and Kartheuser, 1985; Nielsen, 1986) are the significant insights achieved through this technique. In the context of semiconductor heterostructures, where constituents with strain mismatch often result in huge built-in strains, the body of knowledge generated with studies on bulk crystals is proving invaluable (Pollak, 1990). For convenient tabulations of deformation potential constants characterizing the band extrema of the electronic band structure and the zone-center optical phonons, we refer readers to Blacha et al. (1984), Anastassakis (1980, 1991), and Pollak (1990).

The investigation of the effects of hydrostatic stress on the spectra exhibited by semiconductors is obviously a closely related field with great significance for piezospectroscopic studies. We emphasize, however, that the symmetry of the solid is unaltered under hydrostatic stress. The present state-of-the-art of hydrostatic stress techniques allows stresses in the range of megabars to be reached in experimental conditions in which meaningful spectroscopy can be carried out, whereas very few specimens can stand an uniaxial stress in excess of 20 kbar cm^{-2} without being shattered. Larger samples that can be used in a uniaxial stress apparatus and the ease of measurements at cryogenic temperatures with it result in a higher spectroscopic precision; in addition, of course, splittings occur due to the reduced symmetry under uniaxial stress. The advent of the diamond anvil cell has revolutionized the field of hydrostatic pressure effects accessible with spectroscopy (Jayaraman, 1983; Weinstein and Zallen, 1984). The pressure range covered allows the study of important and significant phase transitions besides providing pressure coefficients for band parameters and phonon frequencies. Piezospectroscopy and hydrostatic stress spectroscopy are indeed complementary techniques.

A technique that combines several perturbations simultaneously can provide additional characterizations of an electronic transition, as has been elegantly demonstrated by Freeth et al. (1986, 1987), Fisher et al (1987), and Lewis et al. (1989), who studied the excitation spectrum of Ge(Ga) and Si(B); they combined Zeeman- and piezospectroscopy. Novel level crossing effects and insights into the g-factors of the ground and excited states of the acceptor-bound holes have emerged from these studies. Villeret and Rodriguez (1991) recently carried out a theoretical analysis of the magnetopiezospectroscopy of acceptors in Si and Ge. Jagannath et al. (1985) and Youngdale (1985) studied the $1s(A_1) \rightarrow 1s(T_2)$ transitions in Ge(P) and Ge(As) also under combined magnetic field–uniaxial stress perturbation in four-wave mixing experiments. These studies provide microscopic insights into the ground-state wave functions as affected by the chemical splitting.

The deformation potential constants of electronic transitions in semiconductors are typically one or two orders of magnitude larger than those characteristic of zone-center optical phonons. It is thus possible to observe resonance effects in which an electronic transition has been brought in and out of coincidence with an optical phonon transition with uniaxial stress, resulting in line broadening and the formation of hybrid "phonon-electronic" states. Such effects were observed in the $1s(A_1) \rightarrow 2p_0$ transition in Si(Bi) and line 2 of Si(Ga) (Onton et al., 1967; Butler et al., 1975; Chandrasekhar et al., 1976). Fano resonances in the absorption spectrum of Ge(Zn$^-$) and their piezospectroscopy have been reported by Piao et al. (1990); these resonances arise in the frequency range where the transitions in the $p_{3/2}$ series in combination with the zone-center optical phonon of Ge ~ 305 cm^{-1} occur. With uniaxial stress one can also control the resonance enhancement of a Raman line as in the case of the remarkable multiple-overtone series observed in the Raman spectrum of CdS (Briggs and Ramdas, 1976). Pressure-tuned resonant Raman scattering has been reported by Trommer et al. (1976) and by Yu and Welber (1978) for bulk GaAs and by Kourouklis et al. (1990) for a GaAs–Al$_x$Ga$_{1-x}$As superlattice; in the latter, the electronic transition occurs between quantum confined electronic states.

A fascinating class of defects in Si and Ge is associated with oxygen introduced during the growth. In Si this occurs unavoidably if it is grown in a quartz crucible using the Czochralski technique. When it is dispersed, oxygen is electrically neutral and exhibits a vibrational line at 9 μm in Si (Kaiser, Keck, and Lange, 1956) and at 11.7 μm in Ge (Kaiser and Thurmond, 1961). It is associated with the antisymmetric (B_1) normal mode of Si$_2$O or Ge$_2$O units, oxygen forming bonds with two nearest-neighbor Si or Ge atoms. Hrostowski and Kaiser (1957) observed a rich fine structure at low temperatures indicative of libration and tunneling (Hrostowski and Alder, 1960). The lower symmetry of the Si$_2$O unit implies an orientational degeneracy and the associated latent anisotropy. Corbett et al. (1964) removed this orientational degeneracy by applying uniaxial stress along [111] at an elevated temperature; the equivalence of the potential energy minima occupied by the Si_2O units is destroyed and the population of those in the lower minima increase preferentially. On quenching to room temperature, the preferential orientation established at the higher temperature for the Si_2O units is frozen. The infrared absorption at 9 μm now shows dichroism. Similar results were observed with the 11.7-μm band of Ge. The absence of this preferential alignment for stress along [100] confirms the microscopic model in which the oxygen bridges the nearest-neighbor Si atoms; the annealing behavior of the dichroism provides insights into the diffusion of oxygen in Si and Ge.

Si containing neutral dispersed oxygen becomes n-type following a prolonged anneal at $\sim 450°C$ (Kaiser et al., 1958). These "thermal donors" exhibit excitation spectra with effective mass like excited states similar to these exhibited by shallow donors (Hrostowski and Kaiser, 1958; Aggarwal, 1965). As many as nine distinct donor species have been identified, their ground states reflecting differing central cell corrections. Stavola et al. (1985) have studied the piezospectroscopy of these thermal donors; the absence of splitting of the excitation lines for $F \parallel [001]$ has been explained by postulating that the 1s ground states of these centers are derived from a pair of conduction-band valleys such as those along [001] and [00$\bar{1}$] rather than all six $\langle 100 \rangle$ valleys (see also Stavola, 1987). Piezospectroscopy has also been exploited extensively by Haller and co-workers in the context of the electronic spectra of shallow donor and acceptor complexes in ultrapure Ge (see Haller, 1986).

Acknowledgments

The authors are grateful to the National Science Foundation for its support during the period in which this Chapter was prepared (DMR-89-21717).

References

Aggarwal, R. L. (1965). Ph.D. thesis, Purdue University.
Aggarwal, R. L. (1972). "Semiconductors and Semimetals" (Willardson, R. K. and Beer, A. C. eds.), vol. 9, p. 151. Academic Press, New York.
Aggarwal, R. L., and Ramdas, A. K. (1964). *Proc. 7th Int. Con. on the Physics of Semiconductors* (Hulin, M., ed.), p. 197. Dunod, Paris.
Aggarwal, R. L., and Ramdas, A. K. (1965a). *Phys. Rev.* **137** A602.
Aggarwal, R. L., and Ramdas, A. K. (1965b). *Phys. Rev.* **140**, A1246.
Aggarwal, R. L., and Ramdas, A. K. (1965c). unpublished.
Albert, C., Jullie, A. M., Montiel, E, Jullie, A., and Ranvaud R. (1976). *Proc. 13th Int. Conf. on the Physics of Semiconductors, Rome* (Fumi, F. G., ed.), p. 1039. Tipografia Marves, Rome.

Anastassakis, E. M. (1980). "Morphic Effects in Lattice Dynamics in Dynamical Properties of Solids," (Horton, G. K. and Maradudin, A. A. eds.), vol. 4, pp. 157–375. North-Holland, Amsterdam.
Anastassakis, E. (1981). *J. Raman Spectros.* **10**, 64.
Anastassakis, E. (1991). NATO Advanced Research Workshop on *Light Scattering in Semiconductor Structures and Superlattices*, Mont Tremblant (Young, J. F. and Lockwood, D. eds.). To be published.
Anastassakis, E., and Cardona, M. (1987). *Solid State Commun.* **63**, 893.
Anastassakis, E., Pinczuk, A., Burstein, E., Pollak, F. H., and Cardona, M. (1970). *Solid State Commun.* **8**, 133.
Anastassakis, E., Raptis, Y. S., Hünermann, M., Richter, W., and Cardona, M. (1988). *Phys. Rev. B***38**, 7702.
Anastassakis, E., Cantarero, A., and Cardona, M. (1990). *Phys. Rev. B***41**, 7529.
Auvergne, D., Camassel, J., Mathieu, H., and Cardona, M. (1974). *Phys. Rev. B***9**, 5168.
Baldereschi, A., and Lipari, N. O. (1973) *Phys. Rev. B***8**, 2697.
Balslev, I. (1966). *Phys. Rev.* **143**, 636.
Balslev, I. (1967a). *Rev. Sci. Inst.* **38**, 1528.
Balslev, I. (1967b). *Solid State Commun.* **5**, 315.
Balslev, I. (1974). *Phys. Stat. Sol (b)* **61**, 207.
Bardeen J., and Shockley, W. (1950) *Phys. Rev.* **80**, 72.
Becker, W. M., Ramdas, A. K., and Fan, H. Y. (1961). *J. Appl. Phys.* **32**, 2094.
Bell, M. I. (1972). *Phys. Stat. Sol.* **53(b)**, 675.
Bethe, H. A. (1942). "Theory of the Boundary Layer of Crystal Rectifier." MIT Radiation Laboratory Report No. 43-12.
Bhagavantam, S. (1966). "Crystal Symmetry and Physical Properties." Academic, London.
Bhagavantam, S., and Venkatarayudu, T. (1969). "Theory of Groups and its Applications to Physical Problems." Academic Press, New York.
Bhargava, R. N., and Nathan, M. I. (1967). *Phys. Rev.* **161**, 695.
Bimberg, D., and Rühle, W. (1974). *Proc. 12th Int. Conf. on the Physics of Semiconductors*, Stuttgart (Pilkuhn, M. H., ed.), p. 561. Teubner, Stuttgart.
Binggeli, N., and Baldereschi, A. (1988). *Solid State Commun.* **66**, 323.
Binggeli, N., Baldereschi, A., and Quattropani, A. (1988). "Shallow Impurities in Semiconductors," (Monemar, B., ed.), p. 521. Institute of Physics, Bristol.
Blacha, A., Presting, H., and Cardona, M. (1984). *Phys. Stat. Solidi (b)* **126**, 11.
Bouckaert, L. P., Smoluchowski, R., and Wigner, E. (1936). *Phys. Rev.* **50**, 58.
Bratt, P. R. (1977). "Semiconductors and Semimetals" (Willardson, R. K. and Beer, A. C., eds.), vol. 12, pp. 39–142. Academic, New York.
Brewster, D. (1815). *Phil. Trans.*, p. 46.
Brewster, D. (1816). *Phil. Trans.*, p. 156.
Briggs, R. J., and Ramdas, A. K. (1976). *Phys. Rev. B***13**, 5518.
Briggs, R. J., and Ramdas, A. K. (1977). *Phys. Rev. B***16**, 3815.
Broeckx, P. Clauws, and Vennik, J. (1986). *J. Phys. C. Solid State Phys.* **19**, 511.
Buczko, R., and Bassani, F. (1989). "Shallow Impurities in Semiconductors" (Monemar, B., ed.), p. 107. Institute of Physics, Bristol.
Burstein, E., Bell, E. E., Davisson, J. W., and Lax, M. (1953). *J. Phys. Chem.* **57**, 849.
Burstein, E., Picus, G., Henvis, B., and Wallis, R. (1956). *J. Phys. Chem. Solids* **1**, 65.
Butler, N. R., and Fisher, P. (1976). *Phys. Rev. B***13**, 5465.
Butler, N. R., Fisher, P., and Ramdas, A. K. (1975). *Phys. Rev. B***12**, 3200.
Cady, W. G. (1964). "Piezoelectricity," vol. 1, pp. 54–56. Dover, New York.
Cardona, M. (1969). "Solid State Physics," (Seitz, F., Turnbull, D., and Ehrenreich, H., eds.), Suppl. 11. Academic Press, New York.

Cerdeira, F., and Cardona, M. (1972). *Phys. Rev.* B5, 1440.
Cerdeira, F., Buchenauer, C. J., Pollak, F. H., and Cardona, M. (1972). *Phys. Rev.* B5, 580.
Cerdeira, F., Fjeldly, T. A., and Cardona, M. (1973). *Phys. Rev.* B8, 4734.
Cerdeira, F., Fjeldly, T. A., and Cardona, M. (1974). *Phys. Rev.* B9, 4344.
Cerdeira, F., Pinczuk, A., Bean, J. C., Batlogg, B., and Wilson B. A. (1984). *Appl. Phys. Lett.* 45, 1138.
Chandrasekhar, M., and Pollak, F. H. (1977). *Phys. Rev.* B15, 2127.
Chandrasekhar, H. R., and Ramdas, A. K. (1980). *Phys. Rev.* B21, 1511.
Chandrasekhar, H. R., Fisher, P., Ramdas, A. K., and Rodriguez, S. (1973). *Phys. Rev.* B8, 3836.
Chandrasekhar, H. R., Ramdas, A. K., and Rodriguez, S. (1975). *Phys. Rev.* B12, 5780.
Chandrasekhar, H. R., Ramdas, A. K., and Rodriguez, S. (1976). *Phys. Rev.* B14, 2417.
Chandrasekhar, M., Cardona, M., and Kane, E. O. (1977). *Phys. Rev.* B16, 3579.
Chandrasekhar, M., Renucci, J. B., and Cardona, M. (1978). *Phys. Rev.* B17, 1623.
Chandrasekhar, M., Rössler, U., and Cardona, M. (1980). *Phys. Rev.* B22, 761.
Cohen, M. L., and Chelikowsky, J. R. (1988) "Electronic Structure and Optical Properties of Semiconductors." Springer-Verlag, New York.
Coker, E. G., and Filon, L. N. G. (1957). "A Treatise on Photoelasticity," p. 181. Cambridge University Press, London.
Corbett, J. W., McDonald, R. S., and Watkins, G. D. (1964). *J. Phys. Chem. Solids* 25, 873.
Couture-Mathieu, L., and Mathieu, J. P. (1953). *Comptes Rendus* 236, 371.
Cuevas, M., and Fritzsche, H. (1965). *Phys. Rev.* 137, A1847.
Curie, P. (1894). *J. Phys. (Paris)* 3, 395.
Damen, T. C., Porto, S. P. S., and Tell, B. (1966). *Phys. Rev.* 142, 570.
Duff, K. J., Fisher, P., and Butler, N. R. (1980). *Aust. J. Phys.* 33, 73.
Elliot, J. P., and Dawber, P. G. (1979). "Symmetry in Physics," vol. 1, 2. Oxford University Press, New York.
Elliot, R. J., and Loudon, R. (1960). *J. Phys. Chem. Solids* 15, 146.
Erlbach, E. (1966). *Phys. Rev.* 150, 767.
Faulkner, R. A. (1969). *Phys. Rev.* 184, 713.
Feofilov, P. P., and Kaplyanskii, A. A. (1962). *Soviet Physics Uspekhi* 5, 79.
Feldman, A. (1966). *Phys. Rev.* 150, 748.
Fisher, P., Haak, W. H., Johnson, E. J., and Ramdas, A. K. (1963). *Proc. Eighth Sym. on the Art of Glass Blowing*, p. 135. The American Scientific Glassblowers Society, Wilmington, DE.
Fisher, P., Freeth, C. A., and Vickers, R. E. M. (1987). *Physica* 146B, 80.
Fitchen, D. B. (1968). "Zero-Phonon Transitions in Physics of Color Centers" (Fowler, W. Beall, ed.), p. 329. Academic Press, New York.
Fjeldly, T. O., Cerdeira, F., and Cardona, M. (1973). *Phys. Rev.* B8, 4723.
Flynn, C. P. (1972). "Point Defects and Diffusion." Oxford University Press, London.
Folland, N. O., and Bassani, F. (1968). *J. Phys. Chem. Solids*, 29, 281.
Freeth, C. A., Fisher, P., and Simmonds, P. E. (1986). *Solid State Commun.* 60, 175.
Freeth, C. A., Fisher, P., and Vickers, R. E. M. (1987). *Proc. 18th Int. Conf. on the Physics of Semiconductors, Stockholm* (Engstrom, O., ed.), p. 841. Word Scientific, Singapore.
Gorman, M., and Solin, S. A. (1977). *Phys. Rev.* B16, 1631.
Grimsditch, M. H., Anastassakis, E., and Cardona, M. (1978). *Phys. Rev.* B18, 901.
Grossman, G., Bergman, K., and Kleverman, M. (1987). *Physica B* 146B, 30.
Guidotti, D., Lai, S., Klein, M. V., and Wolfe, J. P. (1979). *Phys. Rev. Lett.* 43, 1950.
Haller, E. E. (1986). Festkörperprobleme: "Advances in Solid State Physics" (Grosse, P., ed.), vol. 26, p. 203. Vieweg, Braunschewig.
Hass, M., and Henvis, B. W. (1962). *J. Phys. Chem. Solids* 23, 1099.
Haas, M. M. C., and Mathieu, J. P. (1954). *J. Phys. Radium* 15, 492.
Hayes, W., and Loudon, R. (1978). "Scattering of Light by Crystals." Wiley, New York.

Ho, L. T., and Ramdas, A. K. (1972). *Phys. Rev.* **B5**, 462.
Hrostowski, H. J., and Kaiser, R. H. (1957). *Phys. Rev.* **107**, 966.
Hrostowski, H. J., and Kaiser, R. H. (1958). *Phys. Rev. Lett.* **1**, 199.
Hrostowski, H. J., and Alder, B. J. (1960). *Phys. Rev.* **33**, 980.
Jagannath, C., and Ramdas, A. K. (1981). *Phys. Rev.* **B23**, 4426.
Jagannath, C., Grabowski, Z. W., and Ramdas, A. K. (1979). *Solid State Commun.* **29**, 355; (1981) *Phys. Rev.* **B23**, 2082.
Jagannath, C., Aggarwal, R. L., and Larsen, D. M. (1985). *Solid State Commun.* **53**, 1089.
Jagannath, C., Koteles, E. S., Lee, J. Chen, Y. J., Elman, B. S., and Chi, J. Y. (1986). *Phys. Rev.* **B34**, 7027.
Jagannath, C., Zemon, S., Norris, P., and Elman, B. S. (1987). *Appl. Phys. Lett.* **51**, 1268.
Jayaraman, A. (1983). *Rev. Mod. Phys.* **55**, 65.
Kaiser, W., and Thurmond, C. J. (1961). *J. Appl. Phys.* **32**, 115 (1961).
Kaiser, W., Keck, P. H., and Lange, C. F. (1956). *Phys. Rev.* **101**, 1264.
Kaiser, W., Frisch, H. L., and Reiss, H. (1958). *Phys. Rev.* **112**, 1546.
Kaplyanskii, A. A. (1964a). *Optics and Spectroscopy* **16**, 557.
Kaplyanskii, A. A. (1964b). *Optics and Spectroscopy* **16**, 329.
Kane, E. O. (1969). *Phys. Rev.* **178**, 1368. [Note that $\Gamma_{25'}$ on line 10 of Table X should read Γ_{25}.]
Klein, M. V. (1975). "Light Scattering in Solids" (Cardona, M., ed.), p. 147. Springer-Verlag, New York.
Kohn, W. (1957). "Solid State Physics" (Seitz, F. and Turnbull, D., eds.), vol. 5, pp. 257–320. Academic Press, New York.
Kohn, W., and Luttinger, J. M. (1955). *Phys. Rev.* **98**, 915.
Koster, G. F., Dimmock, J. O., Wheeler, R. G., and Statz, H. (1966). "Properties of the Thirty-two Point Groups." MIT Press, Cambridge.
Kourouklis, G. A., Jayaraman, A., People, R., Sputz, S. K., Maines, R. G., Sr., Sivco, D. L., and Cho, A. Y. (1990). *J. Appl. Phys.* **67**, 6438.
Kramers, H. A. (1930). *Proc. Acad. Sci. Amsterdam* **33**, 959.
Langer, D. W., Euwema, R. N., Era, K., and Koda, T. (1970). *Phys. Rev.* **B2**, 4005.
Laude, L. D., Cardona, M., and Pollak, F. H. (1970). *Phys. Rev.* **B1**, 1436.
Laude, L. D., Pollak, F. H., and Cardona, M. (1971). *Phys. Rev.* **B3**, 2623.
Lax, M., and Hopfield, J. J. (1961). *Phys. Rev.* **124**, 115.
Lee, J., Jagannath, C., Vassell, M. O., and Koteles, E. S. (1988a). *Phys. Rev.* **B37**, 4164.
Lee, Y. R., Ramdas, A. K., Kolodziejski, L. A., and Gunshor, R. L. (1988b). *Phys. Rev.* **B38**, 13143.
Lewis, R. A., Fisher, P., and McLean, N. A. (1989). "Shallow Impurities in Semiconductors 1988" (Monemar, B., ed.), p. 95. Institute of Physics, Bristol.
Lipari, N. O., and Baldereschi, A. (1978). *Solid State Commun.* **25**, 665.
Long, D. (1968). "Energy Bands in Semiconductor." Interscience, New York.
Ludwig, G. W., and Woodbury, H. H. (1959). *Phys. Rev.* **113**, 1014.
Marieé, M., and Mathieu, J. P. (1946). *Compt. Rendus* **223**, 147.
Martin, A. D., Fisher, P., and Simmonds, P. E. (1979). *Phys. Lett.* **73A**, 331.
Martin, A. D., Fisher, P., and Simmonds, P. E. (1981) *Aust. J. Phys.* **34**, 511.
Mathieu, H., Merle, P., Ameziane, E. L., Archilla, B., Camassel, J., and Poiblaud, G. (1979). *Phys. Rev.* **B19**, 2209.
Mitsuishi, A., Yoshinaga, H., Yata, K., and Manabe, A. (1965) *Jpn J. Appl. Phys.* **4**, Suppl. I, 581.
Möller, K. D., and Rothschild, W. G. (1971). "Far-infrared Spectroscopy." Wiley-Interscience, New York.
Mooradian, A., and McWhorter, A. L. (1967). *Phys. Rev. Lett.* **19**, 849.
Mooradian, A., and Wright, G. B. (1966). *Phys. Rev. Lett.* **16**, 999.
Mott, N. F., and Gurney, R. W. (1948). "Electronic Processes in Ionic Crystals" 2nd edition. Oxford University Press, Oxford.

Musgrave, M. J. P. (1970). "Crystal Acoustics." Holden-Day, San Francisco.
Nielsen, O. H. (1986). *Phys. Rev.* **B34**, 5808.
Nielsen, O. H., and Martin, R. M. (1985). *Phys. Rev.* **B32**, 3792.
Noack, R. (1977). Ph.D. thesis, University of Stuttgart.
Novak, I. I., Baptizmanskii, V. V., and Zhoga, L. V. (1978). *Optics and Spectrosocpy* **43**, 145.
Nye, J. F. (1957). "Physical Properties of Crystals." Clarendon Press, Oxford.
Ohkawa, K., Mitsuyu, T., and Yamazaki, O. (1988). *Phys. Rev.* **B38**, 12465.
Olego, D. J., Shahzad, K., Cammack, D. A., and Cornelissen, H. (1988). *Phys. Rev.* **B38**, 5554.
Onton, A., Fisher, P., and Ramdas, A. K. (1967). *Phys. Rev. Lett.* **19**, 781.
Pantelides, S. T. (1978). *Rev. Mod. Phys.* **50**, 797.
Pearson, G. L., and Bardeen, J. (1949). *Phys. Rev.* **75**, 865.
Peiser, H. S., Wachtman, J. B., and Dickson, R. W. (1963). *J. Res. Natl. Bur. Standards* **67A**, 395.
Perkowitz, S., and Breecher, J. (1973). *Infrared Phys.* **13**, 321.
Piao, G., Lewis, R. A., and Fisher, P. (1990). *Solid State Commun.* **75**, 835.
Picus, G., Burstein, E., and Henvis, B. (1956). *J. Phys. Chem. Solids* **1**, 75.
Platzman, P. M., and Wolff, P. A. (1973). "Waves and Interactions in Solid State Plasmas." Academic Press, New York.
Pollak, F. H. (1990). "Semiconductors and Semimetals." (Willardson, R. K. and Beer, A. C., eds.), vol. 32, p. 17. Academic Press, New York.
Pollak, F. H., and Cardona, M. (1968). *Phys. Rev.* **172**, 816.
Pollak, F. H., and Aggarwal, R. L. (1971). *Phys. Rev.* **B4**, 432.
Price, P. J. (1956). *Phys. Rev.* **104**, 1223.
Putley, E. H. (1964). *Phys. Stat. Solidi* **6**, 571.
Ramdas, A. K. (1990). SPIE 1990 *Int. Symp. on Raman and Luminescence Spectroscopies in Technology*, p. 25.
Ramdas, A. K., and Rodriguez, S. (1981). *Rep. Prog. Phys.* **44**, 1297.
Reuszer, J. H., and Fisher, P. (1965). *Phys. Rev.* **140**, A245.
Reuszer, J. H., and Fisher, P. (1968). *Phys. Rev.* **165**, 909.
Rodriguez, S., and Kartheuser, E. (1985). *Superlattices and Microstructures* **1**, 503.
Rockwell, B., Chandrasekhar, H. R., Chandrasekhar, M., Ramdas, A. K., Kobayashi, M., and Gunshor, R. L. (1991). *Phys. Rev.* **B44**, 11307.
Schawlow, A. L., Piksis, A. H., and Sugano, S. (1961). *Phys. Rev.* **122**, 1469.
Schechter, D. (1962). *J. Phys. Chem. Solids* **23**, 237.
Seiler, D. G., and Addington, F. (1972). *Rev. Sci. Instrum.* **43**, 749.
Singwi, K. S., and Tosi, M. P. (1966). Phys. Rev. **147**, 658.
Sood, A. K., Anastassakis, E., and Cardona, M. (1985). *Phys. Stat. Solidi (b)* **129**, 505.
Stavola, M. (1987). *Physica* **146B**, 187.
Stavola, M., Lee, K. M., Nabity, J. C., Freeland, P. E., and Kimerling, L. C. (1985). *Phys. Rev. Lett.* **54**, 2639.
Stillman, G. E., Larsen, D. M., Wolfe, C. M., and Brandt, R. C. (1971). *Solid State Commun.* **9**, 2245.
Tekippe, V. J. (1973). Ph.D. thesis, Purdue University.
Tekippe, V. J., and Ramdas, A. K. (1971). *Phys. Lett.* **A35**, 143.
Tekippe, V. J., Chandrasekhar, H. R., Fisher, P., and Ramdas, A. K. (1972). *Phys. Rev.* **B6**, 2348.
Tekippe, V. J., Ramdas, A. K., and Rodriguez, S. (1973). *Phys. Rev.* **B8**, 706.
Thomas, D. G. (1961). *J. Appl. Phys.* **32**, Suppl. 2298.
Tinkham, M. (1964). "Group Theory and Quantum Mechanics," p. 80. McGraw-Hill, New York.
Trommer, R., and Ramdas, A. K. (1979). In "Physics of Semiconductors 1978" (Wilson, B. L. H., ed.), p. 585. The Institute of Physics, London.
Trommer, R. Anastassakis, E., and Cardona, M. (1975). "Light Scattering in Solids" (Balkanski, M., Leite, R. C. C., and Porto, S. P. S., eds.) p. 396. Flammarion, Paris.

Udo, M. K., LaBrec, C. R., and Ramdas, A. K. (1991). *Phys. Rev.* **B44**, 1565.
Varga, B. B. (1965). *Phys. Rev.* **137**, A1896.
Venkatarayudu, T. (1938). *Proc. Ind. Acad. Sci.* **A8**, 349.
Venugopalan, S., and Ramdas, A. K. (1973). *Phys. Rev.* **B8**, 717.
Villeret, M., and Rodriguez, S. (1991). *Il Nuovo Cimento* **13D**, 529.
Vogelmann, H., and Fjeldly, T. A. (1974). *Rev. Sci. Instrum.* **45**, 309.
Wachtman, J. B., and Peiser, H. S. (1965). *J. Res. Natl. Bur. Standards* **69A**, 193.
Wardzynski, W., and Sufczynski, M. (1972). *Solid State Commun.* **10**, 417.
Weinstein, B. A., and Cardona, M. (1972). *Phys. Rev.* **B5**, 3120.
Weinstein, B. A., and Zallen, R. (1984). "Pressure-Raman Effects in Covalent and Molecular Solids in Light Scattering in Solids IV," (Cardona, M. and Güntherodt, G., eds.), pp. 463–527. Springer-Verlag, New York.
Whelan, J. M. (1960). "Semiconductors" (Hannay, N. B., ed.), pp. 389–436. Reinhold, New York.
Wickboldt, P., Anastassakis, E, Sauer, R., and Cardona, M. (1987). *Phys. Rev.* **B35**, 1362.
Wilson, D. K., and Feher, G. (1961). *Phys. Rev.* **124**, 1068.
Wilson, E. B., Decius, J. C., and Cross, P. C. (1955). "Molecular Vibrations." McGraw-Hill, New York.
Wintgen, D., Marxer, H., and Briggs, J. S. (1987). *J. Phys. A: Math. Gen.* **20**, L965.
Yokota, I. (1961). *J. Phys. Soc. Japan* **16**, 2075.
Youngdale, E. R. (1985). Ph.D. thesis, MIT.
Yu, P. Y., and Welber, B. (1978). *Solid State Commun.* **25**, 209.

CHAPTER 4

Photoreflectance Spectroscopy of Microstructures

Orest J. Glembocki and Benjamin V. Shanabrook

NAVAL RESEARCH LABORATORY
WASHINGTON, DC

I.	INTRODUCTION	222
II.	MICROSTRUCTURES	223
III.	MODULATION SPECTROSCOPY	228
	1. *Line Shapes*	230
	2. *Modulated Dielectric Function*	230
	3. *Bulk Materials*	231
	4. *Artificially Structured Materials*	233
IV.	EXPERIMENTAL DETAILS	238
V.	EXPERIMENTAL SPECTRA	240
	5. *Quantum Wells*	240
	a. *E_0 Gap*	240
	b. *Line Shape Analysis*	245
	c. *Excited Excitonic Transitions*	249
	d. *Identifying Light and Heavy Holes*	251
	e. *Modulation Mechanisms*	254
	f. *Interference Effects*	257
	g. *E_1 Gap*	259
	h. *Characterization of Quantum Wells*	260
	i. *New Materials Systems*	262
	6. *Coupled Wells*	264
	a. *Asymmetrically Coupled Quantum Wells*	264
	b. *Symmetrically Coupled Wells*	269
	7. *Superlattices*	271
	a. *Wave Function Localization*	271
	b. *Superlattice Band Structure*	273
	c. *Superlattice Effects in Multiple Quantum Wells*	275
	8. *Modulation-Doped Heterojunctions*	277
	a. *The HEMT*	277
	b. *NIPI*	281
	9. *Novel Photoreflectance Techniques*	281
	10. *Applications to Bulk Materials*	284
	a. *Carrier Densities*	284
	b. *Strains*	286
	c. *Elevated Temperatures*	287
VI.	CONCLUSIONS	288
	REFERENCES	289

ISBN 0-12-752136-4

I. Introduction

During the last 15 years, artificially structured materials (ASM), such as multiple quantum wells (MQW), superlattices (SL), and modulation-doped heterojunctions, have been the subject of considerable theoretical (Bastard, 1988; Smith and Mailhiot, 1990) and experimental studies (Weisbuch, 1987). These microstructures are of technological and scientific interest because they exhibit properties that are significantly different from bulk materials. Furthermore, because of the flexibility offered by high-technology growth techniques, the chemical compositions, thicknesses, and doping concentrations of the layers are design parameters. This allows the properties of ASM to be tuned for particular technological or basic science needs. A few examples of the technological impact of ASM include the development of the modulation-doped field-effect transistor (MODFET) permeable base transistor, electro-optic modulators, and quantum well lasers. On the scientific side, the study of these new materials has resulted in a better understanding of the nature and variability of the band structures of solids near the interface of two dissimilar materials. Furthermore, the flexibility of the epitaxial growth techniques has allowed materials to be tailored so they serve as model systems for the investigation into phenomena that are not well understood, such as screening by one-component plasmas, hot-electron transport, and the fractional quantum Hall effect. In addition, ASM also offer us a unique opportunity to reexamine bulk properties under the special conditions of decreased dimensionality.

Many experimental techniques have been applied to the study of microstructures. The optical tools used in these studies probe either the vibrational or electronic properties of the microstructure. These include electronic and vibrational Raman scattering (Pinczuk and Abstreiter, 1989) photoluminescence (PL), photoluminescence excitation spectroscopy (PLE) (R. C. Miller *et al.*, 1981), absorption (Masselink *et al.*, 1985), photoconductivity, and modulation spectroscopy. In order to understand the properties of microstructures, it is important to have as much information as possible about their band structure. This usually involves interband measurements, such as PLE, absorption and reflection spectroscopy, or modulation spectroscopy. In the case of bulk semiconductors, modulation spectroscopy was highly successful in obtaining maps of the band structure of various semiconductors (Cardona, 1969; Seraphin, 1972). In microstructures, however, it has only recently been applied to the study of band structure.

In this chapter, we describe the application and flexibility of modulation spectroscopy, particularly electroreflectance (ER) and photoreflectance (PR), to the study of ASM. We will also show how measurements on microstructures have aided in the understanding of the properties of ASM and of the modulation mechanisms involved in electromodulation spectroscopy.

II. Microstructures

The types of physical systems that we are interested in studying are of the multilayer variety, whose building block is usually composed of two materials having different band gaps. This difference in band gaps can lead to the confinement of carriers in one of the two materials (i.e., the carriers can become confined to a quantum well). Also of interest is the case where the layers are sufficiently thin so that the electronic wave functions can tunnel from one quantum well to the next and synthetic superlattices are formed. We will only present a rudimentary overview of microstructures, which should be sufficient for understanding electromodulation spectroscopy of such systems. Additional details can be found in Weisbuch (1987), Bastard (1988), and Smith and Mailhiot (1990).

Shown in Fig. 1 is a schematic representation of the archetypal multiple quantum well sample composed of GaAs and AlGaAs. We assume that the growth direction is z and that it lies along the (001) crystallographic direction. For this system, the Brillouin zone-center band structure is such that GaAs always has a smaller zone-center gap than AlGaAs, regardless of the alloy composition. Furthermore, the relative band alignments are such that the $k = 0$ electrons (holes) of GaAs occur at lower (higher) energies than those of AlGaAs. In particular, work of others has indicated that 70% of the band-gap difference between AlGaAs and GaAs, Q_e, occurs between the conduction bands of the two materials (Menedez et al., 1986; Wolford et al., 1986) This particular band alignment indicates that when the AlGaAs and GaAs layers are thick (>50 Å) that the low-energy electron and hole states of this system will be, for the most part, confined to the GaAs. The band structure of the quantum well can be described with the effective-mass approximation. The wave function describing the electrons in a GaAs quantum well are given by

$$\psi_n^c = F_n^c(z) u_{c0} \exp(i \mathbf{k} \cdot \mathbf{r}), \tag{1}$$

where n is the subband index, u_{c0} is the $k = 0$ cell periodic Bloch state for the conduction band of the bulk, $\mathbf{k}$ is the wave vector perpendicular to the growth direction, and the envelope function $F_n(z)$ is determined by the solution of the following Schrödinger-like equation:

$$\left[\frac{-\hbar^2}{2m_c(z)} \frac{d^2}{dz^2} + V_c(z) \right] F_n^c(z) = E_n^c F_n^c(z). \tag{2}$$

In this equation, m_c is the effective mass of the electrons in the two materials, $V(z)$ is the potential caused by the relative band alignments of GaAs

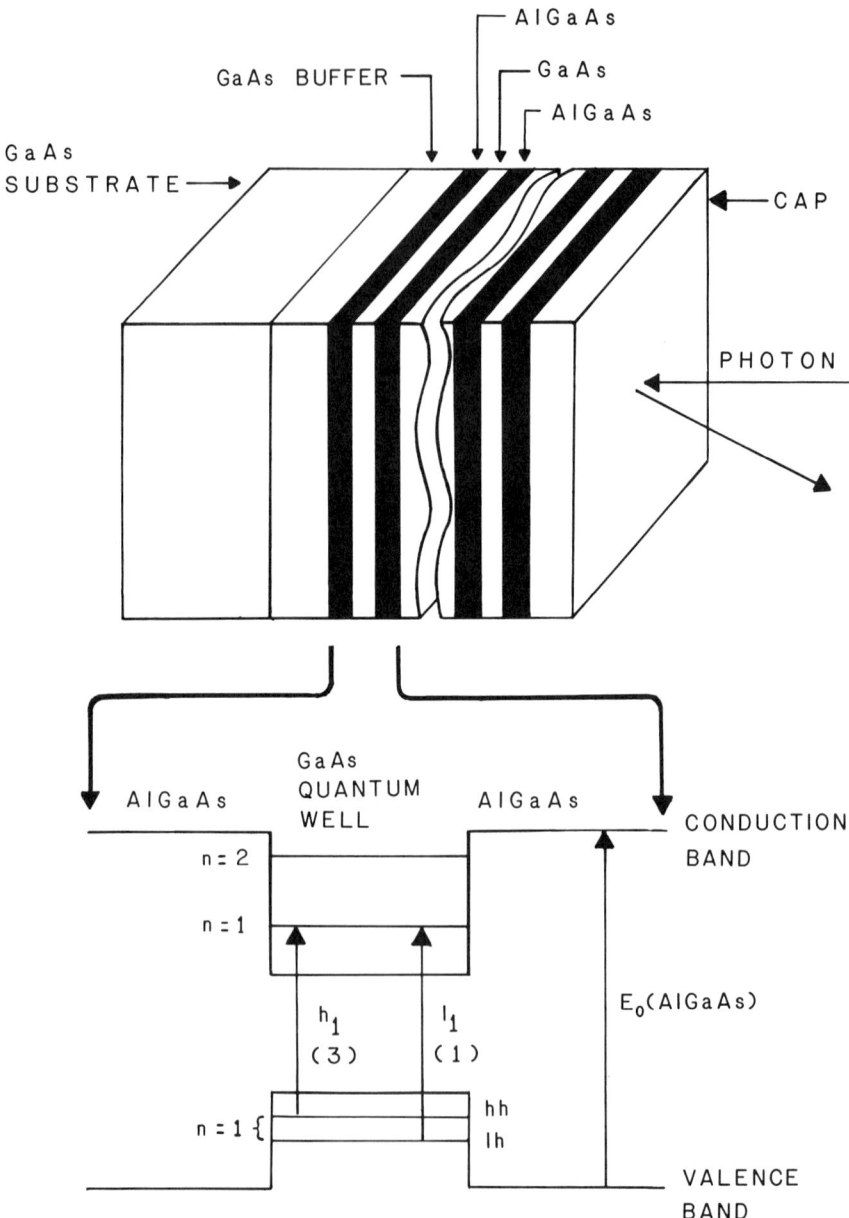

FIG. 1. Schematic representation of a GaAs–AlGaAs MQW structure showing the $n = 1$ electron and hole subbands and the $n = 2$ electron subbands. The allowed light (l_1) and heavy (h_1) interband transitions and are also shown. (After Glembocki et al., 1985.)

and AlGaAs, and E_n^c is the energy of the nth subband at $k = 0$. The current-conserving continuity conditions at the interface are that $F_n(z)$ and $[1/m_c(z)][dF_n(z)/dz]$ be continuous. The wave vector dependence of the energy of the electron in subband n is

$$E_n^c(k) = E_n^c + \frac{\hbar^2 k^2}{2m_c}. \tag{3}$$

For purposes of illustration, assume that $V_c(z) = \infty$ for $z < 0$ or $z > L$, and $V_c(z) = 0$ for $L > z > 0$, where L is the width of the GaAs layer. Then the boundary condition that $F_n(0) = F_n(L) = 0$ requires that

$$F_n^c = A_n \sin\left(\frac{n\pi z}{L}\right), \qquad n = 1, 2, 3, 4 \ldots \tag{4}$$

and

$$E_n^c = \frac{n^2 \pi^2 \hbar^2}{2m_c L^2}. \tag{5}$$

In the zinc blende structure, the interface potential $V_c(z)$ causes a mixing between the heavy- and light-hole character in the valence band of the quantum well and, as a consequence, the procedure for calculating the valence-band wave functions of the quantum well is more complicated than that just described. However, because of the form of the Luttinger Hamiltonian that describes the valence band, the simple one-band model described is accurate for the calculation of the $k = 0$ valence-band wave functions and energies (Sanders and Chang, 1987; Broido and Sham, 1987). In the following discussion, the mixing that occurs in the valence band away from $k = 0$ will be ignored.

The optical absorption for nonexcitonic interband transitions is related to the quantum well wave functions of the electrons from conduction subband n and holes from valence subband m, F_n^c and F_m^v, by

$$\alpha_{nm}(\hbar\omega) \sim \sum_k |\mathbf{e} \cdot \mathbf{P}_{cv}^{nm}|^2 \delta(E_{cv}^{nm}(k) - \hbar\omega), \tag{6}$$

where

$$\mathbf{P}_{cv}^{nm} = \langle u_{c0}|\mathbf{p}|u_{v0}\rangle \int_{-\infty}^{\infty} dz\, F_n^c(z) \cdot F_m^v(z) = \mathbf{p}_{cv} I_{nm}, \tag{7}$$

where $\mathbf{e}$ is the polarization of the incident light, p_{cv} is the interband matrix element of the bulk, I_{nm} is the overlap integral between the valence- and conduction-subband envelope functions, and

$$E_{cv}^{nm}(k) = E_c^n(k) + E_v^m(k) + E_{gap}.$$

The form of $\alpha_{nm}(\hbar\omega)$, shown in Fig. 2, is characterized by an abrupt onset of absorption at $E_{cv}^{nm}(0)$ that continues with constant magnitude to higher energies. This form of the absorption arises from the sum over k space in Eq. (6) and is characteristic of the constant two-dimensional density of states. The form of α_{nm} changes significantly when the excitonic interaction between the electron and hole is included in the model (Shinada and Sugano, 1966). As shown in Fig. 2, sharp peaks occur at energies lower than $E_{cv}^{nm}(0)$ that arise from absorption of light that creates electrons and holes that are bound to each other. Furthermore, there is no longer an abrupt onset for absorption at $E_{cv}^{nm}(0)$, but the high-lying bound excitonic states with large quantum number merge smoothly into the continuum of unbound excitonic transitions. In addition, the correlation between the electron and hole in the unbound excitonic transitions causes the absorption coefficient to be enhanced near $E_{cv}^{nm}(0)$.

In the given illustration (i.e., neglect of the Coulomb interaction and an infinitely deep quantum well for electrons and holes), one would obtain the "allowed" selection rule for interband transitions, i.e., $n = m$, where n, m are the quantum numbers of the conduction and valence subbands. The selection rule is a consequence of the orthogonality of the wave functions F_n^c and F_m^v. These selection rules form a guide to understanding any spectrum from

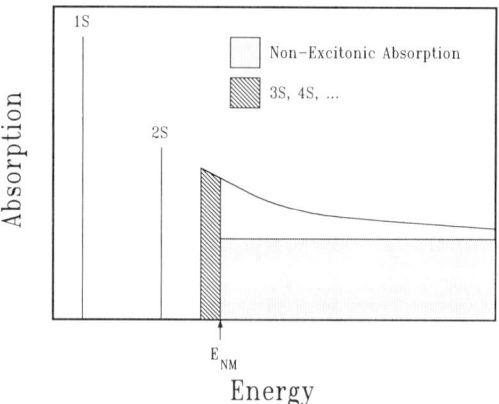

FIG. 2. Schematic representation of the absorption spectrum for a 2D system, with and without excitonic effects. Note that the 3S, 4S,... states and the Coulombic scattering states merge smoothly with the nonexcitonic continuum. The single-particle band edge for electron subband n and hole subband m occurs at E_{nm}.

a single quantum well (SQW) or from MQWs. We also find that (for the infinite well case) any allowed transition originating from a light-hole state will be three times weaker than a corresponding one from a heavy-hole state. This factor of 3:1 is due to the different symmetries of the valence states relative to the conduction band for zinc blende structures (Kane, 1966; Laude et al., 1971).

For quantum wells with finite barrier heights, the situation is somewhat more complex. In the well, F_n will have a sine behavior, while in the barriers (the AlGaAs layers) F_n will be exponentially decaying. Because F_n penetrates into the barrier and $V(z)$ has a well-defined parity, the selection rule is modified to $n + m = 0, 2, 4, 6, \ldots$. However, the transitions with $n = m$ are still the strongest. This selection rule will be broken by a nonsymmetric $V(z)$, by nonparabolicities in the valence band states as well as by electric fields (internal, external), etc.

Of interest to us is the effect of an applied electric field on the wave function in a confined system. Numerous studies have shown that as long as the field is small, then the effect of the field is to change the symmetry of the wave functions and to shift the energies and change the oscillator strengths of the optical transitions (Bastard, 1983; D. A. B. Miller et al., 1985). Shown in Fig. 3 is a schematic representation of a quantum well in an applied field. Note that the electron and hole wave functions polarize and skew in the direction of the field and that there is a red shift of the transition energy. Furthermore, the shape of the confining potential has changed. The shift in the energy levels can be easily calculated from perturbation theory. For weak fields, Bastard (1983) finds that the ground state of a particle in an infinite well will shift by

$$\Delta E_{1,c,v} = -C_{\text{pert}} \frac{m_{c,v} e^2 F^2 L^4}{\hbar^2}, \qquad (8)$$

where e is the electronic charge, F is the electric field, and $C_{\text{pert}} = (15/\pi^2 - 1)/24\pi^2$. More complex forms exist for a quantum well with finite barriers, but this equation is adequate to show how the energy levels shift with applied electric field. These results are important in that they apply qualitatively to any quantum confined system. Although the form of the confining potentials can be quite different, the fact remains that these potentials lead to states that have envelope functions that go to zero in at least two opposing directions. Basically, the field changes the shape of the potential and rigidly shifts the energy of the state by a discrete amount. The magnitude of the shift is related to the carrier mass and the well width. The confined states can still be quasistationary. This result, which can be derived from perturbation theory for any general confining potential, is important and will be used in subsequent discussions.

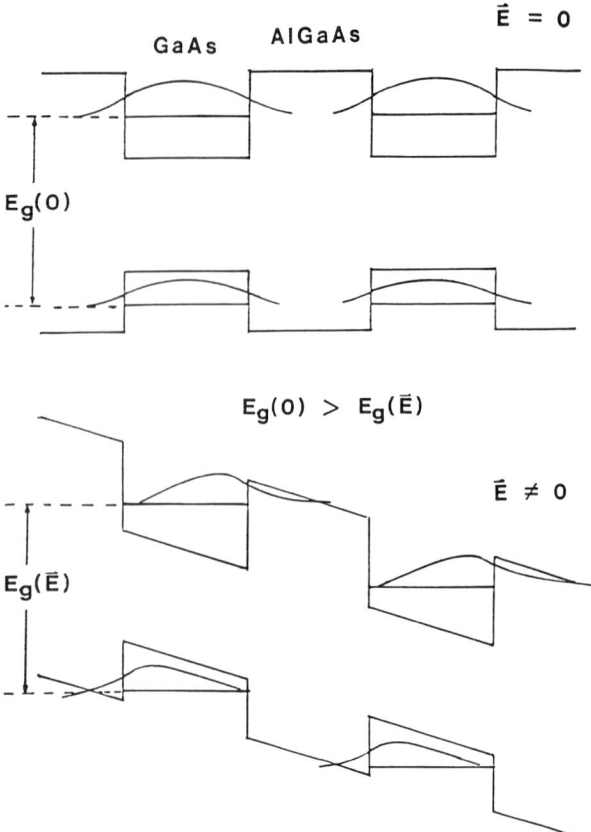

FIG. 3. Schematic representation of the effects of an electric field on the subband states in a quantum well. Note that the wave function polarizes and that the gap shrinks at nonzero electric fields.

III. Modulation Spectroscopy

With modulation spectroscopy, the derivative of the absorptivity or reflectivity with respect to some parameter is evaluated (Cardona, 1969; Seraphin, 1972). This spectroscopy has been shown to be sensitive to critical point transitions in the Brillouin zone, with the resulting spectrum having sharp derivative-like features and little if any featureless background. Also, weak features that may have been difficult to observe in the absolute absorption or reflection spectrum can be enhanced. Because of this derivative-like nature, a large number of sharp spectral features can be observed even at room temperature. In addition, there is also information in the other experimental variables such as the modulation frequency, amplitude, and phase.

The modulation can easily be accomplished by varying some parameter of the sample or experimental system in a periodic fashion and measuring the corresponding normalized change in the optical properties. While it is possible to modulate a variety of parameters, including the wavelength of the incident beam, the sample temperature, an applied stress, or applied magnetic field, we are basically interested in a modulation of either an externally applied electric field or a photoinduced variation in the built-in electric field.

Figure 4 shows a comparison of reflectivity (R) and electroreflectance (ER) spectra for GaAs over the same photon energy range (Phillip, 1963). While the reflectivity is characterized by broad features, the ER spectrum is dominated by a series of very sharp lines with zero signal as a baseline. Notice that much of the broad featureless background of the reflectivity measurement is not present in the ER spectrum. Herein lies the power of modulation spectroscopy: uninteresting background structure is eliminated in favor of sharp lines corresponding to specific transitions in the Brillouin zone. While it is difficult to calculate a full reflectance spectrum, this is not the case for modulation spectroscopy. For well-known critical points in the Brillouin zone, it may be possible to account for the line shapes in a modulation spectrum (Cardona, 1969).

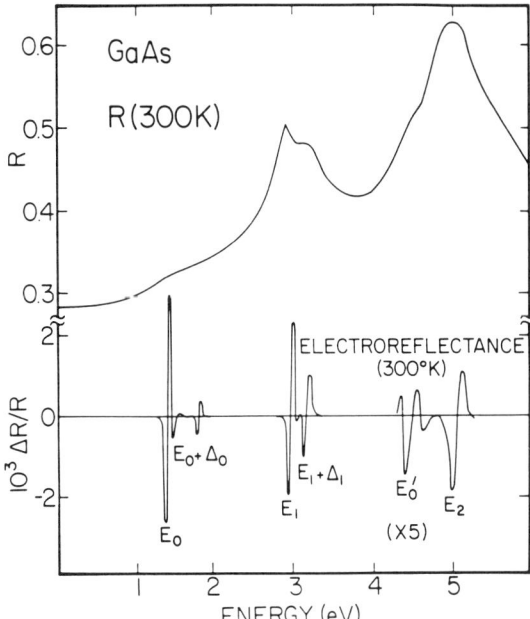

FIG. 4. Room temperature reflectivity (top) and electroreflectance (bottom) spectra of GaAs. (After Phillip and Ehrenreich, 1963.)

1. LINE SHAPES

Differential changes in the reflectivity can be related to the perturbation of the complex dielectric function in a simple manner, as expressed by Seraphin and Bottka (1966):

$$\frac{\Delta R}{R} = \alpha \Delta \varepsilon_1 + \beta \Delta \varepsilon_2, \tag{9}$$

where R is the reflectivity, $\Delta \varepsilon_1$ and $\Delta \varepsilon_2$ are changes in the complex dielectric function, $\varepsilon = \varepsilon_1 + i\varepsilon_2$, and α, β are the Seraphin coefficients, which are related to the unperturbed dielectric function. Near the fundamental gap of bulk materials, $\beta \approx 0$, so that $\Delta R/R \approx \alpha \Delta \varepsilon_1$ is the only important term. However, in multilayer structures interference effects are important so that the Seraphin coefficients are modified, and both $\alpha \Delta \varepsilon_1$ and $\beta \Delta \varepsilon_2$ may have to be considered. The exact functional form of $\Delta \varepsilon_1$ and $\Delta \varepsilon_2$ can be calculated provided that the dielectric function and type of the critical point are known. In the next sections, we will derive expressions for $\Delta \varepsilon_1$ and $\Delta \varepsilon_2$.

In modulation spectroscopy, one often encounters discussions that revolve around the "phase of the line shape." As one can see, from Eq. (9), the modulated reflectivity can be a mixture of $\Delta \varepsilon_1$ and $\Delta \varepsilon_2$. The modulation spectroscopy phase, therefore, represents the fraction of $\Delta \varepsilon_1$ and $\Delta \varepsilon_2$ that constitutes the observed line shape. As we will see, a further complication can arise in systems in which the modulating parameter (e.g., electric and magnetic fields, stress, or temperature) can produce changes in several properties of the system. For example, in quantum wells, the electric field not only shifts the band gap but it also changes the oscillator strength. This type of effect will also mix different line shapes into the modulated reflectivity. Therefore, the modulated reflectivity phase contains information not only about the Seraphin coefficients but also about optical interference, as well as modulation mechanisms.

2. MODULATED DIELECTRIC FUNCTION

The dielectric function of bulk materials has been studied in detail and is well understood. Interband transitions involving absorption with and without Coulomb interactions between the optically created electrons and holes have been considered (Aspnes, 1973, Lederman and Dow, 1976; Blossey and Handler, 1972). If free-electron–hole pairs (without excitonic interactions) are formed in a periodic material, the two particles will be accelerated by the electric field. Because of the position-dependent potential arising from the electric field, translational symmetry is destroyed and k is no longer a good quantum number. Aspnes and Studna (1973) have shown that in the absence of excitonic interactions the electromodulation line shapes for low electric fields are directly related to the third derivative with respect to photon energy of the zero-field dielectric function. At high fields, the Franz–Keldysh effect is operative, and more complex line shapes are obtained (Keldysh, 1958; Franz, 1958; Aspnes, 1973). The distinction between low and high fields is treated in detail by Aspnes and Studna (1973) and will not be discussed here.

On the other hand, if Coulomb interactions between the optically created electrons and holes are present, excitons will form. This case can be qualitatively treated in the effective-mass approximation as an isolated atom in an electric field. The electric field will affect the excitons in two ways. For small electric fields, there will be a Stark shift in the energy levels resulting in a change in the exciton binding energy. At higher fields, the exciton can be field-ionized, leading to a decreased lifetime (Blossey and Handler, 1972). Both of these effects produce first-derivative-like EM line shapes.

In order to better understand electromodulation spectroscopy in microstructures, we will review the salient features of Aspnes' work in bulk materials, particularly the case of nonexcitonic transitions and how low electric fields lead to third derivative functional form (TDFF) EM line shapes. This approach is useful in that it provides a very general expression for the EM line shape and also allows us to learn some details about the transition from three dimensions to two.

For our purpose, the most useful expression for the dielectric function involves the Fourier transform of the time-dependent current. The details of this form are widely documented and are beyond the scope of this chapter. The interested reader is referred to Aspnes and Studna (1973). The general expression for ε is

$$\varepsilon(E, \Gamma) = 1 + \frac{4\pi i e^2 \hbar}{m^2 E^2} \int_{BZ} d^3k \int_0^\infty dt |\mathbf{e} \cdot \mathbf{p}_{cv}|^2$$

$$\times \exp \int_{-t/2}^{t/2} dt' \frac{i[E - E_{cv}(\mathbf{k}) + i\Gamma]t'}{\hbar} \qquad (10)$$

where Γ is a phenomenological Lorentzian lifetime broadening parameter and we have assumed that $\mathbf{p}_{cv}$ is independent of k. Because ε is a complex quantity, Eq. (10) is a general result for the complex dielectric function, which can be related directly to either the absorption or reflection coefficients.

3. BULK MATERIALS

For the case of bulk materials with an applied electric field, the main effect of the electric field is to change the wave vector of the Bloch state from $\mathbf{k}$ to

$\mathbf{k} + e\mathbf{F}t/\hbar$, where $\mathbf{F}$ is the applied electric field. Mathematically, this substitution expresses the fact that the field mixes all wave functions that have components along the field direction. For periodic systems therefore, the component of $\mathbf{k}$ along the field direction is no longer a good quantum number. Because of the continuous nature of the wave vector in periodic systems, even small fields can mix many wave functions. The magnitude of the mixing is determined by the time parameter t. From a free-particle point of view, the electron or hole is driven up its dispersion curve and its energy is changed continuously, giving the appearance that it is accelerated in the electric field. This is just a classical picture of an electron in an electric field. Bloch (1928) and Jones and Zener (1934) showed that a Gaussian wave packet composed of Bloch states would change in time by the preceding prescription. Following Aspnes (1973), we begin by replacing $\mathbf{k}$ with $\mathbf{k} + e\mathbf{F}t/\hbar$ in E_{cv} of Eq. (10), and then expand in a Taylor series around $\mathbf{k}$. With this approximation we get

$$\varepsilon(E, F, \Gamma) = 1 + \frac{iA}{E^2} \int_{BZ} d^3k \int_0^\infty dt \exp\left[\frac{i(E - E_{cv} + i\Gamma)t + i\Omega t^3/3}{\hbar}\right], \quad (11)$$

where $\Omega^3 = e^2 F^2/8\mu\hbar$, $A = 4\pi e^2 \hbar |\mathbf{e} \cdot \mathbf{p}_{cv}|^2/m^2$, and μ is the joint density-of-states effective mass in the direction of the applied field and is related to $\nabla_k^2 E_{cv}$ ($i = x, y, z$). Equation (11) is the general expression for an Airy function.

A simplification occurs when the fields are small or when the homogeneous broadening (Γ) is large; i.e., $\Gamma t \gg \Omega t^3$. In this case, we can expand the exponent containing Ω [$\exp(i\Omega t^3/3) \approx 1 + i\Omega t^3/3$], which results in Aspnes's small-field limit:

$$\varepsilon(E, F, \Gamma) = 1 + \frac{A}{E^2} \int_{BZ} d^3k \int_0^\infty dt \left(1 + \frac{i\Omega t^3}{3}\right) \exp\left[\frac{i(E - E_{cv} + i\Gamma)t}{\hbar}\right]. \quad (12)$$

In Eq. (12), the first term under the integral represents the unperturbed dielectric function, and the second term is the field-dependent perturbation. We immediately notice that the time dependence (t^3) of the perturbation term can be obtained by differentiating the unperturbed term under the integral three times with respect to E (and also Γ or E_{cv}). Integrating Eq. (12) yields the final form

$$\varepsilon(E, F, \Gamma) = 1 + \frac{iA}{E^2} \int_{BZ} d^3k \frac{1}{E - E_{cv} + i\Gamma}$$
$$+ \frac{2iA}{E^2} \int_{BZ} d^3k \frac{(\hbar\Omega)^3}{(E - E_{cv} + i\Gamma)^4}. \quad (13)$$

This result is just the well-known Aspnes third derivative functional form, which can be expressed compactly as $\Delta\varepsilon \sim d^3(E^2\varepsilon(E,0,\Gamma))/dE^3$. The third derivative nature of Eq. (13) is readily apparent from Eqs. (10) and (12). The term in t^3 arises directly from the fact that there is a dispersion curve along the field direction and hence a gradient with respect to **k**. This gradient makes Ω a finite number in Eq. (11). It is interesting to note that for a two-dimensional critical point, characterized by a step density of states, the effective mass is finite along two dimensions and infinite along the third. Therefore, for a field applied along the direction having the infinite mass, there will be no third derivative terms in $\Delta\varepsilon$. This fact is our first hint to the final result for MQWs and other confining potentials.

4. Artificially Structured Materials

We will now apply the results of Aspnes, as developed in the previous section to microstructures and show that for these types of systems, $\Delta\varepsilon$ has just a first derivative functional form (FDFF) and not third derivative as for three-dimensional systems (Glembocki and Shanabrook, 1989). We begin by assuming that the confining potential is along the z-direction. As noted in Eq. (1), the wave function can be written as the product of an envelope function $F_n(z)$ and a Bloch state $u_{co}\exp(i\mathbf{k}\cdot\mathbf{r})$ representing the unconfined directions. Because of the confining potential, k_z is no longer an eigenvalue of the system. Because of this, we must modify Eq. (10):

$$\varepsilon(E,\Gamma) = 1 + \frac{8\pi^2 ie^2\hbar}{m^2 E^2 L}\sum_{nm}\int_{BZ} d^2k \int_0^\infty dt\, |\mathbf{e}\cdot\mathbf{p}_{cv}(\mathbf{k})I_{nm}|^2$$
$$\times \exp\left[\int_{-t/2}^{t/2} dt'\frac{i[E+i\Gamma - E_{cv}^{nm}(\mathbf{k})]t'}{\hbar}\right], \quad (14)$$

where n and m are the subband indices of the conduction and valence bands, L is the width of the confined region, and I_{nm} is the overlap integral between the electron and hole envelope functions (subband states). The sum over the subband states takes the place of the k_z integral in Eq. (10). If an electric field is applied along the direction of confinement, the shape of the confining potential changes, and if the field is small new quasibound states are formed. Furthermore, the new potential redistributes the electronic charge in such a manner that electrons and holes are localized at opposite ends of the confining region. This effect is reflected in both the form of the wave functions and the overlap integral I_{nm}. Unlike the 3D case, however, the new potential does not accelerate the particles and, hence, does not produce a continuously changing energy.

This difference between the free and confined particles in an electric field can be viewed classically in the following manner. Because of the acceleration due to the additional force of the electric field, the energy of an electron becomes time-dependent in that it will change continuously in time. Thus, at any two different times, the energy of the particle is not the same. In the confined case, the electric field acts to change the shape of the potential. Thus, at any two times after the field is applied, the energy of the particle is unchanged because it is in a quasibound state. The main effect of the field is just to change the energy level, the functional forms of the wave functions, and the lifetime of the particle. Many different calculations of quantum wells in electric fields have shown this to be the case (Bastard, 1982; D. A. B. Miller et al., 1985). They find that for small fields, the energy changes by a fixed amount rather than in a continuous fashion as it would in the case of a free particle. The Franz–Keldysh effect, which reduces to the TDFF for low fields, becomes important only when there is considerable penetration of the wave function through the confining barrier.

These results are very important because they indicate that we cannot replace $\mathbf{k}$ by $\mathbf{k} + e\mathbf{F}t/\hbar$ in Eq. (10). Instead, we should change the interband energy $E_{cv}^{nm}(\mathbf{k})$ and the overlap integral between electrons and holes, I_{nm}, by some fixed amount that is not dependent on the time t. The new energy is given by $E_{cv}^{nm}(\mathbf{k}) + \delta E_{cv}^{nm}$, and the new overlap integral is $I_{nm} + \delta I_{nm}$. The quantities δE_{cv}^{nm} and δI_{cv}^{nm} can be calculated from stationary-state perturbation theory.

If δE_{cv}^{nm} is much smaller than any other energy in the problem and $\delta I_{nm}/I_{nm} \ll 1$, we can expand the exponent and the overlap integral to obtain

$$\varepsilon = 1 + \frac{iA'}{E^2} \sum_{nm} |I_{nm}|^2 \left(1 + \frac{2\delta I_{nm}}{I_{nm}}\right) \int_{BZ} dk^2 \int_0^\infty dt \left(1 + \frac{it\delta E_{cv}^{nm}}{\hbar}\right)$$

$$\times \exp\left\{\frac{i[E - E_{cv}^{nm}(\mathbf{k}) + i\Gamma]t}{\hbar}\right\}, \quad (15)$$

where $A' = 8\pi^2 e^2 \hbar |\mathbf{e} \cdot \mathbf{p}_{cv}|^2/m^2 L$. Performing the integral over t and keeping only first-order terms in the perturbation leads to

$$\varepsilon(E, F, \Gamma) = 1 + \frac{iA'}{E^2} \sum_n |I_{nm}|^2$$

$$\times \int_{BZ} dk^2 \left[\frac{i\delta E_{cv}^{nm}}{(E - E_{cv}^{nm}(\mathbf{k}) + i\Gamma)^2} + \frac{(1 + 2\delta I_{nm})/I_{nm}}{E + E_{cv}^{nm}(\mathbf{k}) + i\Gamma}\right]. \quad (16)$$

Equation (16) is the general result for any type of 1D confining potential with the field applied along the confinement direction. In Eq. (16), the first term under the integral represents changes in ε from electric-field-induced changes in the gap, E_{cv}^{nm}, while the second term is a combination of the unperturbed dielectric function and a term due to the modulation of the overlap integral. A most important observation is that the modulated terms are first derivatives and not third derivatives as in the bulk case.

The lack of any particle acceleration is indicated in Fig. 3 by the form of the wave function of a particle confined in a quantum well. Here we show schematically the wave function of the $n = 1$ state with and without an electric field. We notice that the wave function within the well has not gained any extra oscillations (characteristic of acceleration), indicating that no acceleration has ocurred. The peak of the wave function, however, has shifted in accordance with a redistribution of charge in the well. These results are in agreement with the ideas already presented regarding the nature of electric-field effects in confined structures.

In Eq. (16), the integrals over k-space can be performed and for a 2D critical point. Assuming that this is done, we can express the result of Eq. (16) in a compact manner as a first derivative functional form for the modulated dielectric function

$$\Delta\varepsilon = \left[\frac{d\varepsilon}{dI_{nm}}\frac{dI_{nm}}{dF} + \frac{d\varepsilon}{dE_{cv}^{nm}}\frac{dE_{cv}^{nm}}{dF} + \frac{d\varepsilon}{d\Gamma}\frac{d\Gamma}{dF}\right]\Delta F. \quad (17)$$

Equation (17) is the most general form of the modulated dielectric function for confined systems. In it we have included the possibility that the electric field can modulate the lifetime of the state, as in the case of field-induced tunneling. This equation was first proposed by Shanabrook et al. (1987a) to describe PR measurements for excitons in MQWs. Here we see that it also holds for any confining potential. It is interesting to note that Eqs. (16) and (17) apply only when the field is along the direction of confinement. For transverse fields, the TDFF is applicable and for fields along any arbitrary direction, the EM line shapes will be combinations of TDFF and FDFF.

We wish to illustrate the various forms of the derivatives of Eq. (17). Since this equation is general and works for any type of dielectric function, we use the simplest possible, a Lorentzian oscillator sometimes appropriate to excitons. Shown in Fig. 5 are the various derivatives of Eq. (17) for the real part of the dielectric function. Notice the difference in line shape between gap and lifetime modulation. Similar differences will occur for other forms of the dielectric function.

Enderlein et al. (1988) has also treated the dielectric function of quantum wells, but he begins with the integrated form of Eq. (14):

$$\varepsilon(E,\Gamma) = 1 + \frac{iA'}{E^2}\sum_{nm}|I_{cv}^{nm}|^2 \int_{BZ} dk^2 \frac{1}{E - E_{cv}^{nm}(\mathbf{k}) + i\Gamma}. \quad (18)$$

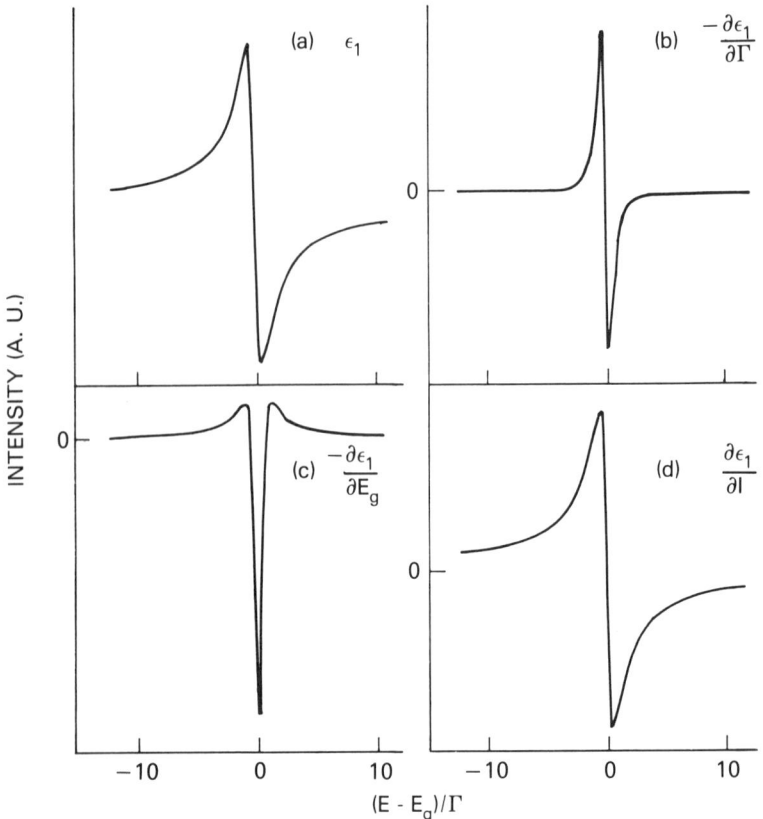

FIG. 5. (a) The real part of the dielectric function for a Lorentzian oscillator and its derivatives with respect to (b) linewidth, (c) energy gap, and (d) intensity or oscillator strength. (After Shanabrook et al., 1987.)

He then calculates the field-induced shifts of the gap and changes in the intensity. He also concludes that the modulated spectrum should be a first derivative. Enderlein also treated the effects of Coulomb interactions on the PR in MQW. It is quite clear from the preceding approach and from Enderlein's work that PR in MQWs is a first derivative spectroscopy.

The results derived here are applicable to isolated, confined systems. If we now let two such systems interact by reducing the barrier thickness between them, we get two coupled states, whose wave functions are shared between the two systems. If many interacting systems are brought together, the wave functions become extended and a miniband is formed, similar to the formation of bands in a bulk material. In this case, we have a 3D-like situation, because there is dispersion in the miniband, as well as a new translational symmetry. For small fields, which do not destroy the miniband, the TDFF results of Eqs. (12)–(13) are applicable to nonexcitonic transitions at the miniband edges. Bleuse and co-workers (1988) have shown theoretically that at moderate fields, the electron and hole wave functions begin to localize to several layers, and finally at higher fields there is complete localization into single layers. These results are important in electromodulation spectroscopy, because they suggest that for moderate and high fields, the TDFF will not be applicable. Rather, we expect that for moderate fields, variations in the overlap integral will be the dominant modulation mechanism. Meanwhile at high fields where complete localization occurs, we expect a modulation of the Stark shifts of the levels to be more important. It is therefore very important in EM of superlattices to determine which electric-field regime is applicable for the chosen experimental conditions.

Until now in our discussion of the modulated dielectric function, we have ignored the Coulomb interaction between the photoexcited electron and hole. Excitons, however, play an important role in the optical properties of semiconductors. They are even more important in the case of microstructures, where the electron and hole are confined in real space. Some very important parameters must be considered before the effects of excitons on the optical properties are ignored. These are as follows: quantum wells tend to be <500 Å in width, while for bulk GaAs the exciton Bohr radius is approximately 200 Å. If quantum well states are formed, the electrons and holes are sensitive to the boundaries of the well. Consequently, it is difficult to imagine that these particles do not sense their mutual Coulomb interaction. Because of this argument, we feel that whenever quantum states are formed for both electrons and holes, one cannot ignore the Coulomb interaction. This indeed has been the experimental case in quantum wells. It is well known that excitons exist in quantum wells even at high temperatures (Dawson *et al.*, 1983) and also in high electric fields (D. Miller *et al.*, 1985). Therefore, excitonic effects must be included in our model of modulation spectroscopy. Because of this, we conclude that even for SLs, in small fields, EM will be sensitive to excitons and hence TDFF will not be applicable.

The case of the exciton can be obtained rather easily by realizing that the Coulomb interaction leads to radial confinement and a set of discrete states for energies less than the binding energy. It is important to note that an exciton acts as a charge neutral unit, containing an electron and hole. Even though the unit can have kinetic energy, it *cannot* be accelerated by the electric field, because of its neutrality. Instead, the field polarizes the exciton and reduces its binding energy (Blossey and Handler, 1972). The general result of Eq. (17) still holds. It is interesting to note that the EM results for excitons in 2D and 3D are the same. In addition, and as indicated in Fig. 2, a very important observation is that whenever excitons play a role in interband transitions, the nonexcitonic band edge is modified so (see Fig. 2) that there is a smooth merging of the exciton states ($n = 1$ to ∞) with the band-edge continuum (Shinada and Sugano, 1966). Therefore, there is no discontinuity in ε at the position of the nonexcitonic band edge, and the use of TDFF to account for the nonexcitonic band to band transitions along with excitons for the same state is inappropriate. Detailed calculations of the form of the interband absorption properties of single quantum wells and superlattices that include light- and heavy-hole mixing and the Coulomb interaction have been reported by Chu and Chang (1989).

IV. Experimental Details

Modulation techniques take advantage of the application of a small periodic perturbation to a physical property of the sample. The change in the optical function (reflectivity or absorptivity) is only a small fraction of its unperturbed value, typically 1 part in 10^4. The perturbation is usually extracted through the use of a lock-in amplifier tuned to the modulating frequency. This aspect of modulation spectroscopy is common to all techniques, including electroreflectance, photoreflectance, wavelength modulation, piezoreflectance, and thermoreflectance. We will describe in more detail electromodulation techniques.

In electroreflectance, an applied electric field is modulated to produce a periodic variation in the dielectric function, while in photoreflectance this modulation is accomplished through the production of photoexcited carriers. The experimental apparatus for photoreflectance spectroscopy is shown in Fig. 6. The probe light is a monochromatic beam obtained from either a tungsten lamp or a quartz halogen source dispersed through a monochromator. For most routine applications, a monochromator having a focal length of 1/4 m is adequate. At low temperatures, excitons can exhibit narrow line widths and a 0.85-m or 1-m monochromator is better. The light impinging on the sample has intensity I_0. The modulation (electric field, temperature, stress, etc.) is applied to the sample at a frequency Ω_m.

The reflected light is detected by a photomultiplier or a suitable photodiode. In the case of most GaAs–AlGaAs microstructures, a Si PIN diode is often used. The light striking the detector contains two signals: a dc or average value, $I_0 R$, and a modulated value, $I_0 \Delta R$, which varies with the frequency Ω_m. Since it is the quantity $\Delta R/R$ that is of interest, it is necessary to eliminate the common factor I_0. This normalization can be accomplished in one of three ways. The simplest, although the most time-consuming, manner is to measure ΔR and R independently and to divide the two quantities. The drawback of this technique is that it requires two measurements and it is always more difficult to rely on ratioing, especially in the case of a light source such

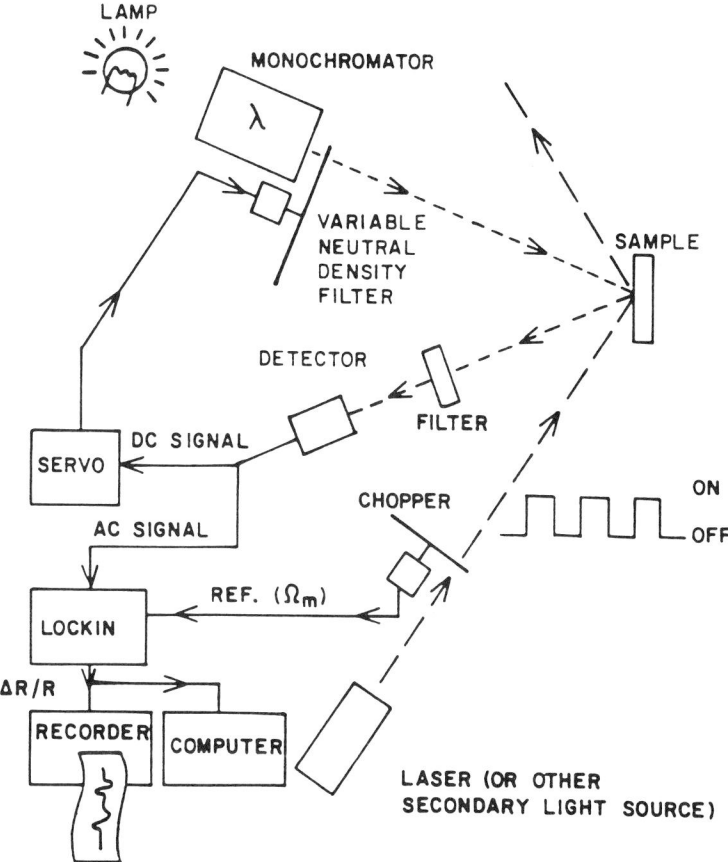

FIG. 6. Schematic diagram of the photoreflectance apparatus. Monochromatic light is used as a probe. Its intensity is kept constant by a variable neutral density filter, which is controlled by a servomechanism. The pump is provided by a chopped laser having a frequency Ω_m. The lock-in is tuned to Ω_m and detects the reflected light at this frequency. (After Shen et al., 1987a,b.)

as a Xe arc lamp, which has many sharp lines. A more elegant method of obtaining $\Delta R/R$ is to make R constant electronically. This can be accomplished in one of two ways. In this technique, the dc output of the photomultiplier tube (PMT) is kept constant through a servomechanism that compares the dc background output to a reference voltage and uses the difference between the two to change the gain of the PMT. This idea can also be applied to a photo-diode through the use of a programmable gain amplifier. A third technique was recently developed and involves keeping the dc intensity of the reflected light constant through the use of a continuously variable neutral density filter placed after the monochromator (Shen et al., 1987a,b). This circular filter is rotated by a servomotor driven by the voltage difference between the dc signal and a fixed reference signal. This is illustrated in Fig. 6. Both of the electronic servo techniques result in the output of the lock-in amplifier being proportional to $\Delta R/R$.

In the case of photoreflectance, it is important for the apparatus to have good filtering of the stray laser light, because it has the same frequency (chopped) as the signal of interest and can easily be detected. Furthermore, laser illumination can produce band-gap photoluminescence, which under certain conditions is greater in intensity than the signal of interest. This problem can be eliminated by using long-focal-length optics or by using a second monochromator running in unison with the probe monochromator (Theis et al., 1988). For a double monochromator, two scans are taken, one with the probe light on and one without it. Subtracting, the two traces effectively eliminates the PL. An alternative technique involves using a dye laser as the probe beam and a detector placed sufficient far away from the sample so as to reduce the PL, which is usually emitted isotropically (Shanabrook et al., 1987a; Glembocki and Shanabrook, 1987b).

V. Experimental Spectra

The experimental results serve as examples relating specifically to the theoretical aspects discussed earlier. In addition, we will attempt to present an overview of the state of photoreflectance in microstructures. Because PR is in many ways similar to electroreflectance, we will take the liberty of including some ER results as needed for illustration.

5. QUANTUM WELLS

a. E_0 Gap

Quantum wells form the basis of our understanding of superlattices and other microstructures. As noted, they represent the simplest circumstance to study experimentally. Because of significant confinement effects that result from the light electron effective masses, the E_0 gap has been the most widely studied band gap of microstructures. In addition, in many systems the fundamental gap is easily accessible by luminescence measurements. Because of the wealth of information available about transitions at the E_0 gap, we will begin our discussion here.

Historically, Mendez et al. performed the first electromodulation measurements on a microstructure system, specifically on superlattices (SL) (Mendez *et al.*, 1981). Three years later, Erman and co-workers, used Schottky barrier electroreflectance to study the interband transitions in a single GaAs–AlGaAs QW grown on a doped (100) GaAs substrate (Erman *et al.*, 1984). Their measurements consisted not only of transitions at the E_0 gap but also those at the E_1 and $E_1 + \Delta_1$ gaps. Despite the work of Mendez *et al.* and Erman *et al.*, modulation spectroscopy of microstructures received little attention. Presumably, this was due to the fact that the configuration for applying the electric field, Schottky contact, or an electrode in an electrolytic solution required doped substrates and special surface preparation techniques. Consequently, the applicability of ER was limited to a certain class of samples, while contactless techniques, such as PL or PLE, did not suffer from such limitations.

Soon after the Erman work, Glembocki *et al.* showed that multiple quantum wells grown on semi-insulating substrates can be studied by the photoreflectance technique (Glembocki *et al.*, 1985). Because the modulation of the dielectric function is accomplished by a chopped laser, the PR technique is totally contactless and applicable to any sample (with a built-in electric field), regardless of the carrier density. Shown in Fig. 7 are the 300 K spectra of Glembocki *et al.* for a GaAs–AlGaAs heterostructure composed of thick layers, and for several MQW samples. The MQW samples consisted of a series of 30 to 40 single quantum wells, grown on semi-insulating (100) GaAs. The interband transitions are denoted by h_n and l_n, where the letters l and h specify the nature of the valence-subband states, light (l) or heavy (h). The number n, is the subband index of the valence and conduction subbands. Note that we are dealing with allowed transitions, for which m (valence) = n (conduction). The first thing to notice is that each spectrum has GaAs and an AlGaAs lines. The GaAs transitions originate from the buffer regions separating the MQW from the substrate. The AlGaAs transitions are from the AlGaAs barriers in the MQW. These two lines are very important because they are directly related to the height of the AlGaAs barrier, relative to the GaAs well. (See Fig. 1). There are two possible situations in relation to these transitions, depending on the layer thicknesses. For thick AlGaAs barriers (> 10 nm), the transition energy is nearly equal to that of the energy gap of the particular alloy concentration. It is also possible to have a circumstance in which Kronig–Penny effects will produce an AlGaAs miniband and thus blue-shift the AlGaAs E_0 transition (Wong *et al.*, 1986). In this case, a Kronig–Penny calculation would be required to get the alloy concentration from the transition energy. In the case of Fig. 7, the AlGaAs barriers are 150 Å thick and Kronig–Penny effects are unimportant. Therefore, from the position of the $Al_xGa_{1-x}As$ line, we find that the samples had x-values near $x = 0.27$.

A further examination of Fig. 7 reveals that the number of transitions increases with increasing well width. This behavior is expected from quantum confinement, for which the number of subbands increases with the width of

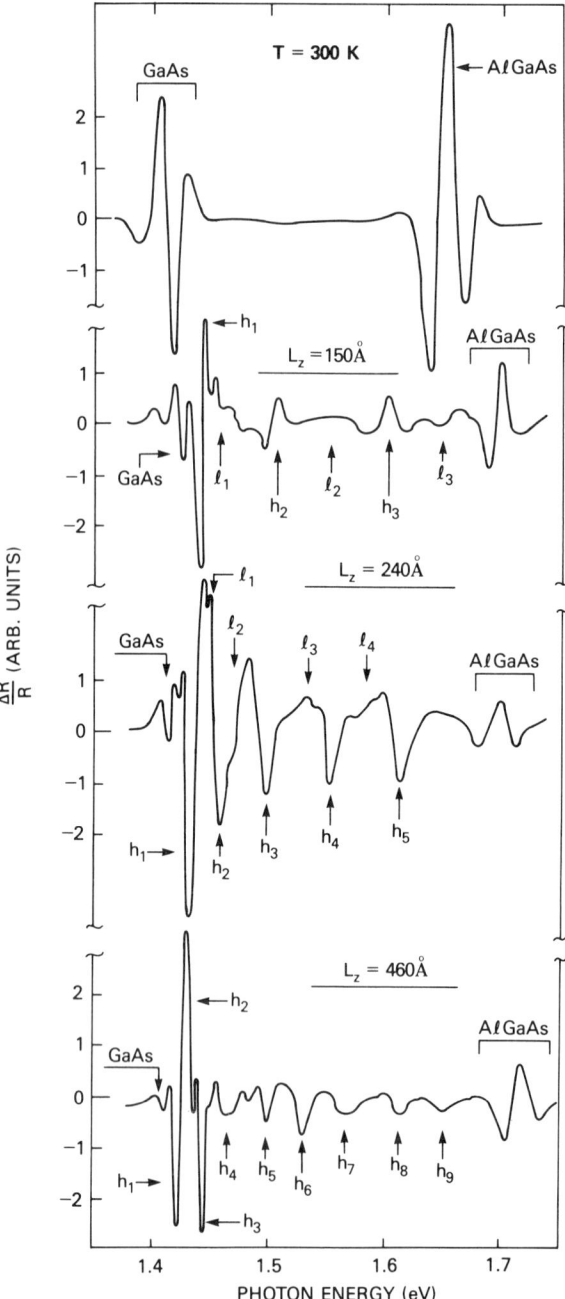

FIG. 7. Photoreflectance spectra for a GaAs–AlGaAs heterojunction (top) and MQW samples with well widths, L_z. The labels $h_n, l_n, n = 1, 2 \ldots$ represent parity-allowed light- and heavy-hole interband transitions. The vertical arrows locate calculated positions of the transitions. (After Glembocki et al., 1985.)

the GaAs layer. Furthermore, note that there is a difference in the intensities of the l_n and h_n transitions, with the h_n excitations being much stronger. The reasons for this are related to differences in oscillator strengths of the light and heavy holes as well as modulation mechanisms. We will discuss this in more detail later.

Glembocki and co-workers compared the transition energies measured in Fig. 7 to ones calculated from a finite square well model, using the following parameters (R. C. Miller, et al., 1984): conduction-band offsets of $Q_e = 0.6$, where $Q_e = [V_c(\text{AlGaAs}) - V_c(\text{GaAs})]/[E_g(\text{AlGaAs}) - E_g(\text{GaAs})]$ conduction-band effective mass of $m_c = 0.0665$, and light- and heavy-hole masses of $m_{lh} = 0.096$ and $m_{hh} = 0.34$. The results from the calculation compared favorably with the experimental results.

Interactions between light- and heavy-hole bands of different indices, n and n' can occur (Chang and Schulman, 1983). This leads to a mixing of light- and heavy-hole wave functions and to a breakdown of the parity selection rule. The mixing produces states whose wave functions have both light-hole and heavy-hole character. Therefore, normally forbidden transitions can now be observed. It has been shown in various other spectroscopies, such as PLE, that parity-forbidden transitions, for which $\Delta n \neq 0, 2, 4\ldots$, can be observed in QW samples. A good example of this is in a study in which PLE experiments in GaAs–AlGaAs MQW were compared with theoretical calculations (Miller, Gossard, Sanders, and Schulman, 1985). This work showed that the intensity of the 21L "forbidden" transition [nmH (L) denotes transitions between the nth conduction subband and the mth heavy (light) hole subband] could be larger than that of an allowed transition.

In addition to the mixing described, an applied electric field breaks the parity selection rule and causes transitions with $n + m = 1, 3, 5\ldots$ to be observed. This type of field induced mixing has been observed in the photocurrent measurements of Collins et al. (1987). Since the signal in photoreflectance comes from regions in which there exists a modulating electric field, "forbidden" transitions should be observed.

In the PR data of Fig. 7 for the 150-Å sample, we see evidence of extra features between l_1 and h_2. In this regard, Glembocki et al. studied the temperature dependence of a MQW sample with 200-Å wells over a limited spectral range (Glembocki et al., 1986). Shown in Fig. 8 is a series of spectra as a function of temperature, between 300 and 125 K. As the temperature is decreased toward 125 K, several forbidden transitions, 12H, 13H, 21H, and 23H emerge from the stronger allowed transitions. In addition, several other features marked by question marks were observed. It was suggested that these transitions might be related to a camel's-back-type warping of the valence band, caused by mixing effects. This assertion, however, was never fully substantiated.

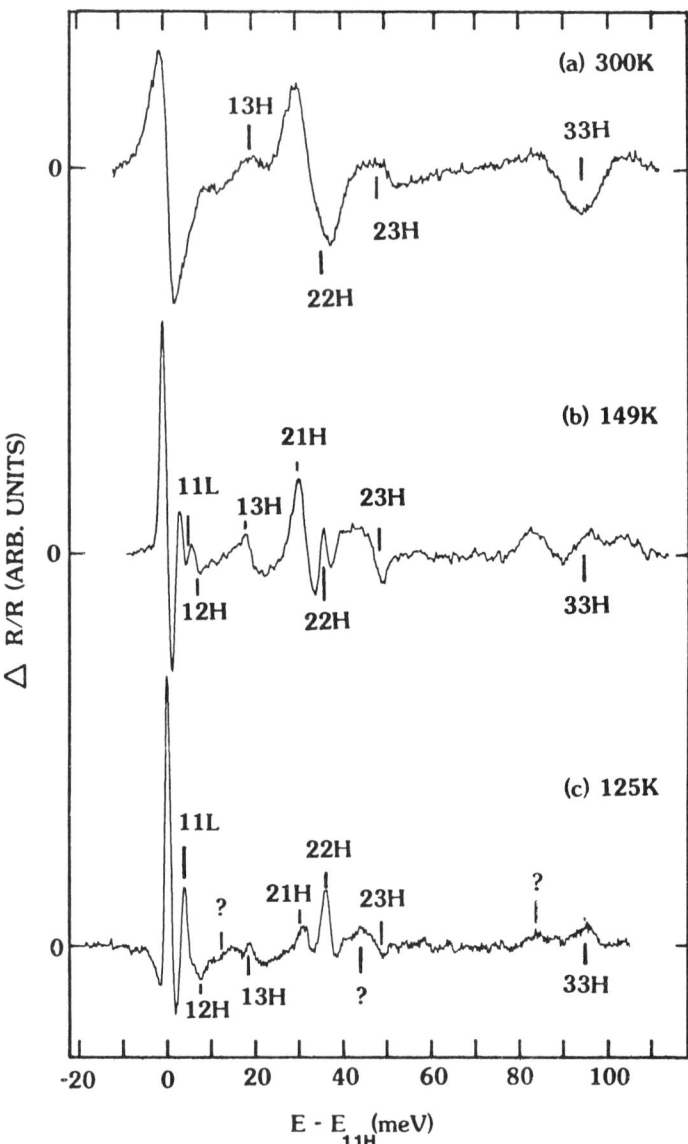

FIG. 8. Temperature dependence of the photoreflectance for a 205-Å well taken at (a) 300 K, (b) 149 K and (c) 125 K. Transitions are labeled as nmH(L), where n and m are the conduction- and valence-subband indices, respectively. The question mark in (c) is an unidentified transition. (After Glembocki et al., 1986.)

Forbidden transitions have also been observed at room temperature by Shen *et al.* (1986a). They performed a least-squares fit to the TDFF of a 2D critical point and obtained a reasonable fit only if forbidden transitions were included in the analysis.

b. *Line Shape Analysis*

Historically, line shape analysis (fitting) has been an integral aspect of electromodulation spectroscopy in bulk materials. These fits allowed complicated ER lines to be compared with theoretical predictions (Blossey and Handler, 1972). This approach survived over the years, because of the complex nature of the lines and the fact that the phases of the ER lines varied with circumstances. Thus line shape analysis would provide valuable information, about the energy gaps and linewidths of transitions.

The first fits to QW data were attempted by Shen *et al.* (1986b), who assumed the TDFF for a two-dimensional critical point and obtained reasonable fits to the room temperature experimental data. This result was surprising in light of the fact that excitons dominate QW absorption spectra at room temperature and at fields considerably larger than those required to ionize bulk excitons (D. A. B. Miller, *et al.*, 1984). Therefore, as we saw previously, the ER or PR spectra of 2D critical points or excitons in QWs should be first-derivative-like. In order to clarify the situation, Shanabrook *et al.* compared PR with PLE of 200-Å GaAs–AlGaAs MQW in the temperature range of 6 to 250 K (Shanabrook *et al.*, 1987a,b]. Experimental difficulties prevented them from making room temperature measurements. The idea was that PLE is related to absorption spectra and hence should exhibit peaks for excitonic transitions. If simultaneously measured PR data agreed in spectral positions with the PLE peaks, then the PR transitions are also excitonic. The data of Shanabrook *et al.* (1987) is shown in Fig. 9. The transitions correspond to 11H, 11L, and 12H. Notice that the spectral positions of the PR lines are in very good agreement with the energies of peaks in the PLE data. This is certainly the case over the entire temperature range. Because the exciton binding energies are approximately 8 meV, one concludes that the PR features are excitonic in nature. Similar results were also obtained by Klipstein and co-workers (1986), who studied the ER spectra of SQWs and MQWs at 15 K and compared ER with both PL and PLE (Klipstein *et al.*, 1987). They also found good agreement between PLE and ER for both types of samples, further justifying the conclusions that the PR transitions are excitonic.

In an attempt to understand the results of Shen *et al.* (1986b), Shanabrook *et al.* (1987a) also tried a line shape analysis for the data of Fig. 9 using an excitonic FDFF and TDFF of a 2D critical point. In the case of the 6 K data the best fit was obtained by using the FDFF and a Lorentzian dielectric

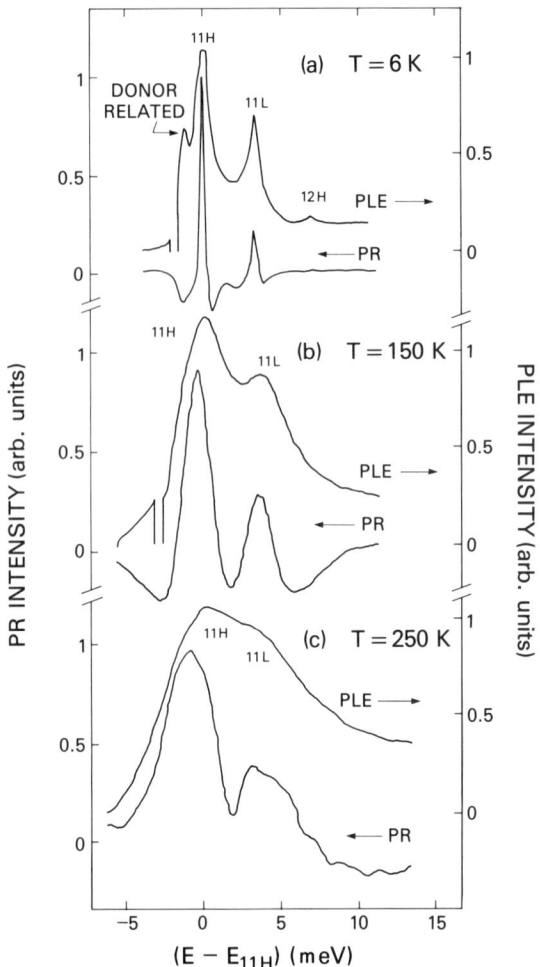

FIG. 9. A comparison of PLE to PR for a 200-Å multiple quantum well at (a) 6 K, (b) 150 K, and (c) 250 K. The energies are measured relative to 11H, where $E_{11H} = 1.5235$, 1.4995, and 1.4537 for 6 K, 150 K and 250 K, respectively. (After Shanabrook et al., 1987.)

function for the exciton. The fit of Shanabrook et al. to Eq. (14) and a Lorentzian dielectric function is shown in Fig. 10. The allowed transitions, 11H and 11L, are best described by a modulation of the exciton gap, while the forbidden transition, 12H, is best fit by an intensity modulation. These results were made possible by a detailed comparison of the PLE and PR data, especially in relation to the linewidths of the transitions. The linewidths used in the FDFF fits were constrained by the widths observed in the PLE data. Because of this, it was possible to fit the 12H transition with an

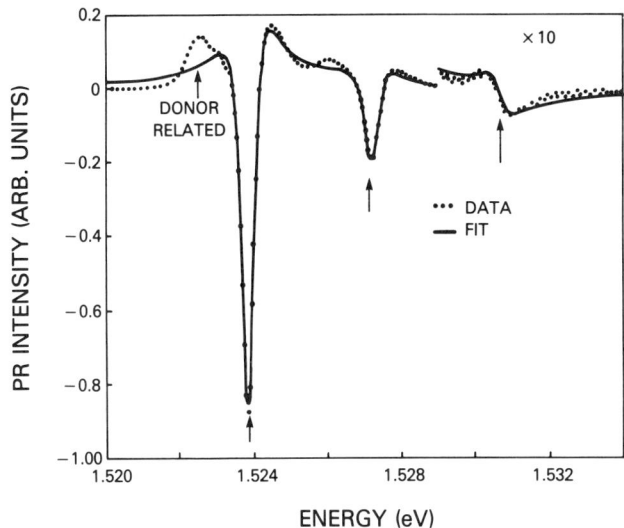

FIG. 10. A fit to the 6 K PR data for a 200-Å MQW. Equation (17) (FDFF) was used along with a Lorentzian oscillator. The transitions 11H, 11L, and 12H, are at 1.524, 1.527 and 1.529 eV respectively. (After Shanabrook et al., 1987a.)

intensity modulation mechanism rather than a lifetime modulation, both of which have similar line shapes. Interestingly, the fits to TDFF were rather poor.

As Shanabrook et al. (1987a) raised the temperature from 6 to 250 K, they found that the Lorentzian excitonic FDFF could not fit the data and that near 250 K, the data was best fit by a third derivative of a 2D critical point, in agreement with earlier room temperature measurements. However, the correspondence in the peak positions measured in PLE and PR proved that the transitions were still excitonic. Glembocki and Shanabrook (1987a) were able to account for this discrepancy by a temperature-dependent inhomogeneous broadening of the excitons. Toyozawa (1958) has shown that this type of broadening, which can be caused by strong exciton–phonon interactions, impurities, defects, and weak exciton–phonon coupling at elevated temperatures, changes the absorption profile of excitons from Lorentzian to Gaussian. In these cases, the Lorentzian dielectric function is inappropriate and Eq. (17) must be used with a "Gaussian" dielectric function. The real part of the Gaussian dielectric function is just a confluent hypergeometric function (Glembocki and Shanabrook, 1987a; Shen et al., 1987c):

$$\varepsilon_1 = Ax\Phi\left(\frac{1}{2}, \frac{3}{2}, \frac{1}{2}, x^2\right)\frac{\exp(-x^2)}{\Gamma}, \qquad (19)$$

where A is a constant containing effective masses, optical matrix elements, and the index of refraction. In Eq. (19), $\Phi(a,b,c,z)$ is a confluent hypergeometric function and x is a dimensionless parameter given by $x = (E - E_{cv})/\Gamma$.

Shown in Fig. 11 is a fit to the 150 K data of Fig. 9. Notice that at this temperature, a Gaussian exciton fits better than a Lorentzian exciton. Glembocki and Shanabrook (1987a) have shown that the TDFF for a 2D critical point mimics the FDFF of a Gaussian dielectric function and can be used as a reasonable approximation to the Gaussian functional form. This result is useful because the evaluation of a confluent hypergeometric function is numerically tedious. Therefore, the energies of the interband transitions in MQWs can be obtained in a simpler fashion. It is important, however, to realize that the transition from Lorentzian to Gaussian is not abrupt and that between 50 and 150 K, the line shapes are of an intermediate form between Lorentzian and Gaussian. Therefore, care must be exercised in using any line shape analysis in quantum well systems. However, meaningful gap energies and relative linewidths can always be obtained from any of the dielectric functions discussed, but actual values for the broadening parameter and details of the line shape may be suspect.

In a related study, Theis and co-workers (Theis et al., 1988) performed a first-principles calculation, including excitonic effects and valence-band mixing of the quantum well states and wave functions and used them to compute the dielectric function. They found good agreement between their theoretical

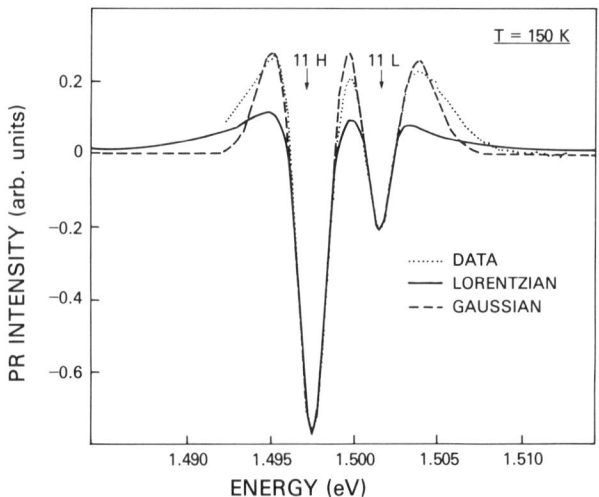

FIG. 11. A comparisons of FDFF fits from Eq. (17), using Lorentzian (solid line) and Gaussian (dashed line) dielectric functions. The data was 150 K PR for a 200-Å MQW (same sample as in Fig. 10). (After Glembocki and Shanabrook, 1987b).

results and experimental data. In addition, they also verified the FDFF and the previously observed Gaussian broadening.

In the previous discussion, we have shown that the line shape depends upon the nature of the dielectric function and the manner in which it is modulated. Fitting to complicated lines can be avoided under special conditions. For isolated lines, there is a simple procedure to *approximate* the energy gap. Aspnes has developed the three-point fitting method for Lorentzian-broadened dielectric functions (Aspnes, 1973). The method involves using two of the most intense extrema in the ER spectrum. Shown in the insets of Fig. 12 are two typical ER spectra and the two extremas need in the three-point fit. The procedure involves knowing the energies of the two extrema, $E(A)$ and $E(B)$, and their intensities, $I(A)$ and $I(B)$. Given these parameters, one then uses the following equations to determine the energy gap and the broadening parameters:

$$\rho = -\frac{I(B)}{I(A)} > 0, \qquad (20)$$

$$E_g = E(A) + [E(B) - E(A)] f(\rho), \qquad (21)$$

$$\Gamma = [E(B) - E(A)] g(\rho), \qquad (22)$$

where the functions $f(\rho)$ and $g(\rho)$ are plotted in Fig. 12. It is easy to see that the energy gap is between the two largest extrema and that for a symmetric line it occurs between A and B regardless of the critical point type. Also note that if one extremum is much larger than the other, then the gap is always near the larger of the two extrema. For example, in the case of an exciton, if $I(A) \gg I(B)$, then $\rho = 0$, $f(\rho) = 0$, and $E_g = E(A)$. What is important to note is that for any given value of ρ the error in obtaining the energy gap by using the wrong critical point type is no more than 20% of the energy difference between the two extrema. This is less than 20% of the linewidth.

c. Excited Excitonic Transitions

As noted, the QW transitions are excitonic in nature. In many cases, we observe only the 1s state of the exciton. In high-quality samples, it is also possible to see 2s excitons in PLE (R. C. Miller, et al., 1981). Such observations are important because they allow the binding energy of the 1s exciton to be estimated. The first evidence of excited excitonic transitions in PR was obtained by Shen and co-workers (Shen, 1987c), who saw at 77 K a weak PR feature located between 11H(1s) and 11L(1s). This feature was also observed in 6 K PLE and 77 K photoreflectance excitation spectroscopy (PRE). Even

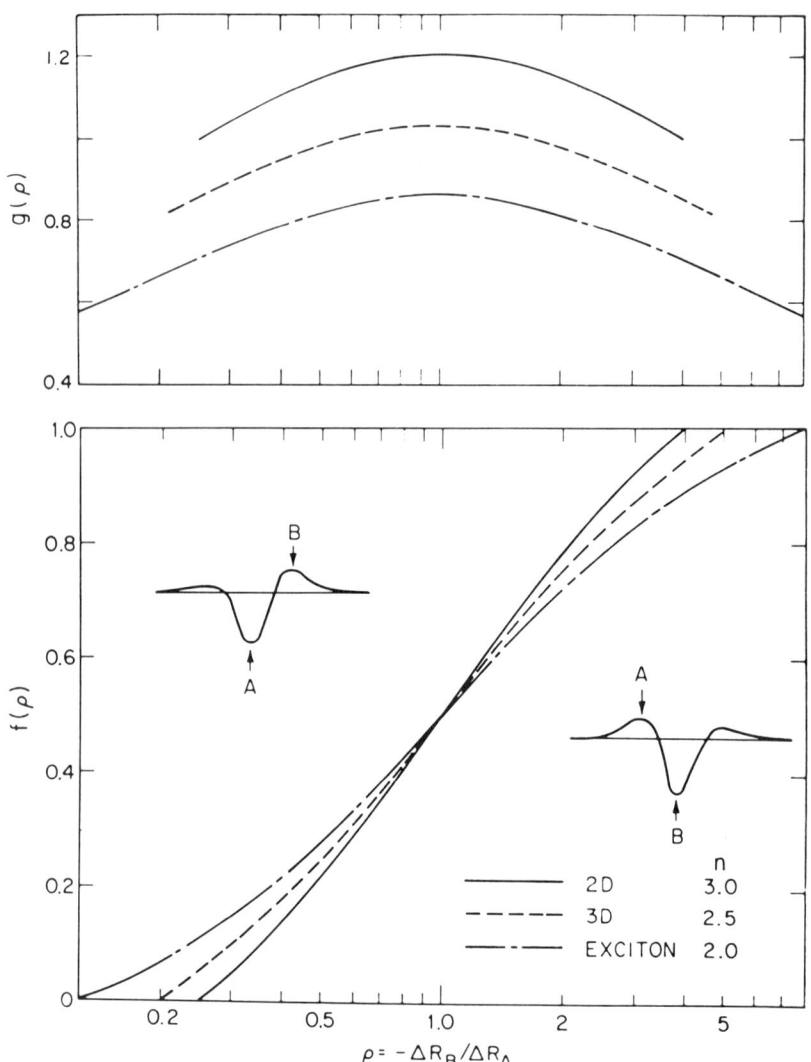

FIG. 12. The function $f(\rho)$ and $g(\rho)$ used in obtaining the gap energy and the broadening parameter. For the exciton, a FDFF analysis was used, while for the 2D and 3D critical points the TDFF was employed. In all cases, a Lorentzian broadening was assumed. (After Aspnes, 1973.)

more compelling evidence for excited states was obtained by Theis et al. (1989a). Shown in Fig. 13 is a comparison of 6 K PLE, PR, and a theoretical calculation of the PR spectrum. Note that there is a well-defined feature corresponding to the 11H(2s) transition. They also observed the transition 11L(2s), which is not seen in this figure. The theoretical calculation also in-

4. PHOTOREFLECTANCE SPECTROSCOPY OF MICROSTRUCTURES   251

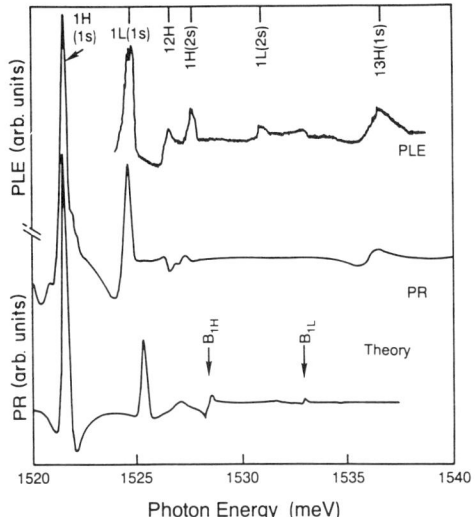

FIG. 13. A comparison of PLE (upper) and PR (middle) for a 225-Å MQW, taken at 6 K. On the top axis are denoted positions of various transitions. Note the presence of 1s and 2s excitons in both the PLE and PR. The lower trace is a calculation of the PR spectrum showing the onset of 1H and 1L band-to-band transitions as B_{1H} and B_{1L}, respectively. (After Theis et al., 1989a.)

cludes two features not seen in PR, namely B_{1H} and B_{1L}. These are the positions of the nonexcitonic band-to-band transitions. The fact that the PR data do not contain these features is proof of the fact that excitonic interactions replace band-to-band transitions by a series of atomic-like transitions that merge with a continuum of states (see the schematic diagram in Fig. 2).

d. *Identifying Light and Heavy Holes*

As we have seen, heavy- and light-hole transitions differ in intensities. In a spectrum that contains many other transitions, it becomes important in being able to distinguish between light- and heavy-hole transitions. While in the GaAs–AlGaAs system, much is known and the identification of PR peaks is simply a matter of comparison with theory, this may not be the case for other systems. For example, in strain-layer systems, the strain may invert the order of the light and heavy holes and modify their optical matrix elements (Voisin et al., 1984). Without detailed knowledge of the strains and band offsets, additional information is required in sorting out the light- and heavy-hole character of a given transition.

In photoreflectance and electroreflectance, we can accomplish this by taking advantage of the modulation of the electric field. As we saw, in the case of MQWs an important modulation mechanism for allowed transitions is simply an electric-field shift of the $n = 1$ quantum well level. For this case, using Eqs. (17) and (18) we can write the intensity of any transition as

$$\frac{\Delta R}{R} = C|\mathbf{e} \cdot \mathbf{P}_{cv}|^2 L(E, E_{cv}, \Gamma) A_g, \tag{23}$$

where C is a constant, $L(E, E_{cv}, \Gamma)$ is a line shape factor, and $A_g = dE_{cv}/dF$ is the modulation coefficient. The coefficient C is treated in detail by Masselink et al. (1985), who evaluate the oscillator strength for excitonic transitions in AlGaAs–GaAs MQWs. In the unmodulated case, Eq. (23) still applies, but without the factor A_g and a different line shape factor. We will describe how to use a comparison of modulated and unmodulated spectroscopy to identify light and heavy holes.

The two most important factors determining the intensity of a given transition are the optical matrix elements and the modulation coefficient. For GaAs–AlGaAs quantum wells, the integrated absorption coefficients of light- and heavy-hole excitons have a ratio of about 1:2.5, depending on well width (Masselink et al., 1985). Therefore, in unmodulated reflectivity, PLE, absorption, etc., we would expect the intensities of light- and heavy-hole excitons to have a ratio of approximately 1:2.5. The coefficient C contains terms involving effective masses, and this modifies the intensity ratio from the 1:3 ratio expected from optical matrix elements involving the Bloch states ($|\mathbf{e} \cdot \mathbf{p}_{cv}|^2$). For our case, the explicit value of C is unimportant.

The modulation coefficient can be obtained from Eq. (8) by differentiating with respect to F. This gives $A_g = -2C_{\text{pert}}(m_c + m_v)e^2FL^4/\hbar^2$. It is important to note that A_g depends on the sum of the electron and hole effective masses. For (100)GaAs–AlGaAs QW we can compute a ratio of heavy- to light-hole modulation factors. Using known masses (R. C. Miller et al., 1984), we find that $A_{g,h}/A_{g,1} = 2.5$. This suggests that in PR, the intensity ratio of heavy-hole to light-hole transitions will be 2.5 times larger than in unmodulated spectroscopy. Shown in Fig. 14 is an expanded view of the comparison of PR to PLE of Fig. (Shanabrook et al., 1987a). Notice that in the PLE trace (Fig. 14a), the ratio of the heavy- to light-hole intensities is approximately 2, which is in agreement with other measurements. In the case of PR (Fig. 14b), we notice that the ratio is closer, to 5, as expected from the additional modulation coefficient.

An interesting experimental procedure of identifying light and heavy holes involves the realization that the polarization selection rules are different for light polarized in the well (growth) plane and perpendicular to it. For polarization in the well plane, both heavy- and light-hole transitions are allowed, while perpendicular to it only light-hole transitions are allowed. Therefore, comparison of these two polarization selection rules will aid in distinguishing between light and heavy holes.

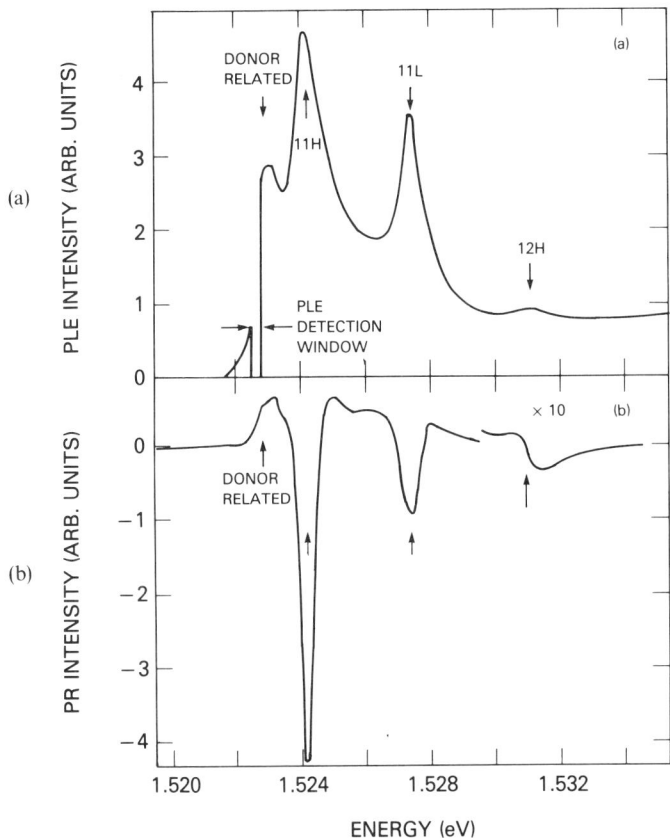

FIG. 14. Comparison of a) PLE and b) PR for a 200-Å MQW taken at 6 K. Note the greater intensity of 11L (relative to 11H) in the PLE spectrum compared with the PR spectrum. (After Shanabrook et al., 1987a.)

In order to obtain polarization perpendicular to the well, the k vector of the incident light has to be near 90° with respect to the growth direction. In most III-V and group IV materials systems, the index of refraction is high, and the refracted beam experiences significant bending toward the normal to the surface. Hence, one cannot get much polarization parallel to the growth direction. Ksendzov et al. have overcome this difficulty for the case of InGaAs quantum wells grown on GaAs (Ksendzov et al., 1990a). Because the InGaAs wells have a smaller band gap than GaAs, it is possible to sandwich them between thick GaAs buffers and to form a total internal reflectance (TIR) arrangement. This allows experiments to be performed with the k vector of light perpendicular to the growth direction. In the work of Ksendzov et al. (1990a), the probe beam was focused onto the thin cleaved side of the sample, while the modulation beam was incident on the normal face of the sample. A schematic comparison of TIR and normal mode PR is shown in Fig. 15. In Fig. 16 are displayed the spectra of Ksendzov et al. (1990a), comparing the normal PR measurement to the TIR mode. The trace with z-polarization corresponds to polarization perpendicular to the wells and shows a substantial decrease in the 11H intensity compared with the xy-polarization scan. Meanwhile the 11L transition intensity is unaffected. This result is consistent with the assignments of the peaks. This clearly is a useful technique for identifying the character of a given transition.

e. Modulation Mechanisms

In photoreflectance, the modulation of the dielectric function is provided by a chopped laser, whose energy is above the gap of the material being studied. In bulk materials, a number of early studies have pointed to the fact that the photoreflectance is simply a modulation of the built-in electric field (Shay, 1970; Aspnes, 1970; Cerdeira and Cardona, 1969). The field modulation mechanism is schematically depicted in Fig. 17, for an n-type semiconductor. Because of the pinning of the Fermi energy at the surface, there exists a space-charge layer. The occupied surface states contain electrons from the bulk. Photoexcited electron–hole pairs are separated by the built-in field, with the minority carrier (holes in this case) being swept toward the surface. At the surface, the holes neutralize the trapped charge, reducing the built-in field. Cerdeira and Cardona (1969) studied the electric-field dependence of

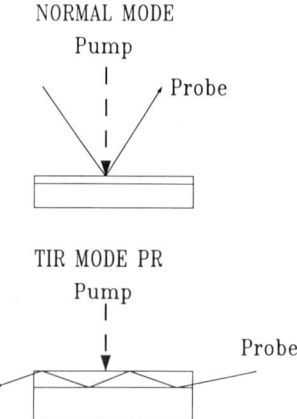

FIG. 15. Comparison of the configurations for normal mode and total internal reflection mode of PR.

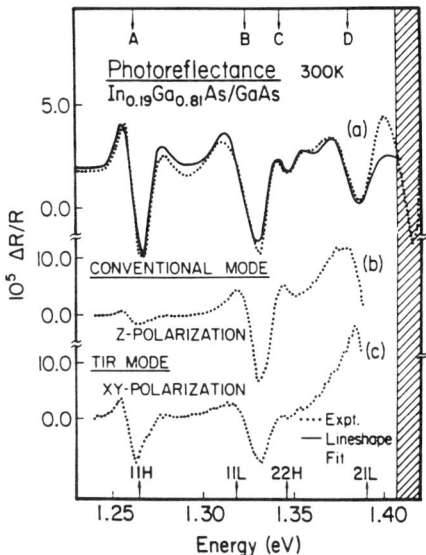

FIG. 16. Room temperature data for InGaAs–GaAs in (a) conventional mode PR and for the two different polarizations of TIR. In (b), the polarization is along the growth direction, and in (c), it is perpendicular to it. (After Ksendzov et al., 1990a.)

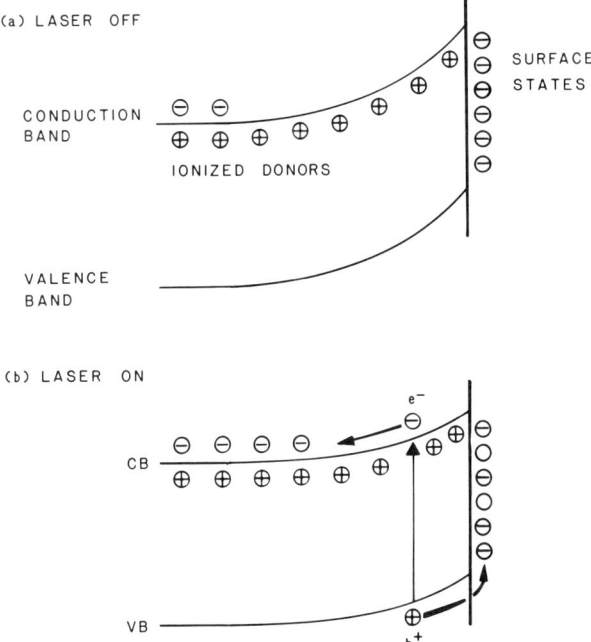

FIG. 17. Schematic representation of the photoreflectance effect for an n-type sample. In (a), the sample is in the dark, and Fermi-level pinning to surface states induces a built-in field. In (b), photoexcited carriers separate in the field, with minority carriers neutralizing charge in surface states. (After Glembocki et al., 1985.)

PR of Si in an electrolyte configuration, while Aspnes (1970) used a Schottky barrier configuration to study the PR and ER of Ge. Both groups systematically changed the electric field and showed that the electric-field modulation mechanism was dominant. In addition, Aspnes, by comparing flatband potentials as measured in ER and PR, also showed that the Dember potential was important in PR.

Because of the preceding work, it is automatically assumed that the modulation mechanism in PR is electric-field modulation. This is not always the case. In early work on PR, several other modulation mechanisms were considered. These include band-filling effects (Gay, 1968), thermoreflectance (Nilson, 1969), and exciton screening by photogenerated carriers (Albers, 1969). While these effects are possible, they were observed to have a minor role in PR of bulk materials.

In QW systems, Theis et al. (1989b) suggested that at low temperatures, linewidth broadening of excitons can play an important role in the PR effect. They performed reflectance measurements and observed that without any pump light the exciton lines narrowed with temperature through the temperature range studied, 6 to 150 K. On the other hand, when they used a HeNe pump beam of modest intensity, 0.5 mW/cm^2, they found that the exciton lines narrowed with temperature only to 125 K. Below 125 K, the exciton linewidths actually increased. In addition, they also found that the PR line of the 11H exciton, measured at 80 K, increased in width by more than a factor of 2.5 as the pump power was increased from 0 to 100 mW/cm^2. These effects were associated with the neutralization of the ionized donors by the pump beam. Above 125 K, all of the donors were ionized, so only LO phonon broadening was important. Below 125 K, the pump light could produce neutral donor-bound excitons (D^0X), whose population was modulated by the pump light. Without any pump light, only free excitons are formed; with a pump light, D^0X can form. The presence of D^0X has been suggested to lead to the observed broadening of 11H.

Theis et al. also observed a below-gap feature similar to the one seen in Fig. 10 and labeled as impurity-related. This feature was shown to be the D^0X transition, and its intensity relative to 11H increased with increasing pump powers. Note that the action of the pump in this case was to modulate the population of D^0X. Its presence in the PR spectrum correlated with the observed exciton broadening. This study clearly shows that population effects of impurities can play an important role in the modulation mechanism for PR.

The foregoing discussion has some important implications. We previously suggested that the intensity ratio of the light- and heavy-hole excitons was affected by an electric-field coefficient, which made this ratio closer to 6 for

(100) QW rather than the 2.5–3 expected from absorption measurements. If the modulation mechanism changes, as suggested by Theis *et al.*, then the light- to heavy-hole intensity ratio would effected. Interestingly, the PR data of Theis *et al.* show a ratio of 2:1. This may be a result of the fact that the linewidth modulation for light and heavy holes is similar in magnitude.

In concluding this section, we wish to caution the reader that because photoreflectance involves photogenerated carriers, there may be many possible mechanisms *other* than just electric-field modulation. We will see that this indeed is the case when we consider modulation-doped heterojunctions.

f. Interference Effects

The line shapes measured in any form of modulation spectroscopy can be profoundly affected by optical interference phenomena. This is especially true for microstructures, which are often clad with protective layers or films to reduce surface electric fields. If the path length difference between the layer of interest and the surface is an odd multiple of half the wavelength of light in the material, destructive interference occurs between light originating from the two regions. This will modify the ER or PR line shape by mixing the real and imaginary parts of the dielectric function in $\Delta R/R$ [see Eq. (9)]. Interference effects are well known for excitonic absorption in bulk samples placed in electric fields. Excitons are ionized near the surface of the sample, where the fields are largest, leaving the front portion of the sample exciton-free (dead layer). Light reflected from the air–sample interface will interfere with light coming from the regions of the sample where excitonic absorption is possible (Silberstein and Pollak, 1980). The ER line shapes changes from dispersive to absorptive, in a cyclic fashion, as an electric field changes the width of the dead layer.

In the case of quantum wells, Klipstein and Apsley have calculated the effect of a thick overlayer on the ER line shapes of a SQW (Klipstein and Apsley, 1986). Shown in Fig. 18 are their results, which illustrate the changes in line shape that are possible simply from different thickness cladding layers. In subsequent work, Thorn and co-workers (1987) studied this effect in a single 100-Å quantum well capped with 2000 Å of GaAs. They changed both the angle of incidence and the temperature of the sample and found line shape changes similar to those shown in Fig. 18, demonstrating the interference effect. Shown in Fig. 19 is the spectra of Thorn and co-workers (1987) for different angles of incidence. Zheng and co-workers (1988) have shown a similar effect for SQWs in reflectivity studies. They saw changes in the reflectance line shapes with thickness of cladding layer, indicating that interference can be important.

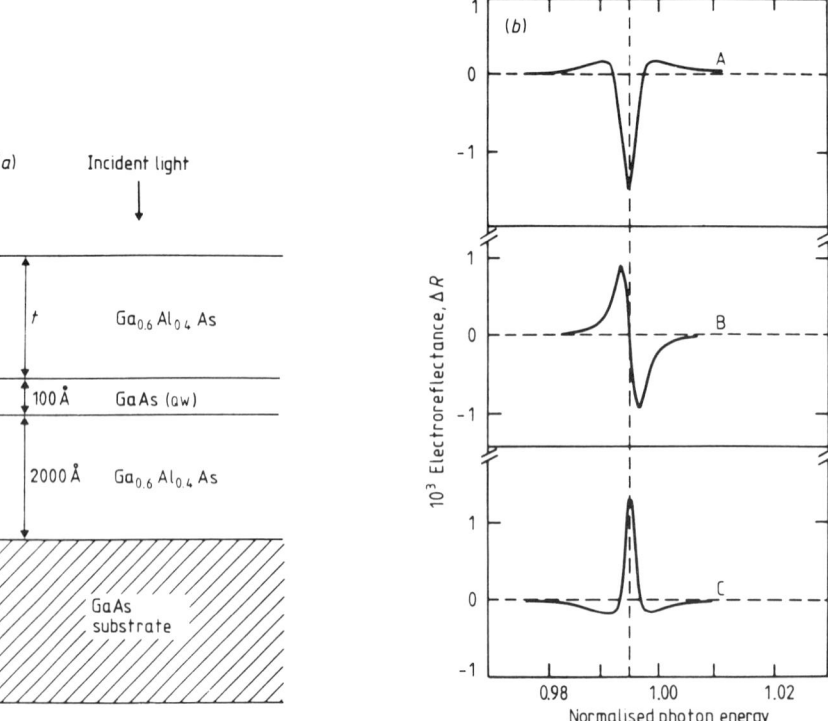

FIG. 18. Calculation of the optical interference effect in electroreflectance of a single quantum well. The cladding layer had thicknesses of (a) 2100 Å, (b) 2400 Å, and (c) 2700 Å. (After Klipstein and Apsley, 1986.)

Interestingly, these effects could also be expected to influence PR line shapes from MQWs. Shanabrook et al. (1987a,b) considered this possibility in their work and found no evidence for it. They noted that even though there were variations in the thickness of their cladding layer, resulting in changes in the reflectance spectrum similar to those described by Zheng et al. (1988), the PR spectra were unaffected. This phenomenon is currently not well understood and may be related to the complicated manner in which light is reflected from a MQW structure.

The interference effects just described must always be considered in the analysis of PR or ER spectra. The line shape analysis of Eq. (5), however, is still valid in a relative fashion. Although interference will change the line shapes, the relative phases of allowed and forbidden transitions will still be preserved, provided that the transitions occur over a small wavelength range. Thorn et al. (1987) have suggested that one can easily correct for optical in-

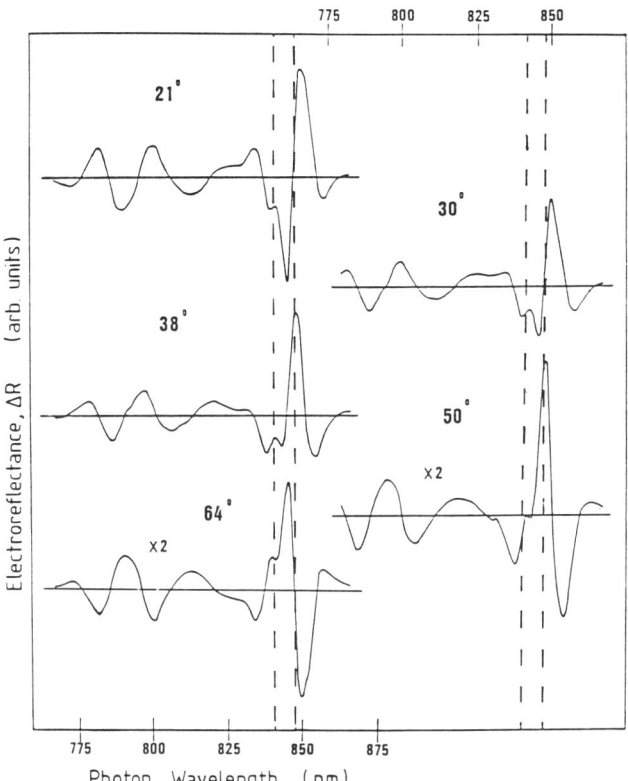

FIG. 19. ER spectra for a 100-Å quantum well with a 2000-Å cladding layer, as a function of angle of incidence. The dashed lines mark the 11H and 11L transition energies. (After Thorn et al., 1987.)

terference through a multilayer reflectance calculation and thus obtain the correct line shape.

g. E_1 Gap

Erman et al. (1984) also observed the effects of confinement on the higher-lying states in the Brillouin zone along L, associated with the E_1 and $E_1 + \Delta_1$ transitions. Figure 20 shows their data in the energy range of 2.5 to 3.5 eV. The data were accounted for by the same model used for the E_0 transitions, but with different masses and band offsets. Interestingly, Erman et al. showed that the valence band wells for the $E_1 + \Delta_1$ transitions were very small, and, therefore, only one transition should be observed. For the E_1 case two transitions were expected, but only one was seen.

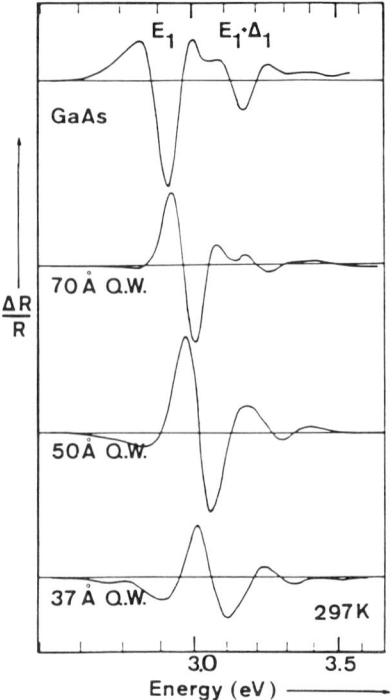

FIG. 20. Room temperature ER spectra for GaAs and three single quantum wells for the E_1 and $E_1 + \Delta_1$ transitions. (After Erman et al., 1984.)

h. *Characterization of Quantum Wells*

The work discussed in the previous section provides a basis for the characterization of quantum wells through the use of modulation spectroscopy. The fact that PR and ER provide information about all of the interband transitions from microstructures, including those from the buffer and the barrier layers, allows us to completely characterize the structure of interest. In this regard, it is important to know several parameters, which have already been mentioned. These include the band offsets, exciton binding energies, and effective masses of the electrons and the light and heavy holes. With these parameters and an appropriate model, one can calculate all of the transitions involved and compare with the experimental data to obtain information about the quantity of interest, e.g., well and barrier widths and barrier heights.

A good example of the type of information that can be obtained from PR is in the work of Parayanthal et al. (1986) in which topographical variations in barrier heights and well widths of a GaAs–AlGaAs MQW were studied. The probe beam used in these experiments was 100 μm in diameter, and the

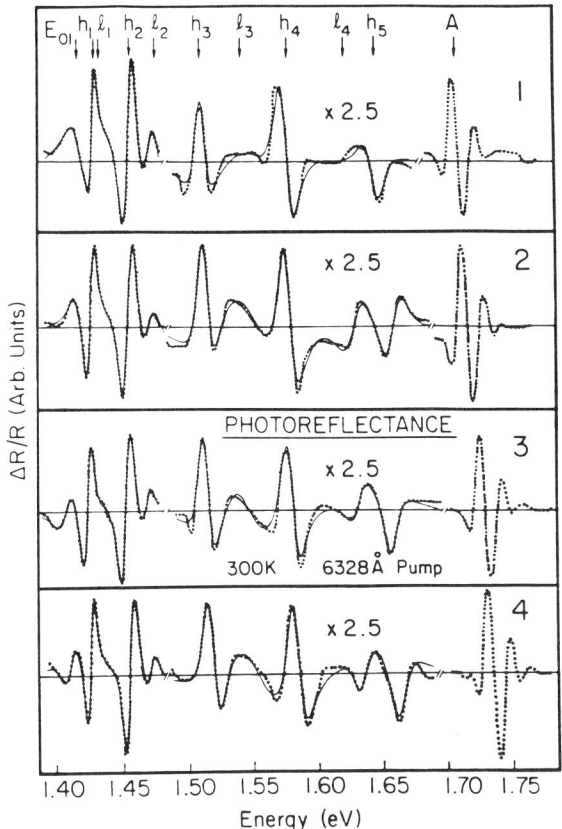

FIG. 21. Room temperature PR spectra for a 200-Å MQW as a function of position along the diagonal of a 1 × 1 cm² sample. The probe spot size was 100 μm. The dotted line is data, and the solid line is a fit to the TDFF. The feature denoted as E_{01} is the substrate GaAs signal, while feature A is from the AlGaAs barriers. (After Parayanthal et al., 1986.)

sample was scanned in both x and y directions. The nominal barrier and well thicknesses were 150 and 220 Å and the Al content was $x = 0.24$. Shown in Fig. 21 is a series of spectra from different positions on the sample. The most prominent thing to note is that the Al content is systematically changing (shifting of feature A). Also notice that there are some subtle changes in the vicinity of the h_1 and l_1 transitions. Parayanthal and co-workers (1986) fit the data to the TDFF and found that in going from position 1 to position 4, the well width changed from 214 to 210 Å and the x-value changed from 0.225 to 0.247. These results are in good agreement with the physical characteristics of the MBE growth in which 10% variations in the cation fluxes were expected over a centimeter. This demonstrates the utility of PR in characterizing microstructures.

i. New Materials Systems

Photoreflectance can also be used to study the optical properties of new materials systems such as InGaAs–GaAs (Ksendzov *et al.*, 1990b) and AlSb–GaSb (Rockwell *et al.*, 1990), as well as the optical properties of materials growth on non-(100) surfaces. For example, Wang has shown that high-quality microstructures can be grown on the (111)B face of GaAs (Wang, 1986). In such situations, the information of interest includes band offsets and carrier effective masses. It is possible to obtain this information by studying quantum wells that have been grown with precise well widths and alloy concentrations (if alloy systems are used). Using PR as a probe, Shanabrook *et al.* studied the optical properties of AlGaAs–GaAs quantum wells grown on the (111)B GaAs face (Shanabrook *et al.*, 1987b). Their measured spectra for samples having well widths in the range of 100 Å to 500 Å are shown in Fig. 22. Transition energies were calculated from a finite square well model, taking into account the nonparabolicity in the conduction band in the [111] direction. The band offsets and hole effective masses were varied. The initial estimates of the effective masses were made from previously determined Luttinger parameters, and the band offsets were assumed to be the same as those of the (100) face (Hess, 1976). This last assumption was based on theoretical work, suggesting that in the AlGaAs–GaAs system, the band offsets were independent of direction (Baldereschi *et al.*, 1988).

Shanabrook *et al.* (1987b) found that if one assumed similar exciton binding energies for the (111) case as for the (100) circumstance, then the measured QW transition energies would agree with calculations using band offsets of $Q_c = 0.55 \pm 0.05$ and effective masses of $m_{lh} = 0.08 \pm 0.01$ and $m_{hh} = 0.8 \pm 0.2$. These values of the masses qualitatively agreed with estimates obtained from the Luttinger parameters. Recent Raman scattering experiments inidcate that the masses used in this fit were very reasonable and lead to a further refinement of the values for the Luttinger parameters of GaAs (Shanabrook *et al.*, 1989). On the other hand, the accuracy of the band offset Q_c is limited by several issues that are not well established, including exciton binding energies of the higher-lying transitions and the nature of the valence-band mixing and the conduction-band nonparabolicities. This study has shown how photoreflectance can provide useful information about new quantum well microstructures.

An interesting aspect of the (111) work comes from a further verification of the ratio of light- to heavy-hole transition intensities, depending on the modulation mechanisms. In the case of (111) QW, the ratio of the light- to heavy-hole modulation coefficients is $A_{g,h}/A_{g,1} = 6$. This would suggest that the heavy- to light-hole transition intensities would be in a ratio of nearly 12:1. Figure 22 c) shows that this indeed is the case.

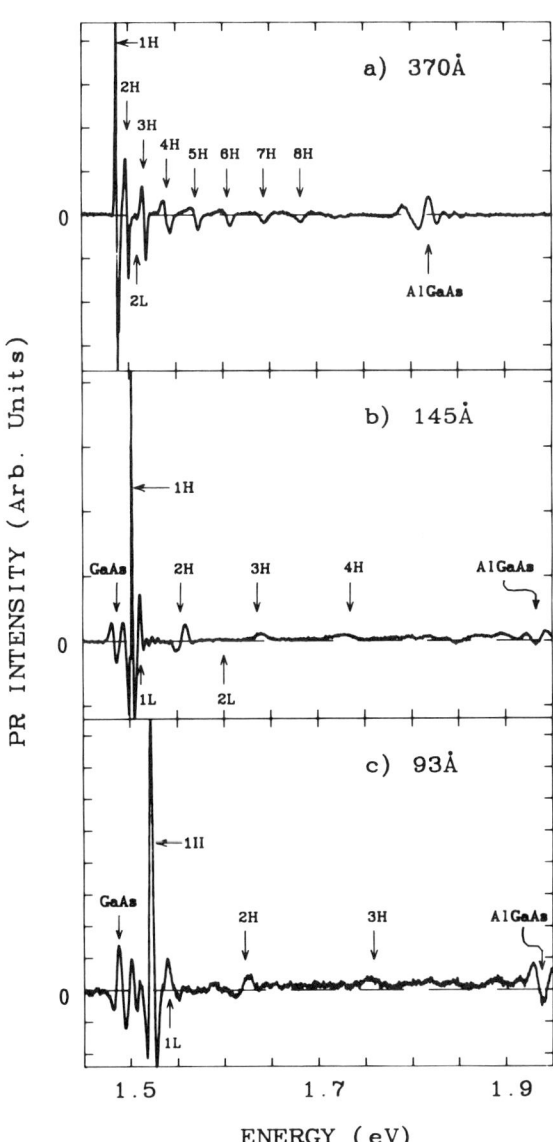

FIG. 22. 150 K PR spectra for three [111] MQW samples. The arrows are the locations of the calculated transition energies using the effective masses given in the text. (After Shanabrook et al., 1987b.)

Another area of considerable interest is strain-layer systems. Ksendzov et al. (1990b) studied the modulated optical properties of $In_{0.19}Ga As_{0.81}-$GaAs single quantum wells having well widths near 110 Å. Their main interest was in the band offsets. In this case, they utilized the fact that the depth of the well determines the number of allowed transitions, and, thus, by knowing the well width, one can determine the well depth from the number and spectral positions of the interband transitions. They found good agreement between the calculated transitions and the experimentally measured ones for a conduction-band offset of $Q_c = 0.45$. Although, exciton binding energies and conduction-band nonparabolicities were not well known, the results agreed with light scattering measurements of Joyce et al. (1988). Interestingly, these results are still somewhat controversial.

In addition to III-V materials system, there has been significant work on Si–Ge superlattices. The interest in this case is in zone-folding the indirect gap and creating a material with a direct gap. Using electroreflectance, Pearsall et al. (1986, 1987) studied Si–SiGe quantum wells and showed that it is important to include the strain in the layers to identify properly the energy states. In addition to the transitions at the fundamental gap, Pearsall et al. (1987) studied higher-lying transitions and observed new interband excitations postulated to be due to ordering effects.

6. COUPLED WELLS

Until now, we have dealt predominately with isolated quantum wells that have thick barriers and states in neighboring wells that do not interact with each other. In the formation of a superlattice, however, the barrier thickness between neighboring wells is reduced so that states in the two wells can interact. Two important effects can occur if we let the barrier thickness decrease to the point where states from adjoining wells interact. First, for just a pair of wells, the interaction between any pair of states in adjoining wells will lead to two new states that are split away from each other, but characteristic of the new coupled system (Koteles et al., 1987; Little and Leavitt, 1989; Little et al., 1986). In other words, any pair of states that were spatially localized in the interacting wells become shared (or "delocalized") in the interacting system. Second, if this is allowed to occur for a large number of interacting wells, then a miniband structure is formed, in a manner similar to the formation of a semiconductor, where bringing together many atoms form semiconductor bands.

a. Asymmetrically Coupled Quantum Wells

As a prerequisite to understanding superlattices, it is instructive to study the nature of interacting states in two quantum wells separated by a thin barrier. As noted, this can be accomplished by studying coupled wells. We begin with a simple case of coupling in only the valence or conduction bands. This can be accomplished by asymmetrically coupled quantum wells (ACQW) (Little and Leavitt, 1989; Little et al., 1986), which are schematically depicted in Fig. 23 (Glembocki et al., 1991). In ACQWs, the wells of interest have different thicknesses and, hence, different subband structures. As can be seen in Fig. 23, at zero applied bias the $n = 1$ conduction subbands are isolated. The conduction subbands are brought into resonance by applying a constant field across the two wells. Resonance couples the $n = 1$ states of the adjoining wells forming two new states that are extended over the coupled wells. However, we notice that at near-resonance conditions for the conduction band, the valence-band states remain uncoupled.

In terms of optical absorption, in the uncoupled system, we can only have one transition per well involving the $n = 1$ states. This is the direct transition.

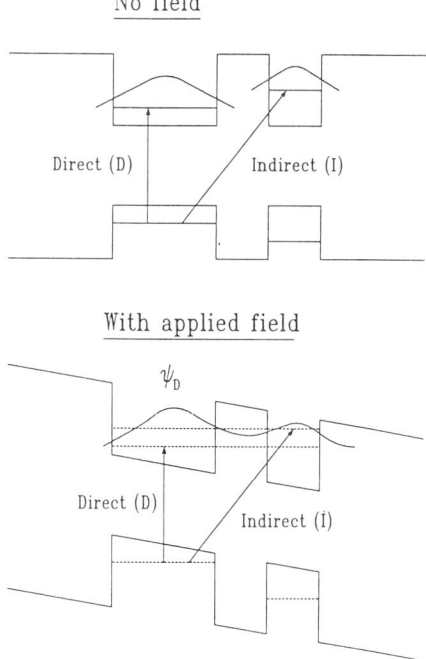

FIG. 23. Schematic representation of asymmetrically coupled quantum wells, showing $n = 1$ conduction and valence subbands. With no field, the conduction subbands are decoupled and the indirect transition (I) is forbidden. A field couples the conduction subbands and allows the indirect transition to occur. Note that the valence subbands are always decoupled (After Glembocki et al., 1991.)

As seen in the top of Fig. 23, the conduction wave functions are isolated in their respective wells and do not extend into the neighboring well. Because of this, there is no overlap between the envelope functions of the valence states with the conduction states that would produce indirect transitions in real space, and, thus, this type of transition is not allowed. Near resonance conditions, on the other hand, the wave functions are shared between the two wells and indirect transitions are allowed. The terminology *indirect* is used because there is a high probability that a photoexcited electron will be found in the well adjoining the one in which the electron–hole pair was created. It is very important to point out that in resonance when the wave function is shared between the two wells, the designation of a transition as direct or indirect is meaningless. In what follows, uppercase (lowercase) letters refer to direct [D, (d)] and indirect [I, (i)] transitions originating from the wide (narrow) well.

Shown in Fig. 24 is a comparison of electroreflectance (ER) and photoconductivity (PC)for a sample containing 10 pairs of InGaAs ($x = 0.1$) quantum wells separated by a 15-Å GaAs barrier (Glembocki et al., 1991). The wells had thicknesses of approximately 45 Å and 90 Å. The sample configuration was a device having a 100-μm-diameter active area. The QWs were located under a thick n^+-GaAs contact layer. The probe light used was focused only on the image of the slit and had an area of 5 mm × 0.5 mm. This is quite large in comparison with the device.

At low fields, the direct transitions are strongest and correspond well to peaks in the PC spectrum. Applying a bias moves the $n = 1$ subband states in the adjoining wells into resonance. As resonance is approached ($F = 45$ kV/cm), the indirect transitions grow in strength and the coupling between the two wells is clearly evident. Finally, beyond resonance, the indirect transitions being to lose intensity, and at sufficiently high fields (not shown) the spectrum would resemble (in terms of the number of transitions) the low-field one.

Based on the FDFF of the quantum well transitions, one might expect to see changes in line shape because the electrons and holes are polarized and some transitions blue-shift while others red-shift. A detailed analysis of ER line shapes was performed by Glembocki et al. for asymmetrically coupled quantum wells (Glembocki et al., 1991).

The ER line shapes in Fig. 24 can be understood in terms of the FDFF of Eqs. (5) and, in particular, from the electric-field dependence of the coefficients of the various derivatives of the dielectric function. A comparison of the ER line shapes in Fig. 24 to the theoretical lines in Fig. 5 suggests that, neglecting interference effects, either the intensity modulation or the lifetime modulation is responsible for the observed ER spectrum. A measurement for a sample having the same overlayer but only 100-Å wells and 150-Å-thick barriers indicates that the 11HD in Fig. 24 is due to an energy-gap modula-

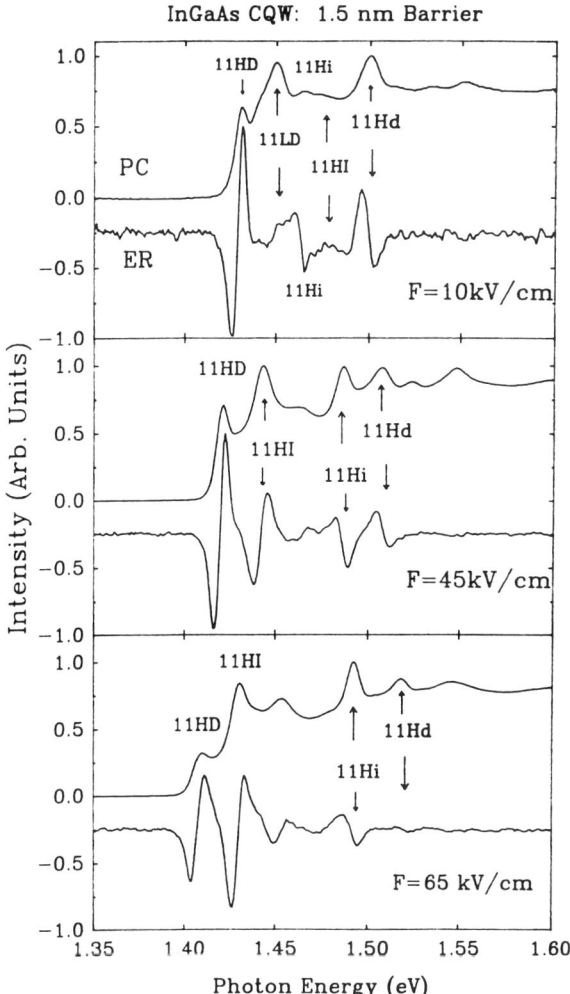

FIG. 24. A comparison of 10 K ER and PC for a sample containing 10 pairs of InGaAs ($x = 0.1$) quantum wells separated by a 15-Å GaAs barrier. The well widths were 45 Å and 90 Å. D (d) marks direct transitions in the wide (narrow) well, and I (i) denotes indirect transitions in the wide (narrow) well. The arrows mark expected positions of the transitions. (After Glembocki et al., 1991.)

tion. This change in line shape is a direct result of the fact that the top n^+ contact layer introduces optical interference effects.

The additional information provided by ER as compared to PC becomes evident upon closer examination of the ER lines and their phases. In all of the spectra, the transitions in the narrow well are opposite in phase from those of the wide well. This occurs because in the vicinity of resonance the $n = 1$ transitions in the wide and narrow wells experience an anticrossing behavior. This repulsion leads to transition energies that red-shift for the wide wells and blue-shift for the narrow ones. Note that this is different from the isolated well circumstance in which case transitions from both the narrow and the wide wells would red-shift. In terms of the ER lines, this suggests that if Eqs. (5) apply, then the phases of 11HD and 11Hd will be opposite provided that the intensity or lifetime terms are not significant. An examination of Fig. 24 indicates that this indeed is the case. A similar result is expected and observed for 11LD and 11Ld, since they also involve the same conduction-band states.

Furthermore, a calculation of the subband energies and the associated wave functions allowed Glembocki *et al.* (1991) to account for the observed line shape, indicating that for the thickness of barrier of this sample the intensity modulation term was small. This occurs because even at zero field the states were coupled, as can be seen from the top PC trace of Fig. 24. Finally, we wish to note that in the top spectra of Fig. 24 the intensity ratio of 11HD to 11HL is much larger for the case of PC than for ER. Once again, this is in agreement with the previous discussion of electric-field effects in identifying whether a transition has light- or heavy-hole character.

The case of AlGaAs–GaAs ACQWs with wider barriers, 25 Å was also studied by Glembocki *et al.* (1991). In this case, it was found that resonance conditions produced changes in the line shape, directly attributable to an intensity modulation produced by field-induced transfer of the wave function from one well to another. This effect is more prominent in thick-barrier samples because the resonance occurs over a narrower field range, resulting in larger differential changes in the electron–hole overlap integrals [see Eq. (17)].

Also note that because of the derivative nature of ER, we are able to identify the source of the observed transitions (wide or narrow well), whereas in PC, which measures the unmodulated absorption coefficient, this is not possible without a detailed mapping of the field dependence of the transitions. This work is a good example of how Eqs. (17) can be used to obtain information about the response of the sample to the applied perturbation. Once the modulation mechanism is identified (gap, intensity, or lifetime), it is possible to probe the details of modulation.

Interestingly, the work in ACQWs is a direct verification of the validity of the first derivative nature of ER in coupled circumstances, where the coupling is between two wells. It is very important to realize that the work of Glembocki and co-workers is independent of any detailed line shape fitting and applies equally well to Lorentzian and Gaussian dielectric functions. This is significant because it suggests that valuable information can be obtained without any need for line shape fitting.

b. Symmetrically Coupled Wells

Cacciatore et al. (1989) used ER to study the coupling between 28-Å GaAs quantum wells separated by 34-Å $Al_{0.3}Ga_{0.7}As$ barriers. This is the case of symmetrically coupled wells, which are in resonance at zero applied bias. Any application of voltage will move the system away from resonance. Shown in Fig. 25 is a schematic representation of their energy-level diagram under a small applied electric field. Notice that both the electron and hole states from neighboring wells are coupled to produce pairs of states E_{11} and E_{12}, H_{11} and H_{12}, and L_{11} and L_{12}. At zero applied field, the wave functions are equally shared between the two coupled wells. As a field is applied, the electrons and holes begin to localize in different wells and we can now have transitions that are "indirect" in real space. This is possible because the localization of the wave function is incomplete. As long as some of the wave function is present in a neighboring well, it is possible to have indirect transitions.

Shown in Fig. 26 are the 300 K spectra of Cacciatore et al. (1989) as a function of dc bias. At zero bias the direct transition $H_{11} - E_{11}$ is strong, but with increasing bias it weakens and red-shifts. Meanwhile, the indirect transition, $H_{12} - E_{11}$ increases in intensity and blue-shifts. These results are a direct consequence of the fact that the electrons and holes are being polarized and

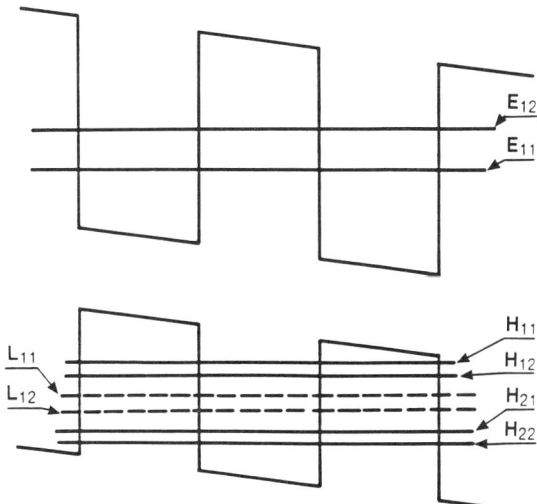

FIG. 25. Schematic representation the $n = 1$ levels in the conduction band and the $n = 1$ and $n = 2$ states in the valence band for a system two coupled (identical) quantum wells. Each level splits into states denoted by $E(H,L)_{n1}$ and $E(H,L)_{n2}$, where n is the subband index of the unperturbed states and E, H, L correspond to electrons, heavy holes, and light holes, respectively. (After Cacciatore et al., 1989.)

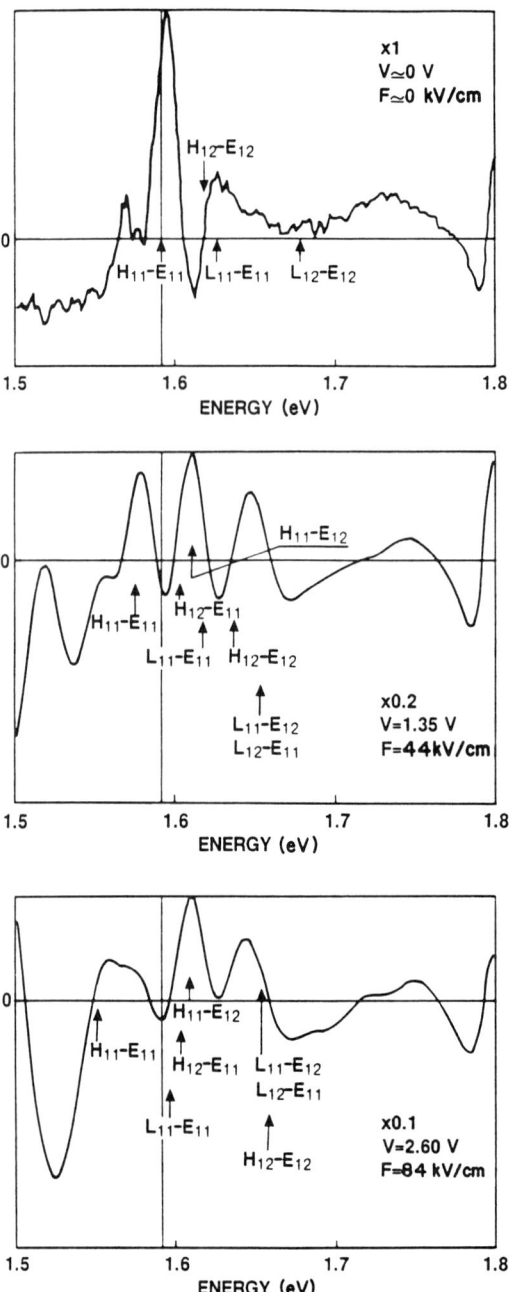

FIG. 26. ER spectra as a function of electric field for 28-Å GaAs quantum wells separated by 34-Å $Al_{0.3}Ga_{0.7}As$ barriers. At zero field only $H_{1m} - E_{1m}$ ($m = 1, 2$) type transitions are seen. At higher fields $H_{1m} - H_{1n}$ ($n, m = 1, 2$) become allowed. (After Cacciatore et al., 1989.)

that the coupling between the wells is breaking down. In this case, the holes are localized in one well. Similar results are evident in Fig. 26 for the transitions $H_{11} - E_{11}$ and $H_{11} - E_{12}$, which involve the electron levels. This study shows the optical behavior of coupled well systems under an applied bias and the manner in which the coupling is destroyed.

7. SUPERLATTICES

Electrolyte electroreflectance has been performed in the case of GaAs–AlGaAs and InGaAs–GaSbAs superlattices over the entire visible wavelength range by Mendez and co-workers (1981). This study showed Kronig–Penny type effects of confinement on the E_0, the $E_0 + \Delta_0$, the E_1, the $E_1 + \Delta_1$ and the E_2 gaps. Shifts of the various energies were accounted for by a Kronig–Penny model. This study considered transitions involving only the first valence- and conduction-subband levels.

Although there is considerable interest in superlattices, there is much less electromodulation work in SLs compared to the amount of work in QWs. The two areas receiving the most attention have been the band structure of short-period GaAs–AlAs superlattices and the question of the collapse of the superlattice under an applied field and the direct-to-indirect crossover for $(GaAs)_m(AlAs)_n$ short-period superlattices. We will consider both of these because they represent some interesting physical problems.

a. *Wave Function Localization*

We now consider the work on wave function localization. As we have seen, the extended nature of the superlattice wave function depends very much on the thickness of the barriers separating neighboring wells. If the barrier thickness is large enough (>2.5 nm in GaAs–AlGaAs), an electric field can be used to bring the states out of resonance. In the case of a superlattice the application of a progressively large field will destroy the extended wave functions and produce localized wave functions whose spatial extent depends upon the magnitude of the field. Ultimately, the wave function will be confined to a single well.

Wave function localization has been studied in GaAs–AlGaAs SLs both theoretically and experimentally (Bleuse *et al.*, 1988; Voisin, 1989). Experimentally, photoconductivity and electroreflectance were employed. A GaAs–AlGaAs superlattice with $x = 0.3$ and 31-Å GaAs wells and 35-Å-wide barriers was used by Voisin. Shown in Fig. 27 are a series of ER spectra of Voisin (1989) as a function of the electric field. The low-field ER spectra at $V = 0$ showed four distinct superlattice features corresponding to the light- and

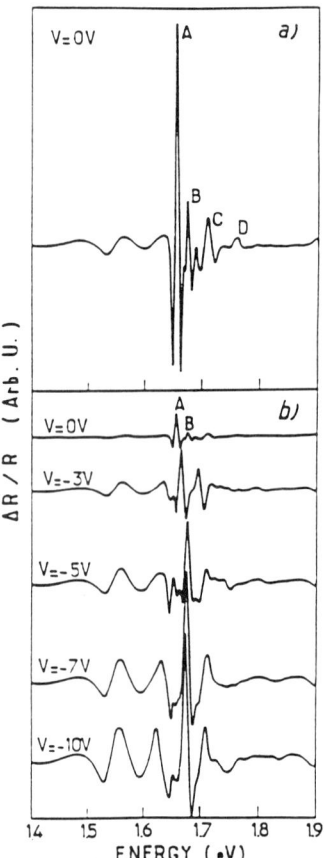

FIG. 27. Electric-field dependence of the ER spectra of a AlGaAs–GaAs superlattice. (After Voisin, 1989.)

heavy-hole excitonic transitions at the minizone center (lower-energy transitions) and a joint density of states feature from the zone edge (higher-energy transitions) (Voisin et al., 1989). From the energy separations of the zone-center and zone-edge transitions, it is possible to determine the minizone bandwidth to within the exciton binding energy. Voisin then increased the AC voltage and observed two effects. The zone-edge transitions decreased in intensity and disappeared, while the zone-center transitions blue-shifted (toward the zone-edge transitions). These results show the collapsing of the minibands, as the wave function is localized, by the disappearance of the zone-edge transitions and the blue shift of the zone-center excitations. The blue shift of the zone-center transitions is expected because the band gap of a SL is lower in energy than the gap of an isolated QW having the same well width.

In a related study of Shen et al. (1987c) for the 10-period GaAs–AlGaAs SL, overlapping transitions were postulated for the 11H, 11L, and 22H interband transitions. The higher-energy transition of a given pair (e.g., 11H) was fit with an ER phase opposite to the lower-energy ones. Because gap modulation dominates this result, it was suggested that the gap of the higher-energy transition moved in the opposite direction (under an applied electric field). This result is similar to that of Voisin's work for the collapse of the minizone.

b. Superlattice Band Structure

The second aspect of superlattice band structure lies in fundamental bandgap studies of short-period SLs, $(GaAs)_m(AlAs)_n$, where m, n specify the number of monolayers. Of interest here is the alignment of the direct and indirect conduction bands. Finkman et al. (1987) have used a combination of PR, PL, and PLE to study the gap of short-period superlattices. In this case, PL occurs from the lowest state, while PR and PLE measure the density of states and thus the direct gap. For indirect gaps, the PL lines would be from the lowest indirect (in real space) transition, while PR would involve the real-space direct gap and would have a feature at higher energies. The lowest energy transition was observed to be from the direct gap by Finkman and coworkers for $m/n = 10/8$, $18/10$, and $29/31$, while the indirect gap was lower in energy for $m/n = 17/5$, $19/19$, and $31/10$. Another study using PR and PL was performed by Fujimoto et al. (1990). They considered the case of $n = 5$, where m was odd and varied from 3 to 11. They found that for $m < 7$, the material was indirect. Shown in Fig. 28 is a 300 K spectrum of Fujimoto

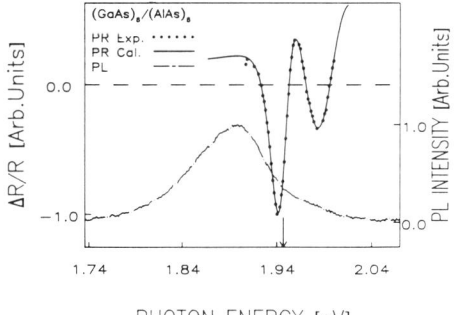

FIG. 28. Comparison of 300 K PR (dashed line) to 300 K PL (solid line) for a $(GaAs)_5$–$(AlAs)_5$ short-period superlattice. The arrow marks the PR transition energy. (After Fujimoto et al., 1990.)

et al. The PL data are observed to peak at a lower energy than the PR transition energy, which is indicated by the arrow. This clearly shows that the gap for the 5/5 SL is indirect. Such work shows how PR along with PL can form a powerful probe of SL band structure.

Recently, Si_8Ge_{32} short-period superlattices have been studied by Dafesh *et al.* (1990). In the notation used, (Si_nGe_m), n, m are the number of monolayers of Si and Ge, respectively. Shown in Fig. 29 is their 87 K spectrum for a 60-period superlattice. The structures B and C and H and I correspond to light- and heavy-hole transitions for the $n = 1$ and $n = 2$ valence minibands at $k = 0$. Meanwhile, E and L also involve the $n = 1, 2$ minibands but for transitions originating from the spin-orbit split valence bands. The excitations labeled K and N are the E_1 and $E_1 + \delta_1$ transitions of the superlattice. Because of the large electron and hole masses along the (111) directions, there is only one subband for both electrons and holes. Finally, the features labeled D, G, M, and P are transitions from a SiGe buffer layer grown between the substrate and the SL layers. The features F, J, and O are unidentified. Dafesh and co-workers (1990) calculated the positions of the subbands, taking into account strain effects, and found good agreement with the data of Fig. 29. This work shows that PR can be used to characterize a complicated band structure, which involves strain.

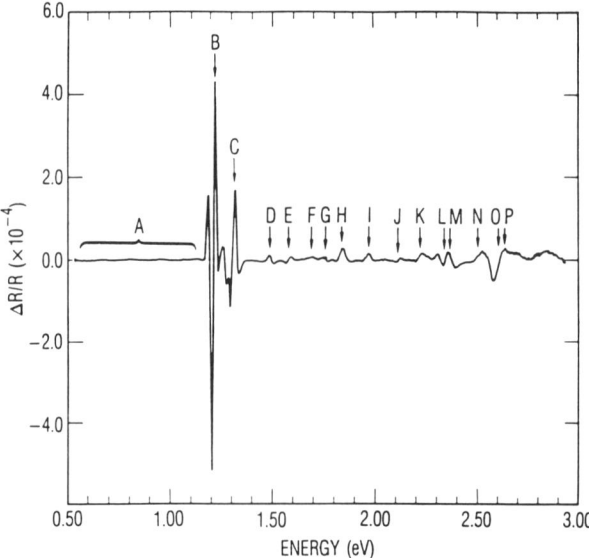

FIG. 29. 87 K PR spectrum of a symmetrically strained Si_8Ge_{12} short-period superlattice. (After Dafesh *et al.*, 1990.)

Also of interest are strain-layer superlattice systems such as InGaAs–GaAs (Reddy et al., 1987a). Recently, there have been PR studies performed in this system. Pan et al. (1988) report the observation of unconfined states, coupling between wells and minizone formation. In addition to this, strain effects had to be included to account for the data. For this system, the strain-induced shifts of the light-hole band lead to a type II superlattice for the light-hole states; i.e., electrons are confined in the InGaAs, while the light holes are confined to the GaAs layers.

c. Superlattice Effects in Multiple Quantum Wells

The preceding discussion focused on multilayer systems that were specifically designed to exhibit superlattice properties, in the ground states. In the case of MQW with finite barriers, the wave functions of higher-lying states penetrate deeper into the barrier than the wave functions of the ground states. In this case, it is possible to have coupling between higher-lying states, even though the ground states are isolated. In a recent study, Shen and co-workers (1988a) considered the effect of coupling on the higher-lying states in the conduction band of a MQW, with sufficiently thick barriers so that the lower-lying states were isolated. The sample was a 39-period GaAs/AlGaAs MQW with well and barrier widths of 71 Å and 201 Å and $x = 0.18$. In this room temperature study, the PR spectra were compared with thermoreflectance (TR), which is a first derivative spectroscopy. Shown in Fig. 30 is the comparison of PR and TR. Notice that the line shapes of the low-lying transitions are observed to be similar, while the line shapes for the 22H transition are markedly different. In Fig. 30 are also shown the line shapes of 22H after two differentiations with respect to energy of the TR spectrum. This operation gives line shapes similar to the PR spectrum. Since TR is a first derivative spectrum, two additional derivatives result in a third derivative line. This suggested the formation of a miniband for these states where excitonic effects are unimportant. Electrons were tunneling out of the well, enabling the field to accelerate the particle, resulting in a third derivative effect. However, at 77 K, where excitonic effects are more important, the TR and PR spectra of 22H are identical; i.e., 22H is just an FDFF due to excitonic effects (Shen et al., 1988a). This work shows that PR can be used as an optical probe of tunneling in superlattices.

In the discussion of MQWs, we considered in detail only transitions from states confined to the well. It can easily be shown that because of the periodicity of the barrier a series of states will form above the AlGaAs barrier in both the valence and conduction bands. These states can form a miniband structure, with a band gap determined by the total thickness of the well and barrier (Wong et al., 1986). Thus, in the absence of Coulomb interactions,

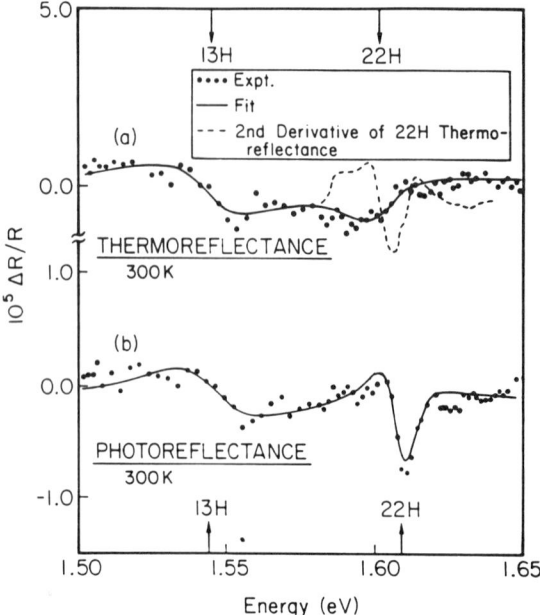

FIG. 30. Comparison of thermoreflectance and photoreflectance for the 22H transitions. Also shown is a spectrum representing two derivatives of the TR data with respect to energy. (After Shen et al. 1988a.)

transitions can occur at critical points at the miniband center (Γ) or edge (π). This effect has been observed in PLE spectroscopy (Song et al., 1987). In PR measurements on a series of GaAs–AlGaAs MQWs with varying barrier and well widths, Reddy et al. (1987b) observed unconfined transitions above the AlGaAs gap for the case of MQW having barrier and well widths of 100 Å and 150 Å, respectively, with $x = 0.15$. A large number of transitions were observed, as many as six. By a comparison with a simple tight-binding calculation of the absorption coefficients, these have been associated with allowed transitions between the valence and conduction minibands. It was noted that the barrier width could be obtained from knowledge of the separation of the Γ and π states for a set of unconfined transitions.

Shen et al. (1988b) have also considered unconfined transitions in GaAs–AlGaAs MQWs and have shown that the energies of the transitions can be accounted for using the simple model of Bastard (1982). This allows for an easy method of determining the barrier thickness.

It is important to note that the aforementioned studies did not take into account how the excitonic interaction affects the form of the dielectric function of the miniband. Additional theoretical calculations would be helpful in clarifying this point.

8. MODULATION-DOPED HETEROJUNCTIONS

Modulation-doped heterojunctions form a very important class of devices that operate on the basis of charge transfer from one layer to another. This class of materials includes the high-electron-mobility FET (HEMT), the NIPI (Ploog and Döhler, 1983), and spike-doped layers. In what follows, we will consider the HEMT and the NIPI structures.

a. The HEMT

The HEMT is formed through a charge transfer from a heavily doped AlGaAs layer to an undoped GaAs layer. This places the electrons in a very pure material, the GaAs, resulting in very high mobilities in the vicinity of 10^6 cm^2/V-sec. Figure 31 shows a schematic representation of such a structure. The material of interest is the undoped GaAs layer, which is separated from the source of the carriers by an insulating AlGaAs spacer layer. The spacer layer thickness controls the number of electrons that are transferred. Once electrons transfer, they are confined to the interfacial region and form a two-dimensional electron gas (2DEG) lying within the triangular potential, as shown in Fig. 31. Just as in the quantum well case, there are subband states in the conduction band. The valence band exhibits no confinement and hence is 3D like. The presence of electrons at the interface is typically determined by Hall effect measurements.

Rather than considering the PR data for this structure, we will start with the electroreflectance work of Snow *et al.* (1988), who not only measured the ER spectrum but also calculated the dielectric function. For the ER measurement, electrical contact was made between the 2DEG and the front surface. In this fashion, it was possible to modulate the carrier density at the interface. In order to model the results, a self-consistent calculation was performed of the conduction-subband states and the bulk-like valence-band states. The resultant energy states and their wave functions were then used to calculate the dielectric function for different 2DEG concentrations. The results of the experiment and the calculation are shown in Fig. 32. Notice the fact that both exhibit broad structures above the gap. This broad structure is a direct result of the nature of the valence band, which allows many valence states to overlap with a single electron state. The calculation shows weak oscillations that are not in the data. The exact source of these oscillations is not clear. They

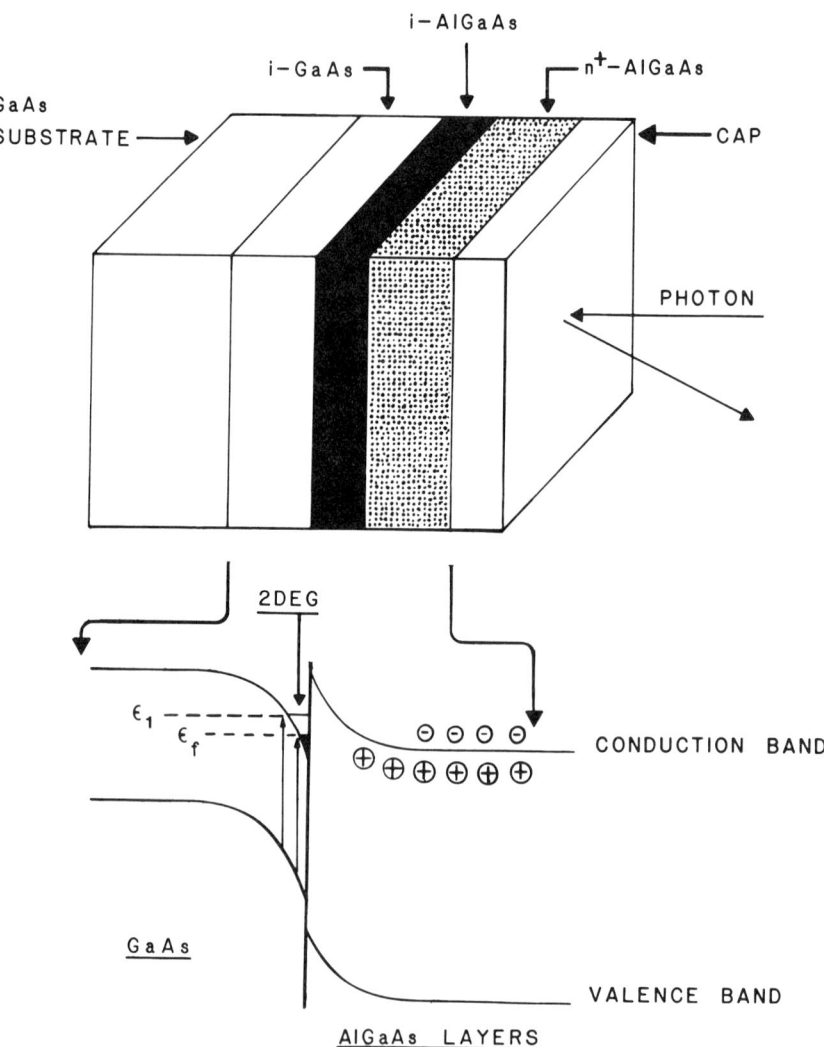

FIG. 31. Schematic representation of a modulation-doped FET structure (MODFET). Electrons from the n^+-AlGaAs layer transfer over to the undoped GaAs layer forming a two-dimensional electron gas (2DEG). Note that there is no confinement in the GaAs valence band. (After Glembocki et al., 1987b.)

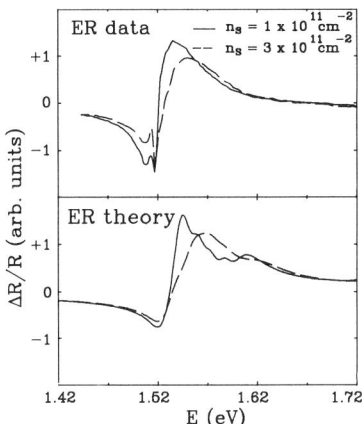

FIG. 32. Comparison of 10 K ER data (top) to calculated spectra (bottom) for a MODFET at two 2DEG densities. The sharp feature in the data located at 1.52 eV is the excitonic transition originating from the low-field regions of the undoped GaAs. (After Snow et al., 1988.)

may arise from the oscillatory nature of the subband wave functions and from their discrete nature. On the other hand, only a limited number of subbands were used in calculating the dielectric function, and the oscillations may be a result of this. A similar experimental result was obtained by Höpfel and co-workers (1985), who concentrated on the large changes in reflectivity at the AlGaAs gap.

In the case of photoreflectance, one might expect that the pump light would introduce carriers into the system, with the largest changes in carrier density and electric fields occurring at the AlGaAs–GaAs interface. Glembocki and co-workers (1985) have shown that the PR technique provides a room temperature fingerprint of the 2DEG. However, the result is very different from the electroreflectance spectrum. Shown in Fig. 33 is a series of spectra of Glembocki et al. (1985), ranging from thick layers to a HEMT that failed and had no 2DEG. The 2D carrier concentrations were obtained by 4 K cyclotron resonance measurements. Notice that there is an extra feature above the GaAs signal (which originates from the back of the GaAs layer containing the 2DEG). This line decreases with intensity, as the carrier density is reduced until it is hardly visible in the sample having no 4 K carrier density. This new feature has also been independently observed by other workers (Tang et al., 1987a). Notice that this feature is very different from the broad structure of Fig. 32. Low-temperature measurements show little change in the shape of the PR feature, indicating that it is probably not a series of Franz–Keldysh oscillations. Although the nature of the PR feature is still unknown, it does serve as a fingerprint of the 2DEG. It should be remembered that in PR, optically created electron–hole pairs modulate the dielectric function and may

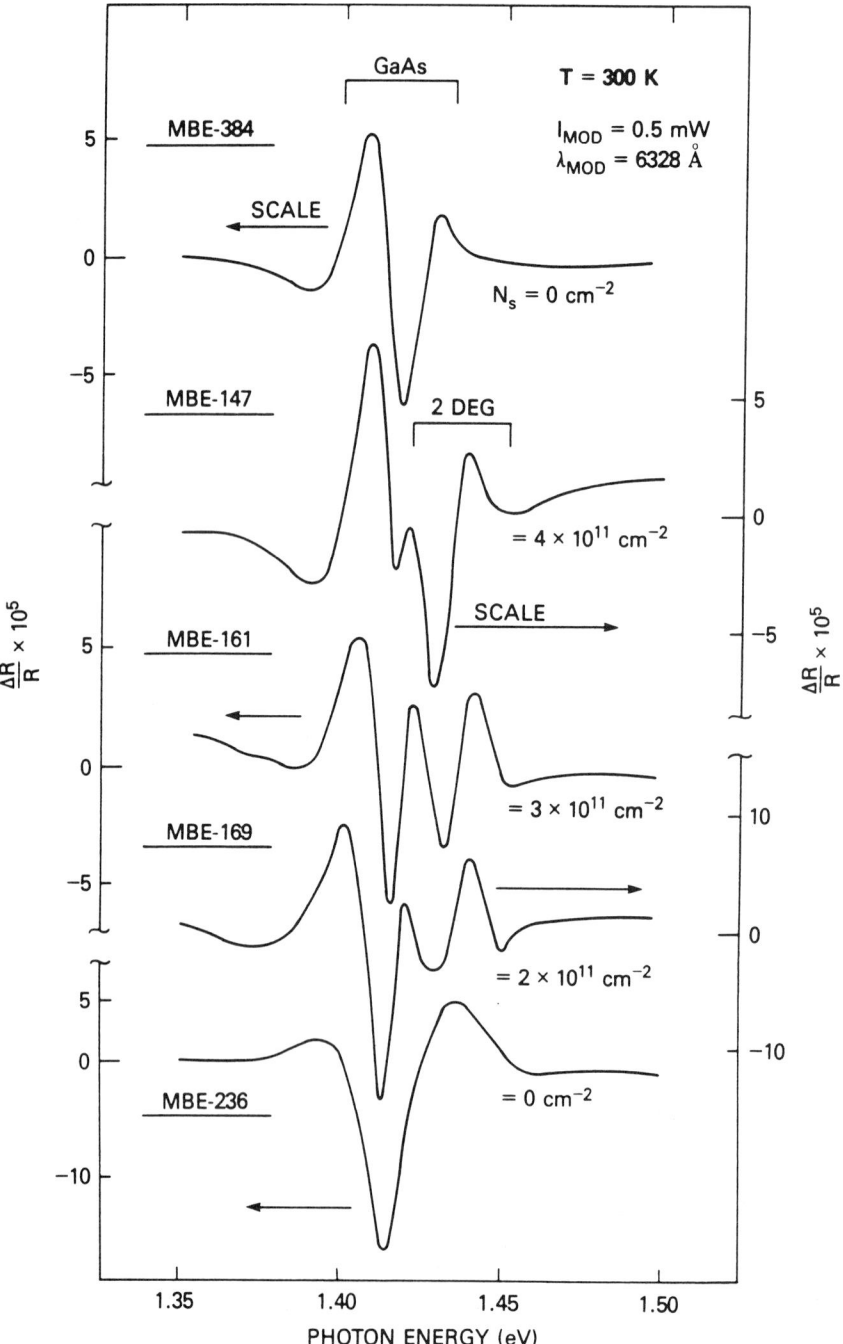

FIG. 33. Room temperature photoreflectance data for an undoped GaAs–AlGaAs heterojunction (top) and four MODFET samples. The carrier densities were obtained from cyclotron resonance measurements. (After Glembocki et al., 1985.)

produce radically different potential changes from simple electric-field modulation. In addition, the ionized shallow acceptors near the 2DEG can capture photoexcited holes.

b. NIPI

Another type of microstructure involving doped heterojunctions is the NIPI, which is just a MQW constructed from alternate layers of n- and p-type GaAs (Ploog and Döhler, 1983). The difference in Fermi energies between the doped layers causes a transfer of charge between the layers, forming a periodic space charge variation. Quantum confinement results where conduction and valence subbands are spatially separated, with electron subbands in the n-region and hole subbands in the p-region. Because the band structure of a NIPI is indirect in real space, any illumination with above-bandgap light results in a spatial separation of electrons and holes that acts to screen the built-in electric fields. This effect has a profound impact on the band structure. In fact, we again expect modulated pump light to lead to a first derivative spectroscopy. The first PR experiments on NIPI structures were performed by Gal and co-workers in the InP doping superlattices (Gal et al., 1986). They interpreted their data in terms of transitions within the n-region originating from the valence band and terminating in the subbands formed in the conduction band. Subsequently, X. Shen et al. (1986) and Tang et al. (1987b) studied GaAs NIPI structures. Both groups considered miniband formation and interpreted their data in terms of transitions between valence subbands in the p-layer and conduction subbands in the n-layer. In addition, X. Shen et al. (1986) changed the frequency of modulation (of the pump beam) from 1 Hz to 3.5 kHz and observed a decrease in the PR signal intensity. This effect was also studied with a background dc laser. Lifetimes were obtained for the carriers ranging between 0.25 and 4.2 msec. The long carrier lifetimes arise because the electrons and holes are separated in real space, and, therefore, recombination is inhibited. The effect of the second laser was to decrease the lifetime, in accordance with a flattening of the bands due to illumination. This effect is shown in Fig. 34.

9. Novel Photoreflectance Techniques

In the experimental section of this chapter, we described the standard photoreflectance technique, which is the most widely used. In what follows, we wish to consider improvements and modifications of the PR technique.

The single most limiting problem with PR is the presence of stray laser light and or band-gap luminescence. Both of these spurious signals have the

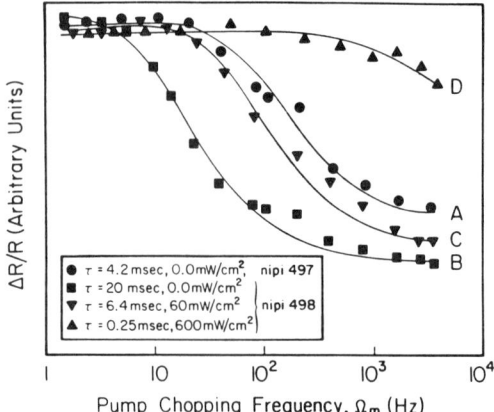

FIG. 34. Frequency dependence of the PR signal for two NIPI samples, as a function of pump laser intensity. (After X. C. Shen, et al., 1986.)

same modulating frequency as the desired PR signal. The spurious light is not optically dispersed and shows up as a large, wavelength-independent background signal. Often, it can be significantly larger than the PR signal, at times making it impossible to perform the measurement. Several methods have been devised to deal with this problem. Theis and co-workers have modified a double monochromator so that the first half of the monochromator provides light for the probe beam. Before the reflected light is detected, it is passed through the second half of the double monochromator, which runs in tandem with the first half. This removes any stray light from the spectrum, except at the wavelengths of the stray light. At these wavelengths, the stray light can be removed by taking a spectrum without the probe light and subtracting.

In another scheme, instead of using a chopper to modulate the intensity of spatially fixed laser spot, it is possible to use a vibrating quartz plate, which displaces the laser spot on the surface of the sample. On one end of the swing, the laser spot overlaps the probe beam, while on the other end it does not. This produces the same effect as a chopper, in relation to field modulation, but causes the stray light and photoluminescence reaching the detector to be unmodulated. This occurs because the lock-in subtracts the signals at the two ends of the swings and because the stray light and photoluminescence are nearly spatially uniform. Pollak (1986) reports limited success with this approach.

Finally, one can take advantage of the fact that both the stray light and photoluminescence are isotropically emitted. Therefore, at large distances from the sample, their intensities will be very weak. Replacing the monochromator with a dye laser, which is a collimated source, allows one to place the detector far away from the sample and to avoid the lenses or mirrors that are routinely placed after the sample to collect the reflected light. These experiments have been successfully carried out in Shanabrook et. al. (1987a) and Glembocki and Shanabrook (1987b).

A very interesting extension of PR was developed by Shen and co-workers (1987b). They realized that the modulation is produced by photoexcited carriers and therefore, just as in PLE the intensity of a PR feature will depend upon the absorption coefficient of the modulating beam. They replaced the pump laser of Fig. 6 with a monochromator. The wavelength of the probe monochromator was fixed at some higher-lying PR feature, and the pump wavelength was scanned through the 11H and 11L regions of a 50-Å GaAs–AlGaAs ($x = 0.33$) MQW at 77 K. This spectroscopy was called photoreflectance excitation spectroscopy. Shown in Fig. 35 is a comparison of the 77 K PRE spectrum to a 6 K PLE spectrum for the same sample. Notice that the PRE and PLE lines are similar in shape and are shifted from each other only by the temperature difference under which the two spectroscopies were performed. The features labeled A and B were identified as 2S excitons. The PR line shape analysis for this sample indicated the presence of these features. In PLE the luminescence from a strong feature is detected, and its changes with the wavelength of the exciting line are monitored (Miller and Kleinman, 1985). This luminescence is usually obtained from the low-energy edge of the free exciton. Therefore, the full line shape of the 11H exciton is difficult to obtain. In PRE on the other hand, any PR feature will do; therefore, one can be chosen far away from the area of interest and the full line shape can thus be defined. Figure 35 illustrates this for the sample studied by Shen et al.

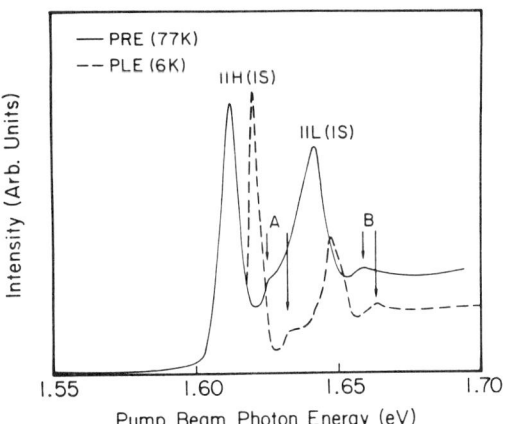

FIG. 35. A comparison of 6 K PLE (solid line) to 77 K PRE. (After Shen et al., 1987b.)

(1987b). Notice how the PLE line is cut off at low energies, where the PRE line is defined.

Finally, an interesting application of PRE would be in the identification of light and heavy holes. One can take a PRE spectrum and compare it with the PR spectrum. All of the previous discussion about the differences in the ratio of the intensities of the light and heavy holes in PLE and PR would apply and could be used to make the identification.

10. APPLICATIONS TO BULK MATERIALS

Thick (>1000 Å) epitaxial layers are of interest because they form the basis for microstructures. The type of information that is important to know includes sample quality, alloy content of alloys, carrier concentrations, and strain in strain-layer systems. Electromodulation techniques can aid greatly in accessing this information, often at room temperature.

In epitaxial growth of alloy compounds, it is important to access the alloy content of the thin film. This can be easily performed from an optical measurement of the band gap, provided that the relationship between the two has previously been established. We will not discuss this now, because there is a good example of determining the alloy content in the section on quantum wells. At our laboratory, we typically calibrate the mole fraction of our MBE AlGaAs material by growing one sample with several 2000-Å-thick layers of $Al_xGa_{1-x}As$ on top of each other. One 300 K PR measurement reveals the different band gaps of the various alloys and calibrates the relative Ga and Al fluxes. In contrast, PL measurements on these samples only exhibit luminescence from the layers near the surface.

a. Carrier Densities

Another interesting property of thick epitaxial layers is the net carrier concentration ($N_D - N_A$) of the film. Electroreflectance or photoreflectance can be used to determine the carrier density. One accomplishes this through the use of the Franz–Keldysh oscillations, whose period can be directly related to the electric field at the surface. Once the field is known, Poisson's equation can be used to obtain the carrier density (Sze, 1981). We describe the theory of the Franz–Keldysh oscillations.

At moderate to high fields, the ER line shapes change and the simple TDFF is not valid. For bulk samples and for photon energies above the band gap, the ER line shapes are characterized by exponentially decaying oscillations. These are known as Franz–Keldysh oscillations (FKO) (Aspnes, 1973; Bhattacharya *et al.*, 1988) and are related to the magnitude of the applied electric field as well as to the interband effective mass along the field direction. Aspnes and Studna have derived a simple approximation for the FKO given by (Aspnes and Studna, 1973)

$$\frac{\Delta R}{R} = B(E - E_{cv})\cos\left[\frac{2}{3}\left(\frac{E - E_{cv}}{\hbar\Omega}\right)^{3/2} + \Theta\right], \quad (24)$$

where E_{cv} is the band-gap energy, $\Theta = \pi(d - 1)/4$, $B(E - E_{cv})$ is an exponentially decaying factor, and d is the critical point dimensionality. Extremas in $\Delta R/R$ occur under the following conditions:

$$n\pi = \Theta + \frac{2}{3}\left[\frac{E_n - E_{cv}}{\hbar\Omega}\right]^{3/2}, \quad (25)$$

where E_n are the energy positions of the nth extrema and $\hbar\Omega$ is the electro-optic energy, which has been defined in Section III. The simple Eqs. (24) and (25) are valid only if the field is constant and $\Omega \gg \Gamma$. Equation (25) can be used to determine the electric field if the effective masses are known. This is a useful result and is routinely used to obtain the net carrier concentration.

Equations (24)–(25) can be used to determine the surface field provided that the penetration depth of light is much greater or much smaller than the skin depth of the field (Bhattacharya, et al., 1988). If this is true, then the energy positions of the extremas in Franz–Keldysh oscillations can be plotted against the extrema number, and Eq. (25) can be used to determine Ω and thus the field. If, however, these conditions are not met, then the measured field is some average field, and a more detailed analysis as described by Bhattacharya et al. is required.

Shown in Fig. 36 are a series of PR spectra of Bhattacharya et al. (1988) for a Zn-doped InP sample having a nominal net carrier concentration of 1.6×10^{16} cm^{-3}. The Schottky contact at the surface was a transparent 650-Å indium–tin oxide layer. The features labeled A–C are associated with the band-gap exciton, while the remainder of the features are FKO. The two types of transitions originate from different parts of the sample. The FKO signal is from the front of the sample, which is under the largest field, while the band-gap exciton originates from the low-field regions that are deeper into the sample. As the bias is increased, the band-gap exciton decreases in intensity, which indicates that the field is penetrating deeper into the sample.

In the case of the FKO, increasing the field shifts the oscillations to higher energies and increases their period. Bhattacharya et al. (1988) plotted the energy positions of the extremas of the FKO (D–J) as a function of their extremum number n [$n(D) = 1, n(E) = 2, n(F) = 3,\ldots$]. From this plot and

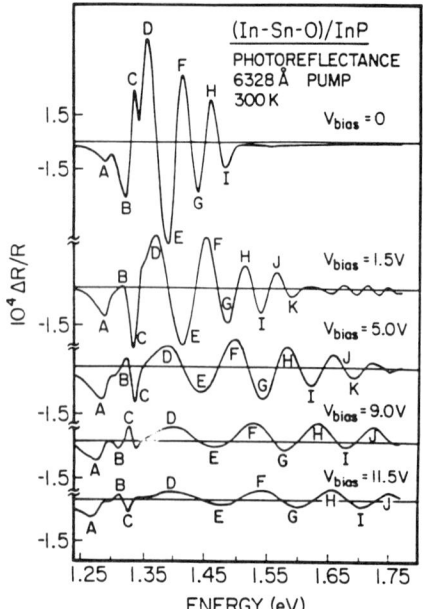

FIG. 36. PR spectra as a function of bias for a Zn-doped InP sample having a nominal net carrier density of 1.6×10^{16} cm^{-3}. The features $A-C$ are related to the band-gap exciton. The features $B-K$ are the extremas in the Franz–Keldysh oscillations. (After Bhattacharya et al., 1988.)

Eq. (25), they were able to determine a built-in potential of -0.45 ± 0.1 V and a carrier density of $(1.55 \pm 0.1) \times 10^{16}$ cm^{-3}, in good agreement with the Hall value. Bottka et al. (1988) used this approach to obtain carrier densities for a wide variety of samples and found a good correlation between the PR-determined carrier concentrations and those obtained from capacitance–voltage measurements.

In PR measurements, the samples are in a built-in surface field. In order to determine the carrier density, one must know the pinning position of the Fermi energy at the surface. For GaAs, this is well known, but in other systems this may not be the case. In such a circumstance, electroreflectance is preferable.

b. Strains

Of interest are strain-layer systems, such as GaAs–Si, InGaAs–GaAs, and InAs–GaSb. For these types of materials, it becomes important to determine the strain in the layer of interest. This can be accomplished by using a modulation technique such as ER or PR. A lattice mismatch between the epitaxial film and the substrate will result in a biaxial strain, which splits the fourfold degenerate valence band into light- and heavy-hole states. There is also a shift of the band gap. There have been many studies utilizing PL (Zemon et al., 1986) and Raman scattering (Jusserand and Cardona, 1988), but few modulation spectroscopy measurements. One PR investigation of GaAs grown on Si reported only the shift of the gap (Bottka et al., 1988). The strain ε was estimated from the shift. Although this approach is useful in the case of binary compounds, it becomes difficult to apply to alloys, where a shift of the gap can be caused by strain as well as alloy content. Therefore, it is much better to look for the splitting of the valence band (Dimoulas et al., 1990). For growth along (100), the biaxial strain produces light- and heavy-hole splittings given by

$$\Delta E_{\text{split}} = \frac{2\varepsilon|b|(C_{11} - C_{12})}{C_{11}}, \qquad (26)$$

where b is the shear deformation potential, C_{11}, C_{12} are the elastic stiffness coefficients, and $\varepsilon = \Delta a_\perp/a_\perp$. In addition, it can be shown that for the light- and heavy-hole interband transitions will have an intensity ratio of 1:3, which facilitates their identification. The utility of PR in this case is clear. For alloy systems, such as InGaAs, this is the only way to determine both the alloy content and the strain: the splitting determines the strain, and the alloy content is obtained from the shifts of the states.

c. *Elevated Temperatures*

Recently, there has been considerable interest in MBE growth mechanisms. In this regard, PR can allow us to study the band gaps and the band structure of semiconducting materials at elevated temperatures. This was accomplished by Shen and co-workers (1988c) for GaAs and AlGaAs in the temperature range of 77 to 900 K and by Hang et al. (1990) for InP over the same temperature range. The E_0 band gap was mapped out as a function of temperature and shown to fit the Varshni equation (Varshni, 1967)

$$E(T) = E(0) - \frac{\alpha T^2}{\beta + T}. \qquad (27)$$

Shown in Fig. 37 are the spectra obtained by Hang et al. (1990) for the temperature dependence of the PR signal up to 600C. The fits to Eq. (27) for the

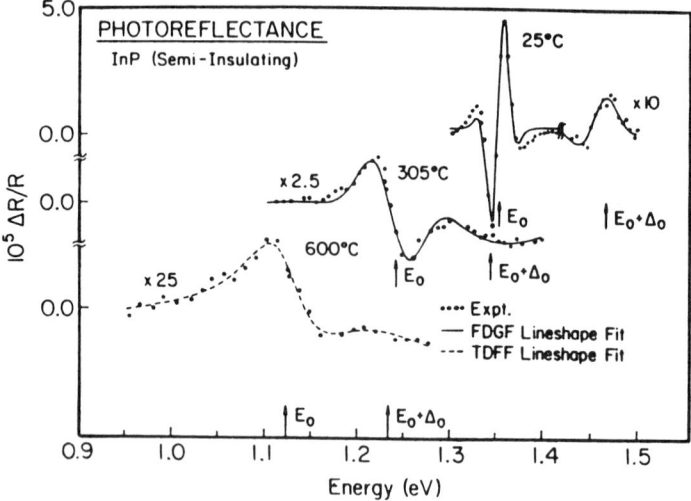

FIG. 37. PR spectra of semi-insulating InP as a function of temperature from 25 to 600°C. The experimental data are represented by a dotted line. (Hang et al., 1990).

E_0 transitions yield the following values for GaAs: $E(0) = 1.512$ eV, $\alpha = 5.1 \times 10^{-4}$ eV K^{-1}, $\beta = 190$ K; for InP: $E(0) = 1.432$ eV, $\alpha = 4.1 \times 10^{-4}$ eVK^{-1}, $\beta = 136$ K. These values should allow PR monitoring of the sample temperature during MBE growth. More interestingly, it may be possible to in situ monitor the development of superlattices and quantum wells.

VI. Conclusions

We have presented an overview of photoreflectance spectroscopy as applied to microstructures. It is clear that the technique is easy to implement and that the information obtained from it can be considerable. The measured line shapes are complicated and require care in analysis. The theoretical treatment, however, should serve as a guide in interpreting photoreflectance data. Despite these complications, the photoreflectance technique is still one of the simplest room temperature optical probes of semiconductors and with care can be routinely used for characterization. We feel that the areas of research in which photoreflectance and electromodulation spectroscopy can have a considerable impact are those involving the effects of electric fields and optically created carriers on microstructures, such as coupled well systems, doping superlattices, and resonant tunneling structures. Here, one may take advantage of the complex impedance nature of the technique to obtain optically and locally (on the sample) information about electric-field behavior of a device. Finally, new systems such as microstructures grown on various crystal faces can be characterized, especially those grown on the (111) face, which have built-in electric fields because of the piezoelectric effect.

References

Albers, W., Jr. (1969). *Phys. Rev. Lett.* **23**, 410.
Aspnes, D. E. (1970). *Solid State Commun.*, **8**, 267.
Aspnes, D. E. (1973). *Surface Sci.* **37**, 418.
Aspnes, D. E., and Studna, A. (1973). *Phys. Rev. B***7**, 4605.
Baldereschi, A., Baroni, S., and Resta, R. (1988). *Phys. Rev. Lett.* **61**, 734.
Bastard, G. (1982). *Phys. Rev. B***25**, 7584.
Bastard, G. (1988). "Wave Mechanics Applied to Semiconductor Heterostructures." Les Editions de Physique, Paris.
Bastard, G., Mendez, E. E., Chang, L. L., and Esaki, L. (1983). *Phys. Rev. B***28**, 3241.
Bhattacharya, R. N., Shen, H., Parayanthal, P., Pollak, F. H., Coutts, T., and Aharoni, H. (1988). *Phys. Rev. B***37**, 4044.
Bleuse, J., Bastard, G. and Voisin, P. (1988). *Phys. Rev. Lett.* **60**, 220.
Bloch, F. (1928). *Z. Phys.* **52**, 555.
Blossey, D. F., and Handler, P., (1972), in "Semiconductors and Semimetals," vol. 9, p. 257. Academic Press, New York.
Bottka, N., Gaskill, D. K., Griffiths, R. J. M., Bradley, R. R., Joyce, T. B., Iito, C., and McIntyre, D. (1988). *J. Cryst. Growth*, **93**, 481.
Broido, D. A., and Sham, L. J. (1987). *Phys. Rev. B***34**, 3917.
Cacciatore, C., Campi, D., Coriasso, C., Rigo C., and Alibert, C. (1989). *Phys. Rev. B***40-II**, 6446.
Cardona, M. (1969). In "Modulation Spectroscopy," p. 15. Academic Press, New York.
Cerdiera, F., and Cardona, M. (1969). *Solid State Commun.* 7, 879.
Chang, Y. C., and Schulman, J. N. (1983). *Appl. Phys. Lett.*, **43**, 536.
Chu, H., and Chang, Y. C. (1989). *Phys. Rev.* **39**, 10861.
Collins, R. T., Viña, L., Wang, W. I., Chang, L. I., Esaki, L., V. Klitzing, K., and Ploog, K. (1987). *Phys. Rev. B***36**, 1531.
Dafesh, P. A., Arbet, V., and Wang, K. L. (1990). *Appl., Phys. Lett.* **56**, 1498.
Dawson, P., Duggan, G., Ralph, H. I., and Woodbridge, K. (1983). *Phys. Rev. B***28**, 7381.
Dimoulas, A., Tzanetakis, P., Georgakilas, A., Glembocki, O. J. and Christou, A. (1990). *J. Appl. Phys.* **67**, 4389.
Erman, M., Theetan, J. B., Frijlink, P., Gaillard, S., Hia, F. J., and Alibert, C. (1984). *J. Appl. Phys.* **56**, 3241.
Enderlein, R., Desheng Jiang, and Yinsheng Tang. (1988). *Phys. Stat. Sol. (b)* **145**, 167.
Finkman, E., Sturge, M. D., Meynadier, M.-H., Nahory, R. E., Tamargo, M. C., Hwang, D. M., and Chang, C. C. (1987). *J. Luminescence* **39**, 57.
Franz, W. (1958). *Z. Naturforsch.* **13**, 484.
Fujimoto, H., Hamaguchi, C., Nakazawa, T., Imanishi, K., Taniguchi, K., and Sasa, S. (1990). *Surf. Sci.* **228**, 206.
Gal, M., Yuan, J. S., Viner, J. N., Taylor, P. C., and Springfellow, G. B. (1986). *Phys. Rev. B***33**, 4410.
Gay, J. G., and Klauder, L. T. (1968). *Phys. Rev.* **172**, 811.
Glembocki, O. J., and Shanabrook, B. V. (1987a). *Superlattices and Microstructures* **3**, 235.
Glembocki, O. J. and Shanabrook, B. V. (1987b). *Proc. SPIE* **794**, 12.
Glembocki, O. J. and Shanabrook, B. V. (1989). *Superlattices and Microstructures* **5**, 603.

Glembocki, O. J., Shanabrook, B. V., Bottka, N., Beard, W. T., and Comas, J. (1985). *Appl. Phys. Lett.* **46**, 970.
Glembocki, O. J., Shanabrook, B. V., and Beard, W. T. (1986). *Surf. Sci.* **174**, 206.
Glembocki, O. J., Snow, E. S., Little, J. W., and Leavitt, R. P., (1991). In press, *Superlattices and Microstructures*.
Hang, Z., Shen, H., and Pollak, F. H. (1990). *Solid State Commun.* **73**, 15.
Hess, K., Binberg, D., Lipari, N. O., Fishbach, U., and Altereli, (1976). Proc. 13th Int. Conf. on Physics of Semincond., Rome (Fumi, F. G., ed.), p. 142. North-Holland, Amsterdam.
Höpfel, R. A., Shah, J., Gossard, A. C., and Wiegmann, W. (1985). *Appl. Phys. Lett.* **47**, 163.
Jones, H., and Zener, C. (1934). *Proc. Roy. Soc* **144**, 101.
Joyce, M. J., Johnson, M. J., Gal, M., and Usher, B. F. (1988). *Phys. Rev.* **B38**, 10978.
Jusserand, B., and Cardona, M. (1989). "Topics in Applied Physics: Light Scattering in Solids V," vol. 66, p. 49. Springer-Verlag, Berlin.
Kane, E. O. (1966). "Semiconductors and Semimetals," vol. 1, p. 75. Academic Press, New York.
Keldysh, L. V. (1958). *Zh. Eksp. Teor. Fiz.* **34**, 1138 [Engl. transl., *Solv. Phys.-JEPT* **7**, 788].
Klipstein, P. C., and Apsley, N. (1986). *J. Phys. C*, **19**, 6461.
Klipstein, P. C., Tapster, P. R., Apsley, N., Anderson, D. A., Skolnick, M. S., Kerr, T. M., and Woodbridge, K. J. (1986). *Phys. C*, **19**, 857.
Koteles, E. S., Chen, Y. J., and Elman, B. S. (1987). Proc. Int. Meeting on Excitons in Confined Systems, Rome.
Ksendzov, A., Shen, H., Pollak, F. H., and Bour, D. P. (1990a). *Solid State Commun.* **73**, 11.
Ksendzov, A., Shen, H., Pollak, F. H., and Bour, D. P. (1990b). *Surf. Sci.* **228**, 326.
Laude, L. D., Pollak, F. H., and Cardona, M. (1971). *Phys. Rev.* **B8**, 2623.
Lederman, F. L., and Dow, J. D. (1976). *Phys. Rev.* **B13**, 1633.
Little, J. W., and Leavitt, R. P. (1989). *Phys. Rev.* **B39**, 1365.
Little, J. W. Whisnant, J. K., Leavitt, R. P., and Wilson, R. A. (1986). *Appl. Phys. Lett.* **51**, 1786.
Masselink W. T., Pearah, P. J., Klem, J., Peng, C. K., Morkoc, H., Sanders, G. D., and Chang, Y. C. (1985). *Phys. Rev.* **B32**, 8027.
Mendez, E. E., Chang, L. L., Ladgren, G., Ludeke, R., Esaki, L., and Pollak, F. H. (1981). *Phys. Rev. Lett.* **46**, 1230.
Menendez, J., Pinczuk, A., Werder, D. J., Gossard, A. C., and English, J. H. (1986). *Phys. Rev.* **B33**, 8863.
Miller, D. A. B., Chemla, D., S., Damen, T. C., Gossard, A. C., Weigmann, W., Wood, T. H., and Burrus, C. A. (1984). *Phys. Rev. Lett.*, **53**, 473.
Miller, D. A. B., Chemla, D. S., T. C. Daman, Gossard, A. C., Wiegmann, W., Wood, T. H., and Burrus, C. A. (1985). *Phys. Rev.* **B32**, 1043.
Miller, R. C., and Kleinman, D. A. (1985). *J. Luminescence* **30**, 520.
Miller, R. C., Kleinman, D. A., Tsang, W. T., and Gossard, A. C. (1981). *Phys. Rev.* **B24**, 1134.
Miller, R. C., Kleinman, D. A., and Gossard, A. C. (1984). *Phys. Rev.* **B29**, 7085.
Miller, R. C., Gossard, A. C., Sanders, G. D., Chang, Y. C., and Schulman, J. N. (1985). *Phys. Rev.* **B32**, 8452.
Nilson, N. G. (1969). *Solid, State Commun.* **7**, 479.
Pan, S. H., Shen, H., Hang, Z., Pollak, F. H., Zhuang, W., Xu, Q., Roth, A. P., Masut, R. A., Lacelle, C., and Morris, D. (1988). *Phys. Rev.* **38**, 3375.
Parayanthal, P., Shen, H., Pollak, F. H., Glembocki, O. J., Shanabrook, B. V., and Beard, W. T. (1986). *Appl. Phys. Lett.* **48**, 1261.
Pearsall, T. P., Pollak, F. H., Bean, J. C., and Hull, R. (1986). *Phys. Rev.* **B33**, 6821.
Pearsall, T. P., Berk, J., Feldman, L. C., Bonar, J. M., Mannaerts, J. P., and Ourmazd, A. (1987). *Phys. Rev. Lett.* **58**, 729.
Phillip, H. R., and Ehrenreich, H. (1963). *Phys. Rev.* **129**, 1550.

Pinczuk, A. and Abstreiter, G. (1989). "Topics in Applied Physics: Light Scattering in Solids V," vol. 66, p. 153, Springer-Verlag, Berlin.
Ploog and Döhler, (1983). *Adv. Phys.* **32**, 285.
Pollak, F. H. (1986). Private communication.
Reddy, U. K., Ji, G., Houdre, R., Unlu, H., Huang, D., and Morkoc, H. (1987a). *SPIE* **794**, 116.
Reddy, U. K., Ji, G., Henderson, T., Morkoc, H., and Schulman, J. N. (1987b). *J. Appl. Phys.* **62**, 145.
Rockwell, B., Chandrasekhar, H. C., Chandrasekhar, M., Pollak, F. H., Shen, H., Chang, L. L., Wang, W. I., and Esaki, L. (1990). *Surf. Sci.* **228**, 322.
Sanders, G. D., and Chang, Y. C. (1987). *Phys. Rev.* **B35**, 1300.
Seraphin, B. O. (1972). "Semiconductors and Semimetals," vol. 9, p. 1. Academic Press, New York.
Seraphin, B. O., and Bottka, N. (1966). *Phys. Rev.* **145**, 628.
Shanabrook, B. V., and Glembocki, O. J. (1987). Proc. 18th Int. Conf. on Phys. of Semicond., Stockholm, (Engstrom, O., ed.), p. 565. World Scientific, Singapore.
Shanabrook, B. V., Glembocki, O. J., and Beard, W. T. (1987a). *Phys. Rev.* **B35**, 2540.
Shanabrook, B. V., Glembocki, O. J., Broido, D. A., Vina, L., and Wang, W. I., (1987b). *J. Phys.* **48**, C5-235.
Shanabrook, B. V., Glembocki, O. J., Broido, D. A., Vina L., and Wang, W. I. (1989). *Phys. Rev.* **B39**, 3411.
Shay, J. L. (1970). *Phys. Rev.* **B2**, 803.
Shen, H., Parayanthal, P., Pollak, F. H., Smirl, A. L., Schulman, J. N., McFarlane, R. A., and D'Haenens, I. (1986a). *Solid State Commun.* **59**, 557.
Shen, H., Parayanthal, P., Pollak, F. H., Tomkiewicz, M., Drummond, T., and Schulman, J. N. (1986b). *Appl. Phys. Lett.* **48**, 653.
Shen, H., Parayanthal, P., Liu, Y. F., and Pollak, F. H. (1987a). *Rev. Sci. Instrum.* **58**, 1429.
Shen, H., Shen, X. C., Pollak, F. H., and Sacks, R. N. (1987b). *Phys. Rev.* **B36**, 3487.
Shen, H., Pan, S., Pollak, F. H., Dutta, M., and AuCoin, T. R. (1987c). *Phys. Rev.* **B36**, 9384.
Shen, H., Pan S. H., Pollak, F. H., and Sacks, R. N. (1988a). *Phys. Rev.* **B37**, 10919; *Proc. SPIE*, **946**, 36.
Shen, H., Pan, S. H., Hang, Z., Pollak, F. H., and Sacks, R. N. (1988b). *Solid State Commun.* **65**, 929.
Shen, H., Pan, S. H., Hang, Z., Leng, J., Pollak, F. H., Woodall, J. M., and Sacks, R. N. (1988c). *Appl. Phys. Lett.* **53**, 1080.
Shen, X. C., Shen, H. C., Parayanthal, P., Pollak, F. H., Schulman, J. N., Smirl, A. L., McFarlane, R. A., and D'Haenens, I. (1986). *Superlattices and Microstructures* **2**, 513.
Shinada, M., and Sugano, S. (1966). *J. Phys. Soc. Japan* **21**, 1936.
Silberstein, R. P., and Pollak, F. H. (1980). *J. Vac. Technol.* **17**, 1052.
Smith, D. L., and Mailhiot, C. (1990). *Rev. Mod. Phys.* **62**, 173.
Snow, E. S., Glembocki, O. J., and Shanabrook, B. V. (1988). *Phys. Rev.* **B33**, 12483.
Song, J. J., Yoon, Y. S., Jung, P. S., Fedotowsky, A., Sculman, J. N., Tu, C. W., Brown, J. M., Huang, D., and Morkoc, H. (1987). *Appl. Phys. Lett* . **50**, 1269.
Sze, S. M. (1981). "Physics of Semiconductor Devices," p. 362. Wiley-Interscience, New York.
Tang, Y., and Desheng, J. (1987a). *Chinese Phys. Lett.* **4**, 283.
Tang, Y., Wang, B., Jiang, D., Zhuang, W., and Liang, J. (1987b). *Solid State Commun.* **63**, 793.
Theis, W. M., Sanders, G. D., Leak, C. E., Bajaj, and Morkoc, H. (1988). *Phys. Rev.* **B37**, 3042.
Theis, W. M., Sanders, G. D., Leak, C. E., Reynolds, D. C., Chang, Y. C., Alavi, K., Colvard, C., and Shidlovsky, I. (1989a). *Phys. Rev.* **B39**, 1442.
Theis, W. M., Sanders, G. D., Evans, K. R., Liou, L. L., Leak, C. E., Bajaj, K. K., Stutz, C. E., Jones, R. L., and Chang, Y. C. (1989b). *Phys. Rev.* **B39**, 11038.
Thorn, A. P., Shields, A. J., Klipstein, P. C., Apsley, N., and Kerr, T. M. (1987). *J. Phys. C* **20**, 4229.

Toyozawa, Y. (1958). *Progr. Theoret. Phys.* **20**, 53.
Varshni, Y. P. (1967). *Physica (Ultrecht)* **34**, 149.
Voisin, P. (1989). *Surf. Sci.* **228**, 74.
Voisin, P., Delahande, C., Voos, M., Chang, L. L., Segmuller, A., Chang, C. A., and Esaki, L. (1984). *Phys. Rev.* **B30**, 2276.
Wang, W. I. (1986). *Surf. Sci.* **174**, 31.
Weisbuch, C. (1987). "Semiconductors and Semimetals," vol. 24, p. 1. Academic Press, New York.
Wolford, D. J., Keuch, T. F., Bradley, J. A., Gell, M. A., Ninno, D. and Jaros, M. J. (1986). *J. Vac. Sci., Technol.* **B9**, 1043.
Wong, K. B., Jaros, M., Gell, M. A., and Ninno, D. (1986). *J. Phys.* **C19**, 53.
Zemon, S., Shastry, S. K., Norris, P., Jagannath, C., and Lambert, G. (1986). *Solid State Commun.* **58**, 196.
Zheng. X. L., Heiman, D., Lax, B., and Chambers, F. A. (1988). *Appl. Phys. Lett.* **52**, 287.

Chapter 5

One- and Two-Photon Magneto-Optical Spectroscopy of InSb and $Hg_{1-x}Cd_xTe$

David G. Seiler*

MATERIALS TECHNOLOGY GROUP
SEMICONDUCTOR ELECTRONICS DIVISION
NATIONAL INSTITUTE OF STANDARDS AND TECHNOLOGY
GAITHERSBURG, MARYLAND

Chris L. Littler

DEPARTMENT OF PHYSICS
UNIVERSITY OF NORTH TEXAS
DENTON, TEXAS

AND

Margaret H. Weiler

LORAL INFRARED & IMAGING SYSTEMS, INC.
LEXINGTON, MASSACHUSETTS

I. INTRODUCTION	294
1. Background and Scope	294
2. Narrow-gap Semiconductors	295
3. Electronic States in Narrow-gap Semiconductors	297
a. Advantages of Using a Magnetic Field	297
b. Landau Levels and Density of States	298
c. Energy-Band Models for Zinc Blende Semiconductors	302
4. Interaction of Laser Radiation with Semiconductors	304
a. Overview	304
b. One-Photon Absorption	307
c. One-Photon Interband Magnetoabsorption	311
d. Two-Photon Absorption	312
e. Two-Photon Magnetoabsorption	319
5. Examples of Magnetoabsorption in InSb	322
II. EXPERIMENTAL METHODS	329
6. Overview	329
7. Photoconductivity	336
8. Photo-Hall Effect	337
9. Photovoltaic Effect	339

*Work initiated at the University of North Texas. It represents the views of the author and is not to be construed as official NIST work.

III. ONE-PHOTON MAGNETOSPECTROSCOPY 339
  10. *Interband One-Photon Magnetoabsorption (OPMA)* 339
    a. *InSb* 339
    b. *HgCdTe*. 345
  11. *Free-Carrier Magneto-Optics* 345
    a. *Conduction-Band Cyclotron Resonance and Variants* 347
    b. *Phonon-Assisted Cyclotron Resonance Harmonics*. 354
    c. *Free-Hole Cyclotron and Combined Resonance* 355
  12. *Impurity- and Defect-Related Laser Spectroscopy* 359
    a. *Shallow Levels* 362
    b. *Deep Levels* 373
IV. TWO-PHOTON ABSORPTION (TPA) AND TWO-PHOTON MAGNETOABSORPTION
(TPMA) SPECTROSCOPY. 389
  13. *Background* 389
    a. *Introduction* 389
    b. *Carrier Lifetimes by TPA Methods*. 390
  14. *Review of TPA and TPMA Work Done on InSb* 391
  15. *Review of TPA and TPMA in Alloys of HgCdTe* 410
    a. *Two-Photon Absorption* 410
    b. *Two-Photon Magnetoabsorption* 413
    c. *Composition and Temperature Dependence of E_g by TPMA Techniques* . . 417
REFERENCES 418

I. Introduction

1. BACKGROUND AND SCOPE

Spectroscopy is a very important area of science since most of our knowledge about the structure of atoms, molecules, and the various states of matter is based on some sort of spectroscopic investigation. Magneto-optical spectroscopy has been shown to be a particularly valuable and powerful means for characterizing and understanding semiconductors. For example, magneto-optical studies have been carried out to investigate various features and properties of semiconductors, such as the conduction or valence bands, excitonic levels, impurity or defect levels, density of electronic states, carrier lifetimes, band symmetries, effective masses, and effects of anisotropy. In addition, magneto-optics has been successfully used to study the influence of externally controlled parameters such as temperature, pressure, electric and magnetic fields, etc., on all of the aforementioned properties.

Although our primary focus in this chapter will be on the magneto-optical properties of the narrow-gap semiconductors InSb and HgCdTe, the principles and techniques can also be used to study magneto-optical effects in wider-band-gap semiconductors and the whole array of new artificially structured materials.

A number of excellent reviews have been previously written on various aspects of magneto-optics in semiconductors: Lax (1963), Palik and Wright

(1967), Smith (1967), Lax and Mavroides (1967), Mavroides (1972), Moss *et al.* (1973), McCombe and Wagner (1975), Pidgeon (1980), Zawadzki (1980), Rigaux (1980), Bauer (1980), Weiler (1981b), Singh and Wallace (1987), Rigaux (1988), and Gaj (1988). The recent *Landau Level Spectroscopy* volume also contains a number of articles that willl be helpful to the interested reader (Landwehr and Rashba, 1991). Our purpose is not to duplicate these works but to augment them. Emphasis will be placed on giving a perspective on the wide variety of magneto-optical phenomena that have been investigated more recently in the narrow-gap semiconductors InSb and HgCdTe. Unique aspects presented here include the review of two-photon magneto-optical studies and of impurity- and defect-related magneto-optical spectroscopy. We will also concentrate on those magneto-optical studies that involved the use of lasers as light sources. The use of lasers with their special characteristics (spectral purity, high power, etc.) has significantly enhanced the ability to study a diversity of magneto-optical effects.

This chapter is divided into four major sections. In Section I, we review the subject of electronic states in narrow-gap semiconductors, summarizing the various band models that have been used to interpret magneto-optical effects. Also included are reviews of various features of both one- and two-photon magnetoabsorption. Section II covers the techniques used to study magneto-optical effects, with illustrations of novel features such as ac magnetic-field modulation techniques for derivative spectroscopy. Section III reviews a wide variety of intra- and interband one-photon absorption phemomena and concentrates on laser-based and impurity- and defect-related magneto-optical results in InSb and HgCdTe. Section IV presents a comprehensive review of all two-photon absorption work in InSb and HgCdTe, with an emphasis on the usefulness of nonlinear magnetoabsorption measurements for characterizing narrow-gap semiconductors.

2. NARROW-GAP SEMICONDUCTORS

Narrow-gap semiconductors (NGS) have long been recognized for their special characteristics that give rise to not only interesting physical effects to study but to useful technological applications as well. Major applications of NGS include infrared detectors for passive imaging and diode lasers for high-resolution spectroscopy. In addition, NGS are used as Hall probes for measuring magnetic fields and thermoelectric devices to monitor power generation and cooling equipment. They are used by radio astronomers as sensitive, reliable, and comparatively fast submillimeter detectors. The specific materials discussed here, InSb and HgCdTe, are used extensively as infrared detectors because of their sensitivity to the regions of the electromagnetic spectrum corresponding to the atmospheric windows at $3-5$ μm and $8-14$ μm. Another area where NGS have shown interesting physical behavior is in nonlinear optical effects. In particular, NGS exhibit very large band-gap resonant optical nonlinearities such as optical bistability and self-defocusing (Craig and Miller, 1986; Miller et al., 1981; Miller and Craig, 1987).

Zawadzki (1972) has defined narrow-gap semiconductors as materials in which the value of the electron energy, as measured from the bottom of the conduction band, can become comparable to the energy gap E_g. According to this broad definition any semiconductor (GaAs, Ge, etc.) can exhibit "narrow-gap" properties under appropriate conditions (e.g., if electrons are optically excited sufficiently high above the band edge). However, a more typical NGS definition is a solid having an energy gap of less than 0.5 eV (Sonowski, 1980). With this last definition it is clear that NGS are found among a broad spectrum of elements, compounds, and alloys as shown in Table I. Improvements of crystal growth techniques have led to the production of many different types of alloys whose energy gap can also be adjusted

TABLE I

SOME REPRESENTATIVE NARROW-GAP SEMICONDUCTORS.

Elements	Compounds					Alloys
	II-V	II-VI	III-V	IV-VI	V-VI	
Tellurium	$BaAs_3$	HgS	InSb	PbS	Bi_2Se_3	$Hg_{1-x}Cd_xSe$
Selenium	$CaAs_3$	HgSe	InAs	PbSe	Bi_2Te_3	$Hg_{1-x}Cd_xTe$
Gray tin	Cd_3As_2	HgTe		PbTe	Sb_2Se_3	$Hg_{1-x-y}Cd_xMn_yTe$
	Cd_3P_2			SnTe	Sb_2Te_3	$Hg_{1-x}Fe_xSe$
	α- and β-EuP_3			GeTe		$Hg_{1-x}Fe_xTe$
	Zn_3As_2					$Hg_{1-x}Mn_xSe$
	Zn_3P_2					$Hg_{1-x}Mn_xTe$
						HgS_xSe_{1-x}
						$Pb_{1-x}Mn_xTe$
	II-IV-V	II-IV				$PbS_{1-x}Se_x$
	$CdSnAs_2$	Mg_2Sn				$Pb_{1-x}Sn_xSe$
						$Pb_{1-x}Sn_xTe$
	Superlattices and quantum wells					$PbSe_xTe_{1-x}$
	(Many combinations from the above lists,					$Cd_{3-x}Zn_xAs_2$
	such as HgTe/CdTe, $InAs/In_{1-x}Ga_xSb$,					$(Cd_{1-x}Mn_x)_3As_2$
	$PbTe/Pb_{1-x}Mn_xTe$, etc.)					$Pb_{1-x}Ge_xTe$
						$InAs_xSb_{1-x}$
						$Pb_{1-x}Cd_xS$
						$Pb_{1-x-y}Eu_xS_ySe$
						$Pb_{1-x-y}Eu_xTe_ySe$
						$Pb_{1-x-y}Cd_xS_ySe$
						$Pb_{1-x}Eu_xTe$
						$Pb_{1-x}Eu_xSe$
						$Pb_{1-x}Sr_xSe$

(From Seiler and Littler, 1990.)

by varying the composition. Among these alloys are the semimagnetic semiconductors like $Hg_{1-x}Mn_xTe$ whose localized magnetic ions lead to significant changes of band structure in the presence of a magnetic field. In addition the emergence of molecular beam epitaxy (MBE) technology has opened up the new and exciting field of band-gap engineering. This MBE growth process refers to a technique in which several atomic (or molecular) beams are incident on a heated substrate material placed in an ultrahigh vacuum environment. Single crystalline layers are grown in registry with the substrate, permitting highly controlled epitaxial growth. With the growth of artifical semiconductor structures by MBE, the optical, transport, and other electron and hole properties can be altered continuously and independently, leading to interesting new physics and new classes of semiconductor devices.

The small energy gap characteristic of these NGS means that the electron effective mass m^* is also small. This small effective mass in turn leads to many interesting physical properties of these materials. The electronic density of states $\rho(E) \propto (m^*)^{3/2}$ and, as a result, a moderate number of conduction electrons will fill the band to relatively high energies above the band edge (i.e., the Fermi energy can become comparable to the gap). Also, the close proximity of the valence band to the conduction band strongly influences the $E(k)$ dependence of the energy bands and produces nonparabolic behavior that is often important and thus must be taken into account. This closeness also produces a mixing of the valence- and conduction-band wave functions; hence, matrix elements for scattering and optical transitions are strongly energy- or k-dependent. As a result, NGS are more sensitive to external influences such as lattice temperature, magnetic field, electric field, or strain than the larger-band-gap semiconductors.

3. ELECTRONIC STATES IN NARROW-GAP SEMICONDUCTORS

a. *Advantages of Using a Magnetic field*

A magnetic field primarily affects the electrons in the solid rather than the lattice, and, hence, the magneto-optical properties of a crystal are largely determined by the electrons alone. Very detailed knowledge about the properties of electrons moving in solids can be obtained from a wide variety of magneto-optical and magnetotransport effects. The fact that the quasicontinuous electron or hole energies of a semiconductor become quantized in a magnetic field results in magneto-oscillatory phenomena involving optical, electronic, thermal, and magnetic properties.

There are a number of important advantages to the field of semiconductor characterization that can be obtained by applying an external magnetic field: (1) External perturbations like a magnetic field are artificially produced and thus can be controlled or continuously varied. This is not the case for intrinsic perturbations like native defects or excitons. (2) The effects of the magnetic field can be predicted on the basis of theoretical models of the electronic structure for the perfect lattice. Comparisons between the theoretical predictions and the experiments then provide a test of the theoretical models themselves. (3) Modulation of the magnetic field and subsequent measurement of the linear or nonlinear optical signals with the same phase and periodicity using lock-in amplifier techniques allows an improvement of up to several orders of magnitude in the signal-to-noise ratio. (4) A magnetic field introduces new quantizations of the electronic states producing a discrete spectrum instead of a continuous one (like the Landau quantization effects). Magnetic fields modify the symmetry of the Hamiltonian and therefore can produce, in addition to energy shifts of the electronic states, splittings of degeneracies, and different selection rules. (5) The application of a magnetic field also allows one to separate and identify individual absorption processes (such as free carrier, impurity-to-band, and nonlinear two-photon). Each process will have a different characteristic spectrum in a magnetic field and so can be identified. Examples of the many different types of magnetoabsorption processes that can be seen in InSb will be given in Section I.5 as a summary overview.

b. Landau Levels and Density of States

Landau (1930) was the first to calculate the energy spectrum of free electrons in a magnetic field in order to explain the diamagnetic behavior of quasi-free electrons in a solid. In contrast, classical theory was unable to predict any such diamagnetism. It is not our intent here to reproduce their derivation (which is done in many texts, for example, Landau and Lifshitz, 1965; Blakemore, 1962), but instead to summarize the results and describe the essential physics involved.

In the presence of a magnetic field, free electrons experience the transverse Lorentz force, which causes them to travel in orbits perpendicular to the magnetic field. Landau's theory, based upon the solution of the Schrödinger equation, shows that the energy of the electrons corresponding to the transverse components of the wave vector are quantized. The resulting energy eigenvalues with the magnetic field parallel to the z-axis are given by

$$E_n^\pm = \left(n + \frac{1}{2}\right)\hbar\omega_c + \frac{\hbar^2 k_z^2}{2m^*} \pm \frac{1}{2} g^* \mu_B B, \tag{1}$$

where n is the Landau level number $(0, 1, 2, \ldots)$. These quantized energy levels in a magnetic field are referred to as Landau levels. The energy of electron motion along the direction of the B-field $(\hbar^2 k_z^2 / 2m^*)$ is unaffected and thus unquantized. On the other hand, the energy of motion in a plane perpendicular to the field $[(n + 1/2)\hbar\omega_c]$ is quantized and related to the cyclotron frequency $\omega_c = eB/m_c^*$, where m_c^* is the cyclotron effective mass. The free-electron mass has been replaced by the effective mass of the electrons in the standard manner, which takes into account the effect of the periodic crystal potential. Table II shows typical values of $\hbar\omega_c$ at 1.0 and 10.0 T for a number of semiconductors and HgCdTe alloy compositions. The effect of the electron's spin is given by the last term $\pm(1/2)g^*\mu_B B$, where g^* is the effective spectroscopic g-factor or spin-splitting factor and the Bohr magneton $\mu_B = e\hbar/2m_0 = 5.77 \times 10^{-2}$ meV/T. This energy term takes into account the spin-orbit interaction. The g-factor in a semiconductor has values quite different from the usual value of two found for atomic systems. It has been shown by Roth *et al.* (1959) that in an InSb-type band structure with a narrow energy gap and a strong spin-orbit interaction, the g^*-value can have large, negative values. This difference is the result of the interaction between the desired band and other energy bands. In the case of multiple degenerate bands that form the valence-band extrema in most semiconductors, this interaction is such that a simple equation of the form of Eq. (1) cannot be applied to the levels near the band edge. For simple bands the Landau energy levels are shown schematically in Fig. 1. The electron moves in real space in the $x + y$-plane with a circular orbit of radius $r_n = \sqrt{(n + 1/2)2\hbar/eB}$ at the cyclotron frequency ω_c. For more complex bands that are anisotropic or nonparabolic, expressions for the Landau levels can still be derived using suitable approximations or be numerically calculated.

The effects of energy quantization become noticeable when the separation between energy levels $\hbar\omega_c$ exceeds kT (otherwise the electron distribution

TABLE II

VALUES OF $\hbar\omega_c$, THE CYCLOTRON RESONANCE ENERGY, FOR VARIOUS SEMICONDUCTORS AT $B = 1.0$ AND 10.0 T.

Semiconductor	m^*/m_0	$\hbar\omega_c$ (1 T) (meV)	$\hbar\omega_c$ (10 T) (meV)
CdS	0.21	0.5	5.5
GaAs	0.067	1.7	17.0
InAs	0.025	4.6	46.0
InSb	0.015	7.7	77.0
$Hg_{0.825}Cd_{0.175}Te$	0.002	58.0	577.0
$Hg_{0.8}Cd_{0.2}Te$	0.007	16.5	165.0
$Hg_{0.7}Cd_{0.3}Te$	0.017	6.8	68.0
$Hg_{0.5}Cd_{0.5}Te$	0.04	2.9	29.0
$Hg_{0.3}Cd_{0.7}Te$	0.067	1.7	17.0

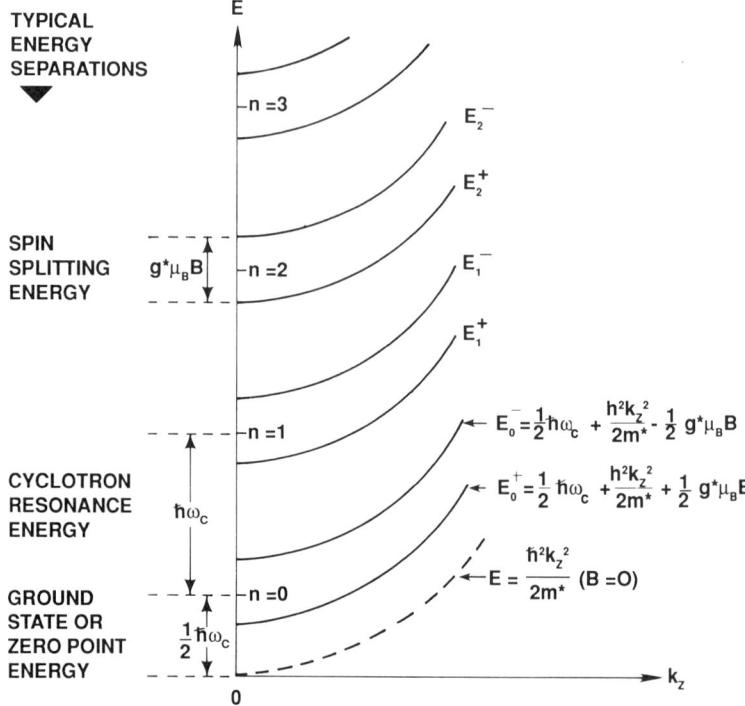

FIG. 1. Landau levels for a simple energy band.

function remains quasicontinuous). It is obvious that at low temperatures, quantization effects are observable even at low magnetic fields. For low-mass HgCdTe samples, quantum effects are still observable at 300 K and low magnetic fields.

Another condition to be satisfied for the observation of the various magneto-oscillatory effects is that the lifetime broadening of the levels $\hbar/\tau$ must be less than the separation of the Landau levels $\hbar\omega_c$; i.e., $\hbar/\tau \ll \hbar\omega_c$ or $\omega_c\tau \gg 1$. This means that the electron must complete at least $1/2\pi$ of an orbit about the field before colliding with the lattice or an impurity. Since $\omega_c\tau \equiv eB\tau/m_c^* = \mu B$, this condition is often expressed in terms of the mobility μ and the field such that $\mu B \gg 1$. Thus at fields of 10.0 T, carrier mobilities must exceed 10^3 cm^2/V-sec; and at 1.0 T they must exceed 10^4 cm^2/V-sec. Values of $\hbar\omega_c$ for various semiconductors at 1.0 T and 10.0 T are given in Table II.

The large changes in the E versus k relation and in the electron wave function when a magnetic field is applied also means large changes in the density of states $[D(E)]$. However, the magnetic field does not alter the total number of states but only alters the distribution of the states. The number of states lying between E and $E + dE$ for a simple conduction band is

$$D(E)\,dE = \frac{1}{4\pi^2}\left(\frac{2m}{\hbar^2}\right)^{3/2} \sum_{n=0}^{n_{\max}} \frac{\hbar\omega_c\,dE}{[E - E_c - (n + 1/2)\hbar\omega_c]^{1/2}}, \quad (2)$$

where the value of $n_{\max}$ is set by the requirement that the denominator be real and E_c is the bottom of the band. This expression reduces to the expression for the density of states in the absence of a magnetic field when B tends to zero, giving the usual result

$$D(E) = \frac{1}{2\pi^2}\left(\frac{2m^*}{\hbar^2}\right)^{3/2} (E - E_c)^{1/2}. \quad (3)$$

The variation of the density of states gives rise to a series of peaks, as shown in Fig. 2, which are easily seen in the spectrum. For large values of B, the function $D(E)$ is zero up to a value of $E - E_c = \hbar\omega_c/2$ and becomes infinite at the points $E - E_c = (n + 1/2)\hbar\omega_c$. However, the energy subbands are usually broadened (Roth et al., 1959), which causes the density of states to be finite at the points of singularity. These periodic increases in the density of states are important for understanding various oscillatory phenomena in a magnetic field.

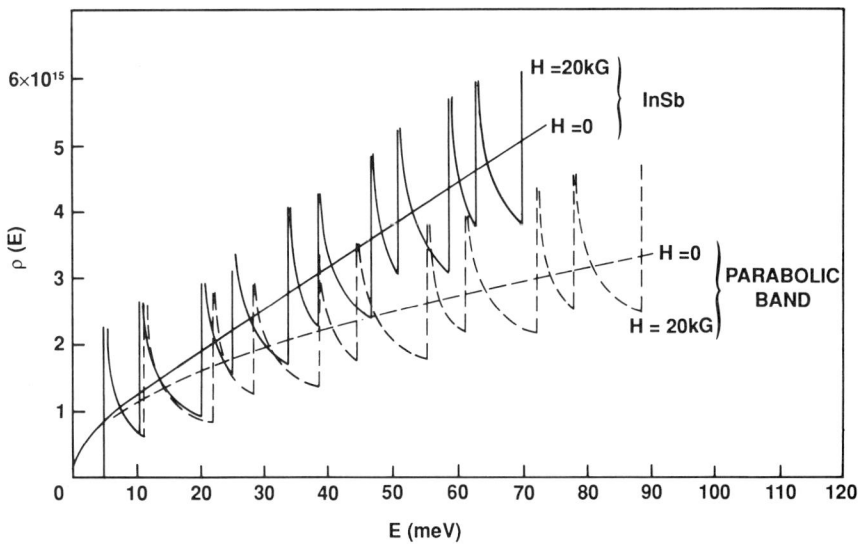

FIG. 2. Density of states in a magnetic field of 2.0 T for InSb.

c. Energy-Band Models for Zinc Blende Semiconductors

The interpretation of any magneto-optical or magnetotransport experiment depends strongly on the model used to describe the energy band structure of a material in the presence of a magnetic field. Fortunately, for zinc blende semiconductors there has been considerable work in this area. There exists a variety of models that can be used to describe the energy-band structure of most NGS to different degrees of accuracy.

In the effective-mass approximation for a spherical, parabolic, nondegenerate energy band, the Landau and spin energies are given in Eq. (1). For cubic semiconductors like InSb, which have zinc blende symmetry, the energy-band extrema near the center of the Brillouin zone are described by one of the three group-theoretical designations Γ_6, Γ_7, or Γ_8, including spin. The conduction band usually has Γ_6 symmetry (S-like, $J = 1/2$, doubly degenerate), the valence band usually Γ_8 (P-like, $J = 3/2$, fourfold degenerate). A second, lower valence band, called the split-off band, has Γ_7 symmetry (P-like, $J = 3/2$, doubly degenerate). The next higher conduction band also consists of another Γ_7, Γ_8 spin-split pair of bands, separated from the lowest conduction band by an energy much larger than that separating the split-off band from the valence band. These bands are illustrated schematically in Fig. 3.

A description similar to Eq. (1) of the degenerate energy band of Γ_8 symmetry gives four sets of levels, of which two are characterized by a light effective mass and two by a heavy effective mass (Luttinger, 1956), due to the different interactions with the other bands. At low quantum numbers n the Landau and spin levels are not uniformly spaced. This is referred to as Luttinger or quantum effects. The Luttinger description in general gives parabolic but nonspherical energy bands for which the magnetic energies depend on the orientation of the magnetic field with respect to the crystal axes.

In narrow-gap semiconductors having InSb-type or inverted (HgTe-type) band structures, the previous descriptions are not adequate, since the energy bands are nonparabolic; that is, the level separations are not linear with magnetic field. The band nonparabolicity in InSb-type materials has been treated in the three-level (3L) model (Bowers and Yafet, 1959), which takes into account the $\mathbf{k} \cdot \mathbf{p}$ interaction between the Γ_6^c, Γ_8^v, and Γ_7^v levels and neglects all other bands. This approach is quite good for the conduction band, the light-hole band (without Luttinger effects), and the spin-orbit split-off band. Using the 3L description, Zawadzki (1963) derived an analytical formula for the energy dependence of the g^*-factor for conduction electrons. Kacman and Zawadzki (1971) developed the 3L model for the inverted band structure of HgTe-type materials and wrote down the resulting wave functions. The energy bands described by the 3L model are spherical.

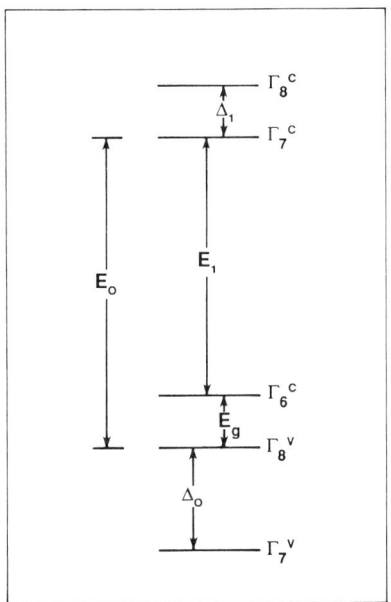

FIG. 3. Important zone-center energy-band positions in a typical zinc blende semiconductor, illustrating the fundamental energy gap E_g, the energy to the highest energy band E_0, and the spin-orbit splittings Δ, Δ_1 of the valence and higher conduction bands, respectively.

The 3L approach results in a cubic equation for the orbital and spin energies. If one is interested in the electron energies satisfying the condition $E \ll E_g + 2\Delta/3$ (where Δ is the spin-orbit energy), the cubic equation can be approximated by a quadratic equation, which can then be solved for the energy eigenvalues in terms of square roots. The following formula has been used often in approximate descriptions of experimental data on narrow-gap materials (cf. Johnson and Dickey, 1970):

$$E(n, k_z, \pm) = -\frac{E_g}{2} + \left[\left(\frac{E_g}{2}\right)^2 + E_g D_{nk_z \pm}\right]^{1/2}, \quad (4)$$

where

$$D_{nk_z \pm} = \frac{\hbar eB}{m_0^*}\left(n + \frac{1}{2}\right) \pm \frac{1}{2}g_0^* \mu_B B + \frac{\hbar^2 k_z^2}{2m_0^*}, \quad (5)$$

in which m_0^* and g_0^* are experimental band-edge values of the effective mass and g-factor.

The 3L model does not describe the band of heavy holes since their large mass comes from the **k** · **p** interaction of the Γ_8^v with the omitted distant bands. This deficiency is removed in the approach of Pidgeon and Brown (1966), who combined the Luttinger and Bowers–Yafet schemes. In other words, the Pidgeon–Brown approach is a 3L model in which the distant bands are incorporated up to the k^2 terms. This corresponds to the Kane model (1957) for the absence of external fields. The bands resulting from this description are both nonparabolic and nonspherical. Mathematically, the Pidgeon–Brown model presents a set of eight differential equations that, under certain conditions, can be factorized into two sets of four equations each and solved in terms of harmonic oscillator functions. The eigenenergy problem reduces then to the diagonalization of two 4×4 matrices. An inclusion of the full bands' anisotropy does not allow the previous factorization and solutions in terms of single-harmonic oscillator functions. A complete version of the Pidegon–Brown approach for **B** $\|$ (001), **B** $\|$ (011), and **B** $\|$ (111) crystal axes has been given by Weiler et al. (1978).

Recently, Pfeffer and Zawadzki (1990) have developed a five-level (5L) model, in which the **k** · **p** interaction between the Γ_7^c, Γ_8^c, Γ_6^c, Γ_8^v, and Γ_7^v levels is taken into account exactly and all other bands are neglected. The energy levels resulting from this model are illustrated schematically in Fig. 3. This description is necessary for medium-gap materials in which the **k** · **p** interaction between the Γ_6^c and Γ_8^v levels is not completely dominant. Within the 5L model the bands' anisotropy is described by a single matrix element of momentum. To solve the eigenenergy problem, one looks for envelope functions in terms of sums of the harmonic oscillator functions, which, according to the procedure first proposed by Evtuhov (1962), leads to a diagonalization of the larger matrices. This approach has been recently used to describe the nonsphericity of the conduction band in GaAs (Pfeffer and Zawadzki, 1990) and InSb (Littler et al., 1990b) (see discussion later in this chapter).

4. Interaction of Laser Radiation with Semiconductors

a. Overview

The interaction of laser radiation with semiconductors is, in general, a complex phenomenon involving many different types of processes occurring before equilibrium is regained. Which particular processes dominate or are important depends upon such parameters as the laser photon energy, laser intensity, laser pulse width, sample temperature, and sample doping. Many of these processes are also important in understanding the interaction of ordinary (nonlaser) electromagnetic radiation with semiconductors. In addition, there are numerous and important semiconductor devices that depend critically upon these processes for their principles of operation. For example, the very large nonlinear optical effects shown by some semiconductors have increased the practical possibility of attaining all-optical switching and signal processing at room temperature for optical computers (Wolfe, 1985). Consequently, this whole area represents a very active and important field of research.

It is well known that illumination of semiconductors with intense laser radiation also leads to carrier heating (Shah, 1978; Moore et al., 1978). Figure 4 shows schematically how photoexcited carriers, created through various absorption processes, can heat the electron gas via electron–electron collisions. For sufficiently low intensities where two-photon or impurity absorption processes can be neglected, free-carrier absorption can take place. There is no increase in the number of carriers n, but their distribution function changes. Electrons excited to high energies by absorption of a photon may undergo energy relaxation through two competing processes: electron-electron scattering and polar-optical phonon emission. This phonon emission transfers the absorbed energy of the photoexcited electrons to the lattice. Electron–Electron scattering, on the other hand, distributes the absorbed photon energy within the electron gas. If this process is sufficiently fast, i.e., if the concentration is high enough, a nonequilibrium carrier distribution will be established, which is characterized by an electron temperature, T_e. It should be noted that the polar-optical phonons emitted by the photoexcited electrons decay to acoustic phonons through a three-phonon interaction and thus have a long lifetime. Some of these optical phonons may therefore be reabsorbed by the electron gas, providing an additional source of heating besides the electron–electron thermalization processes. The ultimate transfer of the absorbed photon energy to the environment surrounding the sample takes place through acoustic phonons, either emitted directly by the gas or created (in pairs) by the decay of optical phonons. But since the rate of emission of acoustic phonons is slow compared with that for emission of optical phonons, this will not become a significant energy loss mechanism until the electron gas has cooled below the point where optical phonon emission can take place. The electron gas will then be cooled by a combination of optical and acoustic phonons. The effects caused by emission of optical phonons prior to thermalization of the photoexcited electrons present additional complications.

Interband and extrinsic absorption processes can also lead to creation of photoexcited carriers and, hence, hot electrons, as discussed previously. In contrast to the free-carrier absorption case however, additional carriers are produced that must then recombine radiatively or nonradiatively. A wide variety of laser-induced heating effects (Moore et al., 1978; Seiler et al., 1978, 1979a,b) and even a laser-induced cooling effect (Seiler and Hanes, 1980; Hanes and Seiler, 1988) in n-InSb have also been observed and investigated.

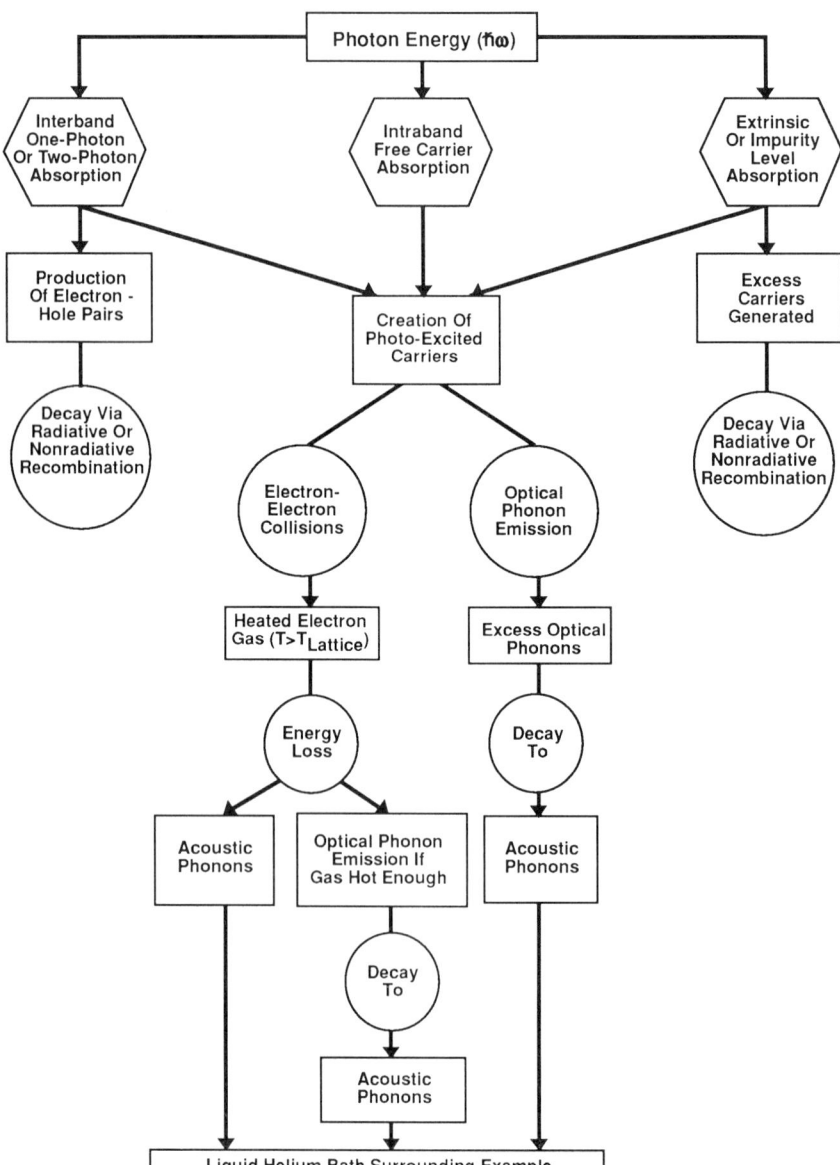

FIG. 4. Photoexcitation, carrier heating, and energy loss processes in semiconductors.

b. *One-Photon Absorption*

The realm of linear optical spectroscopy is quite broad and deals with the various types of absorption processes that can occur by the absorption of only one photon. A tremendous amount of theoretical and experimental information has been accumulated by inducing electronic transitions between energy levels in semiconductors and observing the absorption, emission, or reflection of electromagnetic radiation. (See for example, Bassini and Parravincini, 1975; Pankove, 1971; or Willardson and Beer, 1967.) It is not our purpose to carry out an extensive review but to present a brief overview summarizing the major results.

In general, a large variety of energy states occur in semiconductors, both in the form of bands and in the form of discrete, but usually broadened, energy levels. These include (1) a set of broad valence and conduction electron bands characteristic of pure semiconductors of perfect crystallinity and (2) discrete states that occur in the presence of doped impurities and lattice defects and at low temperatures where excitonic effects become important. There is also an abundance of lattice vibrations that affect the absorption spectrum of a semiconductor. These states are schematically presented in the E versus k diagram of Fig. 5 (Jain and Klein, 1983) for a direct-band-gap semiconductor. The vertical arrows depict the following types of transitions (1) valence- to conduction-band transitions (a) and (a′); (2) free-carrier transitions [intervalence band (b) and (b′), interconduction band (c) and (c′), and intraband (d)]; (3) transitions between the electronic levels of impurities (e); (4) transitions between the impurity levels and the conduction bands (f) and (f′); (5) transitions between the valence bands and impurities (g) and (g′); and (6) transitions between the valence bands and the excitonic levels (h).

The absorption spectrum of a hypothetical direct-band-gap semiconductor with an energy-level diagram similar to that of Fig. 5 is shown in Fig. 6 (Jain and Klein, 1983). This figure depicts qualitatively the spectral location and relative magnitude of the low-intensity absorption coefficients associated with some of these transitions. The value of these coefficients can depend strongly on light intensity (especially when using lasers), lattice temperature, crystal purity, and the application of external parameters such as strain, electric field, and magnetic field.

One-photon transitions between an initial state $|i\rangle$ and the final state $|f\rangle$ are described by Fermi's golden rule, giving the transition probability $\omega_{i \to f}$ per unit time that a perturbation of the form $H_R e^{-i\omega t}$ induces a transition:

$$\omega_{i \to f} = \frac{2\pi}{\hbar} |\langle f|H_R|i\rangle|^2 \, \delta(E_f - E_i - \hbar\omega), \qquad (6)$$

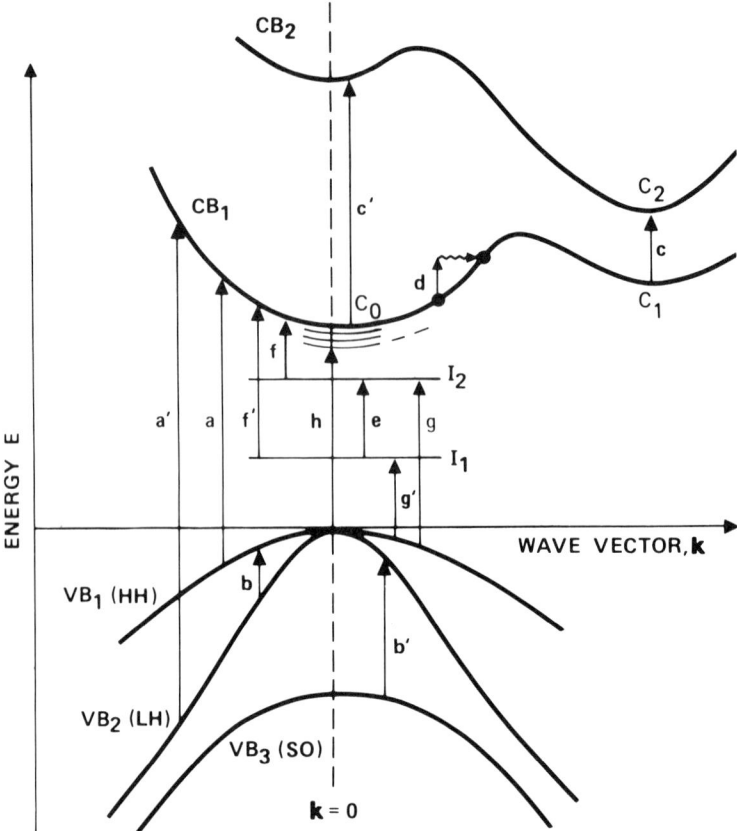

FIG. 5. Schematic representation of the most prominent types of transitions present in a semiconductor. CB_i are the conduction bands and VB_i are the valence bands, with HH, LH, and SO the heavy-hole, light-hole, and split-off bands, respectively. The vertical arrows indicate a variety of transitions that may be induced by optical radiation (From Jain and Klein, 1983.)

where $H_R = (e/m_c)\mathbf{A} \cdot \mathbf{P}$ and $\mathbf{A}$ is the vector potential. Thus, spectroscopy can be carried out such that when $\hbar\omega = E_f - E_i$, a resonant transition occurs. As seen in Fig. 5 and 6, the optical properties of semiconductors are strongly dominated by the form of the valence and conduction bands, as well as the energy gap separating them. The electron energies and wave functions of these bands can be calculated by various approximate methods. It then becomes possible to calculate the total transition rate per unit volume, ω_T, by integrating over all possible vertical transitions in the first Brillouin zone taking into account all contributing bands. Finally, the absorption coefficient α, as defined through Beer's law dependence of intensity on propagation

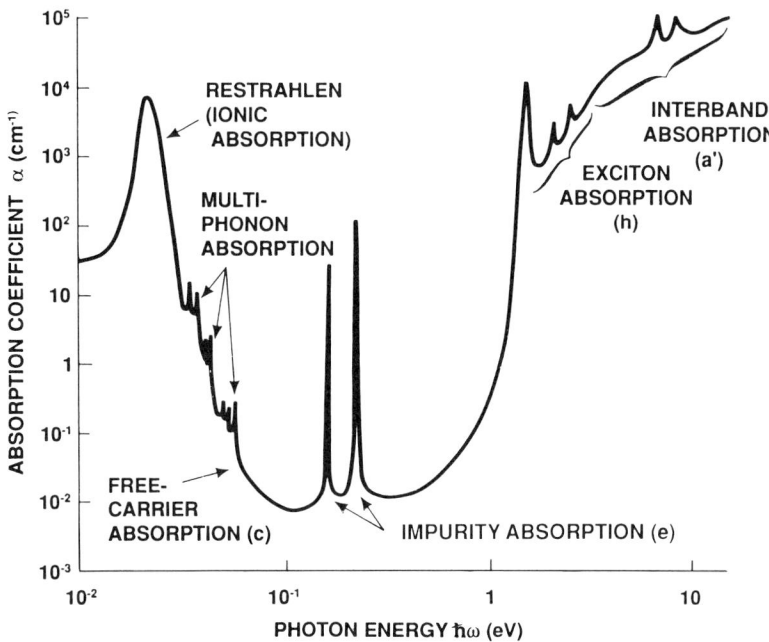

FIG. 6. Tyical low-intensity absorption spectrum for a hypothetical direct-band-gap semiconductor ($E_g \simeq 1$ eV) whose band structure is represented schematically in Fig. 5. The semiconductor is assumed to be at low temperatures and doped with deep-level impurities. The letters in parentheses designate transitions shown in Fig. 5. (From Jain and Klein, 1983.)

distance z, is

$$\alpha = I^{-1}\frac{dI}{dz} = \frac{\hbar\omega}{I}\omega_T. \tag{7}$$

In many cases there are important corrections to the simple interpretation of optical transitions as occurring between two of the electronic states described earlier. One such correction is due to what are called polaron effects: The interaction between an electron and the lattice in a polar material can be shown, to the lowest order, to make its effective mass larger. The corrected effective mass, both without (Fröhlich, 1954; Lee et al., 1953) and with a magnetic field (Larsen, 1964), is given by $m^* = m_0^*(1 - \alpha/6)$, where α is the polaron coupling constant determined by the strength of the electron–phonon interaction. For weakly polar materials like InSb, this coupling is very small, with α estimated to be about 0.02 (Dickey et al., 1967), so the correction to the effective mass is very slight. However, the effect of this interaction can be seen in magneto-optical transitions, where either the initial or final state is separated from another Landau level by an optical phonon energy (Johnson and Larsen, 1966; Dickey et al., 1967; Summers et al., 1968; Dickey and Larsen, 1968; Kaplan and Wallis, 1968; Kaplan et al., 1972). The interaction causes an anticrossing or splitting of the transitions into branches, with a slight offset between the lower-energy branch and an extrapolation of the higher-energy branch. Another effect resulting from the electron–phonon interaction is the occurrence of the phonon-assisted transitions, which are transitions accompanied by the emission or the absorption of a phonon. These have been observed by a number of investigators and will be discussed in detail later.

A second correction is required for the interpretation of interband magneto-optical transitions. These transitions consist of the simultaneous photoexcitation of a hole and an electron; in the final state these may be free or bound in a hydrogen-like state called an exciton. For semiconductors like InSb and HgCdTe, which have narrow gaps, small effective masses, and large dielectric constants, the exciton binding energy is very small, and only in a few instances has there been seen an observable separation between transitions to the bound and free states (see, for example, Ivanov-Omskii et al., 1983a,b; Seiler et al., 1986a; Seisyan and Yuldashev, 1988, as well as Fig. 16 and corresponding discussion). In addition, Kanskaya et al. (1979) reported discrete structure in the absorption edge spectra of pure InSb crystals at 2 K when special measures were taken to eliminate the influence of bound states. The absorption peak at zero magnetic field was attributed to the 1S exciton state. Finally, we point out that an electron–hole liquid has even been reported for InSb by Kalugina and Skok (1984).

Very recent studies by Grober and Drew (1991) have provided a new physical basis reinterpreting the data of Ivanov-Omskii et al. (1983a,b) and Seisyan and Yuldashev (1988). The excitation levels used by Grober and Drew were 300–5000 times smaller than used previously. From a photoluminescence (PL) intensity dependence analysis, PL line shape analysis, and carrier density profile analysis, they concluded that the photoexcited system exists as an electron–hole plasma. They also correlated the intensity-dependent shift of the band edge PL with a universal model for band-gap renormalization proposed by Vashishta and Kalia (1982). At a temperature of 2 K, they extracted a zero-density band gap of 235.3 meV, in excellent agreement with the magneto-optical value of 235.2 meV obtained by two-photon measurements (Littler et al., 1985).

It has been shown for InSb that these exciton effects must be taken into account when fitting both interband and intraband data in order to explain both types of transitions using a single set of energy-band parameters (Weiler, 1979; Goodwin and Seiler, 1983; Littler et al., 1983). A simple approximation for the binding energy for an exciton in a magnetic field B with Landau quantum number n is (Weiler, 1979)

$$E_b(n, B) = 1.6R \left[\frac{\gamma}{2n + 1}\right]^{1/3}, \tag{8}$$

where R is the effective Rydberg $R_0(\mu/m)/\kappa_e^2$, $R_0 = 13.6$ eV, μ is the reduced effective mass for the transition, corrected for nonparabolic effects, κ_e is the dielectric constant, and $\gamma = \hbar eB/2\mu R$.

Other effects may be important when the sample contains impurities. First, transitions involving the donor or acceptor states themselves may be observed, as will be discussed later. Second, collective excitations of the free carriers, called plasmons, can in some circumstances affect the observed magneto-optical spectra. Most of the experiments discussed in this chapter were made on samples with low enough impurity concentrations so that plasmon effects were negligible.

c. One-Photon Interband Magnetoabsorption

The investigation of the properties of electrons and holes in semiconductors has been greatly enriched by the use of magnetic fields. As discussed in the introduction, the presence of a magnetic field influences the motion of electrons in a semiconductor. This is manifested through quantizing the allowed energy states of the electron into subbands known as Landau levels. Over the past several decades, magneto-optical studies of semiconductors have been useful in determining both energy-band parameters and charge-carrier effective masses. This is because sharp optical transitions can be observed between the magnetic-field-quantized Landau levels, thus providing independent measurements of the properties of the charge carriers in both the initial and final states of the transition.

Transitions from Landau-level states in the valence band to states in the conduction band of a semiconductor are known as interband transitions. Interband magnetoabsorption transitions are important to study because they provide much information about the fundamental parameters of a semiconductor that affect its electrical characteristics, such as the energy gap E_g and the effective masses of the electrons and holes.

The strongest allowed transitions for both intraband and interband absorption are those that are proportional to the interband matrix element E_p. The transition energies can be calculated directly from a knowledge of the selection rules for interband transitions and one of a variety of energy-band models. The selection rules for intraband and interband transitions are given by (Weiler et al., 1978)

$$\sigma_R: \quad a(n) \to a(n-1), \quad b(n) \to b(n-1),$$
$$\sigma_L: \quad a(n) \to a(n+1), \quad b(n) \to b(n+1), \qquad (9)$$
$$\pi: \quad a(n) \to b(n+1), \quad b(n) \to a(n-1),$$

where σ_R, σ_L, and π correspond to the light polarizations right circular, left circular, and the linear polarization $\mathbf{e} \parallel \mathbf{B}$. The a and b terms denote the spin state of each Landau level, where a denotes spin-up and b spin-down. Finally, the Landau-level quantum number is denoted by n. In some of the early work (Roth et al., 1959: Pidgeon and Brown, 1966) the selection rules for interband transitions are given as $n \to n - 2$ for σ_R and $n \to n$ for σ_L. In this case the valence states had been renumbered so that $n = -1$ became $n = 0$, etc. This made the label n correspond to the Landau-level quantum number of one of the larger terms for both the conduction and valence bands. The current practice is not to renumber the valence-band states, so all states with the same label n are coupled in the 3L model. This also keeps the selection rules the same for both intraband and interband states.

d. Two-Photon Absorption

The basic theory of two-photon absorption (TPA) processes in atomic systems was first formulated by Maria Goppert-Mayer (1931). Besides TPA processes, her theory includes other two-photon processes such as two-photon emission and absorption–emission processes such as Raman and Rayleigh scattering. Using second-order perturbation theory and the dipole approximation, the two-photon transition probability $W^{(2)}_{i \to f}$ per unit time that a perturbation of the form

$$H_{R1}e^{-i\omega_1 t} + H_{R2}e^{-i\omega_2 t} \qquad (10)$$

induces a transition from the initial state $|i\rangle$ to the final state $|f\rangle$ can be written as

$$W^{(2)}_{i \to f} = \frac{2\pi}{\hbar} \left| \sum_\gamma \left[\frac{\langle f|H_{R2}|\gamma\rangle\langle\gamma|H_{R1}|i\rangle}{E_\gamma - E_i - \hbar\omega_1} + \frac{\langle f|H_{R1}|\gamma\rangle\langle\gamma|H_{R2}|i\rangle}{E_\gamma - E_i - \hbar\omega_2} \right] \right|^2$$
$$\times \delta(E_f - E_i - \hbar\omega_1 - \hbar\omega_2). \qquad (11)$$

This describes the simultaneous absorption of two photons. Note that energy does not have to be conserved for the transitions to the intermediate virtual states γ. The delta function shows the energy conservation for the total process, where the sum of two photons now connects the initial and final states. Once again group theory is important for describing the selection rules for the various possible TPA transitions. The sum over all intermediate states γ is an important distinction that does not exist for the case of one-photon absorption. The interaction Hamiltonian $H_{R1}(H_{R2})$ is given by $(e/mc)\mathbf{A} \cdot \mathbf{p}$, where $\mathbf{A}$ is the vector potential associated with the incident radiation and $\mathbf{p}$ is the momentum operator for the electron.

The transition probabilities depend upon the light polarization. From the usual dipole selection rules it is seen that the parity of final states in TPA is the same as that of the ground state, in contrast to the selection rules for single-photon absorption. There is a wealth of information obtainable by varying the polarizations of the two beams with respect to one another and the crystal axes. There have been several calculations and tabulations of the form of these angular dependences in order that they might be readily available for the analysis of experimental data. Inoue and Toyozawa (1965) gave the angular dependence of two-photon transitions in which either the initial or final state transforms according to the totally symmetric representation of the point group. This work was subsequently extended by Bader and Gold (1968) to (1) the allowed transitions between states belonging to all irreducible representations of the point group and (2) the double-group representations encountered when spin-orbit coupling is included.

The tensor character of $W^{(2)}$ can be seen rather explicitly in an alternative derivation of β from solving Maxwell's equations by treating TPA as a nonlinear polarization source. The TPA coefficient can be expressed in terms of a third-order electric dipole susceptibility tensor $\chi^{(3)}$ by solving the wave equation using the slowly varying amplitude approximation. The explicit expression for the two-photon absorption coefficient β is related to the imaginary part of $\chi^{(3)}$, which depends upon the crystal class and laser electric-field direction. For example, in crystals with $\bar{4}3m$ symmetry (e.g., $Hg_{1-x}Cd_xTe$, GaAs, InSb, etc.) and electric field along the [001] direction (Betchel and Smith, 1976),

$$\beta = \frac{32\pi^2\omega}{n^2c^2}[3\,\mathrm{Im}\,\chi^{(3)}_{1111}(-\omega,\omega,\omega,-\omega)], \tag{12}$$

where the convention used is that of Maker and Terhune (1965).

In noncentrosymmetric media there are electric-field directions for which terms proportional to $\chi^{(2)}$ (the second-order susceptibility tensor) lead to absorption (i.e., direct absorption of the second harmonic). However, with special orientations of the sample, second-harmonic generation is symmetry forbidden and in a measurement of TPA these sample orientations should be used to eliminate these competing absorption processes.

Since $\chi^{(3)}$ is a second-rank tensor there are, in general, nine terms contributing to β with magnitudes that vary with orientation. However, in most systems the crystal symmetry is such that there are relations between certain of the terms. For example, in crystals with $\overline{4}3m$ symmetry there are three possible values of χ for each β, and for crystals with $m3m$ symmetry there are four possible values. These can, in general, be measured by using both linearly and circularly polarized light. It is not always possible to sort out all the different TPA spectra simply by changing the sample orientation or the light polarization because of the competing process of absorption by second-harmonic-produced light.

The TPA coefficient β, the quantity usually measured in an experiment, is related to the two-photon transition rate per unit volume, $W^{(2)}$, by the expression

$$\beta = 2\hbar\omega \frac{W^{(2)}}{I^2} \qquad (13)$$

for the case of two-photon absorption from only one beam. Consequently, the calculation of reliable numerical values for β is conceptually easy but practically difficult since it requires knowledge of the interaction Hamiltonian matrix elements among all the eigenstates of the crystal and summations over all the energy bands. In the past, many approximations and simplifying assumptions regarding the energy bands, momentum matrix elements, and intermediate states were made to reduce the complexity of Eq. (13). Unfortunately, many of the approximations were not justified and the calculated values of the TPA coefficient β were wrong. The work by Pidgeon and co-workers (Pidgeon *et al.*, 1979; Pidgeon, 1981; Johnston *et al.*, 1980) and Weiler (1981) resolved this controversy, except for the role of excitons in the TPA spectra.

Perturbation theory has been applied to crystalline solids where Bloch functions ψ are used to describe the electrons in the conduction and valence bands (see the aforementioned work by Pidgeon *et al.*). The result for the transition probability rate per unit volume of a direct electronic transition from an initial valence band v to a final conduction band c, accompanied by the simultaneous absorption of two photons, each of frequency ω can be written as

$$W^{(2)} = \frac{2\pi}{\hbar} \int \left| \frac{\sum_\gamma \langle \psi_c | H | \psi_\gamma \rangle \langle \psi_\gamma | H | \psi_v \rangle}{E_\gamma - E_v - \hbar\omega} \right|^2$$

$$\times \delta[E_c(\mathbf{k}) - E_v(\mathbf{k}) - 2\hbar\omega] \frac{d^3\mathbf{k}}{(2\pi)^3}. \qquad (14)$$

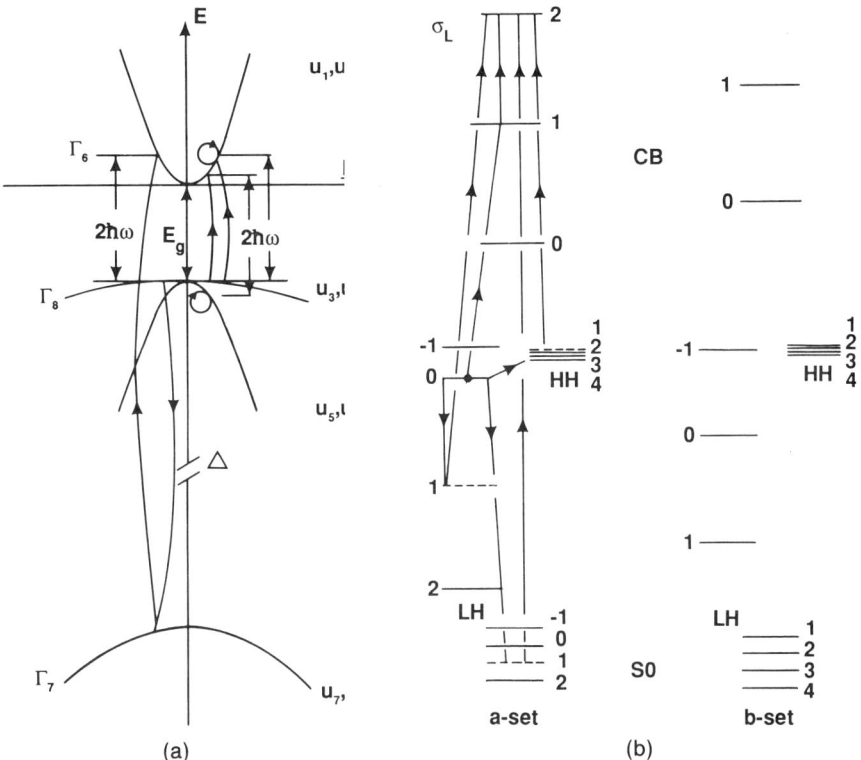

FIG. 7. (a) Conduction and valence bands of InSb near **k** = 0. Typical two-photon transitions are shown. (From Pidgeon et al., 1979.) (b) Splittings of the bands into Landau levels. The a-set are the spin-up levels, and the b-set are the spin-down levels. A two-photon transition $[a^+(0) \to a^c(2)]$ is shown with possible intermediate states enunerated (—•—, initial state, ---, intermediate state, ——— final state.) (From Seiler et al., 1983.)

The summation γ is over all the energy bands, and the energy bands require an integration of the density of states in **k**-space over the entire first Brillouin zone. Figure 7a shows typical two-photon transitions in the 3L model; Figure 7b shows a set of TPMA transitions between the Landau levels in the valence band and conduction band for the $a^+(0)$ and $a^c(2)$ levels. Note the inter–intraband and the intra–interband nature of the transitions. There are two types of processes to be considered: one involves a valence- to conduction-band transition plus either an intraconduction- or intravalence-band transition. The second type involves a valence- to conduction-band transition plus a second interband transition. In all there are 12 final states to be considered, each involving summation over intermediate transitions of these two types. Detailed calculations show that the frequency or photon energy dependence of β for a narrow-gap semiconductor like InSb or HgCdTe is then given by (Weiler et al., 1981)

$$\beta = \left(\frac{4\pi e^4 P}{\hbar n^2 c^2 E_g^3}\right) F(\alpha),$$

where

$$F(\alpha) = \frac{(2\alpha - 1)^{3/2}}{3\alpha^3}\left[\frac{4(3\alpha)^{1/2}}{(3\alpha - 1)^2} + \left(\frac{3\alpha + 3}{2}\right)^{3/2}\left(\frac{9\alpha^4 + 10\alpha^2 + 6}{90\alpha^5}\right)\right]. \quad (15)$$

The function F (whose exact form depends on the band structure) is only a function of the ratio $\alpha = \hbar\omega/E_g$, which determines the states that are optically coupled. Exact, nonparabolic matrix elements were used in this derivation of β. Exciton enhancement effects are not included here because there is, as yet, no clear understanding of their role in TPA processes for NGS. All the TPA experiments on InSb and HgCdTe reported here have been satisfactorily explained without considering excitons. For larger-gap materials like GaAs, two-photon magnetoabsorption experiments must account for the 2P final-state excitons (Neumann et al., 1988).

An alternative to the perturbation theory of TPA was developed by Keldysh (1965), which treated any order of multiphoton absorption and is called the tunneling theory because it reduces for low frequencies and high intensities to the standard theory of interband or Zener tunneling in large electric fields. This theory, as later corrected (Bychkov and Dykhne, 1970), reduced to perturbation theory for photon energies near a submultiple of the energy gap. However, as shown by Weiler (1981), the tunneling theory is less accurate than perturbation theory for transitions well above the energy gap, as is common for narrow-gap semiconductors, because of certain approximations made in calculating the transition matrix elements.

The general case of nonlinear absorption is very complex where I and Δn or Δp are nonlinear functions of time and position. Here, we consider several limiting cases that result in analytical solutions that give physical insight to the TPA processes. These cases are often encountered in the TPA literature.

Neglecting diffusion, excess free-electron or hole absorption, and Auger recombination, one can write the following rate equation for the density of photoinduced electrons (Δn):

$$\frac{d\Delta n}{dt} = \frac{\alpha_e I}{\hbar\omega} + \frac{\beta I^2}{2\hbar\omega} - r_1 \Delta n - r_2 \Delta n(n_0 + \Delta n), \quad (16)$$

where $\Delta p \simeq \Delta n$ and r_2 is the bimolecular or radiative recombination rate. The linear absorption coefficient α_e takes into account electron transitions from impurity or defect levels, with r_1 as the corresponding recombination rate. Here we remind the reader that the reflectivity of the samples should also be taken into account, although, for simplicity, we do not include it here. For low intensities not dominated by TPA processes, the steady-state $(d\Delta n/dt = 0)$ value for Δn is linear in intensity:

$$\Delta n = \left(\frac{\alpha_e}{\hbar\omega r_1}\right) I. \tag{17}$$

However, for higher intensities dominated by TPA $(\alpha_e I/\hbar\omega \ll \beta I^2/2\hbar\omega)$ there are two regions to consider:

(1) Low-density region where $r_1 \Delta n \gg r_2(n_0 + \Delta n)\Delta n$. Here

$$\frac{d\Delta n}{dt} = \frac{\beta I^2}{2\hbar\omega} - r_1 \Delta n. \tag{18}$$

If a long pulse is used so that steady state can be achieved,

$$\Delta n = \left(\frac{\beta}{2\hbar\omega r_1}\right) I^2. \tag{18}$$

Alternatively, for short pulses with durations Δt much less than the decay times $(1/r_1)$

$$\frac{d\Delta n}{dt} = \frac{\beta I^2}{\hbar\omega}$$

or

$$\Delta n = \left(\frac{\beta}{2\hbar\omega}\right)\Delta t \, I^2. \tag{19}$$

In either case, Δn is proportion to I^2.

(2) Higher-density region where $r_1 \Delta n \ll r_2(n_0 + \Delta n)\Delta n$. Here

$$\frac{d\Delta n}{dt} = \frac{\beta I^2}{2\hbar\omega} - r_2(n_0 + \Delta n)\Delta n, \tag{20}$$

and since bimolecular recombination at high densities is fast, $d\Delta n/dt = 0$ can be assumed. For $\Delta n \ll n_0$ we get

$$\Delta n = \left(\frac{\beta}{2\hbar\omega n_0 r_2}\right) I^2, \tag{21}$$

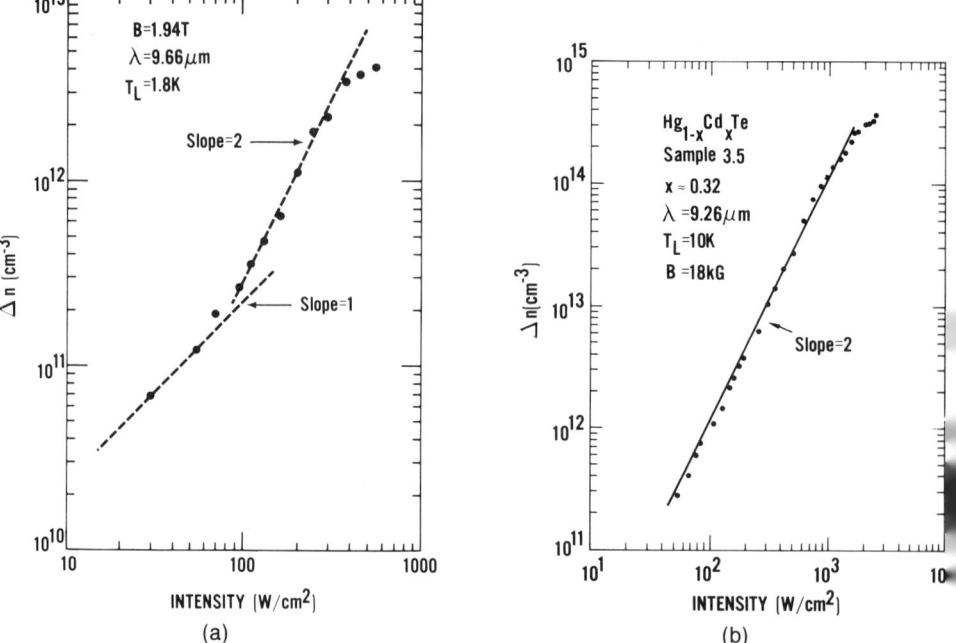

FIG. 8. (a) Intensity dependence of Δn in InSb for a TPMA transition. Three regimes are clearly seen: (1) Δn depends linearly on intensity; (2) Δn is proportional to I^2, and (3) the beginnings of saturation of Δn at the highest laser intensities. (From Seiler and Littler, 1988a.) (b) Intensity dependence of Δn in HgCdTe for TPA. Δn is proportional to I^2. (From Seiler et al., 1986c.)

which is quadratic in I, and for $\Delta n \gg n_0$,

$$\Delta n = \left(\frac{\beta}{2\hbar\omega r_2}\right)^{1/2} I, \tag{22}$$

which is linear in intensity. Thus, depending upon the recombination mechanism and the intensity region, both linear and quadratic intensity dependences of Δn can be obtained. The recombination rates may also depend on intensity.

Examples of two-photon dominated carrier generation can be seen in Fig. 8a and 8b for n-InSb and n-Hg$_{0.68}$Cd$_{0.32}$Te, respectively. In both cases the photo-Hall effect was used to directly measure the number of photoexcited electrons generated by the optical absorption processes. A number of important observations should be noted: (1) The photo-Hall method can be used to detect an extremely small number of photoexcited electrons, $\simeq 7 \times 10^{10}$ cm^{-3} for n-InSb and $\simeq 3 \times 10^{11}$ cm^{-3} for n-HgCdTe. (2) At sufficiently low intensities (< 100 W/cm^2), a linear dependence of the number of photoexcited electrons Δn on intensity I, as predicted by Eq. (17), is seen for n-InSb. This is due to optical absorption resulting from photoexcitation of electrons from impurity or defect levels to the conduction band. (3) At higher intensities, a quadratic dependence of Δn on I is seen for both InSb and HgCdTe, in agreement with the predictions of Eq. (18) or (21) for two-photon dominated absorption processes. In fact, for HgCdTe the quadratic dependence extends over a range of three orders of magnitude. If time-resolved measurements are carried out to determine carrier recombination rates, the TPA coefficient β can be determined (see, for example, Seiler et al., 1986b). (4) At the highest intensities shown, deviations from quadratic behavior occur because of either intensity-dependent carrier lifetimes or TPA coefficients in combination with the high-intensity properties of the rate equations.

The situation after the laser pulse is cut off (at $t = 0$) is given by

$$\frac{d \Delta n}{dt} = -r_1 \Delta n - r_2 \Delta n (n_0 + \Delta n), \tag{23}$$

integration of which gives

$$\Delta n(t) = \frac{(r_1 + r_2 n_0) \Delta n(t = 0)}{[r_1 + r_2 n_0 + r_2 \Delta n(t = 0)] \exp\{(r_1 + r_2 n_0)t\} - r_2 \Delta n(t = 0)}. \tag{24}$$

Thus a hyperbolic decay, characteristic of a quadratic recombination law, can be observed initially after the end of the light pulse (e.g., Schneider et al., 1976). For longer times such that $(r_1 + r_2 n_0)t \to \infty$, we get an exponential decay

$$\Delta n(t) \to \frac{(r_1 + r_2 n_0) \Delta n(t = 0)}{r_1 + r_2 n_0 + r_2 \Delta n(t = 0)} \exp[-(r_1 + r_2 n_0)t]. \tag{25}$$

Alternatively, if $n_0 \gg \Delta n(t = 0)$, the case for small TPA excitation, then

$$\Delta n(t) = \Delta n(t = 0) \exp[-(r_1 + r_2 n_0)t]. \tag{26}$$

e. *Two-Photon Magnetoabsorption*

In order to calculate TPMA interband transition energies in a magnetic field (transitions between Landau levels), both TPMA selection rules and Landau-level energies must be known. Most of the observed TPMA transitions in InSb and HgCdTe can be explained by the usual spherical selection

TABLE III

Selection Rules for Two-Photon Transitions in Zinc Blende Semiconductors for a Magnetic Field $\mathbf{B} \parallel [001]$, $[110]$, and $[111]$ Crystal Axes, both Allowed (A) and Induced by Warping (W) and Inversion Asymmetry (I).

Polarization			$\mathbf{B} \parallel [001]$						$\mathbf{B} \parallel [110]$					$\mathbf{B} \parallel [111]$			
			A		W		I		W		I		W		I	
			Δs	Δn	Δs	Δn	Δs	Δn	Δs	Δn	Δs	Δn	Δs	Δn	Δs	Δn
$\mathbf{E} \perp \mathbf{B}$	σ_L		0	+2	0	$-2, +6$	-1	$+1, +5$	0	$0, -2, +4, +6$	0	$\pm 1, +3, +5$	-1	$0, +6$	0	$-1, +5$
							$+1$	$-1, +3$					$+1$	$-2, +4$	-1	$+3$
															$+1$	$+1$
	σ_R		0	-2	0	$+2, -6$	-1	$+1, -3$	0	$0, +2, -4, -6$	0	$\pm 1, -3, -5$	-1	$+2, -4$	0	$+1, -5$
							$+1$	$-1, -5$					$+1$	$0, -6$	-1	-1
															$+1$	-3
	σ		0	0	0	± 4	-1	$\pm 1, \pm 3$	0	$\pm 2, \pm 4$	0	$\pm 1, \pm 3$	-1	$-2, +4$	0	± 3
							$+1$	$\pm 1, \pm 3$					$+1$	$+2, -4$	-1	$+1$
															$+1$	-1
$\mathbf{E} \parallel \mathbf{B}$	π		0	0	0	± 4	-1	$\pm 1, \pm 3$	0	$\pm 2, \pm 4$	0	$\pm 1, \pm 3$	-1	$-2, +4$	0	± 3
							$+1$	$\pm 1, \pm 3$					$+1$	$+2, -4$	-1	$+1$
															$+1$	-1

(From Goodwin et al., 1982.)

rules ($\Delta n = 0, \pm 2, \Delta s = 0$). The full set of TPA selection rules for zinc blende semiconductors like InSb or HgCdTe in a magnetic field are given in Table III. These selection rules are different for the different light polarizations. The Landau-level energies can then be calculated by using an appropriate energy-band model, and then the TPMA transition energies can be calculated.

The first observation of the interband two-photon absorption in PbTe and InSb (Button *et al.*, 1966) was not understood theoretically at the time. As discussed earlier, a two-photon excitation can be described by the second-order matrix element $\langle f|\boldsymbol{\alpha}_2 \cdot \mathbf{p}|1\rangle\langle 1|\boldsymbol{\alpha}_1 \cdot \mathbf{p}|i\rangle$, where $|i\rangle$ is the initial electron state in the valence band, $|1\rangle$ is the intermediate electron state, which can be either in the conduction or the valence band, and $|f\rangle$ is the final conduction-band electron state. The symbols $\boldsymbol{\alpha}_1$ and $\boldsymbol{\alpha}_2$ are the photon polarizations and $\mathbf{p}$ is the electron momentum number. According to the theories that existed at the time (Zawadzki, private communication), the intermediate states $|1\rangle$ were assumed to be located in a band different from the $|i\rangle$ and $|f\rangle$ bands. It follows that, for a transition of opposite parity between the valence and conduction bands, the intermediate state has to have the same parity as either the initial or the final state, so either the $\langle f|\boldsymbol{\alpha}_2 \cdot \mathbf{p}|1\rangle$ or $\langle 1|\boldsymbol{\alpha}_1 \cdot \mathbf{p}|i\rangle$ matrix element vanishes. Thus, according to this picture, if one-photon transitions between two bands are possible, two-photon absorption between the same bands should not be possible. This confusion was revolved by Zawadzki *et al.* (1967), who observed that, in the presence of a magnetic field, a different Landau level in either the conduction or valence band can also serve as an intermediate state. Thus the two-photon interband transitions have either intra–inter or inter–intra character, in which the intraband part of the total matrix element is the same as for cyclotron-resonance absorption or emission. This perturbation theory approach was further elaborated by Bassani and Girlanda (1970), Nguyen *et al.* (1971), Girlanda (1971), Chaikovskii *et al.* (1972), Bassani (1972), and Bassani and Baldereschi (1973). Hassan (1976a,b) used third-order perturbation to develop the theory of phonon-assisted TPMA processes. There has, to date, been no experimental confirmation of this process. This perturbation theory approach was further used by Zawadzki and Wlasak (1976) (see also discussion in Girlanda, 1977; Weiler, 1982) to derive selection rules and show that most of the intraband–interband (two-photon) resonances observed by various authors for various polarizations can be satisfactorily explained. Figure 7b shows a set of TPMA transitions between the Landau levels in the valence band and conduction band for the $a^+(0) \to a^c(2)$ levels. Note the inter–intraband and the intra–interband nature of the transitions. In principle, other transitions do occur because of band warping, inversion asymmetry, and combinations of these effects.

The tunneling theory of multiphoton absorption was extended to the magnetic field case by Weiler *et al.* (1968); the results appeared to give selection rules different from those derived from perturbation theory. This was later shown to be incorrect (Weiler, 1973); when corrections similar to those of Bychkov and Dykhne (1970) were applied, the results reduced to those of perturbation theory. As discussed for the zero-magnetic-field case, the tunneling theory is less accurate than the perturbation theory for transitions significantly above the energy gap.

Recent theoretical two-photon work in a magnetic field by de Salvo et al. (1985) showed that, in the presence of a static magnetic field, the electronic states in the effective-mass approximation are described by a Hamiltonian containing a nonlocal potential. As a consequence, the electron interactions described using the effective-mass approximation and the radiation fields have to be modified similarly to the case of optical transitions to exciton states in semiconductors. The main difference from the traditional Hamiltonian lies in an effective-mass correction in the interband matrix elements and an additive term proportional to the square of the radiation vector potential.

The effect of external crossed (Hassan and Moussa, 1977) and parallel (Katana, 1983) electric and magnetic fields on the two-photon magnetoabsorption spectra has also been calculated. For crossed fields, the TPMA constant near each Landau level must have an oscillatory behavior due to the presence of discrete Stark levels.

5. Examples of Magnetoabsorption in InSb

An illustration of the wide variety of laser-induced magneto-optical phenomena that are observable in InSb is shown in Figs. 9 and 10 for a number of experimental conditions. In Fig. 9, we show the different spectra seen in n-InSb as a function of externally controlled parameters. In Fig. 9a we see how the spectra change as a function of laser wavelength and applied current or electric field. At $\lambda = 5.22$ μm (obtained from a cw CO laser) OPMA structure is observed (Seiler et al., 1986a), where the two low-field peaks correspond to free-exciton OPMA transitions and the high-field structure is due to bound exciton magnetoabsorption. At 5.63 μm, OPMA is no longer possible, and a series of resonances appear that represent transitions from a deep acceptor level to the conduction-band Landau levels. In this case the acceptor level seen here is the same neutral acceptor participating in the bound-exciton complex. As the laser wavelength is again increased (to 9.20 μm using a cw CO_2 laser), TPMA transitions are clearly seen. In fact, at these CO_2 laser wavelengths, at least three absorption mechanisms are present and compete; the dominant mechanism depends on the laser wavelength and intensity, sample temperature, and bias current. At 9.20 μm and moderate CO_2 laser intensities, TPMA dominates over the weaker impurity- to conduction-band absorption. To delineate these, we must either go to longer wavelengths (as illustrated for $\lambda = 10.65$ μm), where TPMA is no longer energetically possible at these temperatures, or to lower intensities, as seen in Fig. 9b for $\lambda = 9.56$ μm. At $\lambda = 10.65$ μm and $I = 1$ mA in Fig. 9a, a new set of resonances appear that represent unresolved intraconduction-band transitions and transitions from a shallow donor level to the conduction-band Landau levels. At higher fields these transitions become resolved (see Fig. 30). If the current is

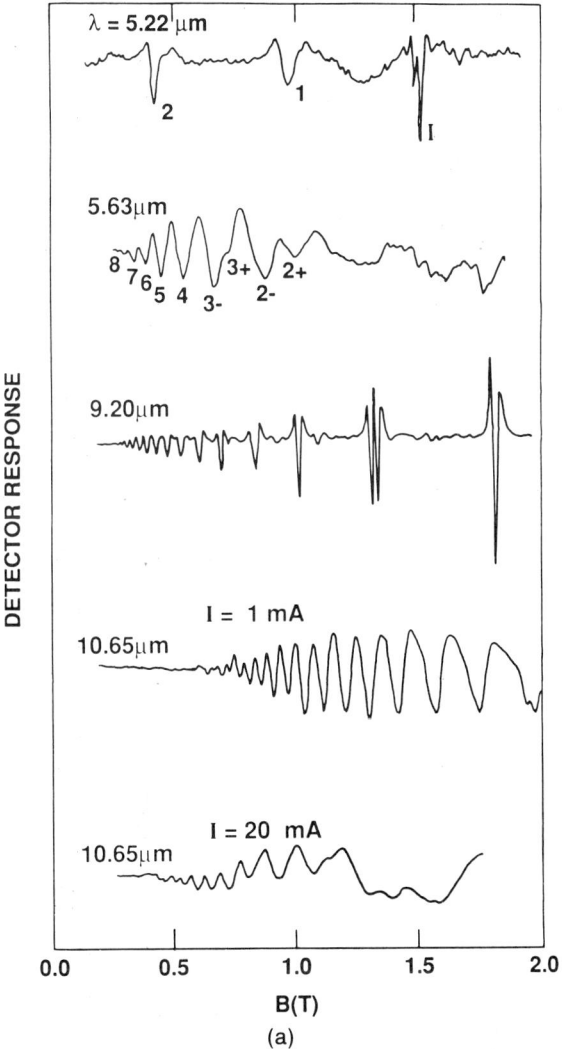

FIG. 9. Magneto-optical effects seen in n-InSb under a variety of conditions. (9e reprinted with permission from Seiler, D. G., and McClure, S. W., *Solid State Comm.* **47**, 17, © 1983 Pergamon Press plc.)

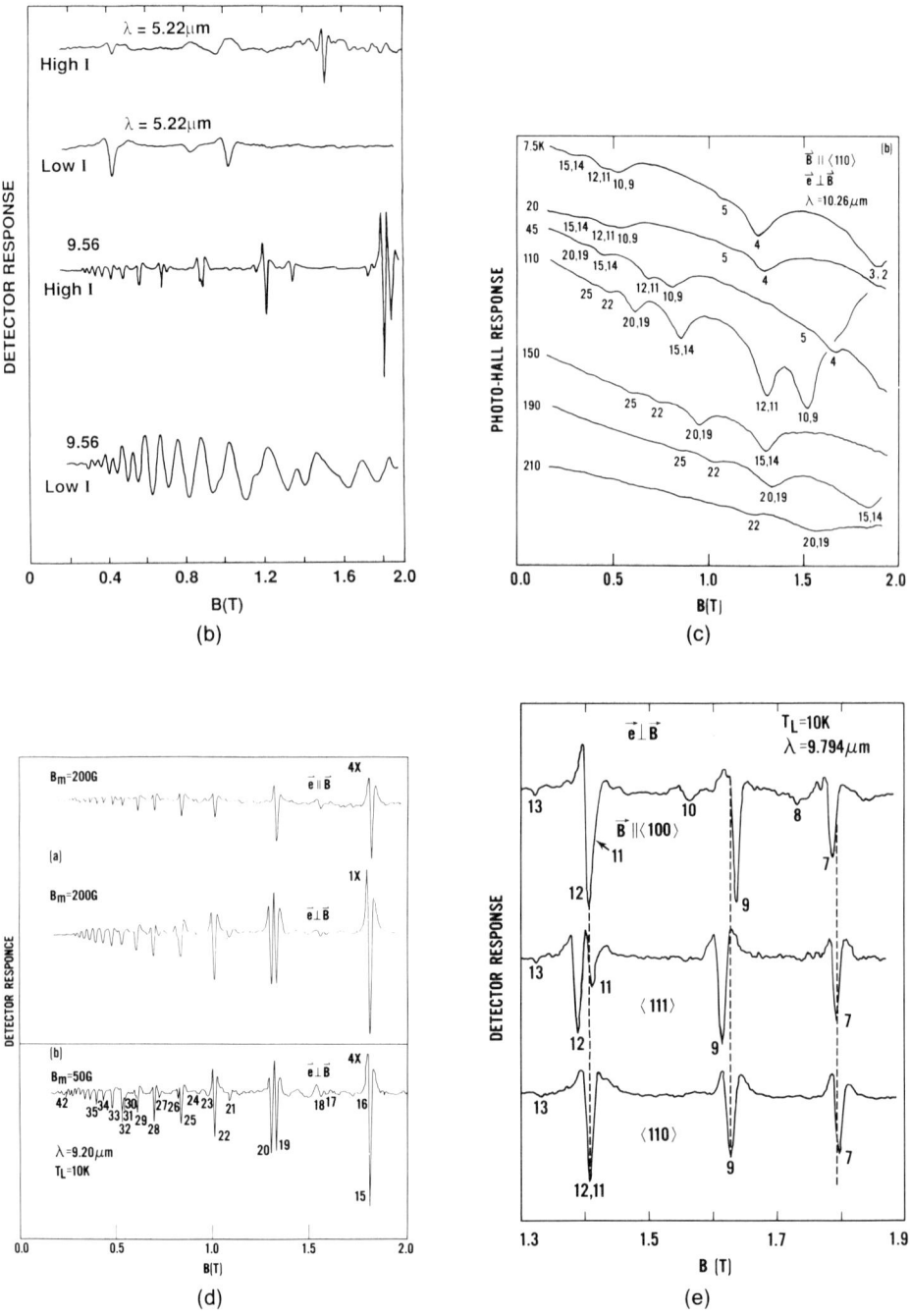

FIG. 9 (Continued)

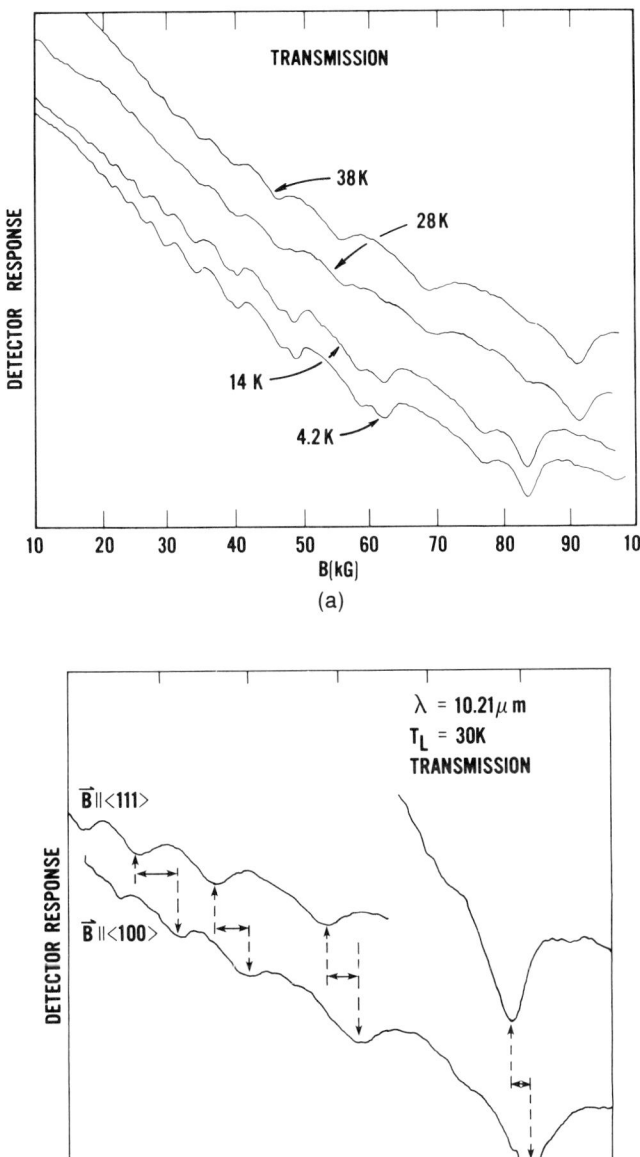

FIG. 10. Magneto-optical effects seen in p-InSb under a variety of conditions.

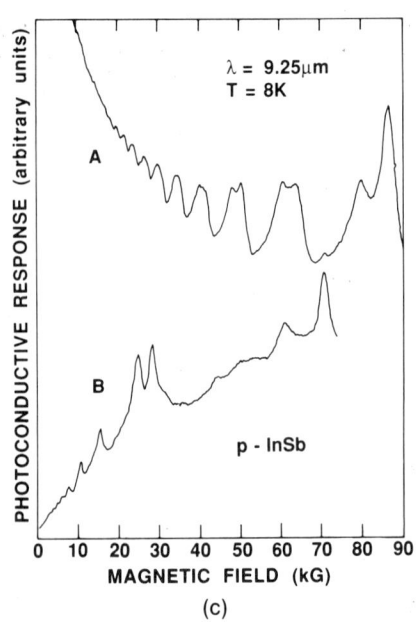

(c)

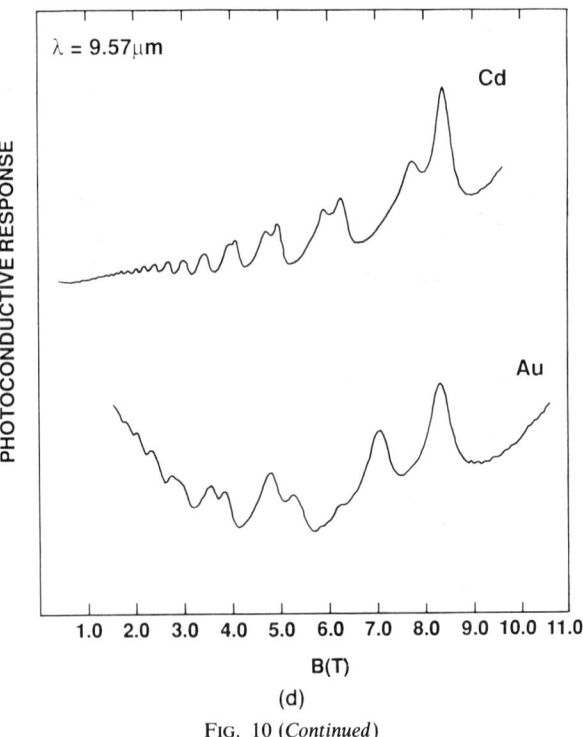

(d)

FIG. 10 (*Continued*)

now increased, ionizing the shallow impurity level, transitions from a deep donor-like level to the conduction-band Landau levels appear (also shown in Fig. 9a). These are the same deep-level transitions that appear at low intensity for $\lambda = 9.56$ μm in Fig. 9b.

Figure 9b represents the magneto-optical spectra observed for different laser intensities in the two spectral regions (5 μm–6 μm and 9 μm–11 μm) under consideration. In the 5–6 μm region only the free excitonic structure is seen at low intensities. As the laser intensity is increased, the free-exciton transitions saturate and the bound-exciton structure appears and becomes dominant. As discussed, in the 9–11 μm wavelength region at low intensities the deep-level to conduction-band transitions dominate the magneto-optical spectra. However, as the laser intensity is increased, the TPMA structure becomes dominant.

Interband magnetoabsorption (either one-photon or two-photon) can be used to monitor the energy-gap dependence on temperature. In other words, if the energy gap changes as a function of temperature, this will change the magnetic-field positions of either OPMA or TPMA transitions for a fixed laser energy. An example of using photo-Hall techniques to determine the temperature dependence of the energy gap E_g is shown in Fig. 9c. As the temperature is increased, the observed TPMA transitions are seen to shift to higher field. This results from an energy gap that decreases as a function of increasing temperature. This measurement has provided the most accurate measurement of the temperature dependence of the energy gap of InSb (Landbolt-Börnstein, 1986) and alloys of HgCdTe (Seiler *et al.*, 1990) to date.

The magneto-optical spectra observed in NGS also depend strongly on light polarization, since there are specific selection rules governing magneto-optical transitions. An example of this for TPMA transitions is seen in Fig. 9d for **e** ∥ **B** and **e** ⊥ **B** polarizations in the Voigt configuration. Note the additional structure and splitting in the **e** ⊥ **B** spectra, resulting from additional allowed transitions for this light polarization.

In Fig. 9e the effects of crystallographic orientation on the position of TPMA transitions are shown. The magnetic-field positions of the TPMA resonances are seen to shift slightly as the field is directed along the three major crystallographic orientations. This shift is due almost entirely to the anisotropy of the heavy-hole Landau levels that serve as the initial states for the TPMA transitions. The conduction band is nearly isotropic; however, this small anisotropy can be observed in another type of experiment based on intraconduction-band absorption (see Fig. 21a and 21b and corresponding discussion).

For p-InSb, similar effects to those discussed can be seen, as well as additional phenomena. Figure 10a shows the temperature dependence of magneto-optical spectra obtained at 9.57 μm. The observed transmission minima seen here result from absorption resonances corresponding to free-hole transitions between heavy- and light-hole Landau levels (high temperature) and bound-hole transitions between ground and excited shallow acceptor impurity states (low temperature). Thus, by changing the temperature we can observe two entirely different sets of optical transitions.

The heavy-hole anisotropy seen in Fig. 9e for TPMA transitions can be seen more dramatically in Fig. 10b. Here the magnetic-field positions of the free-hole transitions shift quite dramatically as a function of crystallographic orientation with respect to the magnetic field. A similar striking anisotropy is also seen for the bound-hole transitions (see Fig. 34) and is discussed later in this chapter.

If the laser intensity is increased we can see an effect similar to that discussed for n-InSb, i.e., impurity-to band transitions giving way to TPMA. This is shown in Fig. 10c. At low intensities ($\simeq 100$ W/cm^2, trace A) the bound-hole transitions dominate the observed magneto-optical data, whereas at higher laser intensities ($\simeq 1000$ W/cm^2, trace B) TPMA is clearly seen.

Finally, in p-InSb, we can see the effect of the dopant on the observed magneto-optical spectra. In Fig. 10d the upper trace represents the bound-hole spectra obtained at $\lambda = 9.57$ μm for shallow Cd acceptors in InSb. The lower spectra was obtained from another sample of InSb doped with Au, which is a deep acceptor impurity. Note how the spectra, although having the same features (doublet structure, etc.) appear entirely different from that obtained from the sample doped with Cd. Thus, one can delineate between dopants in a sample doped with multiple impurities, since each yields its own characteristic spectrum.

These examples in Fig. 9 and 10 of the variety of spectra observed in a single sample under different experimental conditions raise the question of how one actually identifies the origin of each transition or series of transitions, especially considering that more than one type of transition can be occurring in a given magnetic field or photon energy sweep. Of course, the use of different experimental conditions, such as varying the temperature or laser intensity or polarization, is very important in such identification, along with the models of the different electronic states, their occupancy as a function of temperature and sample doping, and the selection rules and intensity dependence of the transitions. In addition, the fundamental tool for the identification of magneto-optical transitions is to plot the photon energies of a group of spectral features as a function of magnetic field, because the energies of different transitions in a given series tend toward the same energy (such as the energy gap for interband transitions) as the magnetic field is reduced to zero, and the energies separate, or fan out, as the magnetic field increases. A plot of such transition energies is known as a "fan chart" and many examples will be given throughout this chapter. A comparison of a given group of experimental features for different laser wavelengths, for example, with such a theoretical fan chart will often reveal the zero-magnetic-field intercept energy, which is an important clue to the origin of the transition or group of transitions. Also, a comparison of the energy spacing between adjacent levels with previously measured cyclotron resonance energies, for example, will provide an important clue to the type of transition (interband or intraband). As is clear from the variety of spectra seen in Figs. 9 and 10, care must always be taken in making any identification of the observed phenomena.

II. Experimental Methods

6. Overview

Magneto-optical effects are generally observed by shining light onto a sample and then observing the resonant response of the sample either as the magnetic field is varied while keeping the photon energy fixed or by varying the photon energy at a fixed magnetic field. Explicit quantitative information about the band structure of a semiconductor can be obtained because of the resonant nature of the sample response. The origin of the various resonances lies in the formation of Landau levels with their associated sharply peaked density of states at the bottom of each level. Peaks occur in the optical constants corresponding to transitions between Landau levels. As described in the Introduction, a wide variety of effects can be observed depending on the nature of the light (intensity, polarization, wavelength), sample properties (doping, crystallographic orientation), temperature, and sample current or electric-field bias. Infrared tunable, monochromatic light sources used in magneto-optical experiments of NGS are now based on either conventional grating or prism monochromators, Fourier transform infrared spectrometers, or lasers. Earlier cyclotron resonance experiments at low magnetic fields necessitated the use of microwave techniques. However, with the advent of high magnetic fields available from superconducting or Bitter magnets, the focus has shifted to magneto-optics performed in the optical region. The choice of which experimental technique to use in observing the sample response is mainly one of convenience and sensitivity.

Both InSb and HgCdTe are narrow-gap semiconductors. Thus, infrared sources must be used to investigate near-band gap and intraband magneto-optical effects in these samples. Table IV lists the laser source (both pulsed and cw) and their wavelengths of operation and photon energies that have often been used to investigate magneto-optical effects in InSb and HgCdTe. We see from this table that only a few photon energy intervals exist in which both interband and intraband magneto-optical studies can be carried out. Lattice temperature changes also affect the energy gap, and this can also be

TABLE IV

LASER SOURCES MOST OFTEN USED TO INVESTIGATE MAGNETO-OPTICAL EFFECTS IN InSb AND HgCdTe.

Laser source	Wavelength range	Photon energy range
CO laser	5–7 μm	177–248 meV
CO_2 laser	9.14–12.1 μm	102–136 meV
Far infrared	37–1222 μm	1–33.5 meV

used to extend the range of samples studied. In addition, there will undoubtedly be further laser and optical technological advances in the near future that will provide new types of light sources in these photon energy ranges. An example of this is given by Cotter, Hanna, and Wyatt (1976), who where able to provide light in the ranges of 2.5–475 μm, 5.67–8.65 μm, and 11.7–15 μm with powers up to 25, 7, and 2 kW, respectively. It is important to point out that the application of a magnetic field allows the "tuning" of the sample's energy levels such that they can be investigated by the preceding laser sources. This extends by a considerable amount the range of HgCdTe samples that can be characterized (e.g., the cutoff wavelength λ_c can be longer than the laser wavelength).

Nonlinear magneto-optical studies can be carried out with a combination of two photons from either the same laser or from two different lasers. The variations of combining two photons from these lasers are given in Table V, where we list the lasers used, the tunability of each of the two photons, and the tuning range of the sum energy. Thus $Hg_{1-x}Cd_xTe$ samples with $0.22 < x < 0.26$ (in the 8–12 μm spectral region) can be investigated with a CO_2 plus a far-infrared (FIR) laser. Samples with $x = 0.27 < x < 0.33$ can be studied with a CO and a FIR laser. Two CO_2 lasers will allow samples with $0.29 < x < 0.32$ to be investigated. Additional coverage for samples with

TABLE V

TWO-PHOTON ENERGIES FROM VARIOUS COMBINATIONS OF INFRARED LASER SOURCES.

Lasers	$\hbar\omega_1$ (meV)	$\hbar\omega_2$ (meV)	$\hbar\omega_1 + \hbar\omega_2$ (meV)
CO + CO	177–248	177–248	354–496
$CO_2 + CO_2$	102–136	102–136	204–272
FIR + FIR	1–33.5	1–33.55	2–67
CO + CO_2	177–248	102–136	279–384
CO + FIR	177–248	1–33.5	178–281.5
CO_2 + FIR	102–136	1–33.5	103–169.5

$x > 0.33$ is apparent from other combinations of these lasers. We again point out that great flexibility comes from being able to apply an external magnetic field and "tune" the energy gap or the Landau levels. Lattice temperature changes also affect the energy gap, and this can also be used to extend sample "tunability."

Alternating current magnetic-field modulation and phase-sensitive detection techniques have been developed by Seiler and co-workers over the past 15 years for the study of laser-induced hot-electron effects and magneto-optical effects in InSb using 20-μsec-wide CO_2 laser pulses. Lattice heating effects are prevented by using a low duty cycle of a few percent. These techniques are based on a variation of the sampling and magnetic-field modulation technique developed by Kahlert and Seiler (1977) for hot-electron quantum transport experiments. The apparatus used is shown in Fig. 11. The voltage drop across the potential probes on the sample is changed by the optical excitation and/or the influence of the magnetic field. The photoconductive response of the samples using mechanically chopped pulses follows the slow rise time of the optical pulse and eventually reaches an equilibrium. The photovoltage can then be amplified and further electronically processed using the sampling unit of an oscilloscope or a boxcar averager. The output of either of these units passes the modulation signal produced by the low-frequency ac magnetic field impressed on the sample photovoltage signal. If the signal is then directed into a lock-in amplifier that is tuned to the modulation frequency or a harmonic thereof, the signal is demodulated into a dc voltage. For a sufficient signal-to-noise ratio the repetition rate of the laser pulses should be at least 10 times the magnetic-field modulation frequency; higher repetition rates substantially improve the signal-to-noise ratio. A schematic display of the photoconductive and magnetic signals is shown in Fig. 12 for a pulse repetition rate exceeding the modulation frequency (43 Hz) by a factor of 10. The uppermost trace shows the modulation magnetic field superimposed on the linearly increasing dc field, and the second trace shows the photoconductive voltage pulses with no ac magnetic field. The sample photoconductive pulses are modulated as shown in the third trace of Fig. 12 if the sample resistance (and hence the photoconductive response) is an increasing function of the magnetic field. The output of the sampling unit is then modulated as illustrated and carries with it the essential information about the dependence of the photoconductive voltage on the magnetic field. If a linear relationship exists between the sample resistance R and the dc magnetic field B, the dc output of the lock-in amplifier will be essentially constant in time and as a function of dc magnetic field, as shown in the bottom trace of Fig. 12. Any nonlinearity in $R(B)$ changes the amplitude of the ac component of the sampling unit output and produces the first derivative of the sample response if the lock-in is tuned to the modulation frequency. But setting the lock-in amplifier to twice the modulation frequency, we obtain the second derivative

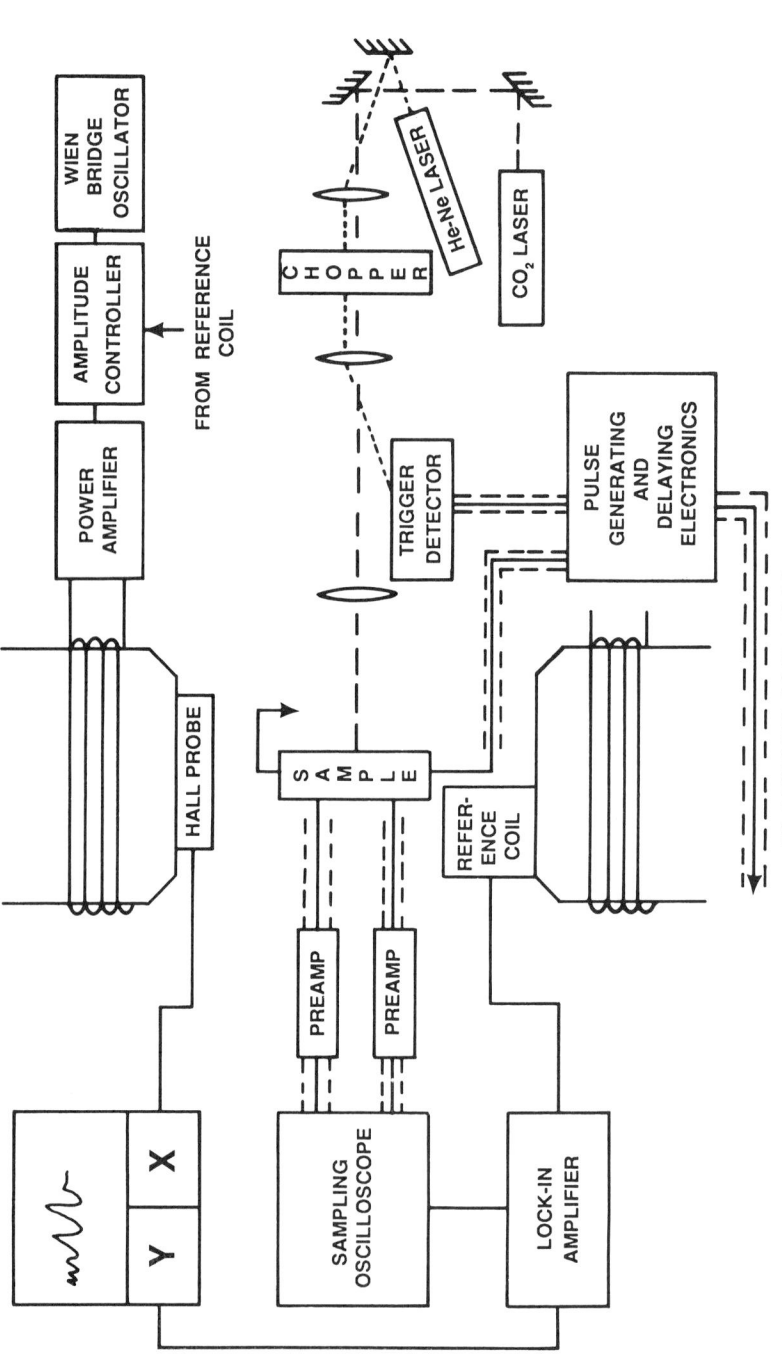

FIG. 11. Block diagram of experimental apparatus used to perform photoconductive measurements using magnetic modulation-sampling-lock-in techniques (From Moore et al., 1978).

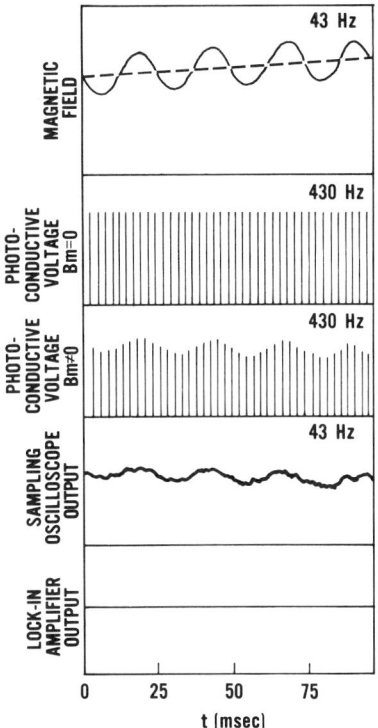

FIG. 12. Schematic time dependence of the magnetic and electronic signals appearing at various components of the apparatus.

of the sample response. Figure 13 demonstrates the previous features applied to the observation of two-photon magnetoabsorption in *n*-type InSb. The top trace shows the sample response as monitored by boxcar average techniques. The lower trace was obtained using the magnetic field modulation with the lock-in amplifier tuned to twice the modulation frequency. It clearly demonstrates how much the resolution improves when employing ac modulation and sampling techniques.

Several experimental methods have been used to investigate the magnetoabsorption properties of semiconductors. The ones most often used in the magnetospectroscopy of NGS are transmission, reflection, photoluminescence, and photoelectronic. Several subcategories fall under the broad category of photoelectronic methods, which include (1) photoconductivity, (2) photo-Hall, and (3) photovoltaic techniques. These techniques are used to monitor, either directly or indirectly, the absorption of light in the sample. Each may be particularly advantageous under different experimental conditions.

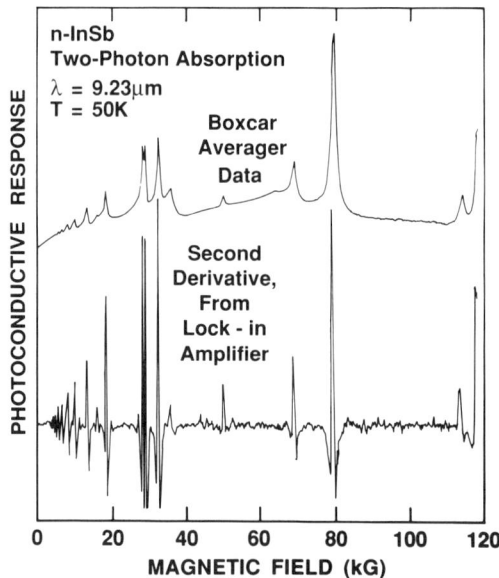

FIG. 13. Comparison of TPMA spectra obtained using conventional boxcar averaging techniques and the magnetic-field modulation–lock-in amplifier method. (From Seiler et al., 1988.)

The attenuation of the laser intensity (I) as the beam propagates through the crystal along the z-direction is described by the expression

$$\frac{dI}{dz} = -\alpha I - \beta I^2 - (\Delta n \sigma_e + \Delta p \sigma_h)I, \tag{27}$$

where α and β are the one- and two-photon absorption coefficients, respectively, σ_e and σ_h are the free-electron and free-hole absorption cross sections and Δn and Δp are the number of photogenerated free electrons and free holes. When the last term can be neglected, the solution to Eq. (27) for the transmittance T is

$$T = \frac{I}{I_0} = \frac{(1-R)^2 \exp(-\alpha l)}{1 + \beta I_0 (1-R)[1 - \exp(-\alpha l)]/\alpha}, \tag{28}$$

where R is the reflectivity as the wavelength of the incident radiation and l is the thickness of the sample. Thus, measurement of T, I, I_0, and R determines α and β. Without simplifying assumptions, only numerical solutions are possible.

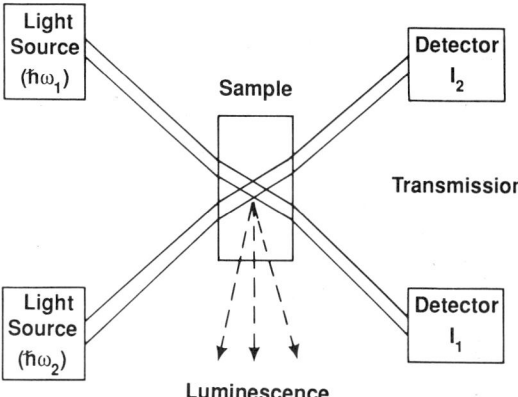

FIG. 14. Schematic diagram of a simple two-photon absorption experiment.

The basic experimental configurations for TPA measurements in semiconductors are summarized in Fig. 14. Two light sources provide beams of light $\hbar\omega_1$ and $\hbar\omega_2$ to be incident on the sample, which is transparent to each of the beams alone. When both beams are simultaneously present (in the same spatial region of the sample and overlapped in time), sufficiently intense, and when $\hbar\omega_1 + \hbar\omega_2$ equals an allowed transition energy, the sample simultaneously absorbs one quantum from each beam. The beam configurations can be parallel, antiparallel, perpendicular, or inclined to one another. To provide sufficient intensity, at least one of the light sources must be a laser, which may or may not be tunable. The other source can be a laser or a tunable beam obtained from a flash lamp and monochromator, or can be obtained by other methods such as stimulated Raman scattering or an optical parametric oscillator.

Gibson *et al.* (1976) have pointed out that in many cases absorption due to TPA photogenerated electrons and holes is not negligible, and thus the absorption contribution of the photogenerated free carriers must be taken into account. Bechtel and Smith (1976) also showed that there may be an apparent laser pulse-width dependence in nonlinear absorption experiments because of this absorption of light by the two-photon created excess carriers. For the assumption that both diffusion and recombination can be neglected (for example, with picosecond pulses) they found that below a critical intensity that is inversely proportional to the laser pulse width, the effect of absorption by excess carriers is small compared with direct TPA. As a final note, we point out that the intensity can also be a function of time and space in most experiments, necessitating integration over these variables. Thus, knowledge of the laser beam's properties both in space and time is needed, and a lack of complete beam characterization can lead to large uncertainties in a quantitative understanding of the transmission and absorption properties (and hence in a determination of β).

Photoelectronic methods exhibit a significantly increased sensitivity over conventional transmission techniques. These photoelectronic effects are caused by the creation of free carriers induced by the absorption of light. These carriers can be detected using photoconductivity, photo-Hall, and photovoltaic effects and are important for understanding materials used as passive sensors or detectors. We will now discuss some of the important features of each of these photoelectronic effects.

7. Photoconductivity

Photoconductivity can be a complex process involving several successive or simultaneous mechanisms: optical absorption, charge-carrier transport, hot-carrier relaxation, and radiative and/or nonradiative recombination. It covers all phenomena by which an increase or decrease in conductivity can take place during and following the absorption of light in the semiconductor. The presence of nonequilibrium carriers generated by the light alters the conductivity of a semiconductor. The general form of the conductivity can be written as

$$\sigma = e[\mu_n(n_0 + \Delta n) + \mu_P(p_0 + \Delta p) + n_0 \Delta \mu_n + p_0 \Delta \mu_P] \quad (29)$$

or

$$\sigma = \sigma_0 + \Delta \sigma,$$

where e is the electronic charge, n_0 (p_0) is the concentration of electrons (holes) at thermal equilibrium in the dark, Δn (Δp) is the excess electron (hole) concentration induced by the light, μ_n (μ_P) is the electron (hole) mobility, $\Delta \mu_n$ ($\Delta \mu_P$) is the change in the electron (hole) mobility caused by the light, $\sigma_0 = e(\mu_n n_0 + \mu_P p_0)$, and $\Delta \sigma = e[\mu_n \Delta n + \mu_P \Delta p + n_0 \Delta \mu_n + p_0 \Delta \mu_P]$. Thus, in general, the conductivity can change if the carrier concentration or the mobility changes. Some detector materials (Putley or hot-electron detectors; Putley, 1964) make use of this mobility change, which can be quite large in the FIR region. Here we are primarily concerned with the changes in concentration produced by the absorption of light. Since in many semiconductors $\mu_P \ll \mu_n$, the change in conductivity arising from nonlinearly produced electron–hole pairs can be written simply as

$$\Delta \sigma = e\mu_n \Delta n. \quad (30)$$

Thus measurements of the photoconductivity are most directly related to changes in electron concentration. It is a very sensitive indicator of small changes in n, particularly if n_0 is small. Thus if $n_0 \simeq 10^{14}$ cm^{-3}, $\Delta n/n_0$ changes of $\simeq 0.001$ (or $\Delta n \simeq 10^{11}$ cm^{-3}) can easily be detected.

8. Photo-Hall Effect

The most common electronic transport method of determining n_0, the electron concentration, is through a measurement of the Hall coefficient R_H, which, in the absence of any light radiation (i.e., in the dark), is given by

$$R_H^D = -\frac{1}{n_0 e}. \tag{31}$$

The Hall effect has been known for nearly 100 years and has been used and written about in many articles. However, if light absorbed by the sample increases the carrier concentration by an amount Δn, then the Hall coefficient is given by

$$R_H \simeq -\frac{1}{ne} = -\frac{1}{(n_0 + \Delta n)e}. \tag{32}$$

Consequently, from the measurement of the photo-Hall coefficient and the dark Hall coefficient the number of photogenerated carriers can be often obtained and is given by

$$\Delta n = -\frac{1}{e}\left[\frac{1}{R_H} - \frac{1}{R_H^D}\right]. \tag{33}$$

The photo-Hall effect is as sensitive as photoconductivity, but has the added feature that it is sensitive only to absorption mechanisms that change the carrier concentration. In contrast, transmission, reflection, and photoconductivity are also sensitive to free-carrier absorption mechanisms. Thus the photo-Hall effect can be used in conjunction with photoconductive measurements to delineate between free-carrier and interband or impurity-to-band transitions (Littler and Seiler, 1985; Littler and Seiler, 1986).

Littler et al. (1984) used photo-Hall (PH) techniques to study TPMA effects in InSb. Representative results are shown in Fig. 15. This represents the first time that TPMA transitions were observed in semiconductors by using a photo-Hall technique. The values of R_H, the Hall coefficient, were calculated in the standard manner. Figure 15 shows the variation of R_H with magnetic

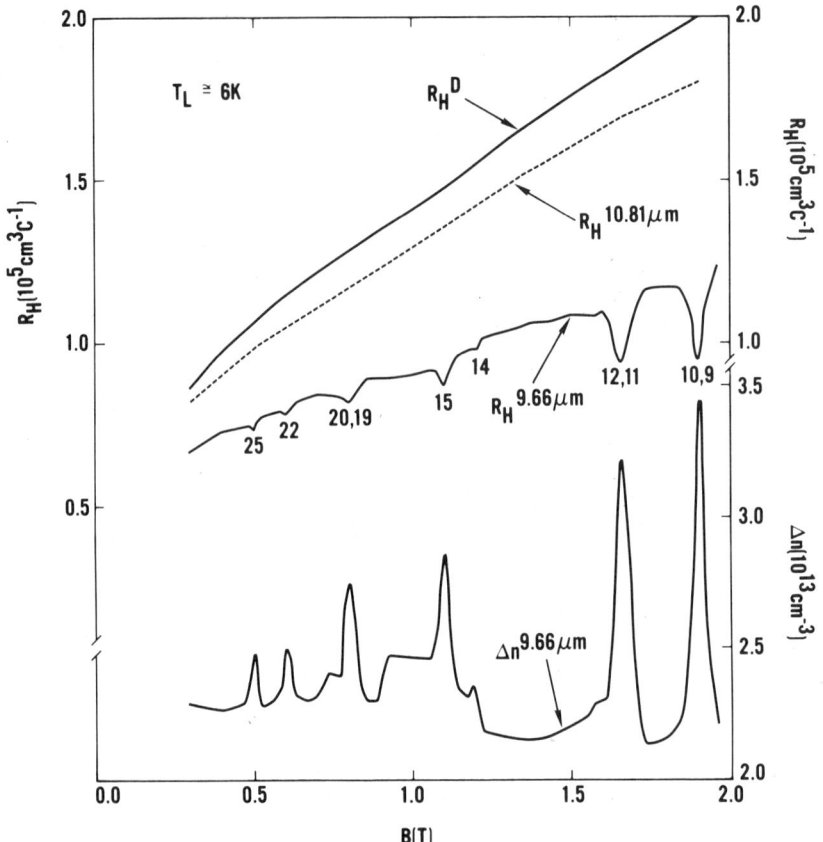

FIG. 15. Plot of the Hall coefficient versus magnetic field under three conditions: (1) in the dark, (2) illuminated with 10.81-μm radiation, and (3) illuminated with 9.66-μm radiation. The resonant decreases in $R_H^{9.66\,\mu m}$ are caused by the resonant increases in the number of two-photon produced carriers. (From Seiler et al., 1984.)

field for the dark case and for two CO_2 laser wavelengths: (1) $\lambda = 10.81$ μm, where no TPMA processes are possible, and (2) $\lambda = 9.66$ μm, where TPA does occur. Magnetic freeze-out effects are important for these pure samples as shown by the increase of the dark Hall coefficient R_H^D with increasing field. The new features shown at 9.66 μm are resonant decreases in R_H due to TPMA processes. Since the electron mobility is much greater than the hole mobility, a one-electron Hall coefficient model can be used to calculate the number of TPA-produced electrons. These results thus demonstrate the use of the PH effect to determine directly the number of free carriers induced by TPMA processes. The dependence of the induced free-carrier concentration on external parameters, such as intensity or lattice temperature, allows direct comparison with the results of rate equation models. In contrast, other methods, such as absorption or photoconductivity, do not allow the direct determination of the number of free carriers; these must be indirectly inferred. In addition, these PH techniques are a very sensitive indicator of even small increases in the free-carrier concentration ($\Delta n \simeq 10^{10}–10^{11}$ cm^{-3} can easily be detected).

9. Photovoltaic Effect

Photovoltaic voltages are generated when $p - n$ junction regions of the semiconductor are illuminated. The exact connection between the voltage signal observed and the amount of absorbed light is difficult to determine. However, it is easily obtained, and has been used by Roth and Fortin (1978) to study OPMA in single crystals and polycrystals of InSb and by Seiler et al. (1983) to detect TPA processes in InSb with a CO_2 laser.

III. One-Photon Magnetospectroscopy

Before the availability of infrared lasers, only one-photon transitions could be observed in NGS. These include interband, intraband or free-carrier, and impurity-related transitions. The available experimental data are reviewed for both InSb and HgCdTe. These results will be combined with the results of two-photon experiments reviewed in Section IV to give a comprehensive picture of the electronic states in both materials.

10. Interband One-Photon Magnetoabsorption

a. InSb

The first interband OPMA measurement for InSb was reported in 1956, and for HgCdTe only a few years later, in 1961. Tables VI and VII give comprehensive compilations of this and subsequent work in these two materials. As the purity of these materials increased, and for HgCdTe, the uniformity of the alloy composition improved, the linewidth of the OPMA signals has improved, and, therefore, the accuracy of the experimental determination of the band-gap energy and other parameters. Also, more detailed observations have been made of such effects as exciton fine structure. The theoretical interpretation has also advanced from simple two-band models to the modified

TABLE VI

REPRESENTATIVE ONE-PHOTON INTERBAND MAGNETOABSORPTION WORK DONE ON InSb.

Authors (year)	Method	Main results
Zwerdling et al. (1956)	Transmission	Saw evidence for a nonlinear shift of the energy gap with magnetic field.
Burstein et al. (1956b)	Transmission	Saw evidence for a nonlinear shift of the energy gap with magnetic field.
Zwerdling et al. (1957)	Transmission	Room temperature observation of OPMA. Obtained energy gap of 0.180 ± 0.002 eV at 298 K. Small effects due to valence-band anisotropy seen.
Wright & Lax (1961)	Magnetoreflection	Obtained energy gap of 228 meV at 80 K.
Zwerdling et al. (1961, 1962)	Transmission	Experimentally established light-hole nonparabolicity, valence-band degeneracy for low quantum numbers, and the existence of high-field excitons
Pidgeon and Brown (1966)	Transmission	Observed direct interband transitions at liquid helium temperatures. Described results using energy-band model that included the interaction between conduction and valence bands exactly, and the effect of higher bands to k^2.
Johnson (1967)	Transmission	Observed exciton fine structure in the interband magnetoabsorption of InSb. Estimated separation between (000) and (002) exciton states.
Pidgeon et al. (1967b)	Electroreflectance	Observed fine structure due to excitons.
Pidgeon and Groves (1968)	Reflection	Linear-k splitting of the valence band deduced. OPMA data for $\mathbf{B} \parallel \langle 111 \rangle$ and $\langle 100 \rangle$.
Pidgeon and Groves (1969)	Reflection	Observation and description of OPMA transitions produced by inversion-asymmetry and warping.
Zhilich (1972)	Transmission	Theoretical study of diamagnetic excitons; comparison with data of Johnson (1967).
Weiler (1979)	Reflection	Described interband data using a modified Pidgeon–Brown energy-band model and included excitonic effects. Introduced a new set of energy-band parameters for InSb.
Efros et al. (1982a,b)	Transmission	Described interband data taking into account excitonic effects and the electron–electron exchange interaction. Reported a new set of energy-band parameters.

TABLE VI (*Continued*)

Authors (year)	Method	Main results
Dennis et al. (1982)	Transmission	Observed saturation of magnetoabsorption spectra. Line shape analysis used to provide information on the coherent electron–hole dephasing time and the carrier relaxation time.
Ivanov-Omskii et al. (1983a,b)	Photoluminescence	Observed evidence for a bound-exciton complex.
Seiler et al. (1986a)	Photoconductivity	Magneto-optical spectra showed presence of free and bound excitons and impurity-to-band transitions where the impurity was identified as the neutral acceptor participating in the bound-exciton complex.

Pidgeon–Brown model, including accurate exciton corrections. Here we give a brief summary of these results.

From the earliest observations in InSb of poorly resolved transitions that shifted with magnetic field (Zwerdling et al., 1956, 1957, 1961; Burstein et al., 1956b, Wright and Lax, 1961), the importance of nonparabolicity was recognized and the results interpreted using a two-band model; simple exciton corrections were also made. The first interpretation of OPMA in InSb incorporating both nonparabolicity and some Luttinger effects was made by Zwerdling et al. (1962); the theory incorporating these effects, neglecting inversion asymmetry and some warping effects, was described by Pidgeon and Brown (1966). Later work, as listed in Table VI, included new observations of exciton structure and effects due to warping and inversion asymmetry. A new group-theoretical analysis (Weiler et al., 1978) introduced a few new parameters to the Pidgeon–Brown model; these were included (Weiler, 1979) in a fit to both inter- and intraband reflection data. Except for the value of the energy gap, which was incorrect due probably either to strain effects or temperature inaccuracies, these parameters have since, with very little modification, been used to describe a large number of experimental data for InSb.

Weiler (1979) showed that excitons must be considered in the quantative interpretation of all one-photon interband magnetoabsorption data, even in the case of narrow-gap semiconductors. Efros et al. (1982a,b) showed that the average deviation between theory and experiment could be improved somewhat by incorporating a more accurate exciton correction and the effects of exchange. The most recent work (Ivanov-Omskii et al., 1983a,b; Seiler et al., 1986a; Seisyan and Yuldashev, 1988) has involved the observation of new transitions at lower energies than previously observed interband

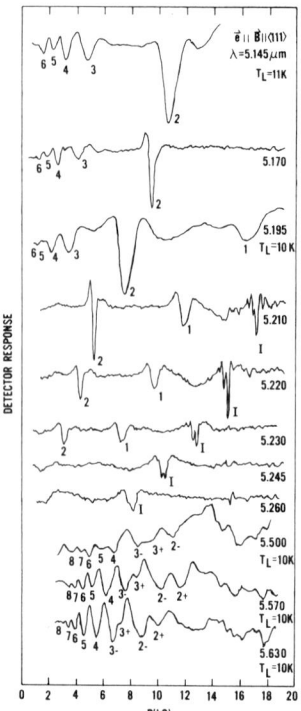

FIG. 16. OPMA spectra obtained for a variety of CO laser wavelengths. Three sets of structure are seen in the data depending on the wavelength employed: (1) free excitons for $\lambda \leq 5.230\,\mu$m (labeled 1–6); (2) bound excitons for $5.210 \leq \lambda \leq 5.260\,\mu$m (labeled I); and (3) impurity-to-band transitions for $\lambda \geq 5.500\,\mu$m (labeled $2\pm$, $3\pm$, 4–8). (From Seiler et al., 1986a.)

transitions, attributed to the creation of excitons bound to impurities, as well as transitions from impurity states to the conduction band. Figure 16 illustrates the large variety of transitions seen by Seiler et al. (1986a) and Fig. 17 the excellent agreement achieved between theory and experiment. We also note that an extensive review of diamagnetic exciton spectra in semiconductors (including InSb and HgCdTe) was recently given by Seisyan and Zakharchenya (1991). These diamagnetic excitons are the states composing the spectrum of exciton states that appear in semiconductor crystals placed in a magnetic field. Thus, rather than speak of a spectrum of oscillatory magnetoabsorption structure, they use the term *diamagnetic exciton spectroscopy*. As discussed in Section 4b the recent photoluminscence work of Grober and Drew (1991) reinterpreted the data of Ivanov-Omskii et al. (1983a,b) and Seisyan and Yuldashev (1988). Thus, consideration must be given to electron-hole plasma and band-gap renormalization effects when interpreting interband data.

The OPMA data have nearly always been interpreted in terms of transitions occurring at $k_h = 0$, i.e., involving states with zero momentum parallel to the magnetic field. Bell and Rogers (1966) carried out a calculation of the energy levels for one orientation of the magnetic field ([100]) including the k_h dependence. Their results showed a complicated dependence on k_h, including maxima away from $k_h = 0$, for the heavy-hole states with low quantum number; these effects are most important in interpreting intravalence-band transitions. The strong minima in the conduction-band levels at $k_h = 0$ forces the interband transitions to occur at this point.

It should be mentioned that interband transitions involving bands other than the lowest conduction band and the highest valence bands have also been observed for InSb. Aggarwal (1972) observed transitions from the spin-

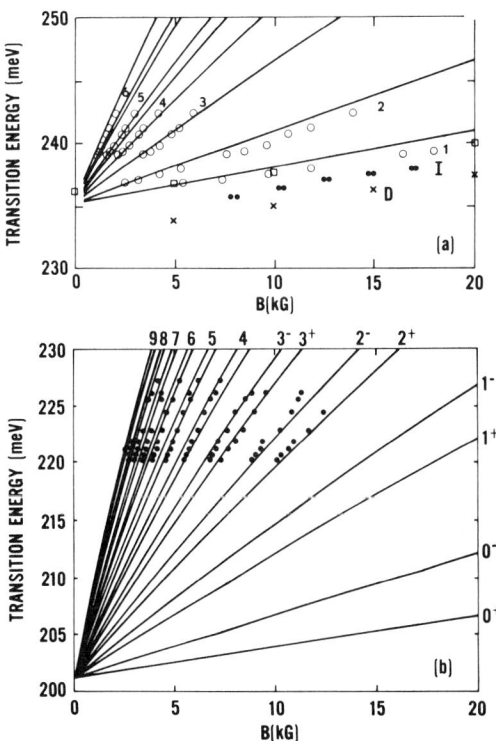

FIG. 17. Transition energies versus magnetic field for the data shown in Fig. 16. Part (a) shows the free-exciton data (open circles) and the bound-exciton data (full circles). Data from Ivanov-Omskii et al., 1983a are also plotted: free-exciton (open squares) and bound excitons (crosses). (From Seiler et al., 1986a.) Part (b) shows the deep-level data due to transitions from a deep acceptor $\simeq 34$ meV above the valence band.

TABLE VII

Compilation of One-Photon Magnetoabsorption Work Done on $Hg_{1-x}Cd_xTe$.

Authors (year)	Experiment	Temp (K)	Alloy comp studied	Comments & Results
Harman et al. (1961)	Magnetoreflection	77	0.14	First interband magneto-optical experiments.
Strauss et al. (1962)	Magnetoreflection	4.2	0.136–0.20	Used Bowers and Yafet model.
Groves et al. (1971)	Magnetoreflection	≃25 & 95	0.161 ± 0.003	Homogeneous sample; first direct determination of heavy-hole mass $[(0.28 \pm 0.1)m_0]$. First study of interband transitions using quasigermanium model; $E_p \simeq 18.5$ eV.
Kim & Narita (1976)	FIR Magnetoabsorption transmission 10-μm samples	4.2	0, 0.07, 0.126, 0.148	Used quasigermanium model; conservation of modified Luttinger parameters with varying of x-value.
Guldner et al. (1977a,b)	Magnetoabsorption transmission	4.2	0.01–0.3	Pidgeon and Brown model used with $F = N_1 = 0$; spherical approximation used; $\Delta = 1.0$ eV used; heavy-hole mass of $(0.4 \pm 0.1)m_0$; polaron effects observed for the first time; determined E_p and E_g as a function of x.
Weiler et al. (1977) Weiler (1981b)	Magnetoreflection	24 & 91	0.175–0.269	Pidgeon–Brown model used; first time exciton corrections applied; $E_p = 19.0 \pm 0.5$, $F = -0.8 \pm 0.3$, $N_1 = 0$; Luttinger parameters $\gamma_1 = 3.3 \pm 0.1$, $\gamma_2 = 0.1 \pm 0.1$, $\gamma_3 = 0.9 \pm 0.1$, $\kappa = -0.8 \pm 0.1$, with $q = 0$; $\Delta = 1$ eV; broad linewidths due to inhomogeneity of alloy composition.
Seiler et al. (1989a,b)	Photoconductivity	7	0.24	Both one-photon and two-photon magnetoabsorption studied with a CO_2 laser. Theoretical description of both done with Pidgeon–Brown model; consistency obtained with a free-exciton binding energy of 2 meV.
Littler et al. (1990a)	Photoconductivity	5	0.22	OPMA studied with a CO_2 laser

orbit split-off band to the conduction band by using electroreflectance and determined values for the spin-orbit splitting and the effective mass of the split-off band. Sari (1971, 1973a,b) observed magneto-optical transitions from the valence band to higher critical points in the conduction band.

b. HgCdTe

Compared with InSb, there have been markedly fewer experimental observations of OPMA in HgCdTe, as shown by the summary in Table VII. After the first observations (Harmon et al., 1961; Strauss et al., 1962), which were interpreted using simple models, Groves et al. (1971) made the first interpretation of HgCdTe OPMA data for one sample with $x = 0.16$, using the Pidgeon–Brown model, followed by Kim and Narita (1976) for samples with different alloy compositions, including the semimetallic case (i.e., $E_g < 0$). Systematic studies of the variation of the parameters with alloy composition were made independently by Guldner et al. (1977a,b) and Weiler (1981b). Ivanov-Omskii et al. (1983a,b) observed magneto-optical spectra in a sample of HgCdTe with $x \simeq 0.3$ and noted that (even though this sample has a gap approximately the same as that of InSb) the magnetoabsorption lines were broader than that in InSb by more than an order of magnitude. Recently Seiler et al. (1989a,b) and Littler et al. (1990a) carried out a unified OPMA and TPMA study of samples with x-values ranging from $x \simeq 0.22$ to $x \simeq 0.30$ and have successfully fit these data with a single set of band parameters. An example of the magneto-optical data for the $x \simeq 0.22$ sample is shown in Fig. 18, with the corresponding comparison of theory and experiment seen in Fig. 19. The $x \simeq 0.24$ sample results are seen in Fig. 38 of the impurity–defect section. The assignment of each transition in Figs. 18 and 19 is given in Table VIII.

11. Free-Carrier Magneto-Optics

Over the past several decades extensive magneto-optical studies of the conduction band of InSb have been carried out in an attempt to determine precisely the energy-band structure (i.e., effective masses, g-factors, anisotropy). In contrast, few studies of this kind have been performed in HgCdTe, largely due to problems of sample purity and inhomogeneity. Even for InSb, there have been few attempts to characterize these processes with theoretical models using the same set of energy-band parameters. Our purpose here is to summarize the major features of free-carrier magnetospectroscopy that have been carried out on InSb and HgCdTe, with an emphasis placed on the more recent laser-based studies. Both conduction- and valence-band data will be presented. We will show how a unified picture including interband

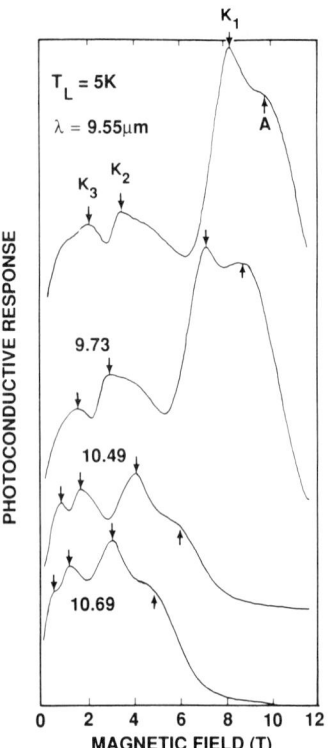

FIG. 18. Wavelength dependence of the photoconductive response, in n-HgCdTe showing both OPMA resonant structure labeled K_1–K_3 and that arising from shallow acceptor impurity-to-conduction band transitions labeled A. (From Littler et al., 1990a.)

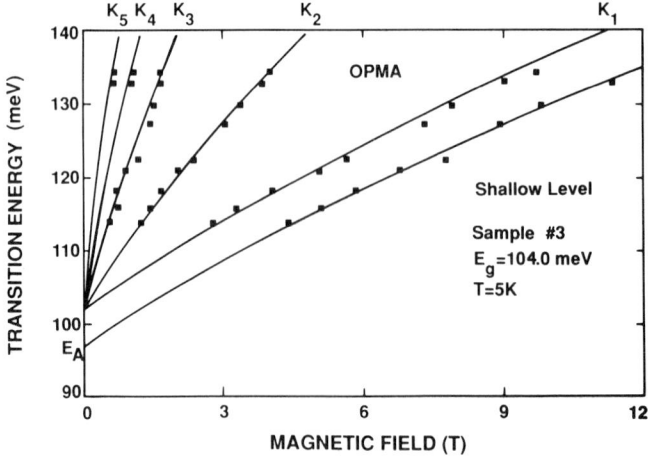

FIG. 19. Transition energy versus magnetic field fan chart for the spectra presented in Fig. 18. (From Littler et al., 1990a.) The lines represent theoretical calculations using the modified Pidgeon–Brown model, and the solid squares represent the magnetic-field positions of the resonances.

TABLE VIII

ONE-PHOTON MAGNETOABSORPTION TRANSITION
ASSIGNMENTS FOR OBSERVED STRUCTURE IN HgCdTe.

Designation	Energy-level transition	Polarization
K_1	$a^-(1) \to a^c(0)$	σ_L
K_2	$b^+(-1) \to b^c(0)$	σ_L
K_3	$a^-(0) \to a^c(1)$	σ_L
K_4	$b^-(0) \to b^c(1)$	σ_L
K_5	$a^+(1) \to a^c(2)$	σ_L

and intraband data is necessary to describe most aspects of the free-carrier data using a modified Pidgeon–Brown model (Weiler et al., 1978); a five-level model (Pfeffer and Zawadzki, 1990) is necessary to describe the full anisotropy of the conduction band of InSb. The transition energies for each free-carrier absorption process will be calculated by using these models in conjunction with the appropriate selection rules that govern each process.

Free-carrier absorption in a NGS can involve a wide variety of optical transitions. In a magnetic field, these transitions can be studied in great detail due to the sharp density of electronic states produced. These transitions include cyclotron resonance, which is seen by absorption of light of energy $\hbar\omega = \hbar\omega_c$, combined resonance involving a spin-flip transition at $\hbar\omega = \hbar\omega_c + \hbar\omega_s$, and spin resonance that takes place at the spin-flip frequency $\hbar\omega = \hbar\omega_s$. In addition various other transitions occur, such as the harmonics of cyclotron resonance, requiring $\hbar\omega = \hbar(n\omega_c)$, combined resonance, requiring $\hbar\omega = \hbar(n\omega_c + \omega_s)$, and the phonon-assisted variants of the transitions, which involve the emission of a longitudinal-optical phonon. We now review various aspects of these absorption processes that have been observed and studied in InSb and HgCdTe.

a. Conduction-Band Cyclotron Resonance and Variants

Cyclotron resonance was first used to investigate the band structure of silicon and germanium (Dresselhaus et al., 1955). Intraband cyclotron resonance occurs whenever the energy between the $n = 0$ and $n = 1$ spin-up conduction-band Landau levels is equal to the incident photon energy. This can be expressed simply as

$$\hbar\omega = \hbar\omega_c = E_a^c(1) - E_a^c(0), \tag{34}$$

where $\hbar\omega$ is the incident photon energy and $E_a^c(n)$ is the energy of the nth spin-up conduction-band Landau level. In addition, the defining equation for the cyclotron resonance effective mass m_c^* is

$$\hbar\omega = \hbar\omega_c, \tag{35}$$

where $\omega_c = eB/m_c^*$ and B is the magnetic-field strength. This equation is traditionally used for cyclotron resonance even for the case of nonparabolic energy bands. Reflecting back to the selection rules discussed earlier in section 4c, we see that cyclotron resonance is a result of the selection rule $a(n) \to a(n + 1)$ and thus requires left circularly polarized light to be observed.

For InSb, extensive studies of cyclotron resonance have been carried out at a variety of temperatures. One of the first measurements of cyclotron resonance in InSb was carried out by Burstein et al. (1956a), who obtained cyclotron resonance signals at room temperature in p-type material. Later, detailed studies of cyclotron resonance followed, such as those carried out by Summers et al. (1968), who investigated the polaron coupling of the optical lattice modes to the magnetic states of both conduction and donor electrons in InSb. The data from a representative set of investigators is presented in Fig. 20 and will be discussed along with other intraband data. In addition to the fixed-temperature measurements, several investigators (e.g., Koteles and Datars, 1974; Matsuda and Otsuka, 1979) have used cyclotron resonance to determine the temperature dependence of the effective mass in InSb. For HgCdTe, cyclotron resonance has been observed and studied in samples with x-values lying in the range $0.15 \le x \le 0.25$ (see review of Dornhaus and Nimtz, 1985). Of particular note was an extensive study by Kinch and Buss (1971), who studied cyclotron resonance in the far-infrared spectral region from 10 to 330 cm^{-1} at low temperatures and observed polaron effects in a sample of $x = 0.204$, yielding a Fröhlich coupling constant $\alpha = 0.037 \pm 0.008$.

Harmonics of cyclotron resonance were first observed by Johnson and Dickey (1970). They observed resonant structure in the transmission spectra of InSb at magnetic field strengths corresponding to the energy conservation condition

$$\hbar\omega = n\hbar\omega_c = E_a^c(n) - E_a^c(0), \tag{36}$$

where $n = 1$ is the condition for cyclotron resonance. Harmonics of cyclotron resonance are thus the result of electrons in the spin-up $n = 0$ Landau level being photoexcited to the spin-up state of some $n > 1$ Landau level. At this point it is well to note that because of the nonparabolicity of the conduction band and quantum effects, the "harmonic" transitions do not occur exactly at the energy $n\hbar\omega_c$. Thus, the notation $\hbar\omega_c$ is being used here merely for label-

5. ONE- AND TWO-PHOTON MAGNETO-OPTICAL SPECTROSCOPY   349

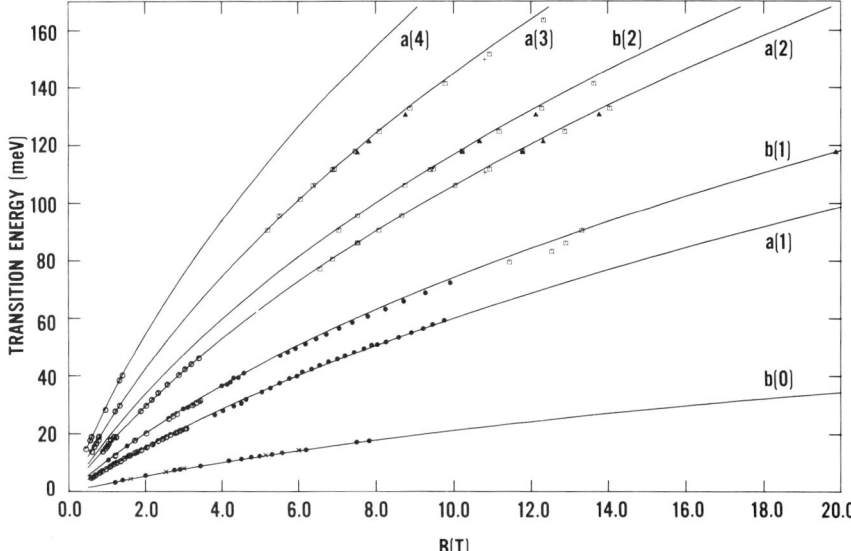

FIG. 20. Fan chart of various intraconduction band processes. The lines represent theoretical calculations obtained using the modified Pidgeon–Brown model and the band parameters listed for Littler et al. (1983). The notation $a(n)$ or $b(n)$ refers to the final spin-up or spin-down level, respectively, of the nth Landau level. The data points are from the following references: solid dots—McCombe et al. (1967), McCombe (1969), crosses—Kuchar et al. (1982); open dots—Johnson and Dickey (1970); open square—Lee (1976); solid triangle—Grisar et al. (1978); plus—Weiler et al. (1974). The labels by the lines represent electron spin resonance $[b(0)]$, cyclotron resonance $[a(1)]$, combined resonance $[b(1)]$, cyclotron resonance harmonics $[a(2), a(3), a(4)]$, and combined resonance harmonics $[b(2)]$. (From Goodwin and Seiler, 1983.)

ing purposes. The origin of the harmonic processes in terms of selection rules cannot be derived in the usual spherical approximation. However, there have been extensive theoretical investigations (Bell and Rogers, 1966; Pidgeon and Groves, 1969; Weiler et al., 1978) of these processes. Selected results from a number of investigators are also shown in Fig. 20.

In addition to cyclotron resonance and its harmonics, other types of intraconduction-band processes have been investigated. These include combined resonance and its harmonics, spin-down cyclotron resonance, and electron spin resonance. Combined resonance is a process in which an electron initially in the $n = 0$ spin-up conduction-band Landau level makes a transition to the $n = 1$ spin-down Landau level. That is, at resonance

$$\hbar\omega = \hbar\omega_c + \hbar\omega_s = E_b^c(1) - E_a^c(0), \tag{37}$$

where $\hbar\omega_s$ is the energy separation between the spin-up and spin-down Landau level and

$$\hbar\omega_s = g_c^* \mu_B B, \tag{38}$$

where g_c^* is the Landé g-factor and μ_B is the Bohr magneton. Harmonics of combined resonance are also possible, obeying the energy conservation condition $\hbar\omega = n\hbar\omega_c + \hbar\omega_s$. These processes require π polarization, and the selection rule is given in Eq. (9). Combined resonance in InSb has been extensively studied by McCombe and co-workers (e.g., McCombe et al., 1967; McCombe, 1969) and Johnson and Dickey (1970), and in HgCdTe (McCombe et al., 1970a; McCombe and Wagner, 1972), including resonant polaron behavior (McCombe, 1969; Swierkowski et al., 1978). In addition to combined resonance in InSb, harmonics of combined resonance have been observed and studied. Among the first observations were those of Grisar and co-workers (Grisar et al.; 1978; Zawadzki et al., 1978).

Spin-down cyclotron resonance can also be observed, and requires the use of left circularly polarized light. This effect is usually only observed at higher temperatures (> 77 K) in InSb where the tail of the probability distribution of carriers is "long" enough to significantly populate the higher-energy spin-down $n = 0$ Landau level. Electron spin resonance is associated with spin-flip processes and can be described by

$$\hbar\omega = \hbar\omega_s = E_b^c(0) - E_a^c(0), \tag{39}$$

where electrons in the spin-up state are photoexcited into the spin-down state of the $n = 0$ Landau level. Bemski (1960) was among the first to directly observe spin resonance using a microwave source of radiation and determined that the value for g^* in InSb decreased for increasing electron concentration for the $10^{14}-10^{15}$ cm^{-3} range and lay between -48.8 and -50.7, depending on the electron concentration. Isaacson (1968) improved on these microwave measurements and performed them over a wider range of concentrations ($\simeq 10^{13}-10^{16}$ cm^{-3}), finding little disagreement between theory and measurement. McCombe and Wagner (1971) used a far-infared laser to investigate spin resonance in order to obtain more precise measurement of g^* and compared their results with the predictions of Johnson and Dickey (1970) and Pidgeon et al. (1967a). Kuchar et al. (1982) found that the strength of the spin-resonance line depended on the amount of uniaxial stress applied to the sample. Recently, electric-dipole-excited spin-resonance experiments by Chen et al. (1985) provided extensive results on the g-factor anisotropy. In HgCdTe, the direct observation of conduction electron spin resonance was made by Bratashevskii et al. (1975, 1977), using a superheterodyne spectrometer. The spin resonance was observed as a single line whose magnetic-field position varied proportionally with the microwave frequency and was independent of sample orientation, excluding the possibility of induced cyclotron or magnetoplasma resonances.

In Fig. 20 are plotted the transition energy data of several authors (McCombe et al., 1967; McCombe, 1969; Johnson and Dickey, 1970; Lee, 1976; Grisar et al., 1978; Weiler et al., 1974; Kuchar et al., 1982; Goodwin and Seiler, 1983). The data represent most of the important work on the conduction band in InSb. The labels correspond to the final state of the transition; the initial state for all transitions shown is the spin-up $[a^c(0)]$ Landau level. The data plotted in Fig. 20 represent a mixture of transition energies obtained for both **B** parallel to the $\langle 111 \rangle$ and $\langle 100 \rangle$ crystallographic direction. The theoretical lines were calculated for the **B** $\| \langle 100 \rangle$ sample orientation by using the band parameters of Littler et al. (1983) listed in Table IX. The transition energies for the two orientations differ only slightly due to spin and orbital anisotropy, for the data presented, this difference is within experimental error. The effects of orbital and spin anisotropy on cyclotron and combined resonance transition energies has recently been observed in high-resolution measurements (Littler et al., 1990b) and will be discussed next.

The energy-band parameters that were used to describe this extensive set of conduction-band data (given in Table IX) were obtained by numerically fitting theoretically calculated transition energies from the Pidgeon–Brown model to the magneto-optical data. In addition to the intraconduction-band data just discussed and the phonon-assisted cyclotron resonance transitions to be discussed later, a set of two-phonon interband and one-photon intravalence-band data were also included in obtaining the band parameter set. The valence-band and two-photon data will also be discussed later.

Even though anisotropy affects the conduction band energies only slightly, effects of both orbital and spin anisotropy on transition energies can be observed (Littler et al., 1990b). An example of the effect of orbital anisotropy on cyclotron resonance harmonic transitions can be seen in Fig. 21a. The high-resolution magneto-optical data shown were obtained from two differently oriented samples illuminated by the same laser, where the photoconductive reponse shown represents the sum of the responses from each sample. The final Landau state quantum numbers of the observed transition are indicated by the numbers. Both the free-electron transition $a^c(0) \to a^c(2)$ [designated 2^+ (free)] and the donor-shifted transition $(000)^+ \to (200)^+$ [designated 2^+ (donor)] are shown for the magnetic-field orientations **B** $\| \langle 111 \rangle$ and **B** $\| \langle 100 \rangle$ crystallographic directions. The donor-shifted transitions will be discussed later in the chapter. Since the value of the conduction-band effective mass is directly related to the transition energy of cyclotron resonance harmonic transition 2^+, any anisotropy in the mass would in turn affect the transition energy. For a fixed excitation energy from the laser, the anisotropy

TABLE IX

COMPARISON OF BAND PARAMETERS.

	One-photon interband					Hole cyclotron & combined resonance			Intraconduction band		Others		
	Zwerdling et al. (1961)	Pidgeon & Brown (1966)	Pidgeon & Groves (1969)	Weiler (1979)	Roth & Fortin (1978)	Efros et al. (1982b)	Bagguley et al. (1963)	Robinson (1966)	Ranvaud et al. (1979)	Johnson & Dickey (1970)	Isaacson (1968) Summers et al. (1968)	Grisar et al. (1978)	Lawaetz theory (1971)	Littler et al. (1983)
E_g (eV)	0.2357 ±0.0005	0.2355 ±0.0005	0.2366	0.2329	0.235 ±0.0001	0.2368	0.236		0.235	0.2367		0.2355	0.237	0.2352
E_p (eV)		21.9	21.2	23.5 ±0.5	21.53 ±0.25	23.42	16.0		26.1	24.0		21.6	23.1	23.2
Δ (eV)		0.9	0.9	0.81	0.81		0.9		0.803			0.803	0.81	0.803
γ_1		1.5	3.51	3.4	1.86	3.44			3.1			0.5	2.59	3.25
γ_2		−1.2	−0.52	−0.3	−0.97	−0.54			−0.4			−1.0	−0.60	−0.20
γ_3		−0.1	0.65	0.9	0.13	0.51			0.7			0.1	0.66	0.90
κ		−2.1	−1.53	−1.2		−1.26			−1.5			−1.4	−1.48	−1.3
F		0.0	0.0	0.0		−1.18							0.0	−0.2
q		0.0	0.39	0.0	0.0 ±0.1	0.31			0.39			0.0	0.15	0.0
N_1				−0.3		−0.33							0.0	−0.55
m_c/m_0	0.0145	0.0144	0.0149	0.0136	0.0145 ±0.0001	0.0139			0.012	0.0139 ±0.0001	0.0137 ±0.0001	0.0145	0.014	0.0136
g_c^*	−48.0	−47.1	−45.3	−51.4	−45 ±0.2	−51.9			−55.3	−51.3	−51.3	−45.3	−48.4	−51.1
m^+/m_0	0.0149	0.0160	0.0157	0.0141	0.0160 ±0.0002	0.0144	0.021 ±0.005		0.013			0.0164	0.015	0.0143
$m^-([111])/m_0$		0.44	0.36	0.44	0.44	0.34	0.45 ±0.03	0.45 ±0.02	0.44			1.18	0.53	0.50
$m^-([110])/m_0$		0.41	0.33	0.40	0.41	0.32	0.42 ±0.03	0.42 ±0.02	0.41			0.95	0.47	0.45
$m^-([100])/m_0$		0.32	0.27	0.31	0.032	0.27	0.34	0.34 ±0.03	0.32			0.60	0.35	0.35

(From Littler et al., 1983.)

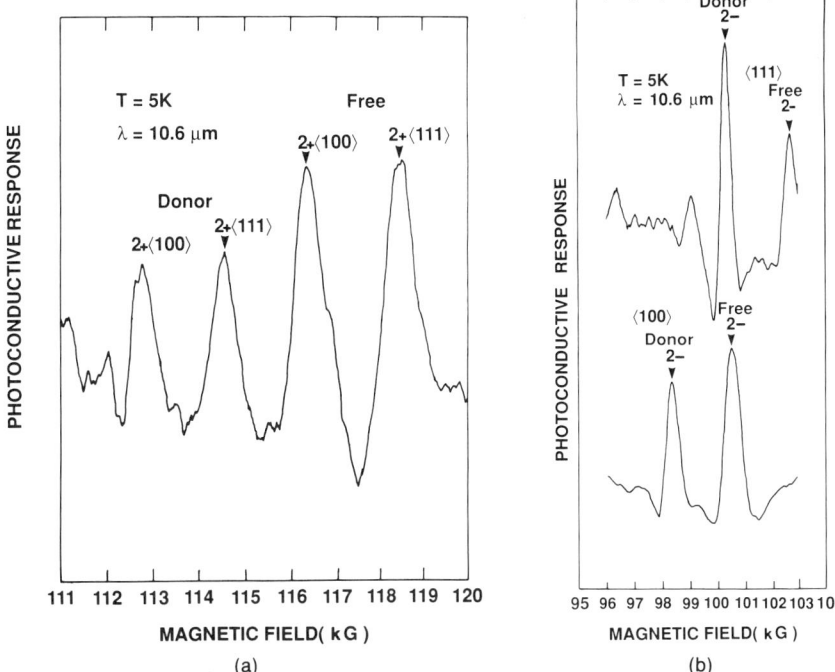

FIG. 21. (a) Photoconductive response obtained from two samples of InSb oriented $\mathbf{B} \parallel \langle 111 \rangle$ and $\mathbf{B} \parallel \langle 100 \rangle$, respectively, showing the effects of orbital anisotropy on the cyclotron resonance harmonic transition $a^c(0) \to a^c(2)$ [labeled 2^+ (free)] and its donor-shifted variant $(000)^+ \to (200)^+$ [labeled 2^+ (donor)]. (b) Combined resonance harmonic transition $a^c(0) \to b^c(2)$ and its donor-shifted variant (labeled 2^-) for $\mathbf{B} \parallel \langle 111 \rangle$ and $\mathbf{B} \parallel \langle 100 \rangle$.

would then manifest itself by a shift in the resonant-field position of the transition. This is indeed seen, as shown in Fig. 21a. Also, from this figure, the anisotropy is seen to be nearly the same for both the free-electron transitions and their donor-assisted variants.

Figure 21b also shows magneto-optical data for the combined resonance harmonic transition $a^c(0) \to b^c(2)$ [designated 2^- (free)] and its donor-shifted varient $(000)^- \to (200)^-$ [designated 2^- (donor)]. Note that this shift in resonance-field position of the free-electron (or donor) combined resonance transition is larger than that of the cyclotron resonance harmonic transitions. This is due to the anisotropy of g^* being larger than m^* and with both m^* and g^* contributing to the anisotropy of the transition energies. Thus, by the technique just described m^* and g^* can be simultaneously measured. The predictions of the five-level model were used to describe the anisotropy data since it describes fully the anisotropy of the effective mass and g-factor. For the 2⁻ transition at $B = 100$ kG, the five-level predictions yield $m^*(100) = 0.02511m_0$, $m^*(111) = 0.02545m_0$, $g^*(100) = -24.96$, and $g^*(111) = -23.74$. Thus, g^* is $\simeq 3.8$ times as anisotropic as m^*. This anisotropy is somewhat less than that predicted by the modified Pidgeon–Brown model (Weiler et al., 1978), but agrees with the results of Chen et al. (1985). The following five-level energy-band parameters for InSb are (Littler et al., 1990b): $E_{P_0} = 23.448$ eV, $E_{P_1} = 4.699$ eV, $E_Q = 13.99$ eV, $C = -0.5$, $C' = -0.05$, $E_1 = 3.11548$ eV, $\Delta_1 = 0.39$ eV, $E_0 = -0.2352$ eV, $\Delta_0 = -0.803$ eV, and $\bar{\Delta} = -0.163$ eV. These parameters yield the band-edge values $m_0^* = 0.01365m_0$ and $g_0^* = -50.9$. The nonresonant polaron correction $(1 + \alpha)/6$, where $\alpha = 0.2$, was included in the calculation of the band-edge effective mass.

b. *Phonon-Assisted Cyclotron Resonance Harmonics*

In addition to the usual cyclotron, combined, and electron spin resonance, phonon-assisted cyclotron resonance and its harmonics (PACRH) have been observed and studied extensively in n-InSb. The theory for PACRH, first presented by Bass and Levinson (1966), showed that a longitudinal-optical phonon could interact with the electron–photon system created by the interaction of light with matter to create absorption resonances. Selection rules for PACRH are determined by a convolution of both the electron–photon and electron–phonon interactions and the transition probability from second-order perturbation theory describing the electron–photon, electron–phonon interactions. The electron–photon interaction for the light polarization σ_L allows a change in Landau-level number from n to $n + 1$ for a transition from the initial to the intermediate or intermediate to the final state for the same spin. In addition, the electron–phonon interaction allows a transition from a state n to any state n', but is strongly spin conserving. Thus, transitions are allowed between any Landau levels n and n' because of the lack of selection rules for the electron–phonon interaction.

Even though transitions for PACRH are allowed from states of quantum n to states of quantum number $n + 2$, etc., due to the large number of electrons in the $n = 0$ spin-up Landau level [$a^c(0)$] at low temperatures, most PACRH transitions would be expected to originate from $a^c(0)$. In light of this, energy conservation for PACRH can be written

$$\hbar\omega = n\hbar\omega_c + \hbar\omega_{LO} = E_a^c(n) - E_a^c(0) + \hbar\omega_{LO}, \quad (40)$$

where $\hbar\omega_{LO}$ is the LO phonon energy.

Figure 22 shows the magnetic-field positions of PACRH absorption resonances obtained at several photon energies from the work of several authors (Johnson and Dickey, 1970; Grisar et al., 1978; Lee, 1976; Enck et al., 1969;

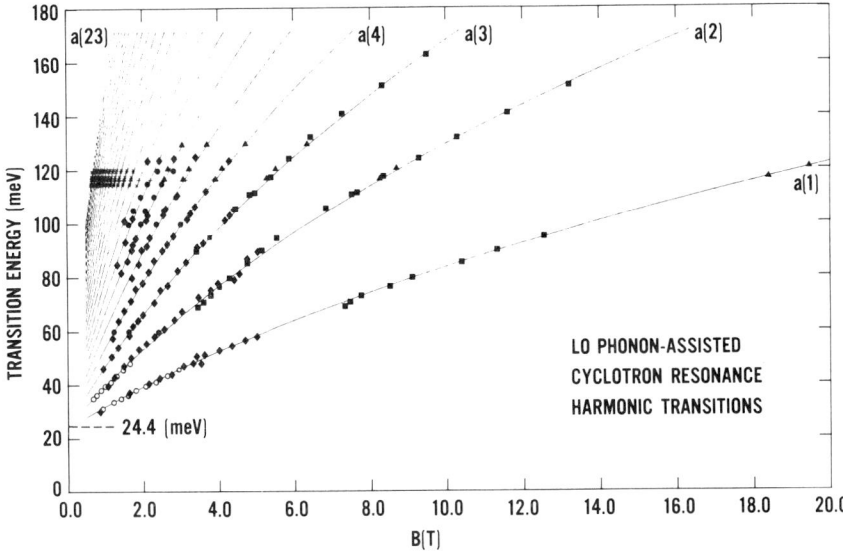

FIG. 22. Fan chart of PARCH for several authors. The labels are the final state of the transition; the initial state for all transitions is the $a^c(0)$ Landau level. The intercept corresponds to the longitudinal phonon energy (24.4 meV). The data points are from the following references: open dot—Johnson and Dickey (1970); solid triangle—Grisar et al. (1978); solid box—Lee (1976); solid dot—Enck et al. (1969); plus—Morita et al. (1975); solid diamond—Ivanov-Omskii et al. (1978); crosses—Goodwin and Seiler (1983). (From Goodwin and Seiler, 1983.)

Morita et al., 1975; Ivanov-Omskii et al., 1978; Goodwin and Seiler, 1983). With the set of band parameters for InSb discussed earlier (Littler et al., 1983), transition energies for PACRH (solid lines) were calculated and compared with the data. From the fit to the data shown in Fig. 22, a value of 24.4 ± 0.2 meV was extracted for the LO phonon energy $\hbar\omega_{LO}$. This agrees very well with the value obtained by Johnson and Dickey (1970), who determined $\hbar\omega_{LO}$ from their measurements to be 24.4 ± 0.3 meV. Excellent agreement of the theory with a wide variety of data is shown.

c. Free-Hole Cyclotron and Combined Resonance

In contrast to the many free-carrier processes observed in the conduction band of InSb, only hole cyclotron and combined resonance have been observed in p-InSb. This is due, in part, to the relative lack of attention that has been paid to the valence band in contrast to the numerous conduction-band studies reported and to the added difficulty in the interpretation of the results obtained due to the greater complexity of the valence bands in InSb.

In addition, the strength of second-order processes, such as PACRH, depend inversely on the mass of the initial state of the transition. For NGS, the heavy-hole states are the lowest lying and thus first populated; thus, transition probabilities would be reduced by m_{hh}^*/m_c^* over those for the conduction band ($\simeq 36$ for InSb).

In order to observe free-carrier absorption effects in p-InSb (or any p-type NGS), a significant free-hole population must first exist. This population can be created either via thermal ionization of shallow acceptor impurities or via the generation of free carriers by light or other radiation sources. The selection rules given in Eq. (9) apply equally for free-hole transitions involving heavy- to heavy-, heavy- to light-, and light- to light-hole Landau levels. Thus, the particular transitions observed depend on the population of the initial states and the range of excitation energies employed.

One of the first cyclotron resonance measurements on p-InSb was carried out by Bagguley *et al.* (1963), using microwaves of 4 mm and 8 mm in wavelength. From these measurements he extracted resonable values for the light- and heavy-hole masses and obtained information on carrier collision times. Later, Button *et al.* (1968) used an HCN laser operating at 337 μm and obtained magneto-optical spectra in which seven cyclotron resonance lines were resolved. In this study, heavy-hole resonances were resolved for the first time and the spectra presented showed the effects of hole anisotropy.

Robinson (1966) extended the cyclotron resonance studies on p-InSb to liquid helium temperatures, using light from a tungsten bulb source to generate free carriers. Employing a microwave spectrometer operating at 35 GH$_3$/sec, he obtained cyclotron resonance spectra for $\mathbf{B} \parallel \langle 100 \rangle$ and $\mathbf{B} \parallel \langle 110 \rangle$. He determined that the differences in the line shapes obtained at low temperatures for these two orientations were due to a splitting of the heavy-hole valence band in the vicinity of $\mathbf{k} = 0$.

Ranvaud *et al.* (1979) and Trebin *et al.* (1979) studied both cyclotron and combined resonance in p-InSb under the application of uniaxial stress. They showed that the application of stress simplifies the hole cyclotron resonance spectra by destroying the cubic symmetry responsible for the degeneracy of the valence bands at $\mathbf{k} = 0$. They used an HCN laser (small photon energies) to study cyclotron and combined resonance transitions near the valence-band edge of p-InSb in detail. From this study a complete set of energy-band parameters for the Pidgeon–Brown model (given in Table IX) for InSb was obtained.

Littler *et al.* (1981, 1983) later used the output of a CO_2 laser to study combined resonance transitions in p-InSb. The higher photon energies of the CO_2 laser allowed these transitions to be observed deep in the valence band, away from the complications the band structure near the band edge resulting from the effects of degeneracy. For these energies the free-hole transitions tend to group around the closely spaced pairs of light-hole Landau

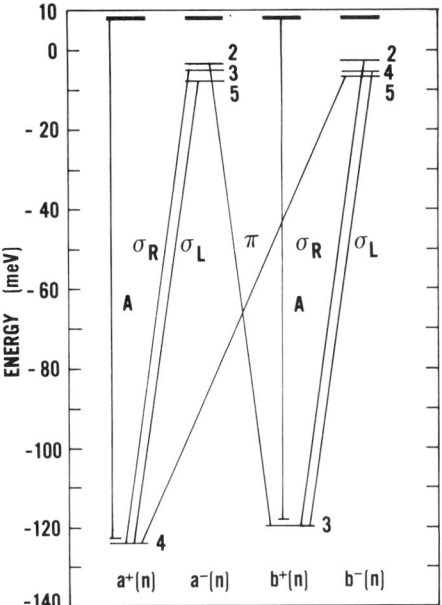

FIG. 23. Diagram of allowed free- and bound-hole transitions involving the $a^+(4)$ and $b^+(3)$ light-hole Landau levels at $B = 6.0$ T for $\mathbf{B} \parallel \langle 111 \rangle$. At low temperatures, the bound-hole transitions (labeled A) originate from the ground-state acceptor level shown. At high temperatures, a number of closely spaced (in transition energy) free-hole transitions between heavy- and light-hole Landau levels [labeled $a^-(n)$, $b^-(n)$ and $a^+(n)$, $b^+(n)$, respectively] are possible. (From Littler et al., 1983.)

levels, as illustrated in Fig. 23. However, as illustrated in Fig. 24, the transition strengths for combined resonance are greater than that for cyclotron resonance in the case where all polarizations of light are present, and the magneto-optical resonances observed can be described as an unresolved pair of combined resonance transitions to the closely spaced light-hole levels.

It is well known that the hole masses of NGS are anisotropic. The anisotropy of the hole masses, in turn, affects the associated Landau-level energies E^{LL}, since $E^{LL} \propto 1/m^*$. Thus, for combined resonance transitions between heavy- and light-hole states, the effects of anisotropy should be observable in the spectra obtained for different sample orientations in a magnetic field. An example of how the crystallographic anisotropy of the heavy- and light-hole masses affects the observed combined resonance spectra is shown in Fig. 25. It is seen that the minima in the $\mathbf{B} \parallel \langle 100 \rangle$ spectra occur at higher magnetic fields than the $\mathbf{B} \parallel \langle 111 \rangle$ minima. In addition, the relative shift between the corresponding minima is seen to decrease with increasing field.

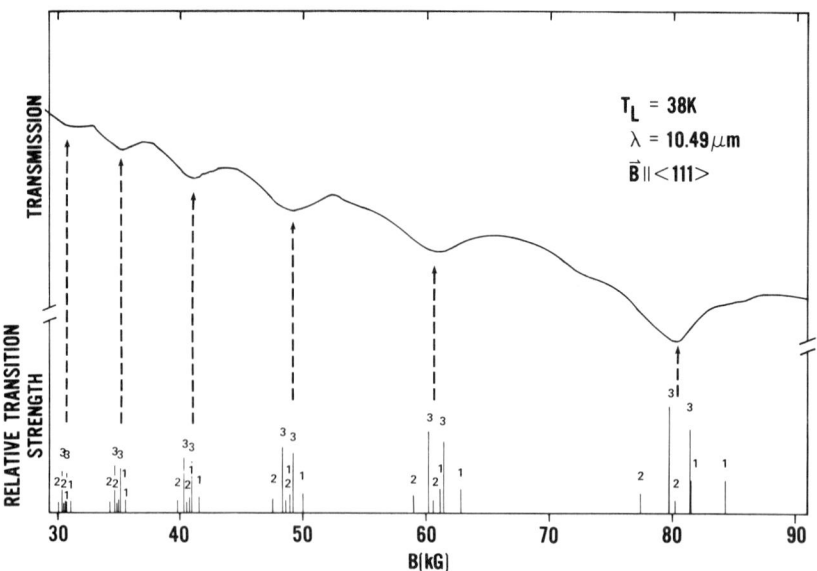

FIG. 24. Comparison of free-hole transmission spectra obtained from p-InSb at λ = 10.49 μm with theoretically calculated field positions and relative transition strengths for different light polarizations. The numbers indicate the light polarization: $\sigma_L = 1$, $\sigma_R = 2$, $\pi = 3$. The arrowed-dashed lines show the one-to-one correspondence between the average magnetic-field position of the combined resonance (π) transitions and the observed transmission minima. (From Littler et al., 1983.)

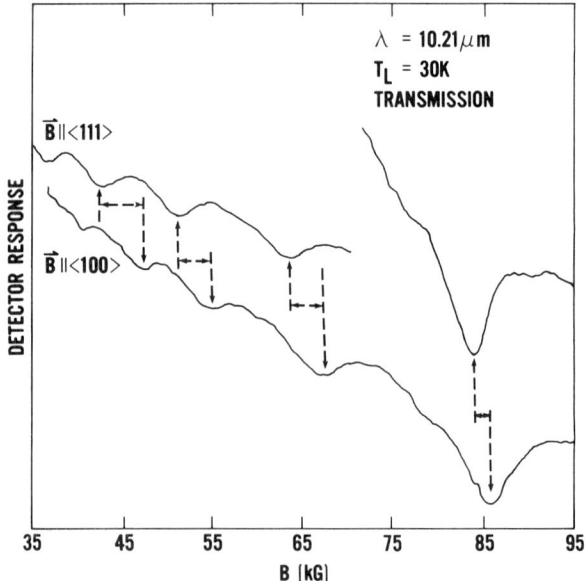

FIG. 25. Free-hole transmission spectra obtained at 10.21 μm for **B** ∥ ⟨111⟩ and **B** ∥ ⟨100⟩. The arrowed-dashed lines show the shift in the resonant minima field positions due to the effects of heavy- and light-hole anisotropy. (From Littler et al., 1983.)

5. ONE- AND TWO-PHOTON MAGNETO-OPTICAL SPECTROSCOPY

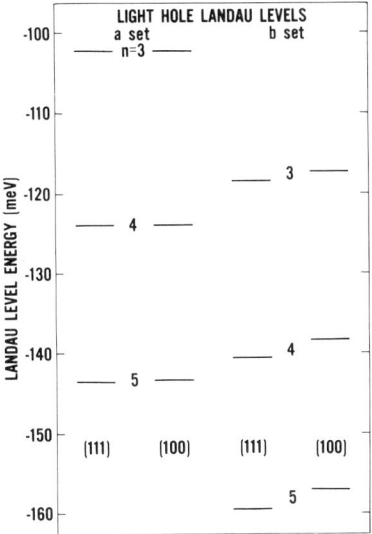

FIG. 26. Diagram of heavy- and light-hole Landau-level energies for $n = 1$ and 2 at 10.0 T and $\mathbf{B} \parallel \langle 111 \rangle$ and $\mathbf{B} \parallel \langle 100 \rangle$, showing the relative anisotropy of each level. (From Littler et al., 1983.)

These features can be explained by looking closely at the anisotropy of the light- and heavy-hole Landau-level energies. As illustrated in Fig. 26, both the a and b set of heavy-hole Landau-level energies are anisotropic, while the b set of light-hole energies displays significantly larger anisotropy than does the a set. Thus, for transitions obeying the selection rule $b^-(n) \to a^+(n-1)$, the shift in corresponding minima is predominantly influenced by the heavy-hole anisotropy, while for transitions obeying $a^-(n) \to b^+(n+1)$ the contributions to the observed shift come roughly equally from the heavy- and light-hole anisotropy.

Figure 27 shows a comparison of both theoretical (solid lines) and experimental (solid dots) combined resonance transition energies $\mathbf{B} \parallel \langle 111 \rangle$. An equally good fit for $\mathbf{B} \parallel \langle 100 \rangle$ is also obtained.

12. IMPURITY- AND DEFECT-RELATED LASER SPECTROSCOPY

Detection and identification of impurities and defects in semiconductor materials have long been topics of technological importance. Of particular interest is the location of the impurity and defect states within the forbidden energy-gap region. For narrow-gap materials, magneto-optics becomes a technique of choice, since resonant transitions from impurity-to-band or

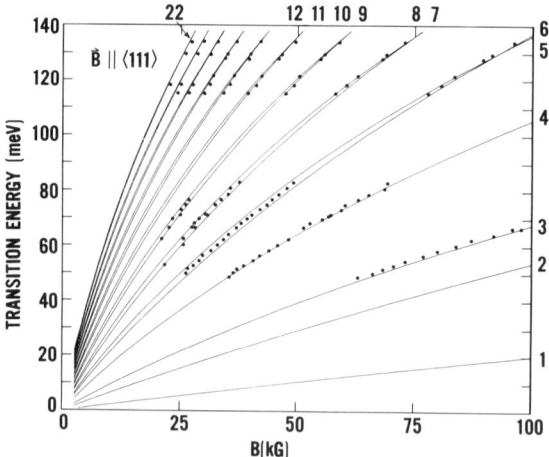

FIG. 27. Fan chart of combined resonance transition energies for **B** ∥ ⟨111⟩. The numbers correspond to the following transition assignments: $1-b^-(1) \to a^+(0)$; $2-a^+(-1) \to b^+(0)$; $3-b^-(2) \to a^+(1)$; $4-b^-(3) \to a^+(2)$; 5, 6, ..., 21, 22 in the ascending sequence $a^-(n) \to b^+(n+1)$ and $b^-(m) \to a^+(m-1)$, where n and m start at 1 and 4, respectively. (From Littler et al., 1983.)

band-to-impurity states allow for the precise determination of just this location. The narrow-gap semiconductors InSb and HgCdTe have been, for the last two decades, the materials of choice for use in infrared detection systems. In these materials, impurities and defects, even in low concentrations, have dramatic effects on device or detector performance. Specifically, detailed information about the energy-band structure and the levels created by the presence of impurities and defects is crucial to understanding and improving the performance of current infrared detectors.

Laser-based magnetospectroscopy has become a very useful technique for the investigation of shallow and deep impurity or defect states in NGS. The use of a laser provides the ability to observe and study impurities and defects even in low concentrations because of the high intensities available. Also, the laser provides essentially single-frequency light, which in turn results in narrower linewidths in the magneto-optical spectra than can be obtained using conventional light sources. In this section, we will detail the impurity or defect work done on InSb and HgCdTe, concentrating on the laser-based measurements that have been performed in the last decade and their contributions to the detection and identification of impurity and defect levels in InSb and HgCdTe.

The origin of impurity- and defect-level magneto-optical transitions can be understood qualitatively by the simplified Landau-level model schematically presented in Fig. 28. The two-photon magnetoabsorption transitions

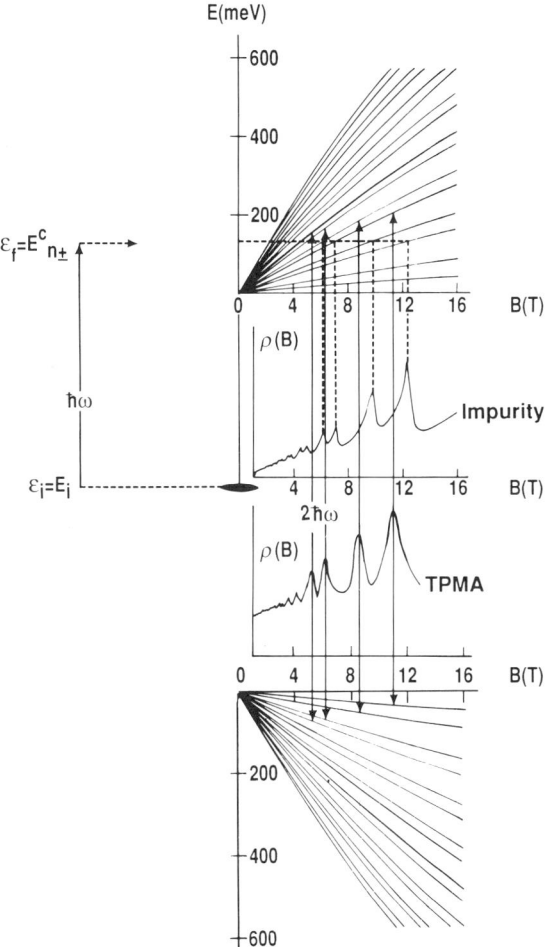

FIG. 28. Schematic representation of the origin of TPMA and impurity- or defect-to-band resonant magneto-optical spectra. The two-photon transitions are represented by the solid arrows and the impurity or defect transitions by the dashed lines. (From Littler et al., 1990a.)

(discussed later) are indicated by the solid arrows and the impurity to defect to conduction-band transitions are indicated by the dashed arrows. Resonant absorption occurs whenever the energy separation between these Landau levels equals the energy prescribed by the particular absorption phenomena. For the two-photon interband transitions, magneto-optical resonances will occur whenever $2\hbar\omega$ is equal to the energy difference between particular valence- and conduction-band Landau levels, in accordance with appropriate selection rules. A sketch of the resulting magneto-optical spectra is shown in the middle of Fig. 28. If, however, the initial state of the magneto-optical transition is a level or state within the forbidden energy gap, then a different magneto-optical spectrum will be seen. These resonances will occur only when $\hbar\omega$ is equal to the energy separation between the impurity or defect level and the conduction-band Landau levels, and follows no particular selection rule. One possible explanation for the absence of selection rules is that the impurity parts of the transition matrix element are nonzero for any initial or final state. This results from the interaction of the free carrier with the Coulomb field of the impurity or defect present (Zawadzki, 1980). Thus, each absorption process posesses its own characteristic spectrum and thus can be so identified.

a. Shallow Levels

Shallow donor states in semiconductors, since their discovery in the magnetic freeze-out effect by Keyes and Sladek (1956) and the pioneering theoretical descriptions of Yafet *et al.* (1956), have been the subject of sustained experimental and theoretical interest. This interest is due, in part, to the fact that the shallow levels control the carrier concentration in the extrinsic regime of the semiconductor, which is important for technological reasons, as well as the fact that the shallow donor system imitates the hydrogen atom. The effect of a magnetic field on shallow donor states is particularly important in narrow-gap materials with small effective masses m^* since, in the absence of the field, the donors are always ionized and cannot be observed. Magneto-impurity investigations have been used primarily to determine such material parameters as the static dielectric constant (Stillman *et al.*, 1969) and its pressure dependence (Zawadzki and W. Lasak, 1984), to identify the chemical nature of impurities (Kuchar *et al.*, 1984), to study screening properties of the electron gas (Fenton and Haering, 1967), and to investigate the metal–nonmetal transition (Robert *et al.*, 1980). More recently, magnetoimpurity investigations have proven to be useful in determining the positions of donors in modulation-doped two-dimensional GaAs–GaAlAs structures (Jarosik *et al.*, 1985; Zawadzki *et al.*, 1987).

i. InSb The first detailed measurements on the optical spectra of shallow donor impurities in InSb were made by Kaplan (1968a), who conducted experimental studies using conventional light sources of the donor-impurity excitation spectra in high magnetic fields. These measurements yielded good qualitative agreement with theoretical predictions with regard to transition energies, selection rules, and absorption coefficients, and quantitatively demonstrated the need for including the effects of nonparabolicity in the theory and indicating that certain corrections to the donor-impurity ground-state energy were required (e.g., central cell corrections). The effect of optic phonons on optical transitions between magnetodonor states in semiconductors was observed by Kaplan and Wallis (1968) in the form of resonant polaron behavior in cyclotron resonance, and by McCombe and Kaplan (1968) in combined cyclotron and spin resonance, yielding information on polaron coupling in InSb.

Far-infrared lasers have also been employed to investigate the properties of shallow donors in InSb. In studies by Kaplan et al. (1978), central-cell structure arising from different residual impurities was observed and investigated. Later, Kuchar et al. (1984) extended the work, identifying specific donor species by the observation of the central-cell structure in the donor excitation spectra of deliberately doped samples. In this study, absorption lines due to Se, Te, and Sn impurities were identified.

Hydrostatic pressure experiments are generally very useful for the detection of impurity levels that are initially resonant with the energy continuum of the host material and/or are associated with a subsidary minimum of the band (Paul, 1968). A study of the pressure coefficients obtained enables one to distinguish between shallow and deep states (Jantsch et al., 1982) and, in the case of shallow levels, to specify the minimum from which they originate. Recently, Brunel et al. (1986), using a far-infrared laser, studied four intraimpurity transitions between the shallow states of residual donors in magnetic fields as high as 19 T and hydrostatic pressures up to 1.1 GPa in nominally undoped samples of n-InSb. The study of the observed chemical shifts for the observed donors (transitions involving the ground state) allowed the determination of the matrix elements of the localized parts of the impurity potentials, yielding information on the localization potential. For the deepest of the donors this matrix element showed anomalous behavior; it increased significantly with pressure, possibly due to a local lattice distortion. In addition, the transitions between excited states (not affected by the central-cell potential) were compared directly with a multiband approach proposed by Trzeciakowski et al. (1986), yielding good agreement.

Transitions between magnetodonor (MD) states where the quantum numbers of the final states are greater than 1 have been observed with far-infrared (Huant et al., 1985) and CO_2 lasers (Grisar et al., 1978; Littler et al., 1989). The larger photon energies provided by the CO_2 laser allow one to probe high-quantum-number transitions (up to $n = 13$) at magnetic fields up to 12 T. Magnetodonor transitions assisted by the emission of optic phonons have also been observed. Due to the phonon assistance, momentum transfer is introduced into the transition, breaking the usual angular momentum selection rules. This in turn allows the study of very high excited states of the magneto-Coulomb system. An example of the magneto-optical spectra using FIR lasers is shown in Fig. 29 and for the CO_2 laser is shown in Fig. 30. For

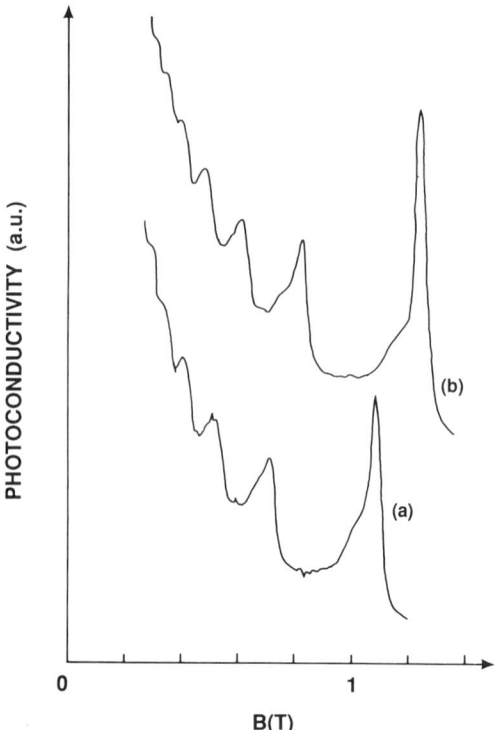

FIG. 29. Photoconductivity spectra obtained at $\lambda = 70.51$ μm for (a) P = 0 MPa and (b) P = 300 MPa. (Reprinted from Huant, S., Brunel, L. C., Baj, M., Dmovski, L., Coron, N., and Dambier, G., *Solid State Commun.* **54**, 131, © 1985 Pergamon Press plc.)

the CO_2 laser results, the observed peaks exhibit a doublet structure in which the higher-field peaks are due to free-electron transistions between Landau levels (LL) and the lower-field peaks are related to the corresponding transitions between MD levels. This assignment is confirmed by the temperature dependence of the structure shown in the insert of Fig. 30, because at higher temperatures thermal excitations depopulate the ground MD state, populate the lowest Landau state, and reverse the relative heights of the peaks in the doublets.

The quantum numbers of the final states involved in the transitions are also detailed in Fig. 30. These have been established by correlating the transition energies with theory, as shown in Fig. 31. Energy conservation for the MD transition yields the resonance condition

$$\hbar\omega = E_f^{MD} - E_i^{MD}(+\hbar\omega_{LO}), \tag{41}$$

5. ONE- AND TWO-PHOTON MAGNETO-OPTICAL SPECTROSCOPY

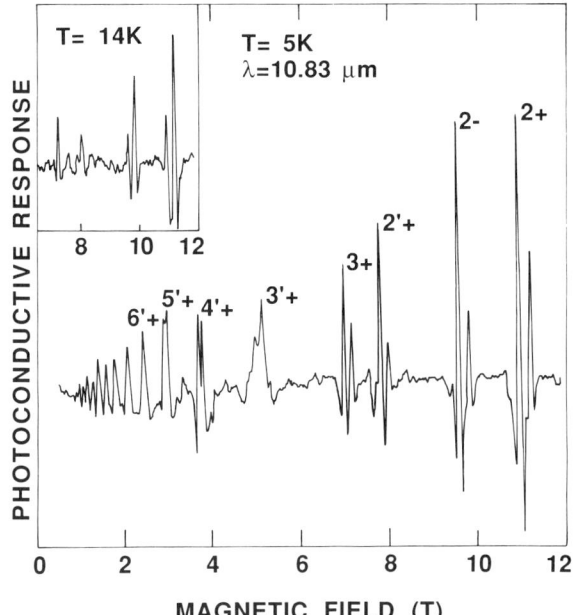

FIG. 30. Photoconductive response versus magnetic field for n-InSb at $\lambda = 10.83$ μm. The numbers indicate the final Landau states associated with the free-electron transitions. The primed numbers indicate the phonon-assisted transitions. The high field peak of each doublet is the free electron resonance, the lower field peak is the corresponding MD resonance. Inset: The magneto-optical spectra at higher temperature. (From Zawadzki et al., 1990.)

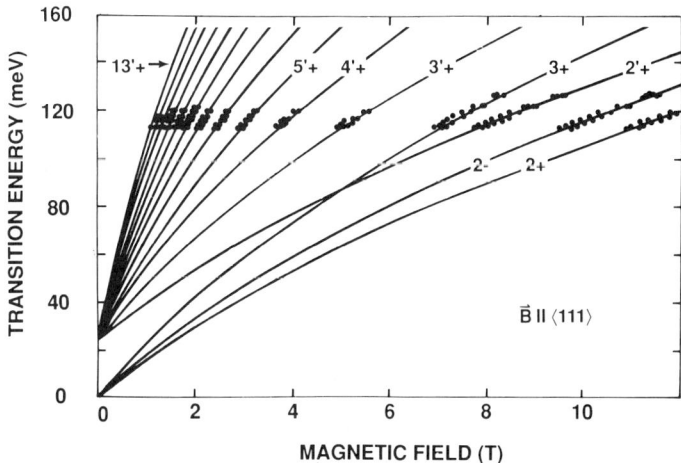

FIG. 31. Fan chart of the observed free-electron and MD transitions, indicating both simple excitations (converging to $\hbar\omega = 0$) and phonon-assisted excitations (converging to $\hbar\omega = \hbar\omega_{LO}$). (From Zawadzki et al., 1990.)

where $\hbar\omega$ is the photon energy, E_f^{MD} is the final magnetodonor state, E_i^{MD} is the initial magnetodonor state, and $\hbar\omega_{LO}$ is the longitudinal-optical energy when phonons are involved in the transition. For the field orientation $\mathbf{B} \parallel (111)$ and the light polarization $\mathbf{E} \perp \mathbf{B}$, the $2\omega_c$ and the $2\omega_c + \omega_s$ transitions are allowed due to the inversion asymmetry of InSb (Weiler et al., 1978). The observed transition $3\omega_c$ is not allowed. The current interpretation is that such forbidden transitions may become allowed due to the presence of impurities in the crystal (Zawadzki, 1980; Huant et al., 1985). Judging by the doublet structure, the corresponding donor-shifted transitions are allowed for the same reasons.

The transitions labeled by the primed numbers are optically induced electron excitations accompanied by the emission of an optical phonon. As seen from the data in Fig. 30, the corresponding phonon-assisted MD excitations are also clearly resolved. Figure 32 compares the high-resolution experimental transition energy data with the theoretical predictions. The solid lines for the LL transitions were calculated by using the Pidgeon–Brown model with the band parameters of Littler et al. (1983) listed in Table IX. The phonon-assisted energies were obtained by adding the LO phonon energy $\hbar\omega_{LO} = 24.4$ meV to the calculated LL energies. The dashed lines for the MD transitions have

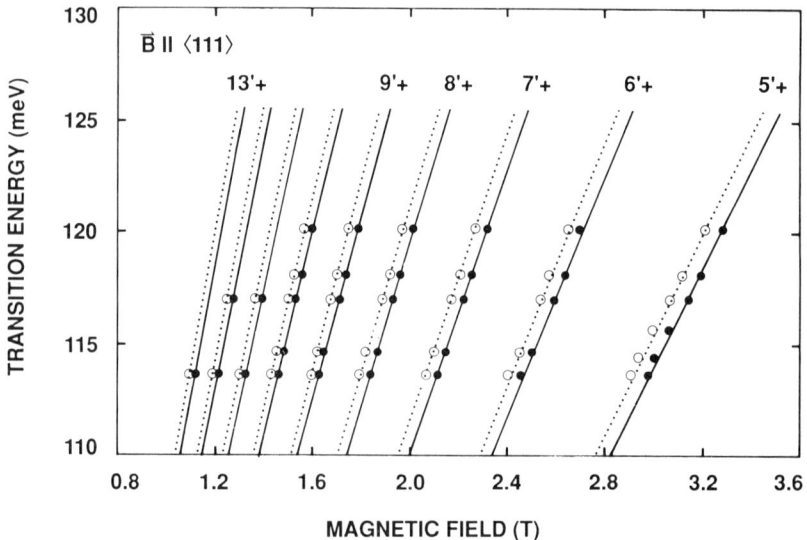

FIG. 32. Energies of the free-electron (solid circles) and MD (open circles) transitions versus magnetic field for the Landau quantum numbers $n = 5$ to 13. The solid lines were calculated by using the Pidgeon-Brown model, and the dashed lines were obtained by adding calculated donor shifts. (From Littler et al., 1989.)

been calculated by adding the corresponding MD shifts to the LL transition energies (up to $n = 13$), and using 0.65 meV for the effective Rydberg.

Impurity transitions from shallow acceptor states in InSb fall into two categories: the impurity to conduction-band transition f' given in Fig. 5 and the valence band to impurity transition g' also shown in this figure. Indirect evidence for the first type of transistion can be seen in the work of D. Miller et al. (1978,1979), who suggested that these transitions were responsible for nonlinear optical effects such as intensity-dependent refractive index and optical bistability, and MacKenzie et al. (1980), who invoked these transitions to explain intensity-dependent Faraday rotation. In addition, Hanes and Seiler (1980), invoked below-band-gap absorption resulting from shallow acceptor- to conduction-band electron transitions in degenerate samples of InSb to explain laser-induced cooling of the conduction-band electron gas.

Several magneto-optical studies of valence band to shallow acceptor-level (bound-hole transitions have been performed in the past 20 years. The first were performed by Kaplan (1968), using monochromator light sources. The results of these studies were later combined with those of Littler et al. (1981, 1983), who employed the use of a CO_2 laser to monitor these transitions deep into the valence band. Figure 33 shows photoconductive and transmission spectra obtained at one monochromator and two laser wavelengths. Note the increased resolution obtainable when lasers are used. This is due to the improved spectral purity that lasers provide. The numbers used to identify the individual transmission minima (or photoconductive maxima) refer to the light-hole Landau levels to which the final acceptor states are associated. Since the hole transitions observed proceed from the acceptor ground state to excited acceptor states closely associated with the light-hole Landau levels, energy conservation for these transitions is given by

$$\hbar\omega = E_{gs}(0) + \Delta E_{gs}(B) + \varepsilon_{lh}^{a,b}(n, B) - E_{es}(B), \tag{42}$$

where $E_{gs}(0)$ is the zero-field acceptor ground-state binding energy, $\Delta E_{gs}(B)$ is the field-induced shift of the ground state, $\varepsilon_{lh}^{a,b}(n, B)$ refers to the nth light-hole Landau level of spin a or b, and $E_{es}(B)$ is the binding energy of the excited light-hole acceptor state. The values of $\Delta E_{gs}(B)$ and $E_{es}(B)$ are small and their difference even smaller; thus, it is sufficient to write

$$\hbar\omega = E_{gs}(0) + \varepsilon_{lh}^{a,b}(n, B), \tag{43}$$

and the observed bound-hole transitions can be described by simply adding the zero-field ground-state binding energy to the light-hole Landau-level energies. Figure 34 shows the fan chart results obtained by using Eq. (43) to describe the shallow acceptor-to-band transitions. Excellent agreement is

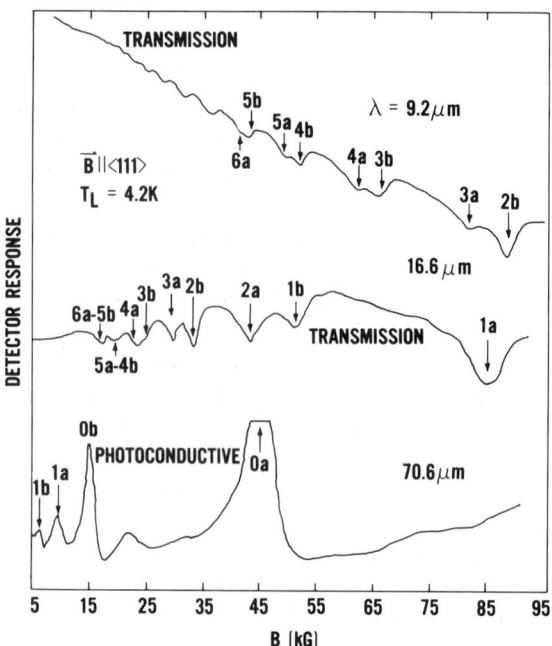

FIG. 33. Wavelength dependence of the bound-hole spectra from p-InSb for $\mathbf{B} \parallel \langle 111 \rangle$. The numbers refer to the light-hole Landau levels to which the excited impurity states are associated. (From Littler et al., 1983.)

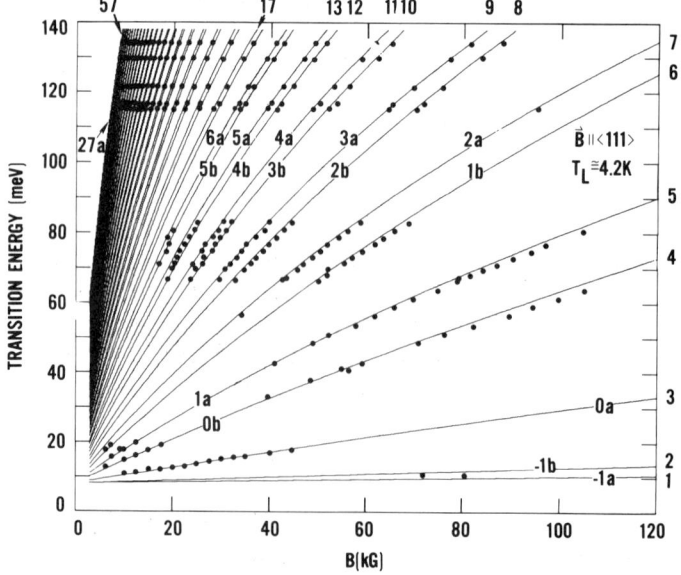

FIG. 34. Fan chart of bound-hole transition energies for $\mathbf{B} \parallel \langle 111 \rangle$. The numbers on the lines identify the light-hole Landau levels, and the numbers on the borders refer to the transition number (e.g., 1—acceptor ground state to $a^+(-1)$ or $-1a$, etc.). (From Littler et al., 1983.)

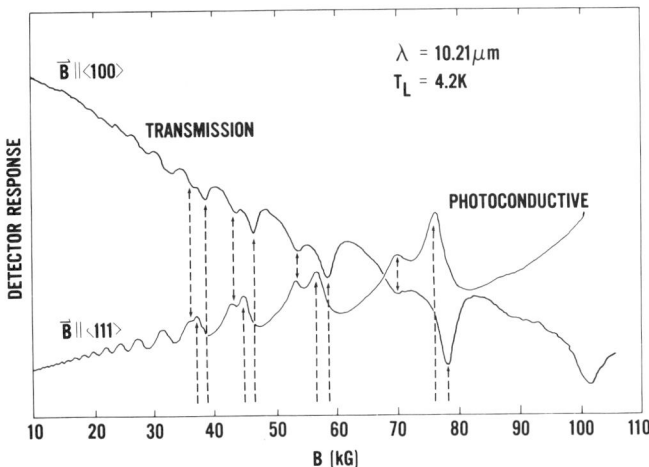

FIG. 35. Bound-hole photoconductive and transmission spectra for **B** ∥ ⟨111⟩ and **B** ∥ ⟨100⟩. The dashed-arrowed lines show how a shift in resonant field position is seen for the higher-field component of each doublet, while no shift is evident for the lower-field component. (From Littler et al., 1983.)

shown over a wide range of photon energies and probing to extremely high quantum numbers ($n = 27$) of the light-hole Landau levels.

A remarkable crystallographic-dependent anisotropy of the bound-hole transition energies in p-InSb is shown in Fig. 35. The high-field component of the doublet structure is seen to shift to higher magnetic-field values upon changing the sample orientation from **B** ∥ ⟨111⟩ to **B** ∥ ⟨100⟩, while the low-field component remains virtually stationary. This behavior is a direct consequence of the anisotropy of the light-hole Landau-level energies (assuming that the binding energies of the excited light-hole acceptor states do not show significant effects of anisotropy) and is predicted faithfully by our energy-band parameter set used in the modified Pidgeon–Brown model.

ii. HgCdTe Until recently, there has been little magneto-optical work done on shallow donors in HgCdTe. This is largely due to the extremely small binding energy of the shallow donor ($\simeq 0.5$ meV; Choi et al., 1988) and the difficulty in obtaining pure, homogeneous samples. However, with the observation of impurity cyclotron resonance by Goldman et al. (1986) came extensive magnetotransport and FIR magneto-optical studies of these shallow donor states by Drew and co-workers (Shayegan et al., 1985a,b, 1986, 1987, 1988; Goldman et al., 1986a,b; Choi et al., 1988). An example of the impurity cyclotron resonance (ICR) for a sample of $Hg_{1-x}Cd_xTe$ with $x = 0.22$ is shown in Fig. 36. The impurity data was analyzed by subtracting the

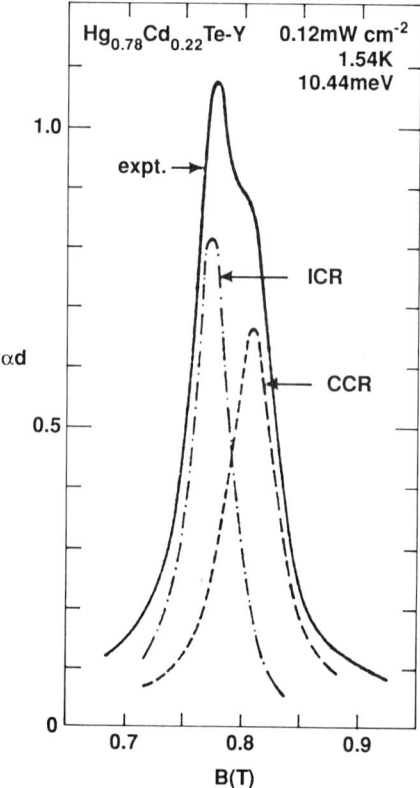

FIG. 36. Example of FIR absorption in $x = 0.22$ n-HgCdTe, showing features of both impurity cyclotron resonance (ICR) and cyclotron resonance (CCR). (From Shayegan et al., 1988.)

Lorentzian line shape obtained at higher temperatures for the free-electron cyclotron resonance (CCR) from the low-temperature experimental absorption to obtain a Lorentzian fit for the ICR transition. Figure 37b shows a plot of the resonant fields for the CCR and the ICR transitions versus photon energy. The lines shows a nonparabolic, Bowers–Yafet (1959) model calculation with the band-edge effective mass used as the fitting parameter. The energy splitting between the CCR and ICR, $(E_{110} - E_{000}) - (E_{1+} - E_{0+})$, was calculated from

$$\Delta B \simeq \left[\frac{d(E_{1+} - E_{0+})}{dB}\right]_B = B_{CR}(B_{CR} - B_{ICR}). \quad (44)$$

and is shown in Fig. 37a. The lines in this figure result from nonparabolic calculations based on a model by Larsen (1968), adapted for $m^* = 5.3 \times 10^{-3}\, m_e$

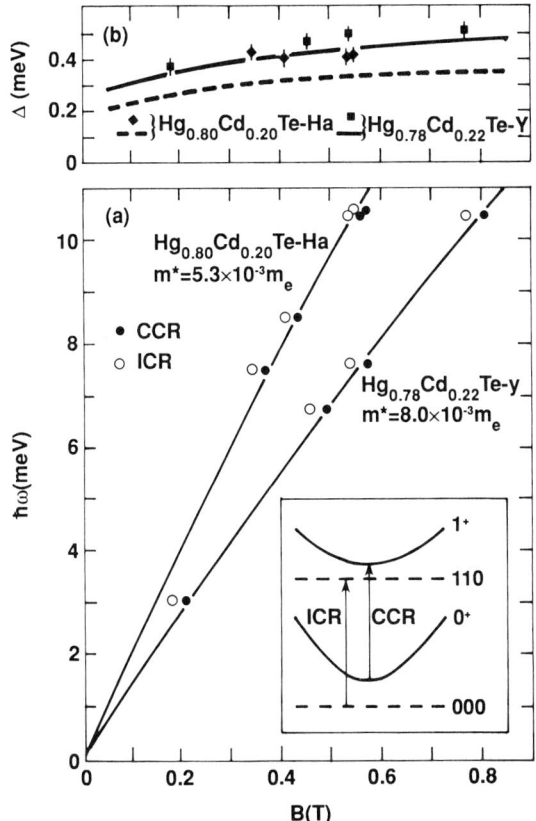

FIG. 37. (a) Resonant magnetic fields at several FIR photon energies. The lines give nonparabolic fits for the $a^c(0) \rightarrow a^c(1)$ transition, with m^* used as the fitting parameter. The inset shows the relevant energy-level scheme. (From Shayegan et al., 1988.) (b) The energy splitting $\Delta = \hbar\omega_{ICR} - \hbar\omega_{CR}$ versus magnetic field. The lines show scaled theoretical predictions of Larsen (1968). (From Shayegan et al., 1988.)

and $m^* = 8.0 \times 10^{-3} m_e$ and a static dielectric constant $\kappa = 17$. This work by Drew's group provided conclusive evidence for the presence of hydrogenic donors in HgCdTe, thus ending the controversy over whether the metal–insulator transition in narrow-gap HgCdTe is due to Wigner condensation or magnetic freeze-out (Shayegan et al., 1985a).

Until recently, shallow acceptor levels in HgCdTe have been studied using only transport techniques in p-type samples such as temperature-dependent Hall measurements (e.g., see C. T. Elliot et al., 1972; Chen and Tregilgas, 1987) or photoinduced mobility effects (Bartoli et al., 1986). Very recently, magneto-absorption has been employed in n-type samples as a new technique to study

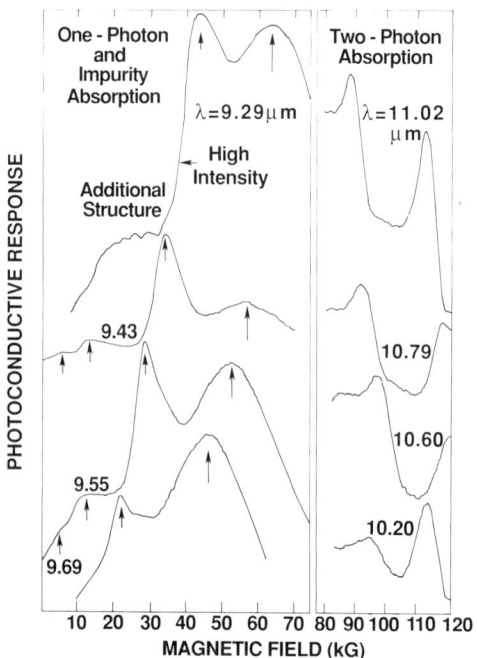

FIG. 38. Magneto-optical spectra at various CO_2 laser wavelengths for an n-HgCdTe sample with $x \simeq 0.24$. At high fields and long wavelengths, TPMA structure is seen. At lower fields and wavelengths, both OPMA (short arrows) and acceptor-to-conduction band transitions (long arrows) are seen. (Reprinted with permission from Seiler, D. G., Loloee, M. R., Milazzo, S. A., Durkin, A. J., and Littler, C. L., *Solid State Comm.* **69**, 757, © 1989 Pergamon Press plc.)

these levels. Figure 38 shows magneto-optical spectra obtained from a sample of n-type $Hg_{1-x}Cd_xTe$ of $x \simeq 0.24$. Three types of resonances are seen in Fig. 38: one-photon magnetoabsorption resonances are indicated by the short arrows, a broad absorption peak by the long arrow, and TPMA peaks occur at the highest fields. In the case of the broad peak, the temperature dependence and time-resolved behavior of the PC response indicates that this absorption peak results from resonant electron transitions from a shallow acceptor (or Hg vacancy level) to the lowest conduction-band Landau level, as does the activation energy of 9.8 meV (Kalisher, 1984; Chen and Dodge, 1986; Chen and Tregilgas, 1987; Bartoli *et al.*, 1986) extracted from the data presented in Fig. 39. To explain properly the OPMA results, an exciton correction (Weiler, 1979) was employed, which amounted to $a \simeq 2$ meV correction to the transition energies at zero magnetic field. Thus, magneto-optics can be successfully used to detect the presence of residual acceptors in n-type samples and yield an accurate value for the activation energy.

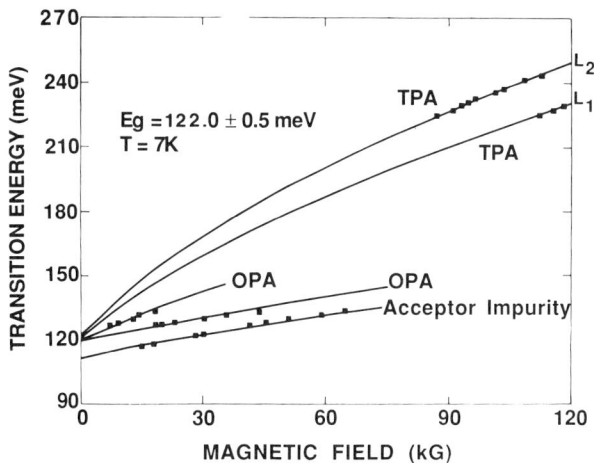

FIG. 39. Fan chart plots of OPMA, TPMA, and shallow impurity transition energies. (From Seiler *et al.*, 1989b.)

b. Deep Levels

Deep levels in semiconductors have been and still are receiving a great deal of attention, mainly because they represent a major limiting factor affecting device performance. Their presence in most cases dramatically affects a semiconductor's electrical and optical properties. Unfortunately, these levels and their origins are still poorly understood. Deep levels are associated with a potential that is short range in nature and hence is not describable by effective-mass theory. In addition, the concentration of these levels is usually small compared with the equilibrium carrier concentration of the host semiconductor, thus making their detection extremely difficult, especially for NGS.

The effects of deep levels usually show up most predominantly in samples with low carrier concentrations. It is this that prevents many optical spectroscopic techniques, notably absorption or reflection, from being useful in studying the properties of deep levels. In contrast, it is just this experimental condition that makes photoconductivity measurements extremely powerful. Because of the low number of thermal equilibrium electrons in the conduction band, it is possible to detect a small percentage (as little as 1 part in 10^4) of photoexcited carriers. In addition, at low temperatures, electrons excited high into the conduction band give rise to enhanced mobilities. Free-carrier absorption, which can obscure the observations of optical transitions from deep levels, is also minimized by low equilibrium carrier concentrations.

TABLE X

SUMMARY OF VARIOUS DEEP-LEVEL MEASUREMENTS IN InSb.

Year	Measurement	Temperature (K)	n- or p-type concentration (cm^{-3} at 77 K)	E_1 (meV)	E_2 (meV)	E_3, E_4, E_5 (meV)	References
1959	Lifetime	77–300	$p, 10^{15}$–10^{18}	170–180			Zitter et al.
1961	Lifetime	15–200	$n, p, 10^{14}$–10^{15}	175	110		Laff and Fan
1962	Lifetime	90–180	$n, p, 10^{12}$–10^{14}	170–180	100–110	70–80	Nasledov et al.
1962	Hall coef., mobility	77–300	$p, 10^{12}$–10^{13}		106		Cunningham et al.
1965	Recombination	4.2	$p, 10^{14}$			27, 42, 50	Pehek and Levinstein
1967	Hall coeff., transmission	55–300	$p, 10^{14}$		120		Baryshev et al.
1967	Hall coeff., lifetime	<150	$n, p, 10^{13}$–10^{15}	159	120		Hollis et al.
1967	Lifetime	77	$n, 10^{12}$–10^{13}	165	110	40	Abduvakhidov et al.
1969	Hall coeff.	78–250	$p, 10^{12}$–10^{13}		120		Galavanov and Oding
1969	Hall coef. Conductivity, mobility, electric-field dependence	4.2–200	$n, 10^{12}$–10^{14}			18, 67	Ismailov et al.
1970	Noise	77–200	$n, 10^{13}$–10^{14}		100	30–40	Heyke et al.
1971	Hall coeff., mobility	4.2–200	$n, 10^{11}$–10^{13}		100–128		Trifonov and Yaremenko
1971	Lifetime	4.2–77	$n, 10^{14}$		125		Guseinov et al.
1973	Noise	77	$p, 10^{12}$	165 ± 5	110 ± 3		Galavanov et al.
1973	Absorption	10	$n, p, 10^{12}$–10^{13}	168, 187	103		Valyashko and Pleskacheva
1975	Lifetime	77–200	$n, 10^{14}$	165 ± 5	110 ± 5	10–20	Blaut-Blachev, Igitsyn et al.
1975	Lifetime, noise	77	$p, 10^{14}$		140		Galavanov and Oding
1975	Hall coeff., stress	77	$p, 10^{13}$		120		Mackey et al.
1975	Lifetime	77–200	$n, 10^{14}$	180	130	22	Blaut-Blachev, Ivleva et al.
1976	Lifetime	30–200	$n, 10^{12}$–10^{14}	163		74 ± 2	Korotin et al.
1982	Mageto-optical	4.2	$n, 10^{14}$	170 ± 2	116		Seiler and Goodwin
1985	Magnetic circular dichroism	230	$n, 10^{14}$				Galanov et al.
1986	Photoconductivity	4.2–80	$p, 10^{12}$			22, 40	Saptsov and Skok
1987	Photoconductivity	45–87	$p, 10^{12}$	144	110	86	Reshchikov and Smetannikova

i. InSb Deep levels in InSb have been investigated by a wide variety of experimental techniques over the past several decades. For *n*-type material, these include lifetime, Hall coefficient, conductivity or mobility, transmission, noise, and photoconductivity. A summary for both *p*- and *n*-type InSb is given in Table X. Examples of magneto-optical spectroscopy of deep levels in *n*-InSb are shown in Fig. 40 and 41. In both figures the detector response is proportional to the second derivative of the photoconductive response of the sample. The sources used to obtain the magneto-optical spectra shown in Fig. 40 and 41 were a CO_2 and CO laser, respectively. The deep-level resonances are extremely weak as compared with PACRH and even TPMA processes, and thus little spectroscopic work has been performed to study these levels. An illustration of just how weak is shown in Fig. 42, which details the dependence of the magneto-optical spectra obtained at 9.56 μm on incident

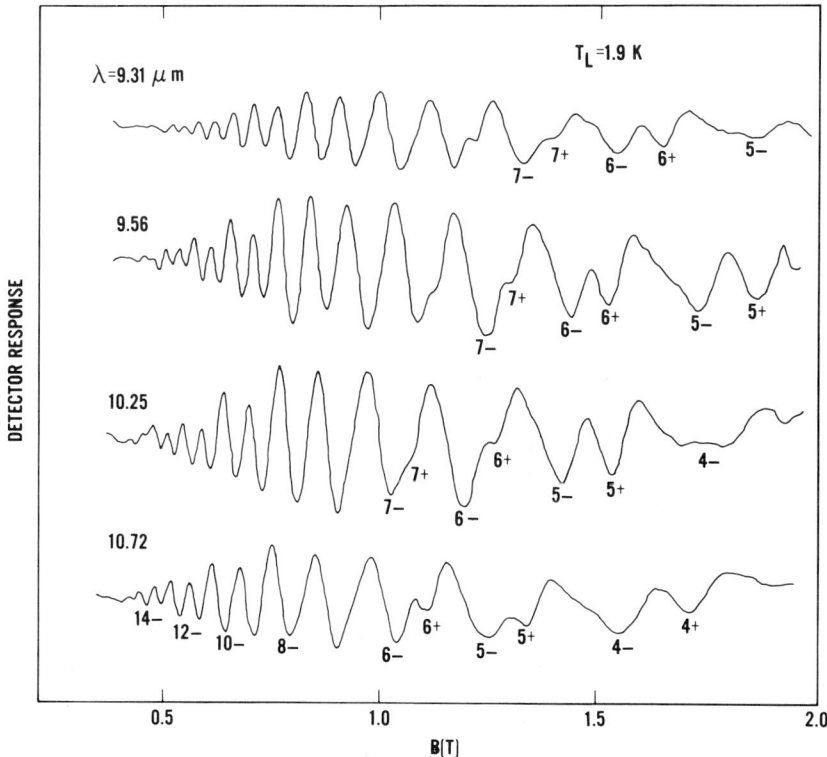

FIG. 40. Wavelength dependence of the deep-level cyclotron resonance harmonic (DLCRH) magneto-optical spectra obtained with a CO_2 laser. The numbers correspond to the final-state conduction-band Landau level of each transition, and the + and − represent the spin-up and spin-down states, respectively. (From Seiler and Goodwin, 1982.)

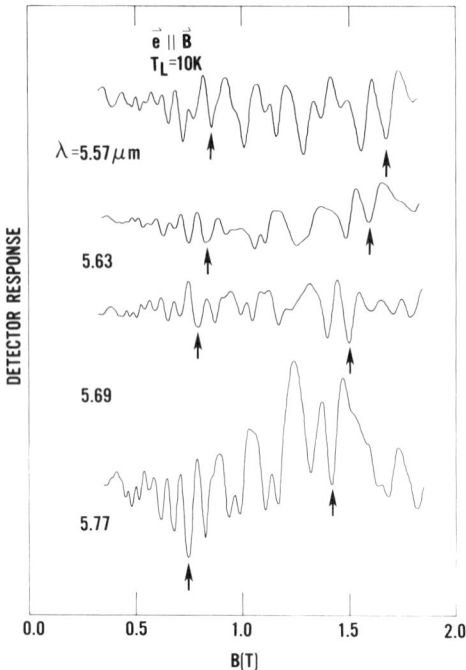

FIG. 41. Wavelength dependence of the CO laser-induced DLCRH spectra from n-InSb. The arrows indicate the shift of the structure to higher magnetic fields for higher photon energies. (From Seiler and Goodwin, 1982.)

laser power. The bottom trace was recorded with the CO_2 laser beam physically blocked. As the laser power P_R is increased ($P_R = 1$ corresponds to a laser power of $\simeq 1.4$ W), a series of weak resonant structure develops; this is the same structure that is presented in Fig. 40. This structure very quickly saturates with increasing laser power and another series of sharper resonances develops. Finally, at the higher laser powers the second series of resonant magneto-optical structure dominates the spectra. This structure is due to two-photon absorption and will be discussed in detail in the next section.

The minimum positions of the laser-induced magneto-optical resonances are plotted for various photon energies of the CO_2 and CO data in Fig. 43a,b. The theoretical lines were calculated by modeling the absorption peaks as resulting from electron photoexcitation from deep levels whose positions are independent of magnetic field, to a Landau level, either spin-up or spin-down, in the conduction band. The energy-band calculations were made by using the energy-band parameters of Littler et al. (1983) listed for InSb in Table IX. Energy conservation for this process is given by

$$\hbar\omega = E^{c\pm}(n) + E_i, \tag{45}$$

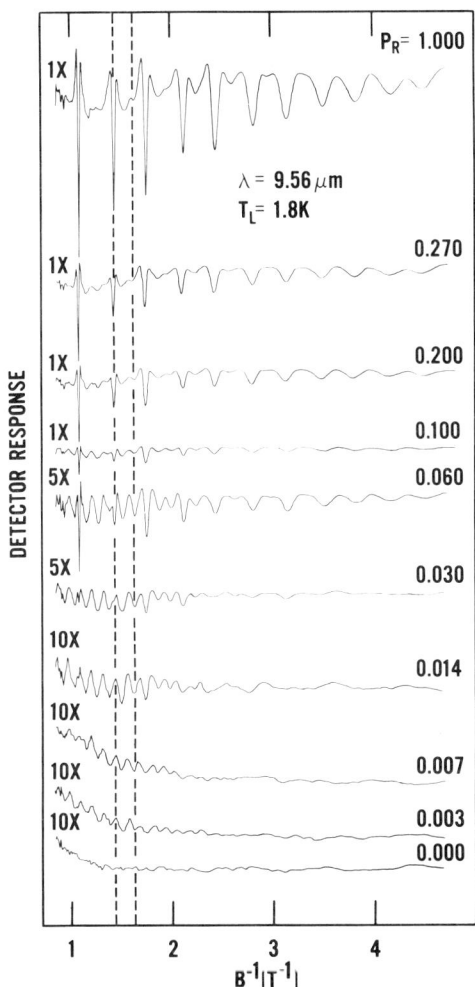

FIG. 42. Photoconductive spectra from n-InSb for different relative peak incident CO_2 laser powers P_R. Note the gain setting on the left. The dashed vertical lines are drawn at the magnetic-field positions of a TPMA and DLCRH resonance, respectively, and illustrate that the magnetic-field positions of each structure are independent of laser power. (From Seiler and Goodwin, 1982.)

where the zero of energy is located at the bottom of the conduction band at zero magnetic field and E_i is the magnitude of the activation of the deep level. The best fit to the data yielded $E_i = E_3 = 74 \pm 2$ meV for the CO_2 laser data and $E_i = E_1 = 170 \pm 2$ meV for the CO laser data (Seiler and Goodwin, 1982). In addition to the previous measurements, Saptsov and Skok (1985) also reported similar laser spectroscopy measurements using a CO laser and

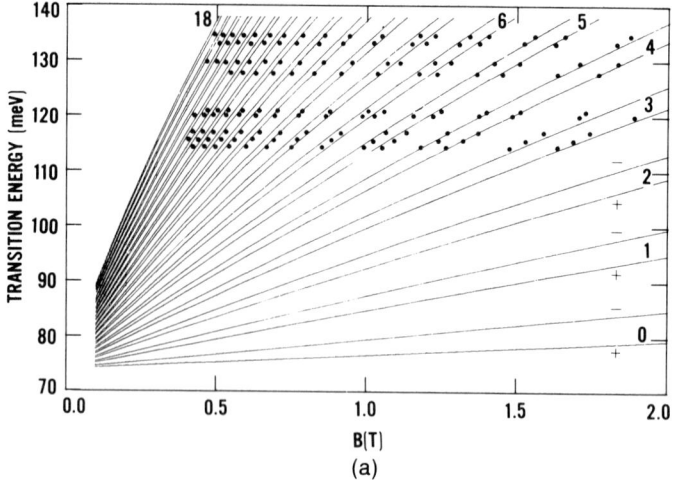

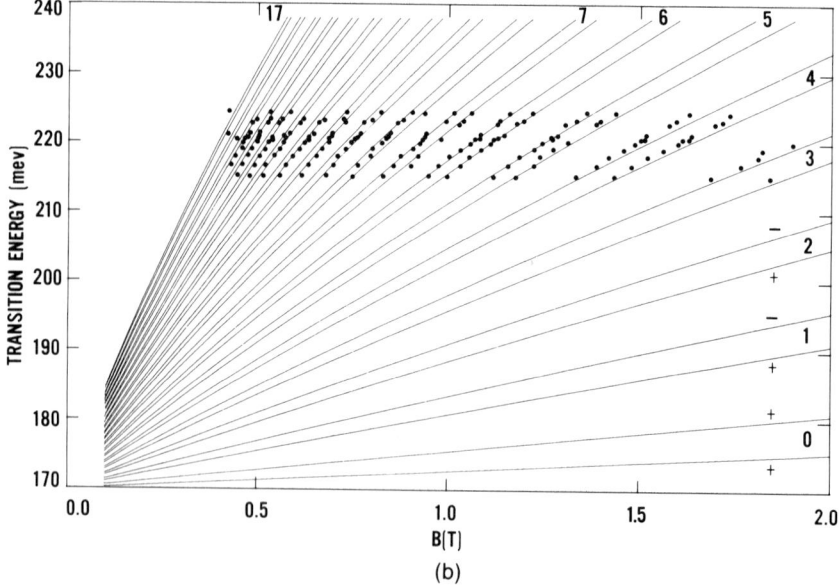

FIG. 43. (a) Fan chart of the DLCRH data shown in Fig. 40. The zero-field intercept is 74 meV. The numbers refer to the final-state conduction-band Landau levels, and the + and − are the spin-up and spin-down states, respectively. (From Seiler and Goodwin, 1982.) (b) Fan chart of the CO laser-induced DLCRH spectra shown in Fig. 41. The zero-field intercept is 170 meV. (From Seiler and Goodwin, 1982.)

derivative spectroscopy techniques. They found a series of magnetoresistance minima whose positions were described as resulting from electron transitions from a deep level to the lowest-lying conduction-band Landau levels. From their analysis, they found the deep level to be 195 ± 1 meV below the conduction-band edge.

From Table X it can be seen that there is quite good agreement on the location E_1, which is located about 170 meV below the conduction-band edge. In addition there seems to be evidence for three other deep levels, labeled E_3, E_4, and E_5, at $\simeq 25$, 40, and 70 meV below the conduction band. The first two are almost always associated with the presence of silver and gold in the sample (Engeler et al., 1961a,b; Pehek and Levinstein, 1965), the 70-meV level not being associated with a specific impurity. The samples with deep levels denoted by E_2 always seem to be present in InSb, regardless of the method of crystal growth or doping level. Thus, it seems reasonable to assume that they are due to native lattice defects that are characteristic of the crystal. From Table X, their position is located approximately in the middle region of the forbidden energy gap.

To date, there have been few magneto-optical studies of deep levels in p-InSb. Referring to Table X, most of the past work has consisted of Hall coefficient measurements, lifetime studies, noise, and mobility experiments. In addition, resonant optical transitions of electrons from the valence band to deep acceptor levels, both with a magnetic field (Littler et al., 1982) and without a magnetic field (Engeler et al., 1961a,b), have been observed and studied. In the latter, the deep-level transitions involved multiple LO phonon emissions. An example of magneto-optical transitions from a deep acceptor level (Au) is shown in Fig. 44. The spectra are complicated, containing many strong and weak resonant peaks. The maxima in the detector response corresponds to a maxima in the conductivity, where the resonant increase in absorption has given rise to an increase in the number of holes. The peaks are labeled according to which acceptor (Au or Cd) the excitation peak has been identified with and to which light-hole Landau level the excited acceptor states are associated. It is seen that the stronger photoconductive peaks result from the excitation of the Au impurities and the weaker peaks are found to result from excitation of shallow Cd acceptors unintentionally introduced during crystal growth.

The same model that was used to explain the Cd-doped InSb samples earlier in this chapter can be employed to analyze the deep-level data. An example of this is shown in Fig. 45. The dashed and solid lines represent calculated transition energies using the Pidgeon–Brown model for the Cd and Au impurities, respectively. The solid dots represent the Au transition energy data. and the open dots are the Cd excitation data. The zero-field intercepts indicated by E_{Au} and E_{Cd} represent the the ground-state binding energy for the

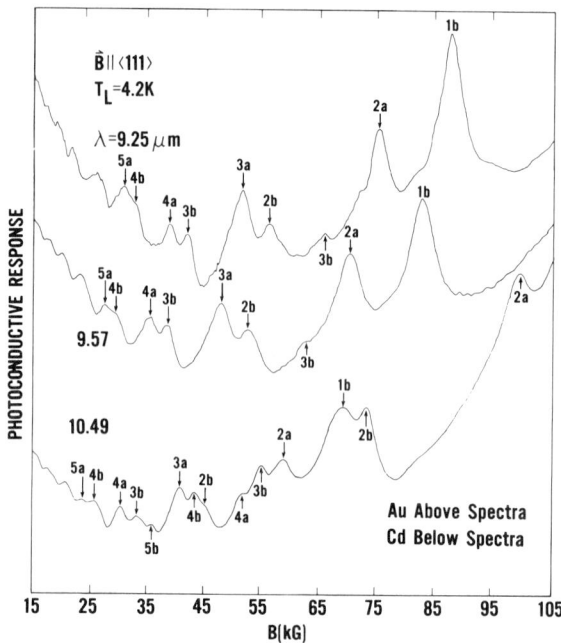

FIG. 44. Photoconductive magneto-optical spectra obtained for three CO_2 laser wavelengths for $\mathbf{B} \parallel \langle 111 \rangle$. The peaks are labeled according to light-hole Landau level that the excited acceptor states are associated with. The Au resonances are indicated by the labels appearing above each spectra (downward-pointing arrows), and the weak Cd peaks are indicated by the labels appearing below each spectra (upward-pointing arrows). (From Littler et al., 1982.)

Au and Cd acceptors. From the analysis the energies 42.5 meV and 8.1 meV are extracted for the Au and Cd ground-state binding energies, in excellent agreement with other work (Engeler et al., 1961a,b).

ii. HgCdTe Only recently have magneto-optical techniques been used for the measurement of deep impurity and defect levels in HgCdTe (e.g., Seiler et al., 1989a,b; Littler et al., 1990a; Kucera et al., 1990). As mentioned previously, magneto-optics allows the accurate determination of impurity and defect levels even at very low concentrations ($\simeq 10^{11}$ out of 10^{15} cm^{-3}). However, in HgCdTe midgap levels dominate, and any optical absorption used must be able to delineate these impurity transitions from two-photon (or TPMA) band-to-band transitions. To delineate between TPMA and midgap-level-to-band transitions, light polarization can be used to determine the origin of the magneto-optical resonances observed. To illustrate the useful-

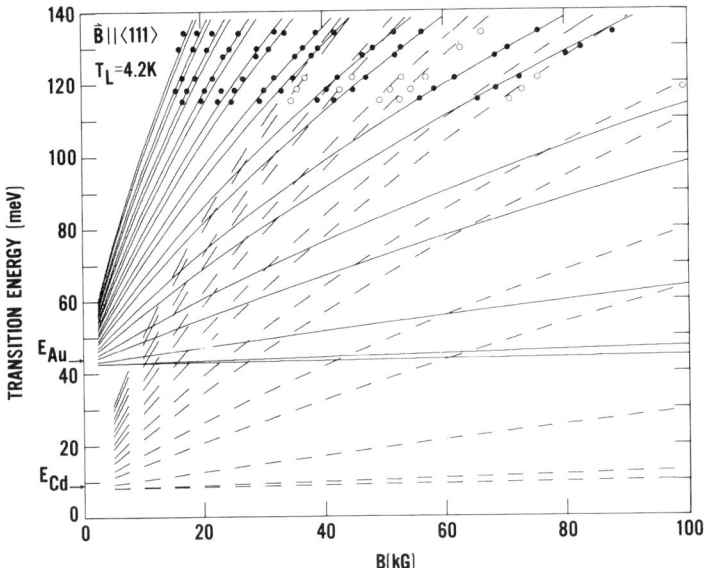

FIG. 45. Fan chart of transition energies for $\mathbf{B} \parallel \langle 111 \rangle$. Dashed lines indicate calculated transition energies for Cd, the solid lines for Au. The solid dots and open dots represent the observed transition energies for the Au and Cd acceptors, respectively. The ground-state binding energies for the Au and Cd are indicated by the zero-field intercepts $E_{Au} = 42.5$ meV and $E_{Cd} = 8.1$ meV. (From Littler et al., 1982.)

ness of light polarization, examples of the TPMA and impurity or defect magneto-optical spectra obtained for left circularly polarized (σ_L) and right circularly polarized light (σ_R) are shown in Fig. 46 for a sample with $x \simeq 0.28$. For σ_L polarization, we see two strong TPMA resonances labeled by the solid arrows L_1 and L_2. The transition assignments for L_1 and L_2 are given in Table XI. The two broader absorption peaks, labeled 1 and 2, are ascribed to electron transitions from two closely spaced impurity or defect levels located at approximately midgap to the lowest-lying conduction-band Landau level. The magnetic-field positions of all arrows indicate the theoretically expected positions of both the TPMA and the impurity or defect resonances. Upon changing the polarization to σ_R, the strong σ_L transitions disappear, being replaced with weaker TPMA resonances labeled by $R_1 - R_5$. It should be noted that R_1 and R_2 are more widely spaced in the magnetic field than the impurity or defect transitions 1 and 2, and some features of 1 and 2 appear for both polarizations. Thus, by the use of circular polarizations, we are able to separate the contributions to the magneto-optical spectra from each absorption process mentioned before.

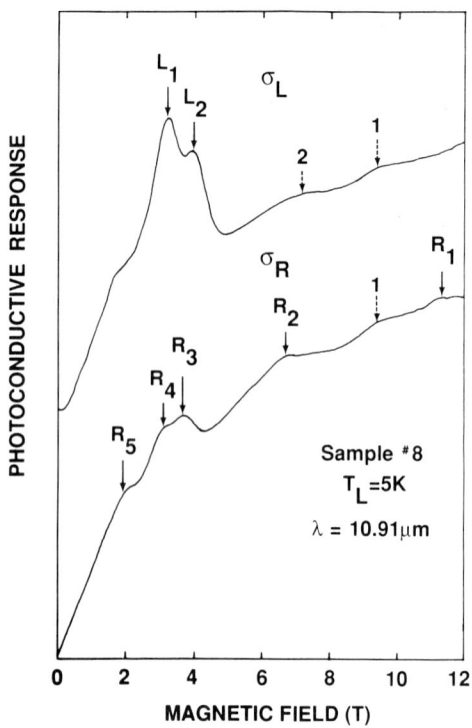

FIG. 46. Polarization dependence of the TPMA and impurity defect transitions in n-HgCdTe. (From Littler et al., 1990a.)

TABLE XI

HgCdTe Two-Photon Magnetoabsorption Transition Assignments.
Assignments for the TPMA Transitions $2\hbar\omega = E_f - E_i$. The
Landau-Level Selection Rules are $\Delta n = +2$ for σ_L Polarization,
$\Delta n = -2$ for σ_R.

Designation	Initial valence Landau level	Final conduction Landau level	Polarization
L_1	$a^-(-1)$	$a^c(1)$	σ_L
L_2	$b^+(-1)$	$b^c(1)$	σ_L
L_3	$a^+(0)$	$a^c(2)$	σ_L
L_4	$b^+(0)$	$b^c(2)$	σ_L
L_5	$a^+(1)$	$a^c(3)$	σ_L
L_6	$b^+(1)$	$b^c(3)$	σ_L
R_1	$a^-(2)$	$a^c(0)$	σ_R
R_2	$b^-(2)$	$b^c(0)$	σ_R
R_3	$a^-(3)$	$a^c(1)$	σ_R
R_4	$b^-(3)$	$b^c(1)$	σ_R
R_5	$a^-(4)$	$a^c(2)$	σ_R

5. ONE- AND TWO-PHOTON MAGNETO-OPTICAL SPECTROSCOPY

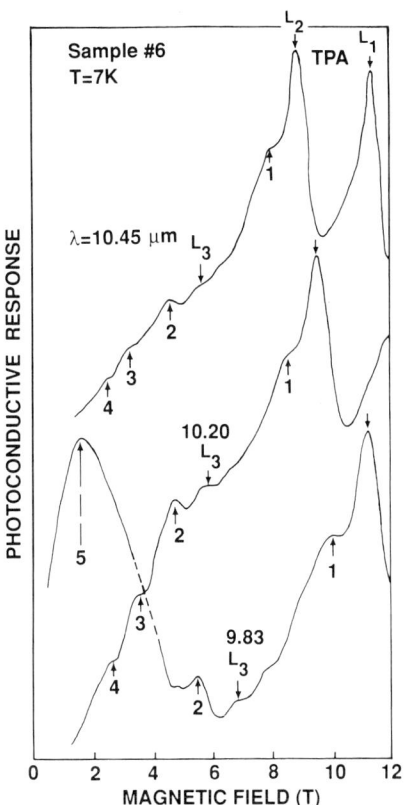

FIG. 47. Wavelength dependence of the magneto-optical spectra for an n-HgCdTe sample with $x \simeq 0.25$, showing both TPMA structure (labeled L_1, L_2, and L_3; see Table XI) and that arising from impurity/defect-to-band transitions. Labels 1–5 refer to the transitions: $1 - E_2 \to a^c(0)$; $2 - E_1 \to a^c(0)$; $3 - E_2 \to b^c(0)$; $4 - E_1 \to b^c(0)$; $5 - E_3 \to b^c(0)$, where E_1, E_2, and E_3 are 63, 68, and 20 meV, respectively. (From Littler et al., 1990a.)

Figures 47 and 48 show two-photon and impurity magneto-optical(IMO) spectra from samples with $x = 0.246$ and 0.296, respectively. For the calculation of the activation energy of the impurity or defect levels, the transition energies from the impurity or defect levels to the conduction band via one-photon absorption can be described by the relationship

$$\hbar\omega = E_c(n, B) - E_i, \tag{46}$$

where E_i is the energy of the identified level (midgap, shallow acceptor, etc.) above the valence band. This analysis procedure is equivalent to that of

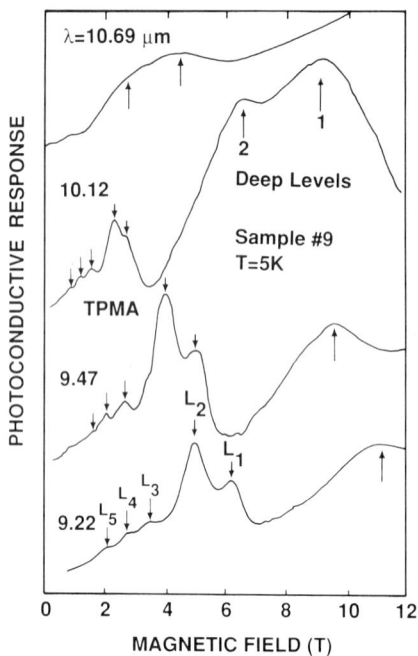

FIG. 48. Wavelength dependence of the magneto-optical spectra for a sample of n-HgCdTe with $x \simeq 0.3$, showing both TPMA resonant structure and that arising from impurity/defect-to-band transitions. The transition assignments for the resonances labeled L_1-L_5 are given in Table XI. The upward-pointing arrows indicate resonances due to deep-level to conduction-band transitions. (From Littler et al., 1990a.)

Eq. (45); the zero of energy in this case has been shifted to the top of the valence band to facilitate the plotting of both interband and impurity-to-band data. As $B \to 0$, $E_c(n, B) \to E_g$ and thus

$$\hbar\omega = E_g - E_i, \qquad (47)$$

which is the $B = 0$ intercept. Figures 49 and 50 show the comparison between theoretical predictions and magneto-optical absorption data, yielding the energy gaps of 136 meV and 222 meV and the activation energies of 63 and 68 meV, 111 meV, and 116 meV at $T = 7$ K.

A survey of past work on impurities and defects in HgCdTe is given in Table XII. This table is not meant to be exhaustive, only representative of past work. From this table, several trends emerge: (1) Shallow acceptor-like levels with activation energies between $\simeq 2$ and 20 meV are seen in most sam-

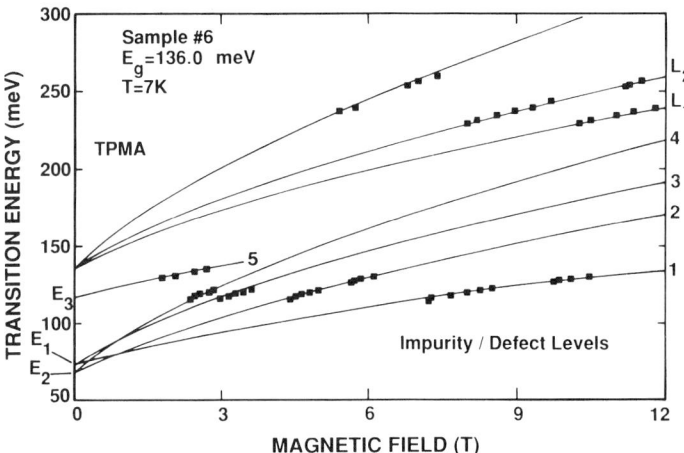

FIG. 49. Fan chart for the magneto-optical data shown in Fig. 47. (From Littler et al. 1990a.)

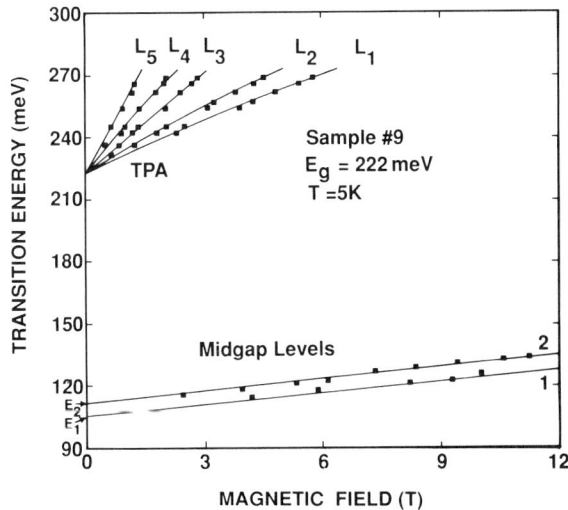

FIG. 50. Fan chart for the magneto-optical data shown in Fig. 48. (From Littler et al., 1990a.)

ples; (2) near midgap levels are also detected in most samples; and (3) there is evidence for a third level at $\simeq (3/4)E_g$ above the valence-band edge. We see that the results of magneto-optics are in good agreement with the results of other techniques and provide a direct, sensitive technique for detecting the presence of impurity and defect levels in HgCdTe.

TABLE XII

SUMMARY OF IMPURITY/DEFECT LEVELS FOR HgCdTe.

Year	Technique	Type	x-value (shallow)	E_1(meV) ($E_g/4$)	E_2(meV) ($E_g/2$)	E_3(meV) ($3E_g/4$)	E_4(meV)	Reference
1972	Hall, PL	p	0.26	15, 16	—	—	—	Elliott et al.
			0.31	22, —	—	—	—	
			0.33	18, 13	—	—	—	
			0.34	19, 10	—	—	—	
1979	Lifetime, PL	n	0.3	18	70	—	—	Andrukhiv et al.
1980a	DLTS	p	0.21	—	—	35, 43	—	Polla and Jones
1980b	AS	Diode	0.22	—	—	46	—	Polla and Jones
			0.31	—	—	160	—	
1980	PL	n	0.32	14 ± 1.5	—	—	—	Hunter et al.
			0.48	4.5 ± 2, 15.5 ± 2	—	—	—	
1981	DLTS	p	0.21	—	43	—	—	Jones et al.
			0.39	—	118	283	—	
1981	DLTS, AS	p	0.207	—	—	37, 44	—	Polla and Jones
			0.215	—	—	35, 46	—	
			0.271	—	—	70, 81	—	
			0.285	—	—	81, 92	—	
			0.305	—	—	161	208	
			0.320	—	—	—	206, 218	
1981a	DLTS	p	0.215	—	—	35, 43	181, 208	Polla et al.
			0.305	—	—	—	206, 218	
			0.320	—	—	—	—	
1981b	Lifetime	p	0.215	—	37	56	—	Polla et al.
			0.220	—	—	—	—	
			0.326	—	—	150	—	
			0.386	—	—	230	—	

Year	Method	Type					Reference
1981	PL	n	0.3	10	—	—	Gelmont et al.
1982	DLTS	n	0.27	—	136	—	Jones et al.
1982	PL	p	0.39	—	150	—	
1982	PL	n	0.32	14 ± 1	70	—	Hunter and McGill
1983	Lifetime	n	0.48	15.5 ± 2	—	—	Pratt et al.
			0.307	—	—	223	
			0.332	—	—	271	
			0.344	—	—	284	
			0.352	12	—	—	
			0.335	30	—	—	
1983	DLTS	p	0.48	—	200, 230	—	Merilainen and Jones
1984	PL	p	0.320	—	$0.4E_g$	—	Polla and Aggarwal
		n	0.340	—	$0.4E_g$	—	
1985	EPR	p	0.3	10	—	—	Jones et al.
1985	Lifetime	p	0.215	15	—	—	Schacham and Finkman
			0.225	15	—	—	
			0.29	20	—	—	
1985	DLTS	n	0.3	5, 12.5	120	—	Cotton et al.
1986	Hall data	p	0.22	7	$E_g/2$	—	Finkman and Nemirovsky
1986	Photo-Hall	p	0.224	—	—	—	Bartoli et al.
1986	Lifetime	n	0.230	10 ± 2	65 ± 5	—	Pratt et al.
			0.234	—	—	124	
			0.245	—	—	200, 232	
			0.31	—	—	233, 253	
			0.32	—	—	—	
			0.33	15 ± 5	70 ± 7	—	

(continued)

TABLE XII (*Continued*)

Year	Technique	Type	x-value (shallow)	E_1(meV) ($E_g/4$)	E_2(meV) ($E_g/2$)	E_3(meV) ($3E_g/4$)	E_4(meV)	Reference
1986	Hall data	p	0.22	11	—	—	—	Chen and Dodge
1987	Lifetime	p	0.24	—	—	—	92	Lacklison and Capper
1987	Hall data	p	0.2	11.5	—	—	—	Chen and Tregilgas
1989a,b	MOS	n	0.23	10	—	—	—	Seiler et al.
			0.25	9.8	—	—	—	
			0.28	—	—	94, 100	—	
1990	PC	p	0.20	7.0	—	45	60	Kucera et al.
1990a	PC	n	0.217	7.0	—	—	—	Littler et al.
			0.225	—	18.0, 26.0	—	—	
			0.225	7.0	—	—	—	
			0.239	9.8	—	—	—	
			0.241	10.0	—	60.0	—	
			0.246	—	—	63.0, 68.0	—	
			0.246	10.2	—	—	—	
			0.278	—	—	90.0, 94.0	—	
			0.296	—	—	111.5, 116.0	—	

PL = photoluminescence; DLTS = deep-level transient spectroscopy; AS = admittance spectroscopy; TSC = thermally stimulated current; MOS = magneto-optical spectroscopy; EPR = electron paramagnetic resonance; PC = photoconductivity.

IV. Two-Photon Absorption (TPA) and Two-Photon Magnetoabsorption (TPMA) Spectroscopy

13. BACKGROUND

a. Introduction

Two-photon absorption processes in semiconductors have been the subject of extensive experimental and theoretical work for more than two decades. This intense interest in the nonlinear TPA properties of materials has been stimulated for both scientific and technological reasons. There are several important scientific features of two-photon absorption. First, it becomes possible to gain information on energy levels in a highly absorbing region while the separate photon energies involved lie in the transparency region of the semiconductor. This is because strong absorption occurs only when the sum of the two-photon energies ($\hbar\omega_1 + \hbar\omega_2$) is close to the difference between two energy levels. Second, one can study transitions that are forbidden to take place in one-photon absorption (OPA) because of the different selection rules for TPA processes. Third, the absorption cross section for TPA is often small, leading to rather uniform absorption throughout the bulk or film volume. Thus, there is no need for ultrathin samples, and there is also a reduced emphasis on surface effects (both of which are important considerations for OPA). Fourth, TPA processes normally involve all the energy bands of the semiconductor as intermediate states in the second-order perturbation theory expression for the TPA transition probability. Thus, a TPA experiment has the potential of giving detailed information about the energy-band structure not obtainable by OPA methods. Interested readers are referred to the following review articles for other information on multi- or two-photon spectroscopy in a wide variety of materials: Peticolas (1967), Bonch-Bruevich and Khodovoi (1965), Fröhlich (1970a,b), Bassani (1972), Worlock (1972); Bredikhin *et al.* (1973), Mahr (1975), Fröhlich and Sondergeld (1977), Salour (1978), Fröhlich (1981), and Nathan *et al.* (1985). Most work that has occurred in the field of multiphoton spectroscopy has taken place in the area of using only two photons. The principle reasons for this are (1) it becomes increasingly more technologically difficult to carry out spectroscopy with more and more photons, and (2) the higher-order processes become weaker and are more difficult to observe.

Interest in TPA processes has also been stimulated by the importance of nonlinear absorption for many aspects of high-power laser technology. Some examples are as follows: (1) Nonlinear absorption limits the transparency of optical window materials and causes laser-induced damage to optical components. (2) TPA processes have been used to produce population inversion in semiconducting laser materials giving high powers. (3) Two-photon absorbers can be used as negative-feedback elements in laser cavities or for controlling laser pulse intensities and pulse durations. (4) TPA techniques can also be used in studying the properties and processes associated with ultrashort laser pulses. (5) Optical power limiters can be constructed by using TPA effects. The purpose of this section is to review how TPA and TPMA techniques have been used to characterize and understand the narrow-gap semiconductors InSb and HgCdTe. Finally, a review of TPA and TPMA work on InSb and HgCdTe is presented, and specific examples are given that demonstrate unique aspects of the two-photon methods.

b. Carrier Lifetimes by TPA Methods

A large amount of effort, both theoretical and experimental, has been expended in the understanding of carrier recombination processes and lifetimes. These processes are especially important in NGS such as InSb and HgCdTe, since the carrier lifetimes in NGS are strongly affected by even small changes to the material's properties. There are numerous methods of measuring carrier lifetimes in semiconductor materials and devices (e.g., see Table I in Milnes, 1973). Unfortunately, as pointed out by Schacham and Finkman (1985), there is a great deal of ambiguity in the interpretation of the data used to determine lifetimes and the corresponding recombination mechanisms in NGS. This ambiguity can very easily be overcome, however, by using TPA methods to generate the photoexcited carriers.

The primary motivation for studying the temperature dependence of carrier lifetimes is to detect and investigate the influence of impurities or defects. Often light sources are used that have large photon energies that create electron–hole pairs via one-photon absorption. Consequently, large absorption coefficients ($\simeq 10^4 - 10^5$ cm^{-1}) result in light penetration depths on the order of a micron. The determination of bulk lifetimes can then become quite complex; the values often depending greatly on the condition of the surface of the material.

The characteristics and advantages of using TPA methods to determine bulk recombination processes has been documented for InSb by Slusher *et al.* (1969), Fossum and Anker-Johnson (1973), Schneider *et al.* (1972), McClure *et al.* (1984), and Loloee *et al.* (1988), and for HgCdTe by Bajaj *et al.* (1983) and Seiler *et al.* (1986b,c). Schneider *et al.* (1972) concluded that the most direct way to determine bulk recombination coefficients is to measure the conductivity decay after excitation via TPA in samples of large volumes. This fact seems to have been recognized by Bajaj *et al.* (1983) in studies of minority-carrier lifetimes in LPE samples of HgCdTe. They carried out minority-carrier lifetime measurements by a photoconductive decay technique using short pulses from a GaAs laser. By illuminating their samples with laser excitation through the CdTe substrate, they found evidence for possible interfacial recombination. However, they concluded that lifetime measurements obtained by volume excitation using TPA methods should be carried out before more definite conclusions could be reached.

Recently, a novel method for the direct determination of true minority-carrier lifetimes in p-InSb samples using TPA techniques was presented by Loloee et al. (1988). They showed that the photoconductive decay method gave complex, difficult to understand, multistage behavior. However, the application of a new method based on the magnetoconductivity tensor components (see also Leslie-Pelecky et al., 1987) allows the direct determination of the time dependence of the minority-carrier concentration, thus allowing the determination of the true minority-carrier lifetime. As a result of the analysis, a trap level of 106 meV was obtained from fits to the data using a Shockley–Read (Shockley and Read, 1952) analysis.

14. REVIEW OF TPA AND TPMA WORK DONE ON InSb

An extensive amount of TPA and TPMA work has been carried out on InSb, both experimental and theoretical. This has been due to the availability of excellent quality material and the fact that the CO_2 laser is ideally suited for TPA studies. At 4 K, one can tune $2\hbar\omega$ from below the band-gap energy of InSb ($\simeq 235$ meV) to well above it. The choice in the experimental investigations seems to have only been whether to use a transverse-electric atmospheric (TEA) or a Q-switched or a cw CO_2 laser. Table XIII summarizes all the known work by the various authors and the year the work was published. Other entries in Table XIII show the two-photon energy used, the temperature, the experimental method used, and a brief summary of the main results obtained. More detailed information can be found in the original articles.

We will now briefly review several experimental configurations used to investigate TPA in InSb. Slusher et al. (1969) used a neon laser at $\lambda = 17$ μm to measure the density and homogeneity of a two-photon-produced plasma from a Q-switched CO_2 laser as shown in Fig. 51. The optical absorption of the beam was primarily caused by the TPA-produced holes that have a large absorption coefficient at 17 μm. Optical measurements that detect and determine the TPA-produced hole concentrations could thus be easily made. Fossum et al. (1973) used an experimental arrangement shown in Fig. 52 to produce and measure the properties of TPA-induced carriers by a Q-switched CO_2 laser. They were able to monitor and study absorption, luminescence, and photoconductivity at the same time. The particular electrical connections shown were able to faithfully reproduce the actual photoconductive transient response of the sample in order to determine carrier lifetimes.

Seiler et al. (1981) carried out the first two-photon experiments using only cw CO_2 lasers in a solid material. They observed resonant TPMA photoconductive structure at certain magnetic fields. Proof of the TPA nature of

TABLE XIII
COMPREHENSIVE COMPILATION OF ALL TWO-PHOTON ABSORPTION WORK DONE ON InSb

Authors (year)	$2\hbar\omega$ (eV)	T_L (K)	Method	Main results
Button et al. (1966)	0.258 0.246 Q	20	PC	Observed multiphoton magneto-optical transitions up to 100 kG in n-InSb.
Gibson et al. (1968)	0.234 Q	30–300	PC	Used p-InSb. PC signal varied as I^2. At high intensities signal saturated. Comparison to response at 5.3 μm.
Weiler et al. (1968)			Theory	Multiphoton interband Landau-level transitions calculated by tunneling theory.
Slusher et al. (1969)	0.234 0.258 Q	4.2 & 77	PC & RR	Density of TPA produced free holes monitored by 17-μm light beam. Density, recombination radiation, and conductivity of TPA produced carriers studied up to 55 kG. Saturation behavior observed.
Danishevskii et al. (1969)	0.234 Q	90	PC	TPA produced concentration varied as I^2 except at high intensities where strong absorption of light by TPA-produced holes occurs. Agreement with theory.
Weiler et al. (1969)	0.258 0.246 Q	15	PC	Experimental & theoretical study of two-photon magnetoabsorption. Tunneling theory.
Weiler et al. (1971)	0.246 0.258 Q	4.2	PC	Magnetic-field studies carried out to see Landau-level transitions. Claims Keldysh theory agrees with data.

Reference		Temp	Method	Description
Nguyen & Strnad (1971)	0.258 Q	1.8	A	Absorption of light by free holes created by TPA processes in a magnetic field. Used n-type (5×10^{13} cm^{-3}) and p-type (3×10^{14} cm^{-3}). Found same TPA spectra in p- as n-type samples.
Nguyen et al. (1971)	0.258 Q	1.8	A	Carried out a perturbation type of theoretical calculation. Theory agreed with TPA experiments in a magnetic field with circular & linear polarized light.
Hongo et al. (1971)	0.234 Q	77	PC	Created electron–hole plasmas by TPA processes. Conductance versus intensity-linear behavior at low intensities and quadratic at high.
Schneider et al. (1972)	0.234 TEA	77	PC	TPA-produced plasma density from $\simeq 10$–10^5 W/cm^2. At high intensities TPA plasma saturates because of increased hole absorption. Electron–hole recombination dominates.
Danishevskii et al. (1972)	0.234 0.258 Q	80–200	PL	Used p-type ($\simeq 10^{14}$ cm^{-3}). Tried to determine polarization dependence of TPA coefficient. Tried to observe optical spin orientation effects.
Danishevskii et al. (1973)	0.234 Q	92–151	A	n-type (1×10^{14} cm^{-3}). Cross section for equilibrium holes at $\lambda = 10.6$ μm depends on T_L and at lower T_L's increases with decreasing T_L. TPA accompanied by intense absorption by excess holes & used to determine hole lifetime and concentration.
Doviak et al. (1973)	0.234 Q	300	A	TPA coefficient measured. Says absorption due to generated free carriers can be neglected.
Manlief & Palik (1973)	0.234 0.268 TEA	15	PC & A	TPMA in n-type (5×10^{13} cm^{-3}). Saturation observed. Strong stimulated recombination at 5.3 μm showing peaks at Landau transitions. Transmitted CO$_2$ laser beam also showed resonant Landau structure.

(*continued*)

TABLE XIII (Continued)

Authors (year)	$2\hbar\omega$ (eV)	T_L (K)	Method	Main results
Fossum & Chang (1973)	0.258 0.234 Q	2 77	PC	Two-photon transition rates calculated and measured.
Fossum & Ancker-Johnson (1973)	0.258 0.234	2 4 77	PC	Radiative recombination of TPA-produced plasma calculated and compared with measurements at 77 K, nonradiative decay dominates, while at 4 & 2 K, bimolecular decay dominates.
Fossum et al. (1973)	0.258 0.234	2 77	PC & A & RR	TPA-produced plasmas studied. Free-hole absorption important for large sample thicknesses. Free-hole cross section $\simeq 5 \times 10^{-15}$ cm^2. Pulse shaping studied.
Lee & Fan (1974)	0.234 TEA	300	A	TPA coefficient measured. Calculations including exciton effects.
Mitra et al. (1975)			Theory	Calculated TPA coefficient at 10.6 μm. Used Braunstein, Basov, and Keldysh models.
Nagao et al. (1974)	0.234 TEA	95	PC & PH	$\Delta n \propto I^2$ at low I and $\Delta n \propto I$ at high I. Plasma density tends to saturation at high I. Recombination is linear at low excitation levels and quadratic at high levels.
Gibson et al. (1976)	0.234 TEA	300	PC & A	TPA coefficient measured. Two orders of magnitude less than previous results. Discusses why large-hole absorption effects need to be accounted for.
Schneider et al. (1976)	0.234 TEA	77	PC	Recombination coefficients determined by two-photon excitation.

Reference		Temp	Type	Description
Zawadzki & Wlasak (1976)			Theory	Matrix elements describing two-photon magneto-optical transitions. Nonsphericity & inversion asymmetry terms included in perturbation theory treatment.
Favrot et al. (1976)	0.468 0.484 TEA	5	A	TPA spectra obtained at high magnetic fields & at 5-μm wavelengths.
Dempsey et al. (1978)	0.258 0.234 TEA	10 77 295	A	TPA coefficient measured at room T_L. Effects due to free-hole absorption important. Pulse shaping observed.
A. Miller et al. (1979)	0.258 0.234	295	Theory	Nonparabolic band dependence on TPA coefficients calculated and compared with data of Dempsey et al. (1978)
Pidgeon et al. (1979)			Theory	Three band, nonparabolic model used for calculating frequency dependence of TPA coefficients.
Holah et al. (1979)	0.234 0.258 TEA	293 77 10	A	TPA coefficients and carrier lifetimes determined.
Vaidyanathan et al. (1980)			Theory	Formulas for describing TPA from Keldysh, Braunstein, and Basov compared to each other and data.
Johnston et al. (1980)	0.234 0.258 TEA	77 300	PC & A	Experimental and theoretical TPA coefficients. Lifetimes measured.
Pidgeon (1980)			Theory	TPA calculations for three-band, nonparabolic model.
Vaidyanathan et al. (1980)			Theory	Corrections to their previous paper, better results.

(*continues*)

TABLE XIII (*Continued*)

Authors (year)	$2\hbar\omega$ (meV)	T_L (K)	Method	Main results
Seiler et al. (1981)	Two tunable cw CO_2 lasers	18	PC	First TPA experiments in solids using only cw lasers. Intensity and polarization measurements carried out. TPMA transition energies versus magnetic field calculated. Sensitive derivative technique used.
Weiler (1981)			Theory	Exact nonparabolic energies and matrix elements used in perturbation theory for calculating TPA coefficients. Frequency dependence and exciton effects calculated.
Seiler et al. (1982)	Tunable cw CO_2 laser	$1.8 \rightarrow 100$	PC	High resolution two-photon magnetoabsorption data presented and analyzed. Temperature dependence of energy gap up to 100 K given.
Pidgeon et al. (1982)	0.234 TEA	77	A	TPA coefficient time independent. Effect of various pulse widths on TPA due to free-hole absorption.
Kwok (1982)	0.234 TEA	300	R	Short pulse experiments (60 psec) at high doping levels to suppress TPA effects. Says electron ionization described by tunneling not TPA.
Goodwin et al. (1982)	Tunable cw CO_2 laser	$1.8 \rightarrow 100$	PC	TPA transition assignments given for 35 transitions in a magnetic field. Selection rules, intensity, and polarization dependence. Discussion of band parameters.
Weiler (1982)			Theory	Selection rules for both TPMA and OPMA in zinc blende materials for various polarizations and orientations.
Hasselbeck & Kwok (1982)	0.234 TEA	300	R	CO_2 laser-induced melting with picosecond pulses consistent with TPA effects.
Kar et al. (1983)	0.234–0.258 TEA	295	A	Observed optical bistability caused by nonlinear refraction induced by TPA.

Reference		Method	Description
Grave et al. (1983)		Theory	TPA by electrons in the conduction band proposed as one mechanism to explain self-induced opacity at high infrared light intensities.
Seiler et al. (1983)	Tunable cw CO_2 laser	PC & PV	Review of TPMA effects in InSb. First report of TPMA in photovoltaic response. Magnetic fields up to 150 kG used. Circular polarization results given and theoretically described. Crystal orientation effects observed.
Seiler & McClure (1983)	Tunable cw CO_2 laser 10	PC	High resolution TPMA spectra for $\langle 100 \rangle$, $\langle 111 \rangle$, and $\langle 110 \rangle$ directions. Anisotropic effects observed and related to anisotropy of initial hole states.
Bresler et al. (1983)	0.258 Q 1.8	PL	TPA luminescence spectra (1.9×10^{14} to 3.3×10^{16} cm^{-3}). Self-absorption and band tails important. Absorption of photons by free electrons in conduction band important. Nonradiative recombination predominated.
Brandi and deAranjo (1983)		Theory	A universal curve for the frequency dependence of the multiphoton absorption coefficient based on a "nonperturbative" approach.
Wherrett (1984)		Theory	Quasi-dimensional analysis for the N-photon interband absorption coefficient of direct-gap semiconductors.
Littler et al. (1984)	Tunable cw CO_2 laser 5	PH	Photo-Hall technique first used to study TPMA. Accurate way of determining number of TPA-produced carriers. Two stage carrier lifetime decay observed and related to both TPA and impurity-level transitions.

(*continues*)

TABLE XIII (Continued)

Authors (year)	$2\hbar\omega$ (meV)	T_L (K)	Method	Main results
McClure et al. (1984)	Tunable cw CO_2 laser	5	PC	Photoconductive lifetime exhibits a two-stage decay because of shallow donor impurities plus TPA effects. Resonant increases in lifetime observed during 1st stage decay at TPMA transitions.
Seiler et al. (1984)	Tunable cw CO_2 laser	5	PC & PH	Nonlinear optical properties at liquid helium temperatures strongly related to the electron population of shallow donor levels. Nonlinear, coupled rate equations describe experimental behavior.
Mathew et al. (1985a)	0.234 TEA	295	A	Detailed study of time resolved self-defocusing from TPA-produced carriers at moderate intensities.
Alekseev et al. (1985)	0.258 Q	1.8	PL	Determined dependence of integrated intensity and polarization of TPA-produced luminescence.
Nathan et al. (1985)			Theory	Reviews theoretical and experimental concepts of TPA in semiconductors. Brief discussion on InSb.
Van Stryland et al. (1985)			Theory	Scaling theory for TPA coefficients of semiconductors involving important material parameters for TPA processes.
Littler & Seiler (1985)	Tunable cw CO_2	2–210	PH	Used TPMA photo-Hall effect to determine accurately the temperature dependence of the energy gap

Ji et al. (1986)	0.234 TEA	295	A	Quasi-steady-state optical bistability in an etalon from free carriers generated by TPA.
Sheik-bahae et al. (1986)	0.234 psec	295	A	TPA coefficient measured with high intensity picosecond pulses. $\beta = 2.5$ cm/MW.
Jarasiunas et al. (1986)	0.234 Q	295	TPG	Studied transient phase grating induced by TPA. Deduced ambipolar diffusion coefficient (26 cm^2 sec^{-1}) and carrier lifetime (54 nsec)
Craig et al. (1986)	0.234 TEA	295	A	Time-resolved measurements of self-defocusing induced by TPA processes. Severe beam distortion observed.
Seiler & Littler (1987)	Tunable cw CO$_2$ laser	1.8–20	PH	Review of TPMA in semiconductors. Eigenstate analysis of n-InSb TPMA data showing how to extract E_g and m^* and g^* for both electrons and holes. Review of TPMA photo-Hall effect data and analysis by a four-level rate equation (conduction and valence bands and two impurity levels). Intensity and time-resolved data analyzed.
Sheik-bahae and Kwok (1987)	0.234 psec	295	A	Self-defocusing of 45-psec pulses caused by TPA-generated free carriers and resulting self-phase modulation. Accurate determination of $\beta = 2.5$ cm/MW.
Sheik-bahae et al. (1987)	0.234 0.258 psec	295	A	TPA coefficient β determined using self-induced refraction and transmission of 45-psec pulses. β different from previous results using nsec pulses. Results consistent with tunneling-assisted ionization theory.

(*continues*)

TABLE XIII (*Continued*)

Authors (year)	$2\hbar\omega$ (meV)	T_L (K)	Method	Main results
Loloee et al. (1988)	CO_2 laser Q	7–190	PMC Tensor components	Magnetoconductivity tensor components used to determine carrier concentrations and mobilities of both majority and minority carriers in p-InSb under TPA conditions. True minority carrier lifetimes determined.
Seiler & Littler (1988)	Tunable cw CO_2	1.8–210	PC & PH	Review of two-photon spectroscopy experiments. Saturation of TPA photo-Hall spectra at 1.8 K.
Seiler et al. (1988)	Tunable cw CO_2 laser	7–150	PC	TPA method for simultaneously and independently determining both E_g and m^*. Both temperature dependencies measured. Use of $k \cdot p$ theory to calculate m^* vs. T inadequate. Electron–phonon interactions important.

PC = photoconductivity; A = absorption (transmission); R = reflectivity; PH = photo-Hall; TPG = transient phase grating; RR = recombination radiation; PL = photoluminescence; PV = photovoltaic; PMC = photomagnetoconductivity.

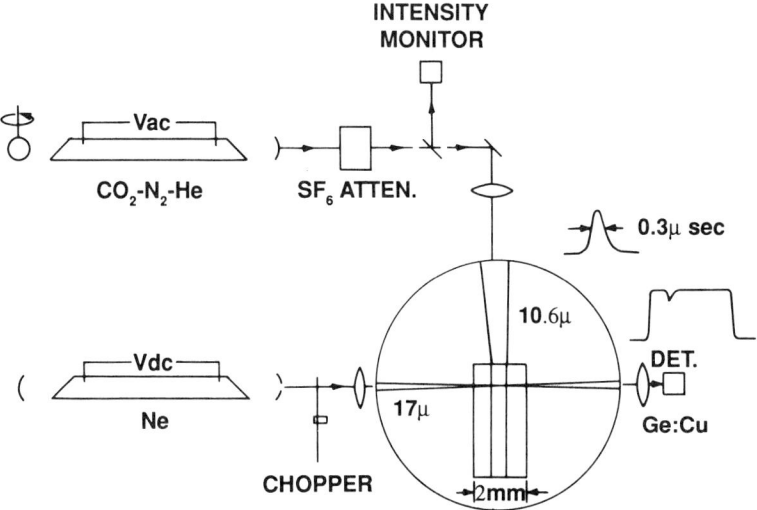

FIG. 51. Diagram of experimental apparatus for measuring density and homogeneity of a two-photon injected plasma in InSb. The enlarged diagram within the circle shows the injecting and measuring laser beams. (From Slusher, et al., 1969.)

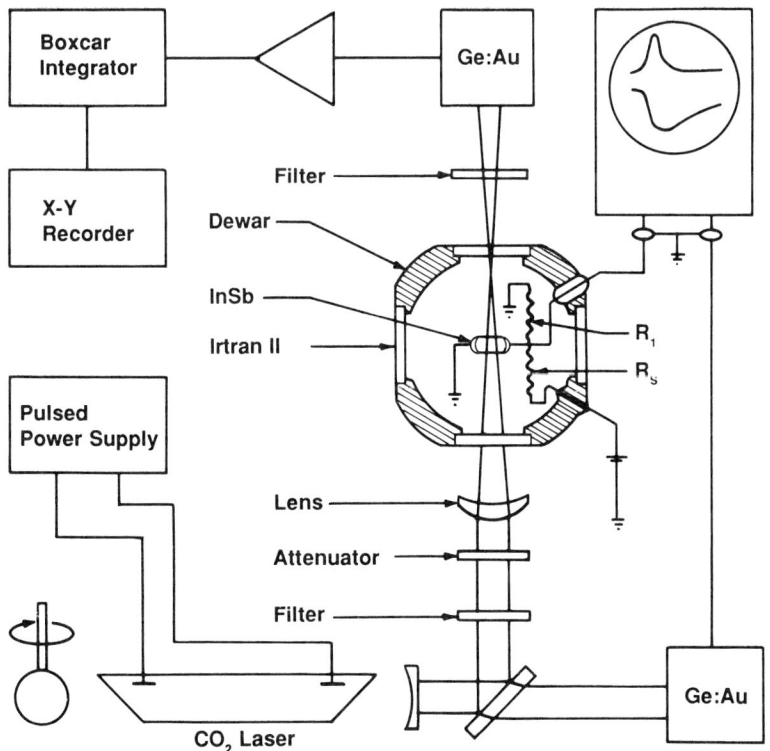

FIG. 52. Experimental arrangement used by Fossum et al. (1973) to produce and measure the properties of TPA-induced carriers in InSb. (From Fossum, et al., 1973.)

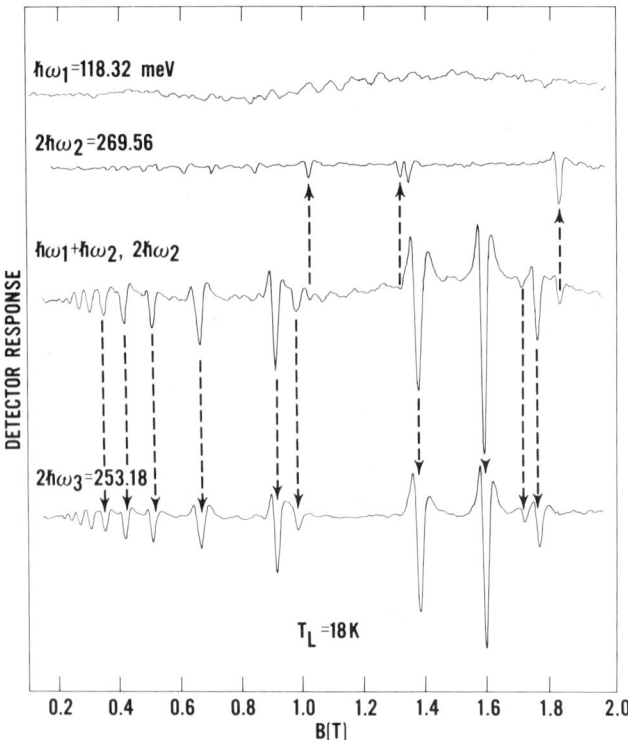

FIG. 53. Photoconductive spectra obtained using various combinations of two-photon energies. The spectrum with $\hbar\omega_1 + \hbar\omega_2$ was obtained with two separate laser beams simultaneously incident on the InSb sample. (From Seiler, et al., 1981.)

the structure was obtained by simultaneously using two individually tunable CO_2 lasers. The results are illustrated in Fig. 53, which shows the second derivative of the PC response versus magnetic field for a high-purity n-InSb sample at 18 K. The top trace shows the sample response obtained while only a single 10.48-μm ($\hbar\omega_1 = 118.32$ meV) cw laser is incident on the sample with a peak power of 1.8 W. To prevent lattice heating, the beams were mechanically chopped at a low duty cycle. The next trace shows weak resonant TPMA structure obtained from a separate cw CO_2 laser operating at 9.20 μm ($\hbar\omega = 134.78$ meV) at a much lower peak power of 0.2 W. Combining the two separate beams produces completely different structure as shown in the next trace ($\hbar\omega_1 + \hbar\omega_2 = 253.10$ meV). Since two-photon processes are the origin of this new structure, the resonant magnetic-field positions are determined by the sum $\hbar\omega_1 + \hbar\omega_2$ from the two lasers and not the individual photon energies. Consequently, the same structure should be observed by using only one laser, but at a different photon energy $\hbar\omega_3$, such that $2\hbar\omega_3 = \hbar\omega_1 + \hbar\omega_2$. The bottom trace shows results where only one laser operating at 9.795 μm ($2\hbar\omega_3 =$ 253.18 meV $\simeq \hbar\omega_1 + \hbar\omega_2$) is incident on the sample. The same structure is observed, proving that the two-photon transition rate [which is proportional to $\delta(E_f - E_i - \hbar\omega_1 - \hbar\omega_2)$] is responsible for the optical transitions and not a single photon transition rate from a deep level to the conduction-band Landau levels.

These high-resolution two-photon magnetoabsorption spectra were obtained with ac magnetic-field modulation and lock-in amplifier techniques as presented in Fig. 54. The boxcar data show that the actual TPMA resonant amplitudes increase with magnetic field because of the increased joint density

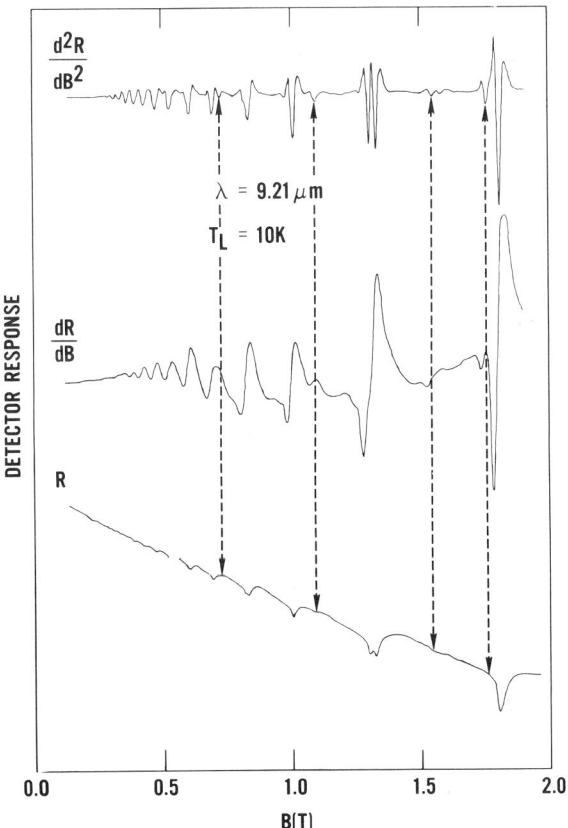

FIG. 54. Comparison of TPMA data obtained using boxcar averaging techniques (lowest curve—R vs. B) and using magnetic modulation techniques, yielding dR/dB (middle curve) and d^2R/dB^2 (upper curve). The dashed lines point to weak transitions in d^2R/dB^2 not found in the R versus B trace. (From Goodwin et al., 1982.)

of states for the higher-field transitions. The TPMA spectra obtained with the ac magnetic field and lock-in amplifier method are modified by a Bessel function envelope to their amplitudes. The derivative techniques clearly allow observation of numerous TPMA transitions that enhance comparison of the TPMA data with theoretical models containing information on selection rules and the energy-band parameters of the sample. Weak resonant structure can also be observed and studied. It is also much easier to observe the small changes in resonant structure due to the crystallographic anisotropy of the sample with respect to the magnetic field (e.g., see Seiler and McClure, 1983) as shown in Fig. 9e.

The wavelength dependence of the TPMA structure is shown in Fig. 55. The numbers in the figure correspond to distinct two-photon transitions that are enumerated in Table XIV. Each photoconductive resonance corresponds to an increase in the conductivity or a decrease in the magnetoresistance of the sample. The TPMA structure shifts toward lower magnetic fields as the

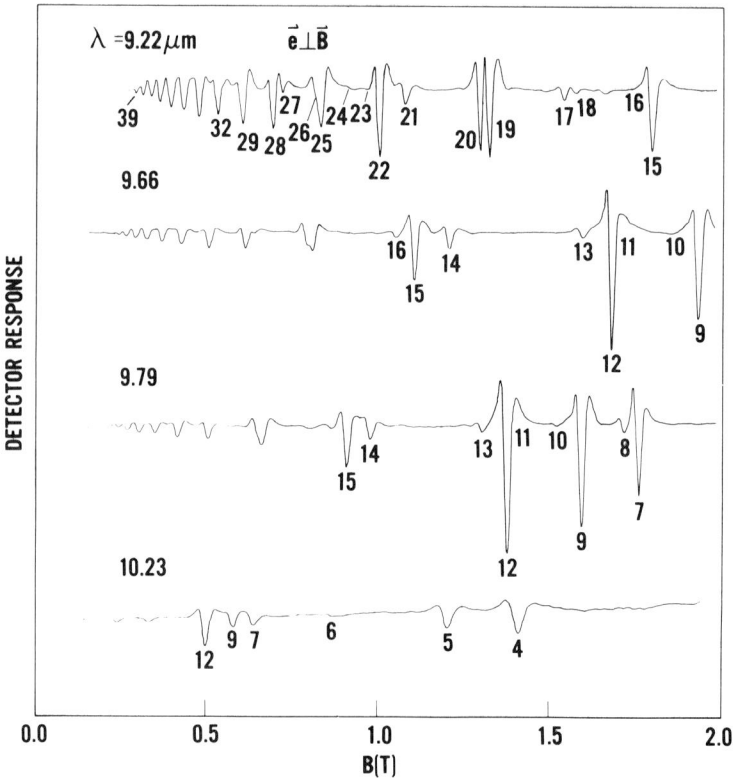

FIG. 55. Wavelength dependence of of the TPMA structure for $\mathbf{B} \parallel \langle 110 \rangle$ and for 0 T to 2 T. (From Seiler, et al., 1983.)

TABLE XIV

Exp. trans. no.	Polarization	Theoretical assignment	Exp. trans. no.	Polarization	Theoretical assignment
*1, *2		$a^-(1)\,a^c(0)$	25	σ_R	$a^-(7)\,a^c(5)$
		$a^+(-1)\,a^c(0)$		σ	$b^+(2)\,b^c(2)$
		$b^-(2)\,a^c(0)$	26	σ_L	$b^+(1)\,b^c(3)$
		$b^-(1)\,a^c(0)$	27	σ_R	$a^+(4)\,a^c(2)$
		$b^+(-1)\,a^c(0)$	28	σ_R	$a^-(8)\,a^c(6)$
3	σ_R	$a^-(2)\,a^c(0)$		σ	$a^+(3)\,a^c(3)$
4	σ	$a^+(0)\,a^c(0)$	29	σ_R	$a^-(9)\,a^c(7)$
5	σ_R	$b^-(2)\,b^c(0)$		σ	$b^+(3)\,b^c(3)$
*6		$a^+(0)\,b^c(0)$	30	σ_L	$b^+(2)\,b^c(4)$
7	σ_L	$a^+(-1)\,a^c(1)$	31	σ_R	$a^+(5)\,a^c(3)$
8	σ	$a^-(1)\,a^c(1)$	32	σ_R	$a^-(10)\,a^c(8)$
9	σ	$b^+(0)\,b^c(0)$		σ	$a^+(4)\,a^c(4)$
10	σ_R	$a^-(3)\,a^c(1)$	33	σ_L	$b^+(3)\,b^c(5)$
11	σ	$b^-(1)\,b^c(1)$		σ_R	$a^-(11)\,a^c(9)$
12	σ_L	$b^+(-1)\,b^c(1)$		σ	$b^+(4)\,b^c(4)$
13	σ_R	$b^-(3)\,b^c(1)$	34	σ_R	$a^+(6)\,a^c(4)$
14	σ_R	$a^+(2)\,a^c(0)$		σ_R	$a^-(12)\,a^c(10)$
15	σ_R	$a^-(4)\,a^c(2)$		σ	$a^+(5)\,a^c(5)$
	σ	$a^+(1)\,a^c(1)$	35	σ_L	$b^+(4)\,b^c(6)$
16	σ_R	$b^-(4)\,b^c(2)$		σ_R	$a^-(13)\,a^c(11)$
*17		$a^+(2)\,b^c(0)$		σ	$b^+(5)\,b^c(5)$
		$a^-(4)\,b^c(2)$	36	σ_R	$a^+(7)\,a^c(5)$
*18		$a^+(1)\,b^c(1)$		σ_R	$a^-(14)\,a^c(12)$
		$a^+(0)\,b^c(2)$		σ	$a^+(6)\,a^c(6)$
	σ_R	$b^+(2)\,b^c(0)$	37	σ_L	$b^+(5)\,b^c(7)$
19	σ_R	$a^-(5)\,a^c(3)$		σ_R	$a^-(15)\,a^c(13)$
	σ	$b^+(1)\,b^c(1)$		σ	$b^+(6)\,b^c(6)$
20	σ_L	$b^+(0)\,b^c(2)$	38	σ_R	$a^+(8)\,a^c(6)$
21	σ_R	$a^+(3)\,a^c(1)$		σ_R	$a^-(16)\,a^c(14)$
22	σ_R	$a^-(6)\,a^c(4)$		σ	$a^+(7)\,a^c(7)$
	σ	$a^+(2)\,a^c(2)$	39	σ_L	$b^+(6)\,b^c(8)$
23	σ_R	$b^-(6)\,b^c(4)$		σ_R	$a^-(17)\,a^c(15)$
*24		$a^+(2)\,b^c(2)$		σ	$b^+(7)\,b^c(7)$

Transition assignments for $\vec{e} \perp \vec{B}$ polarization ($\sigma_R, \sigma_L, \sigma_R + \sigma_L = \sigma$) in the Voigt geometry. The * indicates these transitions which can only be explained by nonspherical transitions. (From Seiler et al., 1983.)

CO_2 laser wavelength is increased. The CO_2 laser is clearly suited to the investigation of TPMA in InSb since its spectral output can be tuned from below to well above the TPMA band edge. Closely spaced transitions near the TPMA band edge were resolved by measurements up to 15 T using boxcar averager techniques, as shown in Fig. 56. At the higher fields, transitions 2 and 3 are well resolved; at the lower fields they are not resolved even when using the derivative techniques.

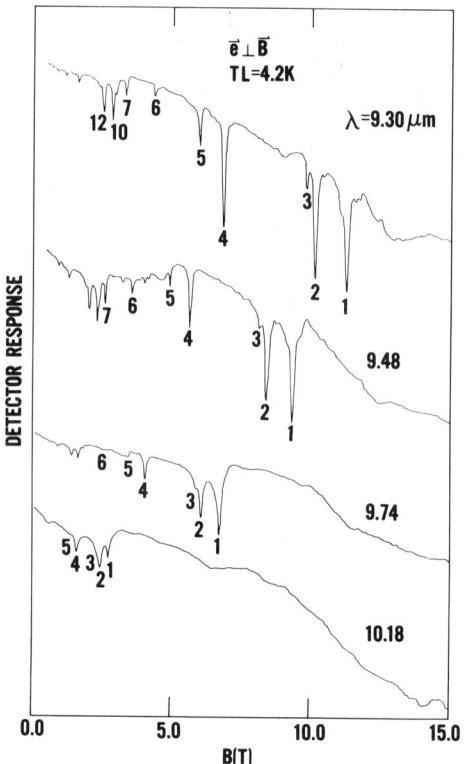

FIG. 56. Wavelength dependence of TPMA for $\mathbf{B} \parallel \langle 211 \rangle$ and for 0 T to 15 T. At the highest magnetic fields the TPMA band edge is directly seen. (From Seiler et al., 1983.)

The effect of light polarization on the TPMA spectra is shown in Fig. 57. In this figure, the light used was slightly elliptical rather than perfectly circular. Thus, for each polarization (σ_R or σ_L), small amounts of the other polarization are present. Clearly, the relative strengths of the transitions depend on the light polarization as expected from the selection rules for TPMA, which differ for each polarization. For example, transitions 11, 13, 14, 17, and 21 are stronger for σ_R, while transitions 12, 16, and 20 are more dominant for σ_L. Some transitions, such as 15 and 19, show no significant dependence on σ_R or σ_L light polarization. These transitions result from σ polarization, which is the sum of σ_R and σ_L light.

The transitions observed using π ($\mathbf{e} \parallel \mathbf{B}$) polarization should match the magnetic-field positions of the σ transitions, because they both obey the same selection rule regarding the initial and final states. However, the strengths of the σ transitions are different from those of the π transitions, because the

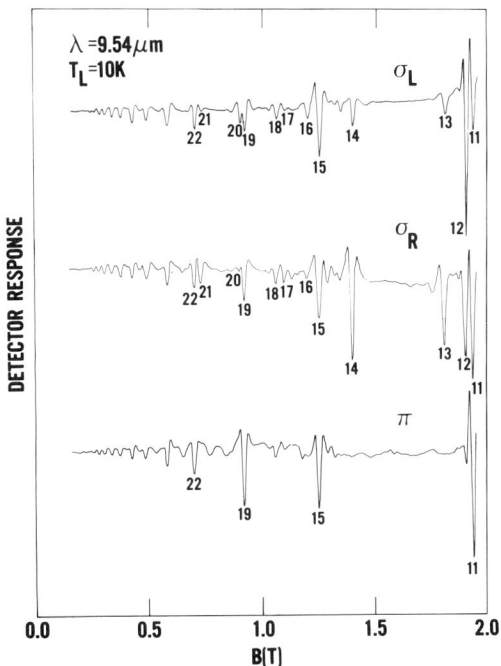

FIG. 57. Polarization dependence of TPMA for **B** ∥ ⟨100⟩. (From Seiler et al., 1983.)

intermediate states for each is different. A schematic presentation of the two-photon magneto-optical transitions for the different light polarizations obeying the spherical selection rules (Seiler et al., 1983) is shown in Fig. 58.

Specific transitions are assigned to the structure by using the polarization dependence and by comparing the theoretically calculated magnetic-field positions of the transitions with the data by using fan charts. Figures 59a,b show fan charts of the experimental (dots) and theoretical (lines) two-photon transition energies as a function of magnetic field. Good agreement is obtained using the theoretical two-photon transition assignments shown in Table XIV. The solid lines result from the use of the spherical selection rules, and the dashed lines result from employing nonspherical selection rules allowed because of warping and inversion asymmetry effects (Goodwin et al., 1982). As mentioned, the band parameters used here are listed in Table IX.

One of the most important applications of TPMA is in the accurate measurement of the energy gap of a NGS and its temperature dependence. This was done for InSb by Littler and Seiler (1985), using the photo-Hall effect to observe the resonant TPMA structure over a wide temperature range (from $\simeq 2$ to 210 K), as shown in Fig. 9c. The shift of the resonant structure to higher

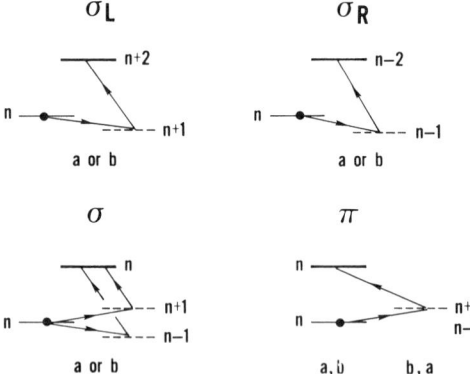

FIG. 58. Schematic diagram of the initial, intermediate, and final states of the two-photon process for the different light polarizations. (From Seiler *et al.*, 1983.)

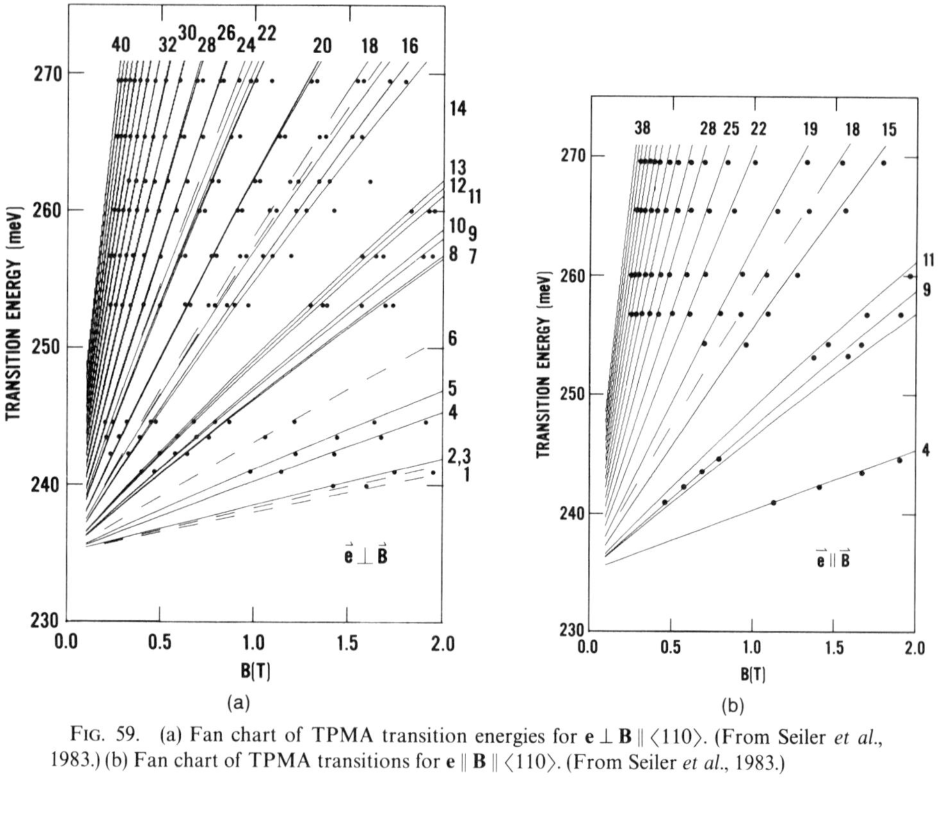

FIG. 59. (a) Fan chart of TPMA transition energies for $\mathbf{e} \perp \mathbf{B} \parallel \langle 110 \rangle$. (From Seiler *et al.*, 1983.) (b) Fan chart of TPMA transitions for $\mathbf{e} \parallel \mathbf{B} \parallel \langle 110 \rangle$. (From Seiler *et al.*, 1983.)

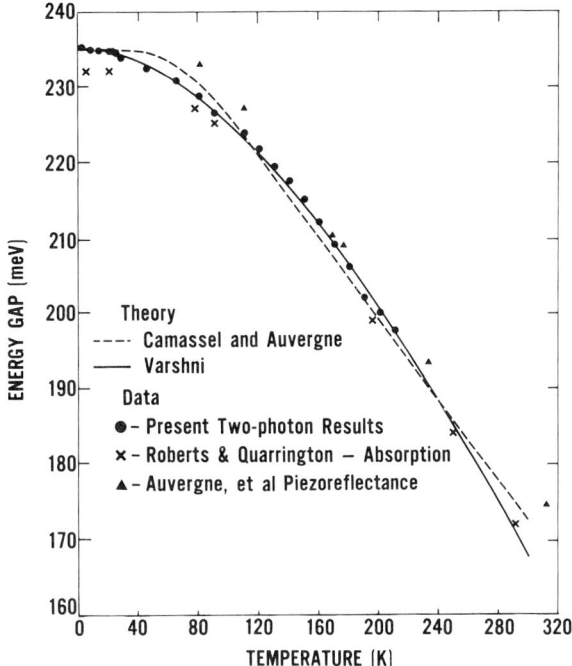

Fig. 60. Temperature dependence of the energy gap of InSb obtained using two-photon photo Hall techniques (dots) compared with selected experimental and theoretical results. (From Littler and Seiler, 1985.)

magnetic fields for larger temperatures is a direct consequence of the decrease of the energy gap since, as E_g becomes smaller, the valence- and conduction-band Landau levels "move" closer together in energy. Thus, larger values of the magnetic field are required to meet the TPMA resonance condition at a given photon energy. Analysis of the wavelength-dependent spectra at each temperature then allows the determination of E_g.

Figure 60 shows the results of the two-photon photo-Hall measurements of E_g versus T compared with various theoretical predictions (Camassel and Auvergne, 1975; Varshni, 1967) and selected experimental results (Roberts and Quarrington, 1955; Auvergne *et al.*, 1974) obtained over the temperature range 0 to 320 K. This temperature dependence is due to the thermal expansion of the lattice and the renormalization of the band energies by the electron–phonon interaction. The two mechanisms compete at low temperatures where few phonons are present and hence cancel each other, giving rise to the nearly flat response below $\simeq 20$ K. The electron–phonon interaction dominates at the higher temperatures, giving rise to the nearly linear decrease of E_g above $\simeq 100$ K. The theoretical predictions shown in Fig. 60 were normalized to our 1.8 K energy gap of 235.2 meV. The pseudopotential calculations of Camassel and Auvergne (1975) do predict the general slope of the data above 110 K, but fail to predict the low-temperature dependence. Varshni's (1967) empirical relationship

$$E_g(T) = E_g(0) - \frac{\alpha T^2}{\beta + T}, \tag{48}$$

with $E_g(0) = 235.2$ meV, $\alpha = 0.6$ meV/K, and $\beta = 500$ K, describes well all features of the two-photon results as well as the 300 K results of Roberts and Quarrington.

15. Review of TPA and TPMA in Alloys of HgCdTe

a. *Two-Photon Absorption*

Relatively little two-photon work has been carried out on HgCdTe alloys, undoubtedly because of the lack of good quality (i.e., homogeneous, high-purity) samples. However, in recent years sufficient quality, homogeneous material has been grown, allowing some interesting TPMA studies to be carried out. Table XV contains a summary of all known TPA and TPMA work on HgCdTe. The earliest TPA work by A. Miller et al. (1979) and Pidgeon et al. (1979) centered around experimentally and theoretically determining the TPA coefficient. Calculations were made using a three-level, nonparabolic energy-band model to explain the magnitude of the experimentally measured results. Bartoli et al. (1983) used TPA methods to uniformly generate carriers in samples with $0.24 < x < 0.27$. They performed photo-Hall measurements to determine the electron mobility as a function of optically generated carrier density at low temperatures where phonon scattering is not important. This photoneutralization technique allows the determination of the compensation density (*n*-type material), which is otherwise very difficult to measure.

Two-photon-induced optical switching and bistability have been investigated at room temperature by Miller, Craig, Matthews, and others (see Table XIV). The TPA-produced carriers cause refractive index changes at wavelengths where the linear absorption of the material is small, allowing higher etalon finesse. Both nonequilibrium and quasi-cw regimes were studied using TEA lasers that produced output pulses of either 30 nsec or 1.75 μsec, respectively. The shorter pulses caused fast switching of an etalon and gave rise to transmission overshoots. The longer pulses were used to give a clear demonstration of optical bistability. Unfortunately, large sample-to-sample variations were also observed and ascribed to inhomogeneities of the

TABLE XV

COMPREHENSIVE COMPILATION OF ALL TWO-PHOTON ABSORPTION WORK DONE ON $Hg_{1-x}Cd_xTe$

Authors (year)	Photon energy (meV)	T_L (K)	Method	Main results
Miller et al. (1979)	117	295	A	TPA coefficients measured and calculated for $x = 0.22$. $\tau = 40$ nsec, and $K_2 = 14.0$ cm/MW
Pidgeon et al. (1979)			Theory	Three-band, nonparabolic model used for calculating frequency dependence of TPA coefficnets.
Johnston et al. (1980)	117 129	77 295	PC & A	TPA coefficients measured and compared to theory. Lifetimes measured.
Pidgeon (1980)			Theory	TPA calculations for 3-band, nonparabolic model.
Yuen (1982)		2	4-wave mixing	Nondegenerate 4-wave mixing. Nonparabolicity for free electrons generated by TPA affected x^3.
Bartoli et al. (1983)	117	10	PC & PH	Photoneutralization technique accurately determined compensation densities using TPA for samples with $0.24 < x < 0.27$.
Miller et al. (1984)	117 TEA	295	A	TPA induced optical switching observed.
Miller & Parry (1984)	117 TEA	295	A	Observation of TPA-induced optical switching (bistability).
Matthew et al. (1985)	117 TEA	295	A	Dynamic optical bistability or optical switching by TPA achieved with $x = 0.23$ samples.
Craig et al. (1985a)	117 TEA	295	A	Study of optical switching by TPA excitation.
Craig et al. (1985b)	117 TEA	295	A	Quasi-steady-state optical bistability in an etalon induced by free carriers generated by TPA.

(continued)

TABLE XV (Continued)

Authors (year)	Photon energy (mev)	T_L (K)	Method	Main results
Craig & Miller (1986)	117 TEA	295	A	Band-gap dependence of free-carrier-induced optical nonlinearities including TPA processes.
Seiler et al. (1986a)	Tunable cw CO_2	10–75	PC	Determination of E_g by TPA techniques for samples with $x \simeq 0.32$.
Seiler et al. (1986b)	Tunable cw CO_2 & Q	8–250	PC & PH	Nonlinear optical characterization by TPA techniques. Determination of number of photoexcited electrons, electron lifetime, TPA coefficient, and TPA cutoff wavelengths (from which E_g can be extracted). Information about recombination mechanisms.
Craig et al. (1986)	117 TEA	295	A	Observed severe beam distortion in line resolved measurements of self-defocusing caused by TPA.
Miller & Craig (1987)	117 TEA	295	A	TPA-induced optical switching and bistability measured and discussed.
Seiler et al. (1989a)	Tunable cw CO_2	7	PC	First report and analysis of TPMA effects in HgCdTe samples ($0.24 < x < 0.28$). Magneto-optical transition from both shallow acceptor and midgap levels to conduction band also seen. OPMA effects also seen.
Seiler et al. (1989b)	Tunable cw CO_2	7–150	PC	Observation and detailed analysis of TPMA spectra in 6 samples ($0.24 < x < 0.30$), both n- and p-type. T_L dependence of E_g accurately measured, revealing a nonlinear variation of E_g versus T_L, not previously reported. Observed impurity (or defect)—to conduction band transitions from midgap and shallow acceptors.
Seiler et al. (1990)	Tunable cw CO_2	7–150	PC	Accurate determination of E_g versus temperature for 3 alloys of HgCdTe with $0.24 \leq x \leq 0.26$ made by analyzing TPMA spectra. New relationship for $E_g(x, T)$ found that more properly accounts for the nonlinear temperature dependence of E_g below 77 K.

PC = photoconductivity; A = absorption (transmission); PH = photo-Hall.

HgCdTe. Finally, Craig, Miller, and Soileau (1986) observed severe beam distortion in time-resolved measurements of self-defocusing caused by TPA.

Using TPA techniques with a cw CO_2 laser, Seiler et al. (1986b,c) measured TPA cutoff wavelengths in a sample with $x \simeq 0.32$. Band-gap energies and hence x-values could be extracted to a precision of $\Delta x \simeq \pm 0.001$. The number of photoexcited electrons was directly determined by the photo-Hall effect, as shown in Fig. 8b. Transient decay measurements of the electron lifetime then allowed the TPA coefficient to be determined.

b. Two-Photon Magnetoabsorption

The TPMA structure has only recently been observed in HgCdTe by Seiler et al. (1989a,b). The TPMA spectra have large linewidths that are caused by spatial variations in the mole fraction (x) of the sample in the region where the photoexcited carriers are generated by the TPA process. These variations in x produce spatial variations in the fundamental energy gap E_g, and a "smearing" or broadening of the Landau-level energies gives rise to broad TPMA spectral features. At low magnetic fields (below 2 T), only the most homogeneous samples (e.g., for $x \simeq 0.3$, $\Delta x < \pm 0.001$) will show resonant TPMA structure, and this structure will be quite broad. For higher magnetic fields (>2 T), a well-defined (i.e., discrete lines, well-separated) resonant structure exists for the homogeneous samples ($\Delta x < \pm 0.001$). We estimate that if access to fields of 10 T is available, TPMA structure can be observed in samples as inhomogeneous as $\Delta x = \pm 0.004$. Consequently, the half-width of the resonant TPMA structure provides a quantitative measure of the sample's compositional inhomogeneity and can be used as a sensitive characterization tool to test the quality of the material grown by various techniques.

In the studies by Seiler et al. (1989a,b) the photoconductive response of a HgCdTe sample, created by illumination using a CO_2 laser, was investigated for the magnetic fields 0–12 T. The characteristic feature of the TPMA spectra obtained is the occurrence of two large resonances, as seen in Fig. 61 for a number of samples with various values of x. This feature is present regardless of whether the sample is n- or p-type, which implies that the absorption process is intrinsic, such as interband TPMA. A comparison of TPMA spectra in an n-InSb sample (representative of TPMA spectra in a homogeneous sample) with that of a similar band-gap sample of HgCdTe with $x \simeq 0.30$ is shown in Fig. 62. Two features are readily apparent: (1) The TPMA resonances are very broad in HgCdTe ($\simeq 10$ larger than in InSb), which indicates the inhomogeneous nature of the HgCdTe samples; and (2) the InSb and HgCdTe spectra themselves are qualitatively and quantitatively different. The latter feature is due to the differences in band-structure parameters that

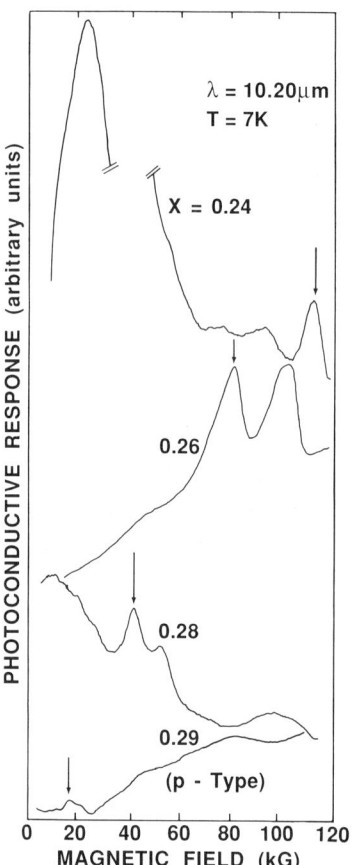

FIG. 61. TPMA spectra for samples of HgCdTe with $x = 0.24, 0.26, 0.28$ (all n-type), and 0.29 (p-type).

produce different magnetic-field and energy dependences of the conduction- and valence-band Landau levels. In addition, the strengths of the transition matrix elements for the same TPMA transitions are different, and so the transition probabilities given by Eq. (14) are different.

The use of left (σ_L) and right (σ_R) circularly polarized light is essential for the identification of the TPMA transitions, as shown in Fig. 46. At the higher magnetic fields two strong TPMA resonances occur, as labeled by the solid arrows L_1 and L_2. The two broader peaks, labeled 1 and 2 (by the dashed arrows), have been ascribed to electron transitions from two closely spaced impurity or defect levels at approximately midgap (Littler et al., 1990a). The magnetic-field positions of all arrows indicate the theoretically expected positions of both the impurity or defect and TPMA resonances. For σ_R polariza-

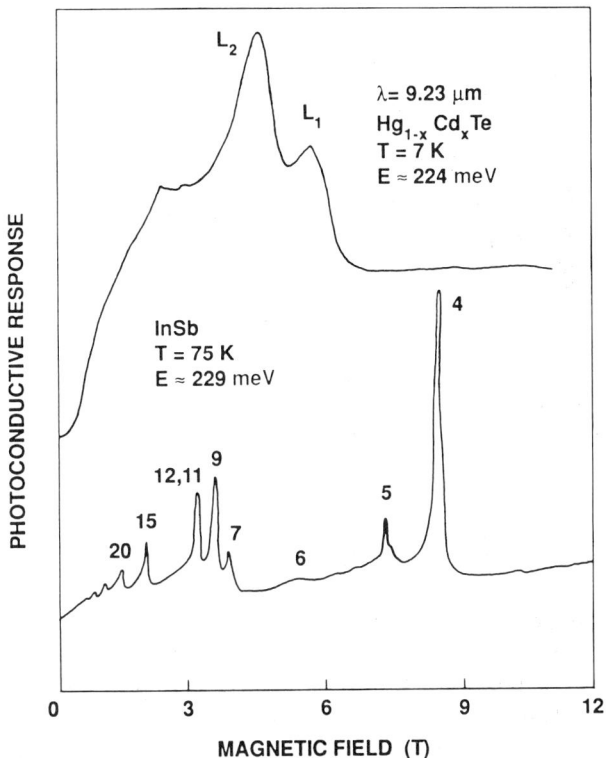

FIG. 62. Comparison of TPMA spectra obtained from InSb and HgCdTe with $x \simeq 0.30$. Note the much broader resonances seen in the HgCdTe sample.

tion, the two strong σ_L transitions disappear and are replaced with weaker resonances $R_1, R_2, \ldots, R_5$. The notation for these transitions is shown in Table XI. The impurity- and defect-related transitions do not depend on the polarization. The excellent agreement between the experimental field positions of the peak TPMA resonances and the theoretically calculated positions as shown by the arrows strongly confirm that the transitions designated by $L_1, \ldots, R_1, \ldots$ arise from TPMA processes.

The wavelength dependence of the PC response for a sample with $x \simeq 0.30$ is shown in Fig. 48. The downward-pointing arrows, labeled $L_1, L, \ldots$ indicate the TPMA transitions as given in Table XI. The two broader peaks at higher fields have been ascribed to one-photon transitions from two closely spaced midgap levels to the lowest conduction-band Landau level. The wavelength dependence of the TPMA transitions can be understood from the two-photon transition energy dependence

$$2\hbar\omega = E_c^{a,b}(n_c, B) - E_v^{a,b}(n_v, B). \quad (49)$$

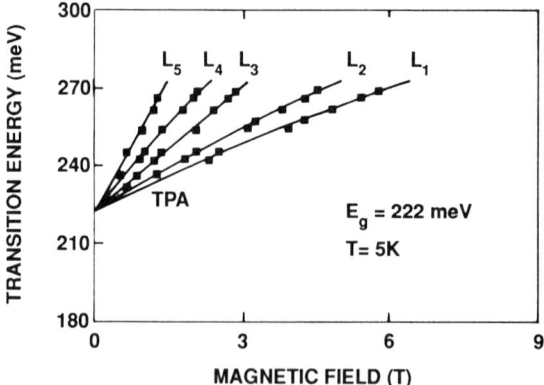

FIG. 63. Fan chart of transition energies for a sample of HgCdTe with $x \simeq 0.30$. The labels L_1-L_5 refer to transition assignments given in Table XI.

Figure 63 shows a comparison of the theoretically calculated and experimentally determined transition energies; good agreement is seen, which verifies the TPMA model used to interpret the observed structure. The values for $2\hbar\omega$ were calculated for each transition using a modified Pidgeon–Brown energy-band model and the transition assignments given in Table XI. Note that as $B \to 0$, $2\hbar\omega \to E_g$. Thus, the value of the energy gap can be accurately determined by using E_g as an adjustable parameter when fitting the model to the data. Weiler's set of energy-band parameters (Weiler, 1981b) accurately describes the data: $E_p = 19.0$ eV, $\Delta = 1.0$ eV, $\gamma_1 = 3.3$, $\gamma_2 = 0.1$, $\gamma_3 = 0.9$, $\kappa = -0.8$, $q = 0.0$, and $N_1 = 0.0$. For the sample shown a value for $E_g = 222.0 \pm 0.5$ meV is obtained. Figure 47 shows the two dominant σ_L TPMA transitions present in a sample with $x = 0.246$. The fan chart plot of $2\hbar\omega$ versus magnetic field for the TPMA structure (Fig. 49) shows excellent agreement between the experimental data and the theoretical calculations. This adds confirmation to the accuracy of Weiler's set of energy-band parameters for describing a variety of HgCdTe alloy compositions.

c. *Composition and Temperature Dependence of E_g by TPMA Techniques*

The primary parameter controlling the operating wavelength range for $Hg_{1-x}Cd_xTe$ intrinsic infrared detectors is the fundamental energy gap E_g. The values of the mole fraction of cadmium, x, and the lattice temperature T critically control the desired values of E_g needed for each device. Several

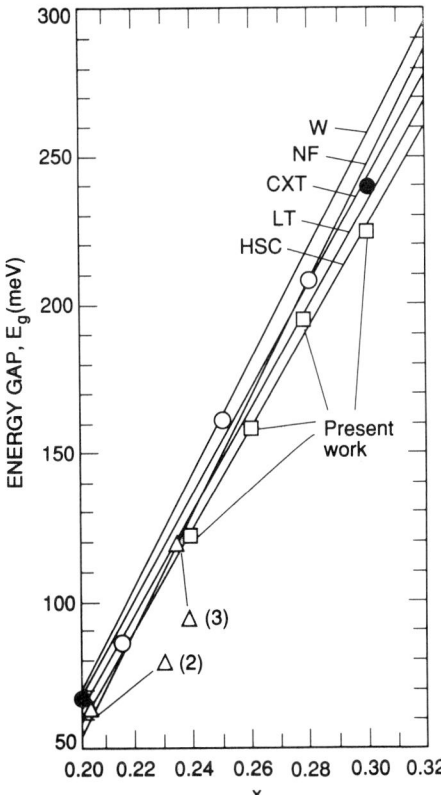

FIG. 64. The low-temperature x-dependence of the energy gap of HgCdTe for $0.2 \leq x \leq 0.3$. The solid lines represent various published relationships describing $E_g(x, T)$: W (Weiler, 1981); NF (Nemirovsky and Finkman, 1979); CXT (Chu et al., 1983); LT (Legros and Triboulet, 1985); HSC (Hansen et al., 1982). The open circles represent the data of Guldner et al. (1977a,b), the solid circles are from the EMIS Datareviews (Brice, 1987), $\Delta(2)$ data from McCombe et al. (1970b), $\Delta(3)$ data from Bridges et al. (1979), and the open boxes are TPMA results of Seiler et al. (1990). (From Seiler et al., 1990.)

empirical relationships $E_g(x, T)$ have been reported by various authors, with considerable disagreement among them. Recently, Seiler et al. (1990) used TPMA techniques to determine E_g accurately for several values of x between $x = 0.22$ and $x = 0.30$. Figure 64 shows a plot of E_g ($T \simeq 0$) over the important alloy composition range $0.2 < x < 0.32$ for several values of E_g determined by one-photon magneto-optical experiments (Guldner et al., 1977a,b) along with the values of E_g determined from TPMA measurements. Values of E_g from the EMIS Datareviews Series No. 3 (Brice, 1987) are also shown at $x = 0.2$ and 0.3. The variation of the five $E_g(x, T)$ relationships are also shown by the solid lines and are identified in the figure caption. The TPMA data support the use of the Hansen–Schmidt–Casselman (HSC) relationship (1982) as representing the best values of $E_g(x)$ at very low temperatures (< 10 K).

Seiler et al. (1990) also used the high accuracy of determining E_g by TPMA techniques to accurately determine the nonlinear temperature variation of $E_g(T)$. By simultaneously fitting three sets of E_g versus T data for samples with $x = 0.239$, $x = 0.253$, and $x = 0.259$, they obtained a new $E_g(x, T)$ relationship that merges smoothly into that of the HSC relationship above 100 K. This relation is

$$E_g(x, T) = -0.302 + 1.93x + 5.35(1 - 2x)(10^{-4})\left[\frac{-1822 + T^3}{255.2 + T^2}\right]$$

$$- 0.810x^2 + 0.832x^3 \qquad (50)$$

for E_g in electron volts and T in kelvins. For samples with $x \simeq 0.2$, the maximum deviation from that of HSC is 3–4 meV at 10–12 K.

References

Abduvakhidov, Kh. M., Volkov, A. S., and Galavanov, V. V. (1967). *Sov. Phys. Semicond.* **1**, 788.
Aggarwal, R. L. (1972). "Semiconductors and Semimetals" (Willardson, R. K., and Beer, A. C., eds.), vol. 9, Ch. 2. Academic Press, New York.
Alekseev, M. A., Bresler, M. J., Gusev, O. B., Merkulov, I. A., and Stepanov, A. O. (1985). *Sov. Phys. Semicond.* **19**, 443.
Andrukhiv, M. G., Maltseva, V. A., Ivanov-Omskii, V. I., Ogoranikov, V. K., and Totieva, T. S. (1979). *Sov. Phys. Semicond.* **13**, 210.
Auvergne, D., Camassel, J., Mathieu, H., and Cordona, M. (1974). *Phys. Rev.* **B9**, 5168.
Bader, T. R., and Gold, A. (1968). *Phys. Rev.* **171**, 997.
Bagguley, D. M. S., Robinson, M. L. A., and Stradling, R. A. (1963). *Phys. Lett.* **6**, 143.
Bajaj, J., Shin, S. H., Pasko, J. G., and Khoshnevisan, M. (1983). *J. Vac. Sci. Technol.* **A1**, 1749.
Bartoli, F. J., Hoffman, C. A., and Meyer, J. R. (1983). *J. Vac. Sci. Technol.* **A1**, 1669.
Bartoli, F. J., Hoffman, C. A., and Meyer, J. R. (1986). *J. Vac. Sci. Technol.* **A4**, 2047.
Baryshev, N. S., Vdovkina, E. E., Martynovich, A. P., Nesmelova, I. M., Tsitsina, N., and Aver'yanov, I. S. (1967). *Sov. Phys. Solid State* **8**, 1800.
Bass, F. G., and Levinson, I. B. (1966). *Sov. Phys.-JETP* **22**, 635.
Bassani, F. (1972). Proc. of Int. School of Phys., Enrico Fermi, Course LII, pp. 592–608. Academic Press, New York.
Bassini, F., and Baldereschi, A. (1973). *Surface Sci.* **37**, 304.
Bassani, F., and Pastori Parravicini, G. (1975). "Electronic States and Optical Transitions in Solids" Pergamon Press, New York.
Bassani, F., and Girlanda, R. (1970). *Optics Commun.* **1**, 359.
Bauer, G. (1980). "Narrow Gap Semiconductors Physics and Applications," p. 427. Springer, New York.

Bechtel, J. H., and Smith, W. L. (1976). *Phys. Rev.* **B8**, 3515.
Bell, R. L., and Rogers, K. T. (1966). *Phys. Rev.* **152**, 746.
Blakemore, J. S. (1962). "Semiconductor Physics." Pergamon Press, London.
Bemski, G. (1960). *Phys. Rev. Lett.* **4**, 62.
Blaut-Blachev, A. N., Iglitsyn, M. I., Iveleva, V. S., and Selyanina, V. I. (1975). *Sov. Phys. Semicond.* **9**, 247.
Blaut-Blachev, A. N., Iveleva, V. S., Korotin, V. G., Krivonogov, S. N., Selyanina, V. I., and Smetannikova, Yu. S. (1976). *Sov. Phys. Semicond.* **9**, 1414.
Bonch-Bruevich, A. M., and Khodovoi, V. A. (1965). *Sov. Phys.-Uspekhi* **85**, 1.
Bowers, R., and Yafet, Y. (1959). *Phys. Rev.* **115**, 1165.
Brandi, H. S., and deAraujo, C. B. (1983). *J. Phys. C: Solid State Phys.* **16**, 5929.
Bratashevskii, Y. A., Tyutyunnik, V. B., Averyanov, K. S., and Nesmelova, I. M. (1975). *Sov. Phys. Semicond.* **9**, 114.
Bratashevskii, Y. A., Tyutyunnik, V. B., Averyanov, K. S., and Nesmelova I. M. (1977). *Sov. J. Low Temp. Phys.* **3**, 586.
Bredikhin, V. I., Galanin, M. D., and Genkin, V. N. (1973). *Sov. Phys.-Uspekhi* **16**, 299–321.
Bresler, M. S., Gusev, O. B., and Stepanov, A. O. (1983). *Sov. Phys. Semicond.* **17**, 755.
Brice, J. C. (1987). "EMIS Datareviews Series No. 3," p. 103. INSPEC, London.
Bridges, T. J., Burkhardt, E. G., and Nguyen, V. T. (1979). *Optics Commun.* **30**, 66.
Brunel, L., Huant, S., Baj, M., and Trzeciakowski, W. (1986). *Phys. Rev.* **B33**, 6863.
Burstein, E., Picus, G. S., and Gebbie, H. A. (1956a). *Phys. Rev.* **103**, 825.
Burstein, E., Picus, G. S., Gebbie, H. A., and Blatt, F. (1956b). *Phys. Rev.* **103**, 826.
Button, K. J., Lax, B., Weiler, M. H., and Reine, M. (1966). *Phys. Rev. Lett.* **17**, 1005.
Button, K. S., Lax, B., and Bradley, C. C. (1968). *Phys. Rev. Lett.* **21**, 350.
Bychkov, Yu. A., and Dykhne, A. M. (1970). *Sov. Phys. JETP* **31**, 928.
Camassel, J., and Auvergne, D. (1975). *Phys. Rev.* **B12**, 3258.
Chaikovskii, I. A., Kovarskii, V. A., and Perlin, E. Yu. (1972). *Sov. Phys.-Solid State* **14**, 620.
Chen, M. C., and Dodge, J. A. (1986). *Solid State Commun.* **53**, 449.
Chen, M. C., and Tregilgas, J. H. (1987). *J. Appl. Phys.* **61**, 787.
Chen, Y. F., Dobrowolska, M., and Furdyna, J. K. (1985). *Phys. Rev.* **B31**, 7989.
Choi, J. B., Kim, L. S., Drew, H. D., and Nelson, D. A. (1988). *Solid State Commun.* **65**, 547.
Chu, J., Xu, S., and Tang, D. (1983). *Appl. Phys. Lett* **43**, 1064.
Cotter, D., Hanna, D. C., and Wyatt, R. (1976). *Optics Commun.* **16**, 256.
Cotton, V. A., Wilson, J. A., and Jones, C. E. (1985) *J. Appl. Phys.* **58**, 2208.
Craig, D., and Miller, A. (1986). *Optica Acta* **33**, 397.
Craig, D., Kar, A. K., Mathew, J. G. H., and Miller, A. (1985b). *IEEE J. Quant. Elec.* **QE-21**, 1363.
Craig, D., Miller, A., Mathew, J. G. H., and Kar, A. K. (1985a). *Infrared Phys.* **25**, 289.
Craig, D., Miller, A., and Soileau, M. J. (1986). *Optics Lett.* **11**, 794.
Cunningham, R. W., Harp, E. E., and Bullis, W. M. (1962). Proc. Int. Conf. on the Physics of Semiconductors, Exeter, p. 732. Institute of Physics, London.
Danishevskii, A. M., Patrin, A. A., Ryvkin, S. M., and Yaroshetskii, I. D. (1969). *Sov. Phys. JETP* **29**, 781.
Danishevskii, A. M., Ivchenko, E. A., Kochegarov, S. F., and Stepanova, M. I. (1972). *Sov. Phys. JETP Lett.* **16**, 440.
Danishevskii, A. M., Kochegarov, S. F., and Subashiev, V. K. (1973). *Sov. Phys. Solid State* **14**, 2754.
Dempsey, J., Smith, J., Holah, G. D., and Miller, A. (1978). *Optics Commun.* **26**, 265.
Dennis, R. B., McKenzie, H. A., Smith, S. D., Voge, D. D., and Wang, W. (1982). *Optics Commun.* **41**, 345.
De Salvo, E., Girlanda, R., and Quattropani, A. (1985). *Nuovo Cimento* **5D**, 63.
Dickey, D. H., and Larsen, D. M. (1968). *Phys. Rev. Lett.* **20**, 65.

Dickey, D. H., Johnson, E. J., and Larsen, D. M. (1967). *Phys. Rev. Lett.* **18**, 599.
Dornhaus R., and Nimtz G. (1985). "Narrow Gap Semiconductors," p. 119ff. Springer, New York.
Doviak, J. M., Gibson, A. F., Kimmitt, M. F., and Walker, A. C. (1973). *J. Phys. C: Solid State Phys.* **6**, 593.
Dresselhaus, G., Kip, A. F., and Kittel, C. (1955). *Phys. Rev.* **48**, 368.
Efros, Al. L., Kanskaya, L. M., Kokanovskii, S. I., and Seisyan, R. P. (1982a). *Phys. Stat. Solidi (b)* **114**, 373.
Efros, Al. L., Kanskaya, L. M., Kokanovskii, S. I., and Seisyan, R. P. (1982b). *Solid State Commun.* **43**, 613.
Elliott, C. T., Melngailis, I., Harmon, T. C., and Foyt, A. J. (1972). *J. Phys. Chem. Solids* **33**, 1527.
Enck, R. C., Saleh, A. L., and Fan, H. Y. (1969). *Phys. Rev.* **182**, 790.
Engeler, W., Levinstein, H., and Stannard, Jr., C. (1961a). *J. Phys. Chem. Solids* **22**, 249.
Engeler, W., Levinstein, H., and Stannard, Jr., C. (1961b). *Phys. Rev. Lett.* **7**, 62.
Evtuhov, V., (1962). *Phys. Rev.* **125**, 1869.
Favrot, G., Aggarwal R. L., and Lax, B. (1976a) Proc. 13th Int. Conf. Physics of Semiconductors. Tipografia Marves, Rome.
Favrot, G., Aggarwal R. L., and Lax, B. (1976). *Solid State Commun.* **18**, 577.
Fenton, E. W., and Haering, R. R. (1967). *Phys. Rev.* **159**, 593.
Finkman, E., and Nemirovsky, Y. (1986). *J. Appl. Phys.* **59**, 1205.
Fossum, H. J., and Ancker-Johnson, B. (1973). *Phys. Rev.* **B8**, 2850.
Fossum, H. J., and Chang, D. B. (1973). *Phys. Rev.* **B8**, 2842.
Fossum, H. J., Chen, W. S., and Ancker-Johnson, B. (1973). *Phys. Rev.* **B8**, 2857.
Fröhlich, D. H. (1970a). "Festkörperprobleme X," pp. 227–255. Pergamon, Vieweg.
Fröhlich, D. H. (1970b). Proc. 10th Int. Conf. on the Physics of Semiconductors, pp. 95–101. U.S. Atomic Energy Commission, Oak Ridge.
Fröhlich, D. (1981). "Festkörperprobleme XXI," pp. 363–381. Pergamon, Vieweg.
Fröhlich, D., and Sondergeld, M. (1977). *J. Phys. E: Sci. Instrum.* **10**, 761–766.
Gaj, J. A. (1988). "Semiconductors and Semimetals," (Willardson, R. K., and Beer, A. C., eds.), vol. 25, p. 275. Academic Press, New York.
Galanov, E. K., Potikhonov, G. N., and Petukhov, I. P. (1985). *Sov. Phys. Semicond.* **19**, 926.
Galavanov, V. V., and Oding, V. G. (1969). *Sov. Phys. Semicond.* **3**, 238.
Galavanov, V. V., and Oding, V. G. (1976). *Sov. Phys. Semicond.* **9**, 1317.
Galavanov, V. V., Ivchenko, E. L., and Oding, V. G. (1973). *Sov. Phys. Semicond.* **7**, 547.
Gelmont, B. L., Ivanov-Omskii, V. I., Maltseva, V. A., and Smirnov, V. A. (1981). *Sov. Phys. Semicond.* **15**, 638.
Gibson, A. F., Kent, M. J., and Kimmitt, M. F. (1968). *Brit. J. Appl. Phys.* **1**, 149.
Gibson, A. F., Hatch, C. B., Maggs, P. N. D., Tilley, D. R., and Walker, A. C. (1976). *J. Phys. C: Solid State Phys.* **9**, 3259.
Girlanda, R. (1971). *Nuovo Cimento* **6B**, 53.
Girlanda, R. (1977). *Nuovo Cimento* **39B**, 593.
Goldman, V. J., Drew, H. D., Shayegan, M., and Nelson, D. A. (1986a). *Phys. Rev. Lett.* **56**, 968.
Goldman, V. J., Shayegan, M., and Drew, H. D. (1986b). *Phys. Rev. Lett.* **57**, 1056.
Goodwin, M. W., Seiler, D. G., and Weiler, M. H., (1982). *Phys. Rev.* **B25**, 6300.
Goodwin, M. W., and Seiler, D. G. (1983). *Phys. Rev.* **B27**, 3451.
Goppert-Mayer, M. (1931). *Ann. Phys.* **9**, 273.
Grave, T., Scholl, E., and Wurz, H. (1983). *J. Phys. C: Solid State Phys.* **16**, 1693.
Grisar, R., Wachernig, H., Bauer, G., Wlasak, J., Kowalski, J., and Zawadzki, W. (1978). *Phys. Rev.* **B18**, 4355.
Grober, R. D., and Drew, H. D. (1991). *Phys. Rev.* **B43**, 11732.
Groves, S. H., Harmon, T. C., and Pidgeon, C. R. (1971). *Solid State Commun.* **9**, 451.
Guldner, Y., Rigaux, C., Mycielski, A., and Couder, Y. (1977a). *Phys. Stat. Solidi* **B81**, 615.

Guldner, Y., Rigaux, C., Mycielski, A., and Couder, Y. (1977b). *Phys. Stat. Solidi B***82**, 149.
Guseinov, E. K., Ibragimov, R. I., Korotin, V. G., Nasledov, D. N., and Popov, Y. G. (1971). *Sov. Phys. Semicond.* **5**, 1549.
Hanes, L. K., and Seiler, D. G. (1980). *Optics Commun.* **34**, 89.
Hanes, L. K., and Seiler, D. G. (1988). *Solid State Electronics* **31**, 493.
Hansen, G. L., Schmit, J. L., and Casselman, T. N. (1982). *J. Appl. Phys.* **53**, 7099.
Harman, T. C., Strauss, A. J., Dickey, D. H., Dresselhaus, M. S., Wright, G. B., and Marvroides, J. G. (1961). *Phys. Rev. Lett.* **7**, 403.
Hassan, A. R. (1976a). *Solid State Commun.* **18**, 437.
Hassan, A. R. (1976b). *J. Phys C: Solid State Phys.* **9**, 2383.
Hassan, A., and Moussa, A. R. (1977). *Nuovo Cimento* **40B**, 354.
Hasselbeck, M., and Kwok, H. S. (1982). *Appl. Phys. Lett.* **41**, 1138.
Heyke, K., Lautz, G., and Schummy, H. (1970). *Phys. Status Solidi A***1**, 459.
Holah, G. D., Dempsey, J., Miller, D. A. B., Wherrett, B. S., and Miller, A. (1979). Proc. 14th Int. Conf. on Physics of Semicond., p. 505. The Institute of Physics, London.
Hollis, J. E. L., Choo, S. C., and Heasell, E. L. (1967). *J. Appl. Phys.* **38**, 1626.
Hongo, S., Panyakeow, S., Shirafuji, J., and Inuishi, Y. (1971). *Jap. J. Appl. Phys.* **10**, 717.
Huant, S., Brunel, L. C., Baj, M., Dmovski, L., Coron, N., and Dambier, G. (1985). *Solid State Commun.* **54**, 131.
Hunter, A. T., Smith, D. L., and McGill, T. C. (1980). *Appl. Phys. Lett.* **37**, 200.
Hunter, A. T., and McGill, T. C. (1982). *J. Vac. Sci. Technol.* **21**, 205.
Inoue, M., and Toyozawa, Y. (1965). *J. Phys. Soc. Japan* **20**, 363.
Isaacson, R. A. (1968). *Phys. Rev.* **169**, 312.
Ismailov, I. M., Nasledov, D. N., Smetannikova, Y. S., and Felitsiant, V. R. (1969). *Phys. Status Solidi* **36**, 747.
Ivanov-Omskii, V. I., Korovin, K. I., and Shereghii, E. M. (1978). *Phys. Stat. Solidi B***90**, 11.
Ivanov-Omskii, V. I., Kokhanovskii, S. I., Seisyan, R. P., Smirnov, V. A., and Yuldashov, Sh. U. (1983a). *Sov. Phys. Semicond.* **17**, 334.
Ivanov-Omskii, V. I., Kokhanovskii, S. I., Seisyan, R. P., Smirnov, V. A., Yukish, V. A., Yuldashov, Sh. U., and Efros, A. L. (1983b). *Solid State Commun.* **46**, 26. .
Jain, R. K., and Klein, M. B. (1983). "Optical Phase Conjugation" (Fisher, R. A., ed.), p. 307. Academic Press, New York.
Jantsch, W., Wünstel, K., Kumagai, O., and Vogl, P. (1982). *Phys. Rev. B***25**, 5515.
Jarasiunas, K., Stonys, S., and Sirmulis, E. (1986). *IEEE J. Quant. Elec.* **QE22**, 1341.
Jarosik, N. C., McCombe, B. D., Shannabrook, B. V., Comas, J., Ralston, J., and Wicks, G. (1985) *Phys. Rev. Lett.* **54**, 1283.
Ji, W., Kar, A. K., Mathew, J. G. H., and Walker, A. C. (1986). *IEEE J. Quantum Elec.* **QE-22**, 369.
Johnson, E. J. (1967). *Phys. Rev. Lett.* **19**, 352.
Johnson, E. J., and Dickey, D. H. (1970). *Phys. Rev. B***1**, 2676.
Johnson, E. J., and Larsen, D. M. (1966). *Phys. Rev. Lett.* **16**, 655.
Johnston, A. M., Pidgeon, C. R., and Dempsey, J. (1980). *Phys. Rev. B***22**, 825.
Jones, C. E., Nair, V., and Polla, D. L. (1981). *Appl. Phys. Lett.* **39**, 248.
Jones, C. E., Nair, V., Lindquist, J., and Polla, D. L. (1982). *J. Vac. Sci. Technol.* **21**, 187.
Jones, C. E., James, K., Merz, J., Braunstein, R., Burd, M., Eetemadi, M., Hutton, S., and Drumheller, J. (1985). *J. Vac. Sci. Technol.* **A3**, 131.
Kacman, P., and Zawadzki, W. (1971). *Phys. Stat. Solidi(b)* **47**, 629.
Kahlert, H., and Seiler, D. G. (1977). *Rev. Sci. Instrum.* **48**, 1017.
Kalisher, M. H. (1984). *J. Crystal Growth* **70**, 369.
Kalugina, N. A., and Skok, E. M. (1984) *JETP Lett.* **38**, 297.
Kane, E. O. (1957). *J. Phys. Chem. Solids* **1**, 249.
Kanskaya, L. M., Kokhanovskii, S. I., and Seisyan, R. P. (1979). *Sov. Phys. Semicond.* **13**, 1420.

Kaplan, R. (1968). *Phys. Rev. Lett.* **20**, 329.
Kaplan, R. (1969). *Phys. Rev.* **181**, 1154.
Kaplan, R. (1973). *Solid State Commun.* **12**, 191.
Kaplan, R., and Wallis, R. F. (1968). *Phys. Rev. Lett.* **20**, 1499.
Kaplan, R., Ngai, K. L., and Henvis, B. W. (1972). *Phys. Rev. Lett.* **28**, 1044.
Kaplan, R., Cooke, R. A., and Stradling, R. A. (1978). *Solid State Commun.* **26**, 741.
Kar, A. K., Mathew, J. G. H., Smith, S. D., Davis, B., and Prettl, W. (1983). *Appl. Phys. Lett.* **42**, 334.
Katana, P. K. (1983). *Phys. Stat. Solidi(b)* **119**, 89.
Keldysh, L. V. (1965). *Sov. Phys. JETP* **20**, 1307.
Keyes, R. W., and Sladek, R. J. (1956). *J. Phys. Chem. Solids*, **1**, 143.
Kim, R. S., and Narita, S. (1976). *Phys. Stat. Solidi* **B78**, 741.
Kinch, M. A., and Buss, D. D. (1971). *J. Phys. Chem. Solids* **32**, 461.
Korotin, V. G., Krivonogov, S. N., Nasledov, D. N., and Smetannikova, Y. S. (1976). *Sov. Phys. Semicond.* **10**, 10.
Koteles, E. S., and Datars, W. R. (1974). *Phys. Rev.* **B9**, 568.
Kucera, Z., Hlidek, P., Höschl, P., Koubele, V., Prosser, V., and Zvara, M. (1990). *Phys. Stat. Solidi(b)* **158**, K173.
Kuchar, F., Meisels, R., and Kriechbaum, M. (1982). Proc. Physics of Narrow Gap Semiconductors, Linz, Austria (Gornik, E., Heinrich, H., and Palmetschofer, L., eds.), p. 197. Springer, New York.
Kuchar, F., Kaplan, R., Wagner, R. J., Cooke, R. A., Stradling, R. A., and Vogl, P. (1984). *J. Phys. C.* **17**, 6403.
Kwok, H. S., (1982). *IEEE J. Quant. Electron.* **QE-18**, 283.
Lacklison, D. E., and Capper, P. (1987). *Semicond. Sci. Technol.* **2**, 33.
Laff, R. A., and Fan, H. Y. (1961). *Phys. Rev.* **121**, 53.
Landau, L. D. (1930). *Z. Phys.* **64**, 629.
Landau, L. D., and Lifshitz, E. M. (1965). "Quantum Mechanics," 2nd ed. Pergamon Press, Oxford.
Landbolt-Börnstein (1987). "Semiconductors: Intrinsic Properties of Group IV Elements and III-V, and I-VII Compounds.," edited by O. Madelung, New Series, Group III, Vol. 22a (Springer, New York), p. 123ff.
Landwehr, G., and Rashba, E. I. (1991). "Landau Level Spectroscopy," vol. 27.1. North-Holland, Amsterdam.
Larsen, D. (1964). *Phys. Rev.* **135**, A419.
Larsen, D. (1968). *J. Phys. Chem. Solids* **29**, 271.
Lawaetz, P. (1971). *Phys. Rev.* **B4**, 3460.
Lax, B. (1963). Proc. Int. School of Physics, "Enrico Fermi" Course XXII, (Smith, R. A., ed.), p. 240. Academic Press, New York.
Lax, B., and Mavroides, J. G. (1967). "Semiconductors and Semimetals," (Willardson, R. K., and Beer, A. C., eds.), vol. 3, p. 321. Academic Press, New York.
Lee, C. C., and Fan, H. Y. (1974). *Phys. Rev.* **B9**, 3502.
Lee, K., thesis, B. S., (1976). Massachusetts Institute of Technology.
Lee, T. D., Low, F. E., and Pines, D. (1953). *Phys. Rev.* **90**, 297.
Legros, R., and Triboulet, R. (1985). *J. Crystal Growth* **72**, 264.
Leslie-Pelecky, D. L., Seiler, D. G., Loloee, M. R., and Littler, C. L. (1987). *Appl. Phys. Lett.* **51**, 1916.
Littler, C. L., and Seiler, D. G. (1985). *Appl. Phys. Lett.* **46**, 986.
Littler, C. L., and Seiler, D. G. (1986). *J. Appl. Phys.* **60**, 261.
Littler, C. L., Seiler, D. G., Kaplan, R., Wagner, R. J., and Zawadzki, W. (1981). *Solid State Commun.* **37**, 783.
Littler, C. L., Seiler, D. G., Kaplan, R., and Wagner, R. J. (1982). *Appl. Phys. Lett.* **41**, 880.

Littler, C. L., Seiler, D. G., Kaplan, R., and Wagner, R. J. (1983). *Phys. Rev.* **B27**, 7473.
Littler, C. L., Seiler, D. G., and McClure, S. W. (1984). *Solid State Commun.* **50**, 565.
Littler, C. L., Zawadzki, W., Loloee, M. R., Song, X. N., and Seiler, D. G. (1989). *Phys. Rev. Lett.* **63**, 2845.
Littler, C. L., Seiler, D. G., and Loloee, M. R. (1990a). *J. Vac. Sci. Technol.* **A8**, 1133.
Littler, C. L., Yoon, I. T., Song, X. N., Zawadzki, W., Pfeffer, P., and Seiler, D. G. (1990b). Proc. 20th Int. Conf. on the Physics of Semiconductors, p. 1763. World Scientific, Singapore.
Loloee, M. R., Seiler, D. G., and Ward, G. B. (1988). *Appl. Phys. Lett.* **53**, 2188.
Luttinger, J. M. (1956). *Phys. Rev.* **102**, 1030.
MacKenzie, H. A., Dennis, R. B., Voge, D., and Smith, S. D. (1980). *Optics Commun.* **34**, 205.
Mackey, H. J., Vaughn, B. J., Rater, L. M., and Seiler, D. G. (1975). *Solid State Commun.* **16**, 997.
Mahr, H. (1975). "Quantum Electronics, vol. I, Nonlinear Optics," Part A, (Rabin, H., and Tang, C. L., eds.), pp. 285–361. Academic Press, New York.
Maker, P. D., and Terhune, R. H. (1965). *Phys. Rev.* **137**, A801.
Manlief, S. K., and Palik, E. D. (1973). *Solid State Comm.* **12**, 1071.
Mathew, J. G. H., Kar, A. K., Heckenberg, N. R., and Galbraith, I. (1985a). *IEEE J. Quantum Elec.* **QE-21**, 94.
Mathew, J. G. H., Craig, D., and Miller, A. (1985b). *Appl. Phys. Lett.* **46**, 128.
Matsuda, O., and Otsuka, E. O. (1979). *J. Phys. Chem. Solids* **40**, 809.
Mavroides, J. G. (1972). "Optical Properties of Solids," (Ables, F., ed.), p. 351. North-Holland, Amsterdam.
McClure, S. W., Seiler, D. G., and Littler, C. L. (1984). *J. Appl. Phys.* **56**, 1655.
McCombe, B. D. (1969). *Phys. Rev.* **181**, 1206.
McCombe, B. D., and Kaplan, R. (1968). *Phys. Rev. Lett.* **21**, 756.
McCombe, B. D., and Wagner, R. J. (1971). *Phys. Rev.* **B4**, 1285.
McCombe, B. D., and Wagner, R. J. (1972). Proc. 11th Int. Conf. Phys. Semicond., p. 321. Pol. Scient. Publ., Warsaw.
McCombe, B. D., and Wagner, R. J. (1975). *Adv. Electrons Electronics* **37**, 1.
McCombe, B. D., Bishop, S. G., and Kaplan, R. (1967). *Phys. Rev. Lett.* **18**, 748.
McCombe, B. D., Wagner, R. J., and Prinz, G. A. (1970a). *Phys. Rev. Lett.* **25**, 87.
McCombe, B. D., Wagner, R. J., and Prinz, G. A. (1970b). *Solid State Commun.* **8**, 1687.
Merilainen, C. A., and Jones, C. E. (1983). *J. Vac. Sci. Technol.* **A1**, 1637.
Miller, A., and Craig, D. (1987), "Optical Properties of Narrow-Gap Low Dimensional Structures," p. 149. Plenum Press, New York.
Miller, A., and Parry, G. (1984). *Phil. Trans. R. Soc. Lond.* **A313**, 277.
Miller, A., Johnston, A., Dempsey, J., Smith, J., Pidgeon, C. R., and Holah, G. D. (1979). *J. Phys. C: Solid State Phys.* **12**, 4839.
Miller, A., Miller, D. A. B., and Smith, S. D. (1981). "Dynamic Non-linear Optical Processes in Semiconductors," *Adv. Phys.* **30**, 697–800.
Miller, A., Mathew, J. G. H., Craig, D., and Parry, G. (1984). *J. Opt. Soc. Am.* **B1**, 475.
Miller, D. A. B., Mozolowski, M. H., Miller, A., and Smith, S. D. (1978). *Optics Commun.* **27**, 133.
Miller, D. A. B., Smith, S. D., and Johnston, A. (1979). *Appl. Phys. Lett.* **35**, 658.
Milnes, A. G. (1973) "Deep Impurities in Semiconductors." Wiley, New York.
Mitra, S. S., Narducci, L. M., Shatas, R. A., Tsay, Y. F., and Vaidyanathan, A. (1975). *Appl. Optics* **14**, 3038.
Moore, B. T., Seiler, D. G., and Kahlert, H. (1978). *Solid State Elec.* **21**, 247.
Morita, S., Takano, S., and Kawamura, H. (1975). *J. Phys. Soc. Japan* **39**, 1040.
Moss, T. S., Burrell, G. J., and Ellis, B. (1973). "Semiconductor Opto-Electronics." Butterworths, London.
Nagao, M., Ishimoto, R., Mukai, T., Hongo, S., and Bo, H. (1974). Memoirs of the Faculty of Engineering, Kobe Univ., No. 20, 207.
Nasledov, D. N., and Smetannikova, Y. S. (1962). *Sov. Phys. Solid State* **4**, 78.

Nathan, V., Guenther, A. H., and Mitra, S. S. (1985). *J. Opt. Soc. Am.* **B2**, 294–316.
Nemirovsky, Y., and Finkman, E. (1979). *J. Appl. Phys.* **50**, 8107.
Neumann, Ch., Nothe, A., and Lipari, N. O. (1988). *Phys. Rev.* **B37**, 922.
Nguyen Van-Tran and Strnad, A. R. (1971). *Optics Comm.* **3**, 35.
Nguyen Van-Tran, Strnad, A. R., and Yafet, Y. (1971). *Phys. Rev. Lett.* **26**, 1170.
Palik, E. D., and Wright, G. B. (1967). "Semiconductors and Semimetals," (Willardson, R. K., and Beer, A. C., eds.), vol. 3, p. 421. Academic Press, New York.
Pankove, J. I. (1971). "Optical Processes in Semiconductors" Dover, New York.
Paul, W. (1968). Proc. 9th Int. Conf. on the Physics of Semiconductors, Moscow, p. 16. Nauka, Leningrad.
Pehek, J., and Levinstein, H. (1965). *Phys. Rev.* **140**, A576.
Peticolas, W. L. (1967). *Ann. Rev. Phys. Chem.* **18**, 233–260.
Pfeffer, P., and Zawadzki, W. (1990). *Phys. Rev.* **B41**, 1561.
Pidgeon, C. R. (1980). "Handbook on Semiconductors, Optical Properties," (Moss, T. S., and Balanskii, M., eds.), vol. 2, p. 223. North-Holland, Amsterdam..
Pidgeon, C. R. (1981). "Theoretical Aspects and New Developments in Magneto-Optics" (Devreese, J. T., ed.), p. 255. Plenum Press, New York.
Pidgeon, C. R., and Brown, R. N. (1966). *Phys. Rev.* **146**, 575.
Pidgeon, C. R., and Groves, S. H. (1968). *Phys. Rev. Lett.* **20**, 1003.
Pidgeon, C. R., and Groves, S. H. (1969). *Phys. Rev.* **186**, 824.
Pidgeon, C. R., Mitchell, D. L., and Brown, R. N. (1967a). *Phys. Rev.* **154**, 737.
Pidgeon, C. R., Groves, S. H., and Feinleib, J. (1967b). *Solid State Commun.* **5**, 677.
Pidgeon, C. R., Wherrett, B. S., Johnston, A. M., Dempsey, J., and Miller, A. (1979). *Phys. Rev. Lett.* **42**, 1785.
Pidgeon, C. R., Johnston, A. M., and Dempsey, I. (1982). Proc. 4th Int. Conf. on the Physics of Narrow Gap Semiconductors, Linz, Austria, p. 123. Springer-Verlag, New York.
Polla, D. L., and Aggarwal, R. J. (1984). *Appl. Phys. Lett.* **44**, 775.
Polla, D. L., and Jones, C. E. (1980a). *Solid-State Commun.* **36**, 809.
Polla, D. L., and Jones, C. E. (1980b). *J. Appl. Phys.* **51**, 6233.
Polla, D. L., and Jones, C. E. (1981). *J. Appl. Phys.* **52**, 5118.
Polla, D. L., Reine, M. B., and Jones, C. E. (1981a). *J. Appl. Phys.* **52**, 5132.
Polla, D. L., Tobin, S. P., Reine, M. B., and Sood A. K. (1981b). *J. Appl. Phys.* **52**, 5182.
Pratt, R. G., Hewett, J., Capper, P., Jones, C. L., and Quelch, M. J. (1983). *J. Appl. Phys.* **54**, 5152.
Pratt, R. G., Hewett, J., Capper, P., Jones, C. L., and Judd, N. (1986). *J. Appl. Phys.* **60**, 2377.
Putley, E. H. (1964). *Phys. Stat. Sol.* **6**, 571.
Ranvaud, R., Trebin, H. R., Rossler, U., and Pollak, F. H. (1979). *Phys. Rev.* **B20**, 701.
Reshchikov, M. A., and Smetannikova, Yu. S. (1987). *Sov. Phys. Semicond.* **21**, 102.
Rigaux, C. (1980). "Narrow Gap Semiconductors Physics and Applications," p. 110. Springer, New York.
Rigaux, C. (1988). "Semiconductors and Semimetals," (Furdyna, J. K., and Kossut, J. eds.), vol. 25, p. 229. Academic Press, New York.
Robert, J. L., Raymond, A., Aulombard, R. L., and Bousquet, C. (1980). *Phil. Mag.* **B42**, 1003.
Roberts, V., and Quarrington, J. E. (1955). *J. Electron.* **1**, 152.
Robinson, M. L. A. (1966). *Phys. Rev. Lett.* **17**, 963.
Roth, A. P., and Fortin, E. (1978). *Phys. Rev.* **B18**, 4200.
Roth, L. M., Lax, B., and Zwerdling, S. (1959). *Phys. Rev.* **114**, 90.
Salour, M. M. (1978). *Ann. Phy.* **3**, 364–503.
Saptsov, V. I., and Skok, E. M. (1985). *Sov. Phys. Solid State* **27**, 2100.
Saptsov, V. I., and Skok, E. M. (1986). *Sov. Phys. Semicond.* **20**, 867.
Sari, S. O. (1971). *Phys. Rev. Lett.* **26**, 1167.
Sari, S. O. (1973a). *Solid State Commun.* **12**, 705.

Sari, S. O. (1973b). *Phys. Rev. Lett.* **30**, 1323.
Schacham, S. E., and Finkman, E. (1985). *J. Appl. Phys.* **57**, 2001.
Schneider, W., Hübner, K., Decker, G., and Röhr, H. (1972). *Phys. Lett.* **41A**, 383.
Schneider, W., Groh, H., and Hübner, K. (1976). *Z. Phys.* **B25**, 29.
Seiler, D. G., and Goodwin, M. W. (1982). *J. Appl. Phys.* **53**, 7505.
Seiler, D. G., and Hanes, L. K. (1980). *J. Phys. Soc. Japan* **49**, Suppl. A329.
Seiler, D. G., and Littler, C. L. (1987). "High Magnetic Fields in Semicond. Phys., Wurzburg," (Landwehr, G., ed.), p. 454. Springer Verlag, New York.
Seiler, D. G., and Littler, C. L. (1988). *Acta. Phys. Polonica* **A73**, 879.
Seiler, D. G., and Littler, C. L. (1990). *J. Research Nat. Int. Standards Tech.* **95**, 469.
Seiler, D. G., and McClure, S. W. (1983). *Solid State Comm.* **47**, 17.
Seiler, D. G., Barker, J. R., and Moore, B. T. (1978). *Phys. Rev. Lett.* **41**, 319.
Seiler, D. G., Barker, J. R., Moore, B. T., and Hansen, K. E. (1979a). Proc. 14th Int. Conf. Phys. Semiconductors, Edinburgh, p. 501. The Institute of Physics, London.
Seiler, D. G., Hanes, L. K., Goodwin, M. W., and Stephens, A. E. (1979b). *J. Mag. Magn. Mater.* **11**, 247.
Seiler, D. G., Goodwin, M. W., and Weiler, M. H. (1981). *Phys. Rev.* **B23**, 6806.
Seiler, D. G., Goodwin, M. W., and Weiler, M. H. (1982). Proc. 4th Int. Conf. on Physics of Narrow Gap Semiconductors, Linz, Austria, p. 192. Springer-Verlag, New York.
Seiler, D. G., Goodwin, M. W., McClure, S. W., and Veilleux, L. A. (1983). "The Application of High Magnetic Fields in Semiconductor Physics," p. 297. Springer-Verlag, New York.
Seiler, D. G., Littler, C. L., and McClure, S. W. (1984). *Optics Commun.* **50**, 359.
Seiler, D. G., Littler, C. L., and Heiman, D. (1985). *J. Appl. Phys.* **57**, 2191.
Seiler, D. G., Littler, K. H., and Littler, C. L. (1986a). *Semicond. Sci. Technol.* **1**, 383.
Seiler, D. G., McClure, S. W., Justice, R. J., Loloee, M. R., and Nelson, D. A. (1986b). *Appl. Phys. Lett.* **48**, 1159.
Seiler, D. G., McClure, S. W., Justice, R. J., Loloee, M. R., and Nelson, D. A. (1986c). *J. Vac. Sci. Technol.* **A4**, 2034.
Seiler, D. G., Milazzo, S. A., Durkin, A. J., Loloee, M. R., and Littler, C. L. (1988). Proc. 19th Int. Conf. on Phys. of Semicond., Warsaw (Zawadzki, W., ed.), p. 913. Institute of Physics, Warsaw.
Seiler, D. G., Loloee, M. R., Milazzo, S. A., Durkin, A. J., and Littler, C. L. (1989a). *Solid State Commun.* **69**, 757.
Seiler, D. G., Littler, C. L., Milazzo, S. A., and Loloee, M. R. (1989b). *J. Vac. Sci. Technol.* **A7**, 370.
Seiler, D. G., Lowney, J. R., Littler, C. L., and Loloee, M. R. (1990). *J. Vac. Sci. Technol.* **A8**, 1237.
Seisyan, R. P., and Yuldashev, Sh. U. (1988). *Sov. Phys. Solid State* **30**, 6.
Seisyan, R. P., and Zakharchenya, B. P. (1991). "Modern Problems in Condensed Matter Sciences" (Landwehr, G., and Rashba, E. I., eds.), vol. 27.1, p. 347. North-Holland, New York.
Shah, J. (1978). *Solid State Elec.* **21**, 43.
Shayegan, M., Drew, H. D., Nelson, D. A., and Tedrow, P. M. (1985a). *Phys. Rev.* **B31**, 6123.
Shayegan, M., Goldman, V. J., Drew, H. D., Nelson, D. A., and Tedrow, P. M. (1985b). *Phys. Rev.* **B32**, 6952.
Shayegan, M., Goldman, V. J., Drew, H. D., Fortune, N. A., and Brooks, J. S. (1986). *Solid State Commun.* **60**, 817.
Shayegan, M., Goldman, V. J., and Drew, H. D. (1987). Proc. 18th Int. Conf. on the Physics of Semiconductors, p. 1205.
Shayegan, M., Goldman, V. J., and Drew, H. D. (1988). *Phys. Rev.* **B38**, 5585.
Sheik-bahae M., and Kwok, H. S. (1987) *IEEE J. Quantum Elec.* **QE-23**, 1974.
Sheik-bahae, M., Mukherjie, P., and Kwok, H. S. (1986). *J. Opt. Soc. Am.* **B3**, 379.
Sheik-bahae, M., Rossi, T., and Kwok, H. S. (1987). *J. Opt. Soc. Am.* **B4**, 1964.
Shockley, W., and Read, W. T. (1952). *Phys. Rev.* **87**, 835.

Singh, M., and Wallace, P. R. (1987). *J. Phys. C.* **20**, 2169.
Slusher, R. E., Giriat, W., and Brueck, S. R. J. (1969). *Phys. Rev.* **183**, 758.
Smith, S. D. (1967). "Handbuch der Physik", XXV/2A, p. 234.
Sonowski, L. (1980). *Proc. Narrow Gap Semiconductors, Physics and Applications, Nimes*, p. 1. Springer-Verlag, New York.
Stillman, G. E., Wolfe, C. M., and Dimmock, J. O. (1969). *Solid State Commun.* **7**, 921.
Strauss, A. J., Harman, T. C., Mavroides, J. G., Dickey, D. H., and Dresselhaus, M. S. (1962). Proc. Int. Conf. on the Physics of Semiconductors, Exeter (Institute of Physics, London), p. 703
Summers, C. J., Dennis, R. B., Wherrett, B. S., Harper, P. G., and Smith, S. D. (1968). *Phys. Rev.* **170**, 755.
Swierkowski, L., Zawadzki, W., Guldner, Y., and Rigaux, C. (1978). *Solid State Commun.* **27**, 1245.
Trebin, H. R., Rossler, U., and Ranvaud, R. (1979). *Phys. Rev.* **B20**, 686.
Trifonov, V. I., and Yaremenko, N. G. (1971). *Fiz. Tekh. Poluprovod.* **5**, 953 [*Sov. Phys. Semicond.* **5**, 839].
Trzeciakowski, W., Baj, M., Huant, S., and Brunel, L. (1986). *Phys. Rev.* **B33**, 6846.
Vaidyanathan, A., and Guenther, A. H. (1980). *Phys. Rev.* **B22**, 6480.
Vaidyanathan, A., Walker, T., and Guenther, A. H. (1980). *Phys. Rev.* **B21**, 743.
Valyashko, E. G., and Pleskacheva, T. B. (1973). *Sov. Phys. Semicond.* **7**, 573.
Van Stryland, E. W., Vanherzeele, H., Woodall, M. A., Soileau, M. J., Smirl, A. L., Guha, S., and Boggess, T. F. (1985). *Opt. Eng.* **24**, 613.
Varshni, Y. P. (1967). *Physica* **34**, 149.
Vashishta, P., and Kalia, R. K. (1982). *Phys. Rev.* **B25**, 6492.
Weiler, M. H. (1973). *Phys. Rev.* **B7**, 5403.
Weiler, M. H. (1979). *J. Magn. Magn. Mater* **11**, 131.
Weiler, M. H. (1981a). *Solid State Commun.* **39**, 937.
Weiler, M. H. (1981b). "Semiconductors and Semimetals" (Willardson, R. K., and Beer, A. C., eds.), vol. 16, p. 119. Academic Press, New York.
Weiler, M. H. (1982). *Solid State Comm.* **44**, 287.
Weiler, M. H., Aggarwal, R. L., and Lax, B. (1974). *Solid State Commun.* **14**, 299.
Weiler, M. H., Aggarwal, R. L., and Lax, B. (1977). *Phys. Rev.* **B16**, 3603.
Weiler, M. H., Aggarwal, R. L., and Lax, B. (1978). *Phys. Rev.* **B17**, 3269.
Weiler, M. H., Bierig, R. W., and Lax, B. (1969). *Phys. Rev.* **184**, 709.
Weiler, M. H., Bierig, R. W., and Lax, B. (1971). Proc. 3rd Int. Conf. on Photoconductivity, p. 145. Pergamon Press, New York.
Weiler, M. H., Reine, M., and Lax, B. (1968). *Phys. Rev.* **171**, 949.
Wherrett, B. S. (1984). *J. Opt. Soc. Am.* **B1**, 67.
Willardson, R. K., and Beer, A. C. eds. (1967). "Semiconductors and Semimetals: Vol. 3, Optical Properties of III-V Compounds." Academic Press, New York.
Wolfe, A., (1985). *Electronics Week*, June 10, pp. 24–27.
Worlock, J. M. (1972). "Laser Handbook," pp. 1323–1369. North-Holland, Amsterdam.
Wright, G. B., and Lax, B. (1961). *J. Appl. Phys. (Suppl.)* **32**, 2113.
Yafet, Y., Keyes, R. W., and Adams, E. N. (1956). *Phys. Chem. Solids* **1**, 137.
Yuen, S. Y. (1982). *Appl. Phys. Lett.* **41**, 590.
Zawadzki, W. (1963). *Phys. Lett.* **4**, 190.
Zawadzki, W. (1971). "New Developments in Semiconductors," p. 442. Nordhoff, Leyden.
Zawadzki, W. (1972). Proc. 11th Int. Conf. Phys. of Semiconductors, Warsaw, p. 87. Polish Scientific Pub. Warsaw.
Zawadzki, W. (1980). "Narrow Gap Semiconductors Physics and Applications," p. 85. Springer, New York.
Zawadzki, W., and Wlasak, J. (1976). *J. Phys. C: Solid State Phys.* **9**, L663.

Zawadzki, W., and Wlasak, J. (1980). "Theoretical Aspects and New Developments in Magneto-Optics" (Devreese, J. T., ed.), p. 367. Plenum New York.
Zawadzki, W., and Wlasak, J. (1984). *J. Phys C.* **17**, 2505.
Zawadzki, W., Hanamura, E., and Lax, B. (1967). *Bull. Am. Phys. Soc.* **12**, 100.
Zawadzki, W., Grisar, R., Wachernig, H., and Bauer, G. (1978). *Solid State Commun.* **25**, 775.
Zawadzki, W., Kubisa, M., Raymond, A., Robert, J. L., and Andre, J. P. (1987). *Phys. Rev.* **B36**, 9297.
Zawadzki, W., Song, X. N., Littler, C. L., and Seiler, D. G. (1990). *Phys. Rev.* **B42**, 5260.
Zhilich, A. (1972). *Sov. Phys. Solid State* **13**, 2425.
Zitter, R. N., Strauss, A. J., and Attard, A. E. (1959). *Phys. Rev.* **115**, 266.
Zwerdling, S., Keyes, R. W., Foner, S., Kolm, H. H., and Lax, B. (1956). *Phys. Rev.* **104**, 1805.
Zwerdling, S., Lax, B., and Roth, L. M. (1957). *Phys. Rev.* **108**, 1402.
Zwerdling, S., Kleiner, W. H., and Theriault, J. P. (1961). *J. Appl. Phys. (Suppl.)* **32**, 2118.
Zwerdling, S., Kleiner, W. H., and Theriault, J. P. (1962). Proc. Int. Conf. Semicond., Exeter, p. 455. Inst. of Phys. and Phys. Soc., London.

Index

A

Airy function, 232
Anisotropy
  effects on magneto-optical transitions, 327, 328, 347, 351, 353, 357, 359, 369
  latent anisotropy, 139, 143

B

Band nonparabolicity, 297
Band structure parameters
  HgCdTe, 342
  InSb, 341, 351, 352, 354
Binding energies
  exciton, 310
  impurities
    InSb
      deep levels, 374, 375–379
      shallow acceptor, 380
      shallow donor, 367
    HgCdTe
      deep levels, 383–385, 386–388
      shallow donor, 371–372
Bound magnetic polarons
  acceptor, 58, 61
  donor
    below, T = 1K, 66–70, 73
    mechanism, 5
    spin–flip energy, 56–58

C

Carrier densities, measurement of in photoreflectance, 284ff

CdMnSe, 115–116
CdSe
  Fe-doped
    Intra-ion Raman transitions, 74, 76–77
    spin–flip Raman scattering, 75
    Van Vleck induced moment, 74–75
  Mn-doped
    bound magnetic polaron, 56–58
    d-d exchange energy, 63–64
    magnetization steps, 63–64
    metal–insulator transition, 69–73
    spin–flip Raman scattering, 63–64, 66–73
CdTe
  d-d exchange energy
    nearest neighbor, 64
    next-nearest neighbor, 55
  magnetic-ion triplets, 64–66
  spin alignment (magnetization), 54–55, 58–61
Combined resonance, 355–359
Conduction bands of semiconductors, 151–156, 307, 327, 347–354
  dilational and shear deformation potential constants, 152
Coupled quantum wells
  asymmetrically coupled wells, photoreflectance of, 264ff
  symmetrically coupled wells, electroreflectance of, 269
Curie Principle, 142
Cu_2O, 112–113
Cyclotron energy, 19–20
  radius, 18
Cyclotron resonance, 299, 355–359

D

Deformation potential
  optical phonons, 210
  potentials, 145
  theory, 144
Density of states, 301
Dielectric function, 231
  bulk materials, 231
  confined systems, 233

E

E_1 and $E_1 + \Delta_1$ transitions in photoreflectance, 259
Effective mass, InSb, 297, 299
  electron, 348, 351
    crystallographic anisotropy of, 347
    hole crystallographic anisotropy of, 327, 328, 357, 359, 369
Elasticity, 139
  compression, 141
  elastic compliance constants, 140
  elastic stiffness constants (elastic moduli), 140
  generalized Hooke's Law, 140
  strain tensor, 140
  stress tensor, 140
  tension, 141
Electronic states (*See also* Landau levels)
  conduction band, 347–354
  density of states, 298–301
  exciton, 310
  impurity, 311, 359–388
  spin–orbit split-off, 345
  valence, 355–359
Electro-optic energy, Ω, 234, 284
Energy band models, 302–304
  discussion of, 302–304
  exciton corrections, 341
  five-level, 304, 347
  four-level Luttinger, 302
  Kane $\mathbf{k} \cdot \mathbf{p}$, 304
  Pidgeon–Brown, 304, 341, 347, 366, 416
  three-level, 302–303
  two-level, 341
Energy gap
  engineering by molecular beam epitaxy (MBE), 297
  impurity and defect states, 359
  temperature dependence of
    HgCdTe, 416–418
    GaAs, 287
    InSb, 324, 327–328
Exchange interaction
  d-d exchange, 6, 53–55, 61–64
    (Cd, Mn)Se, 7, 63–64
    (Cd, Mn)Te, 7, 54–55
  further neighbor, 65, 68–69
  magnetic polaron, 5, 56–58, 66–70, 73
  sp-d exchange, 5–6, 55–56
    (Cd, Fe)Se, 75
    (Cd, Mn)Se, 66–69
Exciton
  binding energies, 47, 48, 310
  diamagnetic exciton, 343
  inhomogenous broadening of transitions, 247
  localization, 48–50, 115
  magneto-, 44–50
  photoluminescence (Cd, Mn)Te, 45
Experimental
  apparatus
    dilution refrigerator, 7–9
    magnet
      Bitter, 9
      hybrid, 9–10
      pulsed, 10
      superconducting, 9
    optical dewar, 8, 11
  techniques, 329–339
    data analysis, fan chart, 328–329
    hydrostatic pressure, for detection of impurity levels, 363
    infrared laser sources, 329–331
    modulation, 149
      magnetic field modulation, 331
    lattice temperature, 329–330
    non-linear magneto-optical studies, 330
    one-photon absorption, 339
    photoconductivity, 333, 336–337, 413
    photo-Hall, 333, 337–339
    photovoltaic, 333, 339
    transmission, 333
    two-photon absorption (TPA), 335, 339
    two-photon magnetoabsorption (TPMA), 337

F

Fan diagram, 328–329
FIR (far infrared) studies
  of HgCdTe, 370–371
  of InSb, 363, 367
Faraday Rotation (Cd, Mn)Te
  ion clusters, 64–65
  spin alignment, 58–61
Fermi energy, 17, 21–22
  level, 38–39
Fiber optics, 11–13
Forbidden transitions, photoreflectance line shapes of, 246
Franz–Keldysh effect, 234, 284

G

GaAs
  cyclotron energy, 20
  multiple quantum wells, 15
    confinement energy, 17
    photoluminescence
      excitons, 45
      $v \geq 1$, 22, 24–26
      $v \simeq 1$, 26–28
      $v \simeq 2/3$, 30–32
    Raman scattering
      electrons, 39–41
      holes, 35–39
      $v = 2$, 41–43
      phonons, 48–50
  single interfaces, 15
  single quantum well, photoluminescence
    $v \simeq 1$, 28–30
    $v = 1/3, 2/3$, 32
    time-resolved, 32–34
  time-resolved spectroscopy of bulk GaAs, 108–114
  g-factor, 299
    of InSb, 350–354

H

HEMT, photoreflectance transitions in, 279$f\!f$
Heterostructures, 175–178
  built-in anisotropic strain due to lattice mismatch, 176–178
  quantum wells, 175

HgCdTe
  band parameters, 342
  deep levels, 380–388, 386$f\!f$
  energy gap dependence on
    composition, 416–418
    temperature, 418
  shallow levels, 369–373
Holes, in InSb, 355–359, 369
  heavy, 327, 328, 356
  light, 356, 367, 369
Hot electron relaxation
  in GaAs/GaAlAs quantum wells, 119

I

Impurity and defect related magneto-optical studies, 328, 359–388, 390
  deep levels, 373–388, 374
    HgCdTe, 380–388, 386$f\!f$
    InSb, 375–380, 374
  shallow levels, 362–373
    HgCdTe, 369–373
    InSb, 362–369
InSb
  band parameters, 352
  combined resonance, 355–359
  cyclotron resonance, 347–359
  deep levels, 375–380, 374
  energy gap dependence on temperature, 327
  g-factor, 350–354
  phonon-assisted resonances, 354–355
  shallow levels, 362–369
  spin resonance, 349
Intraband magneto-optical studies
  combined resonances, 355–359
  cyclotron resonance, 347–359
  phonon-assisted resonances, 354–355
  spin resonance, 349
Interband transitions, 345–359
  in magneto-optics
    absorption coefficient, 308, 308
    exciton correction, 339, 341
    HgCdTe band parameters, 342
    InSb band parameters, 352
    one-photon, 307–312
      HgCdTe, 345
      InSb, 339–345, 340–341

Interband transitions (*Continued*)
  two-photon, 312–322, 389–391
    HgCdTe, 410–418
    InSb, 391–410
  in photoreflectance
    (111) GaAs quantum wells, 262
    band to band, 251
    allowed transitions in multiple quantum wells, 241
    forbidden transitions in multiple quantum wells, 243
    in HEMTS, 277*ff*
    superlattices, 273
  in piezospectroscopy
    direct transitions, 166–172
    indirect transitions, 172–175
  in transient luminescence, 99

L

Landau filling factor, 18
Landau levels, 18–25, 35–39, 298–301, 315
Laser radiation, interaction of, 304–322
  absorption coefficient, 308, 334, 390
  beam characterization, 335–336
  carrier heating, 305
  laser-induced cooling, 305, 367
  light polarization, 313, 327, 350, 380–381, 414
Latent anisotropy, 139, 143
Lifetimes, 390–391

M

Magnetic field modulation, 297–298
Magnetic freeze-out effects, 338
Magnetic length, 18, 41–43
Magnetic semiconductors, mechanisms, 5–7
Magnetization (spin alignment), 58–66
  (Cd, Mn)Te, 58–61
  steps
    pairs, 61–64, 64–65
    triplets, 64–65
Magneto-donors, 362–366
Magneto-optical phenomena, 21–28, 35–41, 294*ff*
  (*See also* Interband magneto-optical studies, Intraband magneto-optical studies, Impurity magneto-optical studies, Two photon magnetoabsorption)
  absorption coefficient, 308, 334
  combined resonances, 355–359
  examples of, 322–329
  exciton correction, 339, 341
  g-factor, 299, 350
  HgCdTe band parameters, 342
  impurity and defect, 328, 359–388, 390
    deep levels, 373–388, 374
      HgCdTe, 380–388, 386*ff*
      InSb, 375–380, 374
    shallow levels, 362–373
      HgCdTe, 369–373
      InSb, 362–369
  InSb band parameters, 352
  one-photon, 307–312
    HgCdTe, 345
    InSb, 339–345, 340–341
  two-photon, 312–322, 389–391
    HgCdTe, 410–418
    InSb, 391–410
  cyclotron resonance, 347–359
    conduction-band, 347–354
    free-hole, 355–359
    phonon-assisted, 354–355
    phonon-assisted resonances, 354–355
    spin resonance, 349
Magnetoroton, 41–43
Metal–insulator transition, (Cd, Mn)Se, 69–73
Modulated dielectric function, 232*ff*
  excitons, 237
  first derivative functional form (FDFF), 235*ff*
  Gaussian line shapes in FDFF, 248
  Lorentzian line shapes in FDFF, 236
  low electric field approximation, 232
  third derivative functional form (TDFF), 232
  three point filling technique, 249
Modulation doping, 15–16
Microstructures, 223
  effective mass approximation, 223
  electric field, confined systems, 227
  excitonic effects, absorption, 226
  optical absorption, 225

N

Narrow gap semiconductors, 295–297, 296
  HgCdTe, 342, 345

INDEX **433**

InSb, 339–345, 340–341
  technological applications of, 295, 360
NIPI, photoreflectance transitions in, 281
Nonlinear optical effects, 295–296, 305, 367, 389
  experimental methods, 330
  optical generation of ultrashort pulses, 92–93

O

One-photon absorption, 307–312, 339–345, 340–341
  absorption coefficient, 308, 334
  experimental methods, 339
  HgCdTe, 243(table), 345, 347
Optical phonons (zone-center), 197–206, 305
  α-quartz, piezospectroscopy of Raman spectrum, 138, 198
  coupled LO phonon–plasmon modes, 207
  deformation potentials, 201
  Fano-interference, 213
  in the presence of free carriers, 210
  interaction with electronic states, 206
  polarizability (Raman) tensors, 200, 203–204
  strain Hamiltonian, 200
  symmetry under uniaxial stress, 199
Optical Stark effect, 129, 130
Orientational degeneracy, 143–144

P

Phase, in modulation spectroscopy, 230
  in relation to optical interference, 257
Phase relaxation, 112–115
Photoluminescence (*See also* GaAs and CdTe)
  circular polarization, 23
  time-resolved, quantum Hall effect, 32–34
  two-dimensional electron system, 14–22
Photoluminescence excitation spectroscopy, compared to photoreflectance, 245, 283
Photoconductive switching, 93–94
Photoconductivity, 333, 336–337
Photo-Hall, 333, 337–339
Photoluminescence spectroscopy
  circular polarization, 23
  time-resolved
    quantum Hall effect, 32–34
    transient, spectrum, 98–101

two-dimensional electron system, 14–22
GaAs
  excitons, 45
  phonons, 48–50
  $v \geq 1$, 22, 24–26
  $v \simeq 1$, 26–28
  $v \simeq 2/3$, 30–32
Photomodulation spectroscopy, 102–105
(*See also* Excite-probe spectroscopy)
  characteristic lineshapes, 104
Photoreflectance
  experimental apparatus
    photoreflectance excitation spectroscopy (PRE), 283
    spurious light reduction, 240, 281ff
    total internal reflection, 253
  line shapes (*See also* Modulated dielectric function)
    coupled quantum wells, 267ff
    Franz–Keldysh oscillations, 284ff
    interference effects, 257
    modulation coefficient, 252
    oscillator strengths, 225, 234, 243, 252
    self consistent calculation of, 248, 250
    superlattice effect in quantum wells, 275
  modulation mechanisms
    excitons, 237
    HEMT, 277ff
    NIPI, 281
    photovoltage modulation, 254
    population of donor bound excitons, 256
    superlattices, 236, 275
Photovoltaic, 333, 339
Piezospectroscopy, 138ff
  Raman spectrum of α-quartz, 138, 198
Polarons, 309, 350

Q

Quantum confinement
  magnetic, 3, 18, 49–50
  potential well, 15–17
Quantum Hall effect
  fractional, 4, 16, 30–32
  integer, 3, 28–30, 32–34
  optical, 28–30, 30–32
Quantum wells, 15, 43–46, 118–126, 240–262
  carrier capture, 118–119
  hot electron relaxation, 119–120
  recombination kinetics, 121

Quantum wells (*Continued*)
  resonant tunneling, 122–123
  time-resolved spectroscopy in, 117–126
  vertical transport, 124–126

R

Raman scattering
  Fe intra-ion transitions, 76–77
  inter-Landau transitions
    electrons, 39–43
    holes, 35–39
  intersubband transitions
    electrons, 39–40
    holes, 35–39
  phonons, 48–50
  plasmons, 41–43
  spin–flip
    bound magnetic polaron, 66–69
    donor electrons
      (Cd, Fe)Se, 75
      (Cd, Mn)Se, 63–64, 66–69
    free electrons, 69–73

S

Scattering
  electron–electron, 305
  impurity, 300
Selection rules
  absence of, in impurity level transitions, 362
  dipole, 313
  magnetoabsorption
    one-photon, 311–312
    two-photon, 319–321, 320, 382, 389
  PACRH, 354
Semimagnetic semiconductors, 297
Shallow impurities
  HgCdTe, 369–373
  InSb, 362–369
  Lyman spectra of, 179–197
    acceptors, 191–197
    chemical splitting of donors, 183
    donors, 179–191
    valley–orbit splitting of donors, 184
Silicon
  effect of [111] compression, 205–206

excitation spectrum
  of arsenic, 187–188
  of boron, 191–197
  of phosphorus, 181
wavelength-modulated reflectivity, 172–174
Spectroscopy
  Faraday rotation, (Cd, Mn)Te
    ion clusters, 64–65
    spin alignment, 58–61
  magneto-optical
    intraband
      InSb
        combined resonances, 355–359
        cyclotron resonance, 347–359
        phonon-assisted resonances, 354–355
        spin resonance, 349
    interband
      absorption coefficient, 308
      exciton correction, 339, 341
      HgCdTe band parameters, 342
      InSb band parameters, 352
      one-photon, 307–312
        HgCdTe, 345
        InSb, 339–345, 340–341
      two-photon, 312–322, 389–391
        HgCdTe, 410–418
        InSb, 391–410
  photoluminescence
    circular polarization, 23
    time-resolved
      quantum Hall effect, 32–34
      transient, spectrum, 98–101
      two-dimensional electron system, 14–22
      GaAs
        excitons, 45
        $v \geq 1$, 22, 24–26
        $v \simeq 1$, 26–28
        $v \simeq 2/3$, 30–32
        phonons, 48–50
  photoreflectivity, 46–48
  Raman Scattering
    CdSe
      Intra-ion Raman transitions, 74, 76–77
      spin–flip Raman scattering, 75
    GaAs
      electrons, 39–41
      holes, 35–39
      phonons, 48–50

$v = 2$, 41–43
time-resolved
　bulk semiconductors, 106–116
　　Cu_2O, 112–113
　　CdMnSe, 115–116
　　GaAs, 108–112
　detectors, 94–96
　excite-probe, 102–106
　general configuration for, 96–97
　instrumentation
　　dye laser, 91–92
　　nonlinear optical generation, 92–93
　　solid state, 89–91
　modulators, 93–94
　quantum wells and superlattices, 117–126
Spin diffusion, 70–71
Strain layer systems, photoreflectance of
　bulk, 286
　InGaAs/GaAs quantum wells, 264
　SiGe superlattices, 274
Streak camera, 94
Superlattice, 50–52, 118, 271–276
　strain-layer, 50–52
Symmetry, 142
　classification of acceptor states, 194
　classification of donor states, 191
　Curie principle, 142
　under uniaxial stress, 143

T

Topographical variation in multiple quantum wells (MQW), 260

Two-dimensional system
　electrons, 3–5, 14–21
　holes, 35–39, 50–52
Two photon absorption, 312–319, 389–391, 392–400
　absorption coefficient, 334, 390
　experimental techniques, 335, 339
　HgCdTe, 410–413
　tunneling theory, 321
Two-photon magnetoabsorption (TPMA), 319–322
　InSb, 391–410
　　crystallographic orientation, 327
　　experimental techniques, 337
　　temperature dependence of, 327
　HgCdTe, 382(table), 413–418
　　temperature dependence of, 418

U

Uniaxial stress apparatus, 146
　optical cryostat, 146

V

Valence bands of semiconductors, 157–160
　deformation potential constants, definition of, 157–158
　spin–orbit interaction, 158

Contents of Volumes in this Series

Volume 1 **Physics of III–V Compounds**

C. Hilsum, Some Key Features of III–V Compounds
Franco Bassani, Methods of Band Calculations Applicable to III–V Compounds
E. O. Kane, The $k \cdot p$ Method
V. L. Bonch-Bruevich, Effect of Heavy Doping on the Semiconductor Band Structure
Donald Long, Energy Band Structures of Mixed Crystals of III–V Compounds
Laura M. Roth and Petros N. Argyres, Magnetic Quantum Effects
S. M. Puri and T. H. Geballe, Thermomagnetic Effects in the Quantum Region
W. M. Becker, Band Characteristics near Principal Minima from Magnetoresistance
E. H. Putley, Freeze-Out Effects, Hot Electron Effects, and Submillimeter Photoconductivity in InSb
H. Weiss, Magnetoresistance
Betsy Ancker-Johnson, Plasmas in Semiconductors and Semimetals

Volume 2 **Physics of III–V Compounds**

M. G. Holland, Thermal Conductivity
S. I. Novkova, Thermal Expansion
U. Piesbergen, Heat Capacity and Debye Temperatures
G. Giesecke, Lattice Constants
J. R. Drabble, Elastic Properties
A. U. Mac Rae and G. W. Gobeli, Low Energy Electron Diffraction Studies
Robert Lee Mieher, Nuclear Magnetic Resonance
Bernard Goldstein, Electron Paramagnetic Resonance
T. S. Moss, Photoconduction in III–V Compounds
E. Antončik and J. Tauc, Quantum Efficiency of the Internal Photoelectric Effect in InSb
G. W. Gobeli and F. G. Allen, Photoelectric Threshold and Work Function
P. S. Pershan, Nonlinear Optics in III–V Compounds
M. Gershenzon, Radiative Recombination in the III–V Compounds
Frank Stern, Stimulated Emission in Semiconductors

Volume 3 **Optical of Properties III–V Compounds**

Marvin Hass, Lattice Reflection
William G. Spitzer, Multiphonon Lattice Absorption
D. L. Stierwalt and R. F. Potter, Emittance Studies
H. R. Philipp and H. Ehrenreich, Ultraviolet Optical Properties
Manuel Cardona, Optical Absorption above the Fundamental Edge
Earnest J. Johnson, Absorption near the Fundamental Edge
John O. Dimmock, Introduction to the Theory of Exciton States in Semiconductors
B. Lax and J. G. Mavroides, Interband Magnetooptical Effects

CONTENTS OF VOLUMES IN THIS SERIES

H. Y. Fan, Effects of Free Carries on Optical Properties
Edward D. Palik and George B. Wright, Free-Carrier Magnetooptical Effects
Richard H. Bube, Photoelectronic Analysis
B. O. Seraphin and H. E. Bennett, Optical Constants

Volume 4 Physics of III–V Compounds

N. A. Goryunova, A. S. Borschevskii, and D. N. Tretiakov, Hardness
N. N. Sirota, Heats of Formation and Temperatures and Heats of Fusion of Compounds $A^{III}B^{V}$
Don L. Kendall, Diffusion
A. G. Chynoweth, Charge Multiplication Phenomena
Robert W. Keyes, The Effects of Hydrostatic Pressure on the Properties of III–V Semiconductors
L. W. Aukerman, Radiation Effects
N. A. Goryunova, F. P. Kesamanly, and D. N. Nasledov, Phenomena in Solid Solutions
R. T. Bate, Electrical Properties of Nonuniform Crystals

Volume 5 Infrared Detectors

Henry Levinstein, Characterization of Infrared Detectors
Paul W. Kruse, Indium Antimonide Photoconductive and Photoelectromagnetic Detectors
M. B. Prince, Narrowband Self-Filtering Detectors
Ivars Melngailis and T. C. Harman, Single-Crystal Lead-Tin Chalcogenides
Donald Long and Joseph L. Schmit, Mercury-Cadmium Telluride and Closely Related Alloys
E. H. Putley, The Pyroelectric Detector
Norman B. Stevens, Radiation Thermopiles
R. J. Keyes and T. M. Quist, Low Level Coherent and Incoherent Detection in the Infrared
M. C. Teich, Coherent Detection in the Infrared
F. R. Arams, E. W. Sard, B. J. Peyton, and F. P. Pace, Infrared Heterodyne Detection with Gigahertz IF Response
H. S. Sommers, Jr., Macrowave-Based Photoconductive Detector
Robert Sehr and Rainer Zuleeg, Imaging and Display

Volume 6 Injection Phenomena

Murray A. Lampert and Ronald B. Schilling, Current Injection in Solids: The Regional Approximation Method
Richard Williams, Injection by Internal Photoemission
Allen M. Barnett, Current Filament Formation
R. Baron and J. W. Mayer, Double Injection in Semiconductors
W. Ruppel, The Photoconductor-Metal Contact

Volume 7 Application and Devices
PART A

John A. Copeland and Stephen Knight, Applications Utilizing Bulk Negative Resistance
F. A. Padovani, The Voltage-Current Characteristics of Metal-Semiconductor Contacts
P. L. Hower, W. W. Hooper, B. R. Cairns, R. D. Fairman, and D. A. Tremere, The GaAs Field-Effect Transistor
Marvin H. White, MOS Transistors
G. R. Antell, Gallium Arsenide Transistors
T. L. Tansley, Heterojunction Properties

CONTENTS OF VOLUMES IN THIS SERIES

PART B

T. Misawa, IMPATT Diodes
H. C. Okean, Tunnel Diodes
Robert B. Campbell and Hung-Chi Chang, Silicon Carbide Junction Devices
R. E. Enstrom, H. Kressel, and L. Krassner, High-Temperature Power Rectifiers of $GaAs_{1-x}P_x$

Volume 8 Transport and Optical Phenomena

Richard J. Stirn, Band Structure and Galvanomagnetic Effects in III–V Compounds with Indirect Band Gaps
Roland W. Ure, Jr., Thermoelectric Effects in III–V Compounds
Herbert Piller, Faraday Rotation
H. Barry Bebb and E. W. Williams, Photoluminescence 1: Theory
E. W. Williams and H. Barry Bebb, Photoluminescence II: Gallium Arsenide

Volume 9 Modulation Techniques

B. O. Seraphin, Electroreflectance
R. L. Aggarwal, Modulated Interband Magnetooptics
Daniel F. Blossey and Paul Handler, Electroabsorption
Bruno Batz, Thermal and Wavelength Modulation Spectroscopy
Ivar Balslev, Piezooptical Effects
D. E. Aspnes and N. Bottka, Electric-Field Effects on the Dielectric Function of Semiconductors and Insulators

Volume 10 Transport Phenomena

R. L. Rode, Low-Field Electron Transport
J. D. Wiley, Mobility of Holes in III–V Compounds
C. M. Wolfe and G. E. Stillman, Apparent Mobility Enhancement in Inhomogeneous Crystals
Robert L. Peterson, The Magnetophonon Effect

Volume 11 Solar Cells

Harold J. Hovel, Introduction; Carrier Collection, Spectral Response, and Photocurrent; Solar Cell Electrical Characteristics; Efficiency; Thickness; Other Solar Cell Devices; Radiation Effects; Temperature and Intensity; Solar Cell Technology

Volume 12 Infrared Detectors (II)

W. L. Eiseman, J. D. Merriam, and R. F. Potter, Operational Characteristics of Infrared Photodetectors
Peter R. Bratt, Impurity Germanium and Silicon Infrared Detectors
E. H. Putley, InSb Submillimeter Photoconductive Detectors
G. E. Stillman, C. M. Wolfe, and J. O. Dimmock, Far-Infrared Photoconductivity in High Purity GaAs
G. E. Stillman and C. M. Wolfe, Avalanche Photodiodes
P. L. Richards, The Josephson Junction as a Detector of Microwave and Far-Infrared Radiation
E. H. Putley, The Pyroelectric Detector–An Update

Volume 13 Cadmium Telluride

Kenneth Zanio, Materials Preparation; Physics; Defects; Applications

CONTENTS OF VOLUMES IN THIS SERIES

Volume 14 Lasers, Junctions, Transport

N. Holonyak, Jr. and M. H. Lee, Photopumped III–V Semiconductor Lasers
Henry Kressel and Jerome K. Butler, Heterojunction Laser Diodes
A. Van der Ziel, Space-Charge-Limited Solid-State Diodes
Peter J. Price, Monte Carlo Calculation of Electron Transport in Solids

Volume 15 Contacts, Junctions, Emitters

B. L. Sharma, Ohmic Contacts to III–V Compound Semiconductors
Allen Nussbaum, The Theory of Semiconducting Junctions
John S. Escher, NEA Semiconductor Photoemitters

Volume 16 Defects, (HgCd)Se, (HgCd)Te

Henry Kressel, The Effect of Crystal Defects on Optoelectronic Devices
C. R. Whitsett, J. G. Broerman, and C. J. Summers, Crystal Growth and Properties of $Hg_{1-x}Cd_xSe$ Alloys
M. H. Weiler, Magnetooptical Properties of $Hg_{1-x}Cd_xTe$ Alloys
Paul W. Kruse and John G. Ready, Nonlinear Optical Effects in $Hg_{1-x}Cd_xTe$

Volume 17 CW Processing of Silicon and Other Semiconductors

James F. Gibbons, Beam Processing of Silicon
Arto Lietoila, Richard B. Gold, James F. Gibbons, and Lee A. Christel, Temperature Distributions and Solid Phase Reaction Rates Produced by Scanning CW Beams
Arto Lietoila and James F. Gibbons, Applications of CW Beam Processing to Ion Implanted Crystalline Silicon
N. M. Johnson, Electronic Defects in CW Transient Thermal Processed Silicon
K. F. Lee, T. J. Stultz, and James F. Gibbons, Beam Recrystallized Polycrystalline Silicon: Properties, Applications, and Techniques
T. Shibata, A. Wakita, T. W. Sigmon, and James F. Gibbons, Metal-Silicon Reactions and Silicide
Yves I. Nissim and James F. Gibbons, CW Beam Processing of Gallium Arsenide

Volume 18 Mercury Cadmium Telluride

Paul W. Kruse, The Emergence of $(Hg_{1-x}Cd_x)Te$ as a Modern Infrared Sensitive Material
H. E. Hirsch, S. C. Liang, and A. G. White, Preparation of High-Purity Cadmium, Mercury, and Tellurium
W. F. H. Micklethwaite, The Crystal Growth of Cadmium Mercury Telluride
Paul E. Petersen, Auger Recombination in Mercury Cadmium Telluride
R. M. Broudy and V. J. Mazurczyck, (HgCd)Te Photoconductive Detectors
M. B. Reine, A. K. Sood, and T. J. Tredwell, Photovoltaic Infrared Detectors
M. A. Kinch, Metal-Insulator-Semiconductor Infrared Detectors

Volume 19 Deep Levels, GaAs, Alloys, Photochemistry

G. F. Neumark and K. Kosai, Deep Levels in Wide Band-Gap III–V Semiconductors
David C. Look, The Electrical and Photoelectronic Properties of Semi-Insulating GaAs
R. F. Brebrick, Ching-Hua Su, and Pok-Kai Liao, Associated Solution Model for Ga-In-Sb and Hg-Cd-Te
Yu. Ya. Gurevich and Yu. V. Pleskov, Photoelectrochemistry of Semiconductors

CONTENTS OF VOLUMES IN THIS SERIES

Volume 20 Semi-Insulating GaAs

R. N. Thomas, H. M. Hobgood, G. W. Eldridge, D. L. Barrett, T. T. Braggins, L. B. Ta, and S. K. Wang, High-Purity LEC Growth and Direct Implantation of GaAs for Monolithic Microwave Circuits
C. A. Stolte, Ion Implantation and Materials for GaAs Integrated Circuits
C. G. Kirkpatrick, R. T. Chen, D. E. Holmes, P. M. Asbeck, K. R. Elliott, R. D. Fairman, and J. R. Oliver, LEC GaAs for Integrated Circuit Applications
J. S. Blakemore and S. Rahimi, Models for Mid-Gap Centers in Gallium Arsenide

Volume 21 Hydrogenated Amorphous Silicon
Part A

Jacques I. Pankove Introduction
Masataka Hirose, Glow Discharge; Chemical Vapor Deposition
Yoshiyuki Uchida, dc Glow Discharge
T. D. Moustakas, Sputtering
Isao Yamada, Ionized-Cluster Beam Deposition
Bruce A. Scott, Homogeneous Chemical Vapor Deposition
Frank J. Kampas, Chemical Reactions in Plasma Deposition
Paul A. Longeway, Plasma Kinetics
Herbert A. Weakliem, Diagnostics of Silane Glow Discharges Using Probes and Mass Spectroscopy
Lester Guttman, Relation between the Atomic and the Electronic Structures
A. Chenevas-Paule, Experiment Determination of Structure
S. Minomura, Pressure Effects on the Local Atomic Structure
David Adler, Defects and Density of Localized States

Part B

Jacques I. Pankove, Introduction
G. D. Cody, The Optical Absorption Edge of a-Si:H
Nabil M. Amer and Warren B. Jackson, Optical Properties of Defect States in a-Si:H
P. J. Zanzucchi, The Vibrational Spectra of a-Si:H
Yoshihiro Hamakawa, Electroreflectance and Electroabsorption
Jeffrey S. Lannin, Raman Scattering of Amorphous Si, Ge, and Their Alloys
R. A. Street, Luminescence in a-Si:H
Richard S. Crandall, Photoconductivity
J. Tauc, Time-Resolved Spectroscopy of Electronic Relaxation Processes
P. E. Vanier, IR-Induced Quenching and Enhancement of Photoconductivity and Photoluminescence
H. Schade, Irradiation-Induced Metastable Effects
L. Ley, Photoelectron Emission Studies

Part C

Jacques I. Pankove, Introduction
J. David Cohen, Density of States from Junction Measurements in Hydrogenated Amorphous Silicon
P. C. Taylor, Magnetic Resonance Measurements in a-Si:H
K. Morigaki, Optically Detected Magnetic Resonance
J. Dresner, Carrier Mobility in a-Si:H

CONTENTS OF VOLUMES IN THIS SERIES

T. Tiedje, Information about Band-Tail States from Time-of-Flight Experiments
Arnold R. Moore, Diffusion Length in Undoped a-Si:H
W. Beyer and J. Overhof, Doping Effects in a-Si:H
H. Fritzche, Electronic Properties of Surfaces in a-Si:H
C. R. Wronski, The Staebler-Wronski Effect
R. J. Nemanich, Schottky Barriers on a-Si:H
B. Abeles and T. Tiedje, Amorphous Semiconductor Superlattices

Part D

Jacques I. Pankove, Introduction
D. E. Carlson, Solar Cells
G. A. Swartz, Closed-Form Solution of I–V Characteristic for a-Si:H Solar Cells
Isamu Shimizu, Electrophotography
Sachio Ishioka, Image Pickup Tubes
P. G. LeComber and W. E. Spear, The Development of the a-Si:H Field-Effect Transitor and Its Possible Applications
D. G. Ast, a-Si:H FET-Addressed LCD Panel
S. Kaneko, Solid-State Image Sensor
Masakiyo Matsumura, Charge-Coupled Devices
M. A. Bosch, Optical Recording
A. D'Amico and G. Fortunato, Ambient Sensors
Hiroshi Kukimoto, Amorphous Light-Emitting Devices
Robert J. Phelan, Jr., Fast Detectors and Modulators
Jacques I. Pankove, Hybrid Structures
P. G. LeComber, A. E. Owen, W. E. Spear, J. Hajto, and W. K. Choi, Electronic Switching in Amorphous Silicon Junction Devices

Volume 22 Lightwave Communications Technology
Part A

Kazuo Nakajima, The Liquid-Phase Epitaxial Growth of InGaAsP
W. T. Tsang, Molecular Beam Epitaxy for III–V Compound Semiconductors
G. B. Stringfellow, Organometallic Vapor-Phase Epitaxial Growth of III–V Semiconductors
G. Beuchet, Halide and Chloride Transport Vapor-Phase Deposition of InGaAsP and GaAs
Manijeh Razeghi, Low-Pressure Metallo-Organic Chemical Vapor Deposition of $Ga_xIn_{1-x}As_yP_{1-y}$ Alloys
P. M. Petroff, Defects in III–V Compound Semiconductors

Part B

J. P. van der Ziel, Mode Locking of Semiconductor Lasers
Kam Y. Lau and Amnon Yariv, High-Frequency Current Modulation of Semiconductor Injection Lasers
Charles H. Henry, Spectral Properties of Semiconductor Lasers
Yasuharu Suematsu, Katsumi Kishino, Shigehisa Arai, and Fumio Koyama, Dynamic Single-Mode Semiconductor Lasers with a Distributed Reflector
W. T. Tsang, The Cleaved-Coupled-Cavity (C^3) Laser

CONTENTS OF VOLUMES IN THIS SERIES

Part C

R. J. Nelson and N. K. Dutta, Review of InGaAsP/InP Laser Structures and Comparison of Their Performance
N. Chinone and M. Nakamura, Mode-Stabilized Semiconductor Lasers for 0.7–0.8- and 1.1–1.6-μm Regions
Yoshiji Horikoshi, Semiconductor Lasers with Wavelengths Exceeding 2 μm
B. A. Dean and M. Dixon, The Functional Reliability of Semiconductor Lasers as Optical Transmitters
R. H. Saul, T. P. Lee, and C. A. Burus, Light-Emitting Device Design
C. L. Zipfel, Light-Emitting Diode Reliability
Tien Pei Lee and Tingye Li, LED-Based Multimode Lightwave Systems
Kinichiro Ogawa, Semiconductor Noise-Mode Partition Noise

Part D

Federico Capasso, The Physics of Avalanche Photodiodes
T. P. Pearsall and M. A. Pollack, Compound Semiconductor Photodiodes
Takao Kaneda, Silicon and Germanium Avalanche Photodiodes
S. R. Forrest, Sensitivity of Avalanche Photodetector Receivers for High-Bit-Rate Long-Wavelength Optical Communication Systems
J. C. Campbell, Phototransistors for Lightwave Communications

Part E

Shyh Wang, Principles and Characteristics of Integratable Active and Passive Optical Devices
Shlomo Margalit and Amnon Yariv, Integrated Electronic and Photonic Devices
Takaaki Mukai, Yoshihisa Yamamoto, and Tatsuya Kimura, Optical Amplification by Semiconductor Lasers

Volume 23 Pulsed Laser Processing of Semiconductors

R. F. Wood, C. W. White, and R. T. Young, Laser Processing of Semiconductors: An Overview
C. W. White, Segregation, Solute Trapping, and Supersaturated Alloys
G. E. Jellison, Jr., Optical and Electrical Properties of Pulsed Laser-Annealed Silicon
R. F. Wood and G. E. Jellison, Jr., Melting Model of Pulsed Laser Processing
R. F. Wood and F. W. Young, Jr., Nonequilibrium Solidification Following Pulsed Laser Melting
D. H. Lowndes and G. E. Jellison, Jr., Time-Resolved Measurements During Pulsed Laser Irradiation of Silicon
D. M. Zehner, Surface Studies of Pulsed Laser Irradiated Semiconductors
D. H. Lowndes, Pulsed Beam Processing of Gallium Arsenide
R. B. James, Pulsed CO_2 Laser Annealing of Semiconductors
R. T. Young and R. F. Wood, Applications of Pulsed Laser Processing

Volume 24 Applications of Multiquantum Wells, Selective Doping, and Superlattices

C. Weisbuch, Fundamental Properties of III–V Semiconductor Two-Dimensional Quantized Structures: The Basis for Optical and Electronic Device Applications
H. Morkoc and H. Unlu, Factors Affecting the Performance of (Al, Ga)As/GaAs and

Contents of Volumes in This Series

(Al, Ga)As/InGaAs Modulation-Doped Field-Effect Transistors: Microwave and Digital Applications
N. T. Linh, Two-Dimensional Electron Gas FETs: Microwave Applications
M. Abe et al., Ultra-High-Speed HEMT Integrated Circuits
D. S. Chemla, D. A. B. Miller, and P. W. Smith, Nonlinear Optical Properties of Multiple Quantum Well Structures for Optical Signal Processing
F. Capasso, Graded-Gap and Superlattice Devices by Band-gap Engineering
W. T. Tsang, Quantum Confinement Heterostructure Semiconductor Lasers
G. C. Osbourn et al., Principles and Applications of Semiconductor Strained-Layer Superlattices

Volume 25 Diluted Magnetic Semiconductors

W. Giriat and J. K. Furdyna, Crystal Structure, Composition, and Materials Preparation of Diluted Magnetic Semiconductors
W. M. Becker, Band Structure and Optical Properties of Wide-Gap $A_{1-x}^{II}Mn_xB^{VI}$ Alloys at Zero Magnetic Field
Saul Oseroff and Pieter H. Keesom, Magnetic Properties: Macroscopic Studies
T. Giebultowicz and T. M. Holden, Neutron Scattering Studies of the Magnetic Structure and Dynamics of Diluted Magnetic Semiconductors
J. Kossut, Band Structure and Quantum Transport Phenomena in Narrow-Gap Diluted Magnetic Semiconductors
C. Riqaux, Magnetooptics in Narrow Gap Diluted Magnetic Semiconductors
J. A. Gaj, Magnetooptical Properties of Large-Gap Diluted Magnetic Semiconductors
J. Mycielski, Shallow Acceptors in Diluted Magnetic Semiconductors: Splitting, Boil-off, Giant Negative Magnetoresistance
A. K. Ramdas and S. Rodriquez, Raman Scattering in Diluted Magnetic Semiconductors
P. A. Wolff, Theory of Bound Magnetic Polarons in Semimagnetic Semiconductors

Volume 26 III–V Compound Semiconductors and Semiconductor Properties of Superionic Materials

Zou Yuanxi, III–V Compounds
H. V. Winston, A. T. Hunter, H. Kimura, and R. E. Lee, InAs-Alloyed GaAs Substrates for Direct Implantation
P. K. Bhattacharya and S. Dhar, Deep Levels in III–V Compound Semiconductors Grown by MBE
Yu. Ya. Gurevich and A. K. Ivanov-Shits, Semiconductor Properties of Superionic Materials

Volume 27 High Conducting Quasi-One-Dimensional Organic Crystals

E. M. Conwell, Introduction to Highly Conducting Quasi-One-Dimensional Organic Crystals
I. A. Howard, A Reference Guide to the Conducting Quasi-One-Dimensional Organic Molecular Crystals
J. P. Pouget, Structural Instabilities
E. M. Conwell, Transport Properties
C. S. Jacobsen, Optical Properties
J. C. Scott, Magnetic Properties
L. Zuppiroli, Irradiation Effects: Perfect Crystals and Real Crystals

CONTENTS OF VOLUMES IN THIS SERIES

Volume 28 Measurement of High-Speed Signals in Solid State Devices

J. Frey and D. Ioannou, Materials and Devices for High-Speed and Optoelectronic Applications
H. Schumacher and E. Strid, Electronic Wafer Probing Techniques
D. H. Auston, Picosecond Photoconductivity: High-Speed Measurements of Devices and Materials
J. A. Valdmanis, Electro-Optic Measurement Techniques for Picosecond Materials, Devices, and Integrated Circuits
J. M. Wiesenfeld and R. K. Jain, Direct Optical Probing of Integrated Circuits and High-Speed Devices
G. Plows, Electron-Beam Probing
A. M. Weiner and R. B. Marcus, Photoemissive Probing

Volume 29 Very High Speed Integrated Circuits: Gallium Arsenide LSI

M. Kuzuhara and T. Nozaki, Active Layer Formation by Ion Implantation
H. Hashimoto, Focused Ion Beam Implantation Technology
T. Nozaki and A. Higashisaka, Device Fabrication Process Technology
M. Ino and T. Takada, GaAs LSI Circuit Design
M. Hirayama, M. Ohmori, and K. Yamasaki, GaAs LSI Fabrication and Performance

Volume 30 Very High Speed Integrated Circuits: Heterostructure

H. Watanabe, T. Mizutani, and A. Usui, Fundamentals of Epitaxial Growth and Atomic Layer Epitaxy
S. Hiyamizu, Characteristics of Two-Dimensional Electron Gas in III–V Compound Heterostructures Grown by MBE
T. Nakanisi, Metalorganic Vapor Phase Epitaxy for High-Quality Active Layers
T. Mimura, High Electron Mobility Transistor and LSI Applications
T. Sugeta and T. Ishibashi, Hetero-Bipolar Transistor and Its LSI Application
H. Matsueda, T. Tanaka, and M. Nakamura, Optoelectronic Integrated Circuits

Volume 31 Indium Phosphide: Crystal Growth and Characterization

J. P. Farges, Growth of Discoloration-free InP
M. J. McCollum and G. E. Stillman, High Purity InP Grown by Hydride Vapor Phase Epitaxy
T. Inada and T. Fukuda, Direct Synthesis and Growth of Indium Phosphide by the Liquid Phosphorous Encapsulated Czochralski Method
O. Oda, K. Katagiri, K. Shinohara, S. Katsura, Y. Takahashi, K. Kainosho, K. Kohiro, and R. Hirano, InP Crystal Growth, Substrate Preparation and Evaluation
K. Tada, M. Tatsumi, M. Morioka, T. Araki, and T. Kawase, InP Substrates: Production and Quality Control
M. Razeghi, LP-MOCVD Growth, Characterization, and Application of InP Material
T. A. Kennedy and P. J. Lin-Chung, Stoichiometric Defects in InP

Volume 32 Strained-Layer Superlattices: Physics

T. P. Pearsall, Strained-Layer Superlattices
Fred H. Pollack, Effects of Homogeneous Strain on the Electronic and Vibrational Levels in Semiconductors

J. Y. Marzin, J. M. Gerárd, P. Voisin, and J. A. Brum, Optical Studies of Strained III–V Heterolayers
R. People and S. A. Jackson, Structurally Induced States from Strain and Confinement
M. Jaros, Microscopic Phenomena in Ordered Superlattices

Volume 33 Strained-Layer Superlattices: Materials Science and Technology

R. Hull and J. C. Bean, Principles and Concepts of Strained-Layer Epitaxy
William J. Schaff, Paul J. Tasker, Mark C. Foisy, and Lester F. Eastman, Device Applications of Strained-Layer Epitaxy
S. T. Picraux, B. L. Doyle, and J. Y. Tsao, Structure and Characterization of Strained-Layer Superlattices
E. Kasper and F. Schaffler, Group IV Compounds
Dale L. Martin, Molecular Beam Epitaxy of IV–VI Compound Heterojunctions
Robert L. Gunshor, Leslie A. Kolodziejski, Arto V. Nurmikko, and Nobuo Otsuka, Molecular Beam Epitaxy of II–VI Semiconductor Microstructures

Volume 34 Hydrogen in Semiconductors

J. I. Pankove and N. M. Johnson, Introduction to Hydrogen in Semiconductors
C. H. Seager, Hydrogenation Methods
J. I. Pankove, Hydrogenation of Defects in Crystalline Silicon
J. W. Corbett, P. Deák, U. V. Desnica, and S. J. Pearton, Hydrogen Passivation of Damage Centers in Semiconductors
S. J. Pearton, Neutralization of Deep Levels in Silicon
J. I. Pankove, Neutralization of Shallow Acceptors in Silicon
N. M. Johnson, Neutralization of Donor Dopants and Formation of Hydrogen-Induced Defects in n-Type Silicon
M. Stavola and S. J. Pearton, Vibrational Spectroscopy of Hydrogen-Related Defects in Silicon
A. D. Marwick, Hydrogen in Semiconductors: Ion Beam Techniques
C. Herring and N. M. Johnson, Hydrogen Migration and Solubility in Silicon
E. E. Haller, Hydrogen-Related Phenomena in Crystalline Germanium
J. Kakalios, Hydrogen Diffusion in Amorphous Silicon
J. Chevallier, B. Clerjaud, and B. Pajot, Neutralization of Defects and Dopants in III–V Semiconductors
G. G. DeLeo and W. B. Fowler, Computational Studies of Hydrogen-Containing Complexes in Semiconductors
R. F. Kiefl and T. L. Estle, Muonium in Semiconductors
C. G. Van de Walle, Theory of Isolated Interstitial Hydrogen and Muonium in Crystalline Semiconductors

Volume 35 Nanostructured Systems

Mark Reed, Introduction
H. van Houten, C. W. J. Beenakker, and B. J. van Wees, Quantum Point Contacts
G. Timp, When Does a Wire Become an Electron Waveguide?
M. Büttiker, The Quantum Hall Effect in Open Conductors
W. Hansen, J. P. Kotthaus, and U. Merkt, Electrons in Laterally Periodic Nanostructures